Perioperative Medicine – Current Controversies

Karen Stuart-Smith
Editor

Perioperative Medicine – Current Controversies

Springer

Editor
Karen Stuart-Smith
Waimairi Anaesthesia Ltd
Christchurch
New Zealand

ISBN 978-3-319-28819-2 ISBN 978-3-319-28821-5 (eBook)
DOI 10.1007/978-3-319-28821-5

Library of Congress Control Number: 2016934461

Springer Cham Heidelberg New York Dordrecht London

Printed on acid-free paper

Springer International Publishing AG Switzerland is part of Springer Science+Business Media (www.springer.com)

For Pepsi

Preface

The World Health Organization estimates that 11 % of the global burden of disease can be treated with surgery, and yet there are no specific charitable organisations nor government departments anywhere in the world that view high-quality perioperative care as a public health priority. Surgery costs money and carries both benefits and risks, and the anaesthetist is an integral part of this complex and expensive way of managing disease. Careful preoperative assessment, attention to fluid balance, pain control and depth of anaesthesia, as well as meticulous post-operative management, would seem intuitively to be beneficial for both the individual patient and society as a whole. It might even appear obvious that a well-managed perioperative course reduces complications and readmissions to hospital. However, there is very little hard evidence at a population level to determine whether these intuitions are valid in practice, and thus it becomes extremely difficult to persuade national budget holders to prioritise perioperative care over other disease burdens which appear more pressing. It is up to clinicians to provide this evidence. No other group has the necessary knowledge to make the case.

The role of the anaesthetist in the twenty-first century has expanded well beyond the operating room and now effectively encompasses the entire journey of the surgical patient, from the moment the decision to operate is made to patient discharge from hospital and, in the case of chronic pain management, for some time after that. In the last 5 years or so, it has become increasingly clear that the choices anaesthetic clinicians make in the operating room, including such fundamental details as 'how much' and 'what' to give of everything, may have profound short- and long-term implications for the patient. Anaesthesia has long been regarded as simply a relatively safe way of rendering the patient amenable to surgery, but the transition from 'anaesthetist' to 'perioperative medicine specialist' now seems inevitable. Indeed, many professional bodies around the world, including most notably the Royal College of Anaesthetists in the United Kingdom, advocate this transition as a vital component of patient care that is both clinically effective and cost-effective.

In spite of this recent interest, the concept of perioperative medicine remains somewhat ill defined. The purpose of this book is to try and gather together at least some of the subject areas relevant to perioperative medicine in one place. This volume covers topics relevant to adult surgical patients. The perioperative management of paediatric patients and the management of obstetric anaesthesia are both substantial areas of interest in their own right and will require their own books—possibly a project for either myself or others in the future. This book describes 'current controversies' for the obvious reason that nothing is settled and most of the topics covered here arouse strong feelings one way or the other at major anaesthetic meetings worldwide. There is no intention to be prescriptive or attempt to establish one type of care over another. The invited authors have given a comprehensive overview of their own areas of special interest. I have no doubt the reader will be able to think of many topics that they feel should have been covered and have not. I am rather in that position myself. All suggestions for future editions are welcome.

Christchurch, New Zealand Karen Stuart-Smith, MD, FRCA
November 2015

Biography

Karen Stuart-Smith was born and raised in Glasgow, Scotland, and graduated in physiology and medicine from the University of Glasgow. She is a Fellow of the Royal College of Anaesthetists and has enjoyed a long career as a clinical anaesthetist. Early training in cardiorespiratory research led to an enthusiasm for applying scientific methodology to clinical practice, and she is also a passionate believer in the role of clinical audit in assessing the quality of patient care. Karen presently resides in New Zealand, where she continues to practice as an anaesthetist and enjoys the outdoor lifestyle in her spare time.

Contents

Chapter 1
Perioperative Medicine: Defining the Anesthesiologist's Role in Shaping Perioperative Outcomes

Ottokar Stundner and Stavros G. Memtsoudis

Abstract The practice of anesthesia has long been regarded as an assistive service for surgery rather than one that, by itself, represents an entity of medical interventions with significant outcome consequences. Only relatively recently, researchers have supplied convincing evidence that anesthesia-related factors impact on outcomes beyond intraoperative care parameters and postoperative pain measures. Specifically, the choice of anesthetic technique may influence major morbidity and mortality, recovery time, length of hospital stay, functional outcomes, long-term quality of life, and economic aspects of care. However, with research in the field evolving, the magnitude of these effects is poorly defined and thus considerable controversy persists. This chapter seeks to introduce the reader to the current literature and to highlight areas where anesthesia related factors have been suggested to significantly impact perioperative outcomes. The presentation focuses largely on the orthopedic patient population, a group that has been most extensively studied. A view of how anesthetic practice may impact on the wider public health system evolves from this presentation.

Keywords Perioperative outcomes • Anesthesia type • Regional anesthesia • General anesthesia • Population-based outcomes

O. Stundner, MD
Department of Anesthesiology, Perioperative Medicine and Intensive Care Medicine, Paracelsus Medical University, Salzburg, Austria

S.G. Memtsoudis, MD, PhD (✉)
Department of Anesthesiology, Hospital for Special Surgery, Weill Medical College of Cornell University, New York, NY, USA
e-mail: MemtsoudisS@hss.edu

K. Stuart-Smith (ed.), *Perioperative Medicine – Current Controversies*,
DOI 10.1007/978-3-319-28821-5_1

1.1 The Evolution of Anesthesiology in Perioperative Medicine

The administration of anesthesia has long been regarded as a medical intervention without immediate diagnostic or curative benefit. Rather, it has been described as an assistive service allowing surgical procedures to take place. With time, the increasingly complex task of providing anesthesia resulted in the need for specialists, who were often surgeons in the early years of the speciality. Expanding knowledge of human physiology and pharmacology, advances in airway, respiratory and hemodynamic management and the advent of regional anesthesia revolutionized perioperative medicine and transformed the field of anesthesiology. The evolution of surgery progressed in parallel, with innovations in anesthesia allowing for procedures to increase in length, complexity and invasiveness, and to be extended to increasingly morbidity-ridden patient populations. Anesthesiologists' responsibilities quickly expanded to include the maintenance of vital organ functions, and consequently comprehensive medical care of the patient during, before and after surgical procedures, including the care for critically ill. Moreover, preoperative medical evaluation became an important constituent of anesthetic management, as it allowed for the stratification of overall surgical risk. The development of the American Society of Anesthesiologists Classification system to predict perioperative risk is only one of the most visible contributions. Along with these developments, the field of anesthesiology has been shifting from a surgery-enabling service to a comprehensive perioperative science.

Following is a discussion of relevant literature regarding various aspects of anesthetic care and their influence of perioperative outcomes. The concept of anesthesia related perioperative outcomes research is introduced, followed by examples in various clinical settings. A special focus on orthopedic surgical patients is presented, as the majority of available evidence concerns itself with this particular population.

1.2 Anesthesia Related Perioperative Outcomes Research

As much of our knowledge in the field of anesthesiology's impact on outcomes is derived from large population-based dataset research, it is necessary to understand basic concepts related to this type of analyses. Many end points of interest occur rather infrequently, thus necessitating large patient samples to conduct valid analyses and derive meaningful conclusions [1]. Methodological advances have allowed for the use of extremely large datasets encompassing information from vast numbers of patients, hospitals and regions, representing "real world practice", thus allowing researchers to overcome the constraints of single-institutional studies usually performed in academic settings. The latter have been criticized in the past, partially because of the limited external validity of the data generated by these studies. Further, randomized controlled trials conducted in such settings are often constrained by unrealistic inclusion and exclusion criteria.

Although associated with many advantages, the analysis of large database constructs has a number of limitations. It must be noted that commonly used databases for clinical research have not been designed for such purposes and lack pertinent clinical detail. In this context, sampling of data with the correct scope, sufficient granularity and integrity requires thoughtful planning, rigorous execution, coordination and time. Frequently, many data contributing centers deliver data of variable quality, thus making analysis and interpretation challenging. In addition, statistical analysis of such datasets – often exceeding millions of discharge entries – is disproportionately more complex than traditional statistical data handling, as computational technology, knowledge and advanced methodology is necessary. Finally, while associations can be established between interventions and outcomes, clear causalities cannot. This is in part due to the issue of residual cofounding, as many data points were collected for administrative purposes and lack important clinical covariates. Readers need to evaluate results of such studies accordingly, realizing their value and their limitations [2].

1.3 Regional Versus General Anesthesia

The choice of anesthetic technique and its impact on outcomes has been on the forefront of anesthesia related outcome research in recent years. In this context, studies evaluating the use of general versus regional anesthesia have shown how the choice of anesthetic approach can impact on numerous perioperative outcomes measures. General anesthesia is defined by a state similar to deep unconsciousness, frequently necessitating securing of the airway and mechanical ventilation due to loss of protective reflexes and respiratory drive. Regional anesthesia comprises blockade of nerve conduction utilizing local anesthetics, either centrally at the level of the spinal cord or at more peripheral locations, at the level of nerve bundles or singular nerves. Advantages of regional anesthesia include the preservation of higher central nervous system functions such as those affecting the respiratory system, while negating the need for invasive procedures including intubation and ventilation. Not all types of surgery lend themselves equally well to regional anesthetic approaches. It has become widely utilized particularly in procedures involving the extremities and lower part of the trunk, either as the sole anesthetic or in combination with sedation or general anesthesia. Basic and clinical science have suggested numerous benefits of regional anesthesia: better and longer-lasting analgesia with less need for systemically administered analgesics, higher patient comfort with lower incidence of postoperative nausea and vomiting, and improved distal tissue perfusion through sympatholysis. The following is a discussion of the literature as it pertains to the choice of anesthetic technique and perioperative outcomes for a number of different procedures. As much of the literature focuses on orthopedic patients, data regarding this population is featured more prominently below.

1.4 Orthopedic Surgery: Complications and Mortality

Orthopedic surgery is a prime target for comparative outcomes research, as it concerns a large and increasing collective of patients, many of whom are advanced in age and suffer from a number of comorbidities.

1.4.1 Unilateral, Elective Total Joint Replacement

A number of publications, both meta-analytical as well as population based, support a benefit of regional over general anesthesia, however with a varying degree of agreement. Over a decade ago, Rodgers et al. performed one of the first meta-analyses on the topic suggesting that the use of neuraxial anesthesia was associated with improved outcomes including decreased blood loss and lower risk of thromboembolic events in a heterogeneous patient population [3]. Although orthopedic patients made up the largest subgroup in this pooled sample, statistically, no concrete conclusion could be drawn regarding this group of patients, pointing to the problem that even pooled datasets have with achieving sufficient power. More recently, Macfarlane et al published a systematic review synthesizing 28 studies with approximately 1,500 patients undergoing knee arthroplasty in 2009; the authors found a better analgesic profile with lower incidence of opioid-related side effects, decreased length of stay and facilitated rehabilitation in patients undergoing the procedure under neuraxial anesthesia [4]. However, they failed to detect a difference in incidence of perioperative death or major cardiovascular complications, deep vein thrombosis or pulmonary embolism. Examining a cohort of patients undergoing total hip arthroplasty from 21 randomized controlled trials, Hu et al suggested that regional anesthesia positively influenced operating time (OR –0.19; 95 % CI –0.33––0.05), blood transfusion need (OR 0.45; 95 % CI 0.22–0.94) and pulmonary embolism event rates (OR 0.46, 95 % CI 0.29–0.80) [5].

Using population based data, Memtsoudis and colleagues carried out a database analysis involving 382,236 patients undergoing total hip or knee replacement, comparing general to neuraxial anesthesia. Approximately 11 % of patients received neuraxial (epidural or spinal anesthesia), 75 % received general anesthesia, and 14 % a combination of neuraxial and general anesthesia. Differences in age and individual comorbidity burden between groups were only moderate. Thirty-day mortality rates were significantly lower among patients undergoing their procedures under neuraxial or combined neuraxial-general anesthesia, compared to general anesthesia only (0.10, 0.10 and 0.18 %, $p<0.001$). While lower crude incidences of many major complications (including pulmonary complications, pulmonary embolism, cerebrovascular events and acute renal failure) as well as lower incidence of transfusion were observed in the regional anesthesia groups, there was no difference in the incidence of myocardial infarction (0.24 % vs. 0.26 % vs. 0.28 %; $P=0.47$) and other cardiac complications (6.20 % vs. 6.61 % vs. 6.42 %; $P=0.07$). Length of stay differed only slightly, but significantly (2.6 (2.2–3.2) days (neuraxial) versus 2.6 (2.2–3.1) days (neuraxial-general) versus 2.7 (2.2–3.5) days (general), $P<0.001$). Similarly, the incidence of binary prolonged hospital stay (exceeding the 75th percentile) was highest in the general anesthesia group (28.7, 27.4, and 35.4 %; $P<0.001$).

After controlling for covariates, adjusted odds ratios for 30-day mortality were significantly higher for groups receiving general anesthesia, when compared to neuraxial blockade alone (odds ratio (OR) = 1.83 (CI 1.08–3.1; $p=0.0211$)) or combined general-neuraxial anesthesia: OR = 1.70 (CI 1.06–2.74; $p=0.0228$)) in patients undergoing knee arthroplasty. There was no significant difference in adjusted mortality risk for patients undergoing hip arthroplasty. However, the odds for numerous major in-hospital complications were lowered by the choice of neuraxial techniques in both surgical groups [6]. In the groups receiving a combination of general and neuraxial anesthesia, the odds for complications were frequently found to fall between those for neuraxial anesthesia and general anesthesia alone. The authors conclude from this finding a potential intrinsic beneficial effect of neuraxial anesthesia, and not just a result of avoiding general anesthesia. Other population-based studies utilizing different datasets have been able to confirm the beneficial association of the use of regional anesthesia and improved outcomes in joint arthroplasty patients. Utilizing data collected for the American College of Surgeons National Surgical Quality Improvement Program, authors were able to show decreased complication rates with the use of regional over general anesthesia and, in the case of total hip replacements, even a decrease in operating room time [7, 8].

1.4.2 Influence of Surgical Invasiveness: Bilateral Elective Total Knee Replacement

Studies evaluating outcomes among unilateral joint arthroplasty recipients have provided a relatively solid literature base. Studies addressing the impact of the type of anesthesia among higher-invasiveness procedures such as bilateral knee replacements, which have been suggested to carry higher morbidity and mortality, are rare [9]. Stundner et al analyzed data from 22,253 patients undergoing bilateral knee arthroplasty. Interestingly, the majority of patients received general anesthesia (80 % general anesthesia, 7 % neuraxial, 13 % combined neuraxial and general anesthesia). The authors reported that neuraxial anesthesia was associated with a lower incidence and lower odds ratios for blood transfusions (44.7, 28.5 and 38.0 %, $p<0.0001$) and an overall reduction of the odds of major in-hospital complications (−16.0 % for neuraxial, −6.0 % for combined neuraxial-general anesthesia) [10]. While more research is needed, these results suggest that regional anesthesia may positively affect outcomes in patients undergoing interventions of higher surgical invasiveness.

1.4.3 Influence of Comorbidities and Age

Intuitively, many anesthesiologists choose regional anesthesia techniques in elderly patients or those with comorbidities in an attempt to blunt surgical stresses in a presumably more risk prone patient collective. In the setting of sleep apnea, the American Society of Anesthesiology has published guidelines suggesting that regional anesthesia should be used in an attempt to reduce complication risk [11]. However, data supporting this approach remain rare. A population based study on outcomes after total lower extremity joint arthroplasty in patients with a confirmed diagnosis of sleep apnea found an almost 20 % reduced risk for major complications when neuraxial was chosen over general anesthesia (OR 0.83 [0.74–0.93]; $P = 0.001$) [12]. In addition, the use of peripheral nerve blocks lead to a reduced need for intensive care services in the same study.

In an attempt to determine whether the benefits seen in previous studies of joint arthroplasty patients were uniform among patient groups of various age and comorbidity burden, the same study group evaluated a national joint arthroplasty cohort. They concluded that although absolute rates of complications were higher in older individuals with cardiopulmonary disease, all patients within their respective groups had lower complication risk with neuraxial techniques [13].

1.4.4 Influence of Anesthesia Type in Non-elective Surgery

In the studies described above, evidence for benefit of regional anesthesia was proven in patients undergoing elective total lower extremity joint replacement. However, for other types of orthopedic surgery, large-scale database studies show mixed results when regional versus general anesthesia was compared. In acute surgical settings, individual interventions may have limited ability to influence outcomes and complication risk, compared to the patient's overall condition and injuries. Hip fracture surgery represents a well-researched example; while Neuman et al. reported a mortality benefit in patients with intra-trochanteric fractures when undergoing surgery under neuraxial anesthesia, patients having surgery after femoral neck fracture did not derive such a result, suggesting that the surgical indication plays an important role in determining outcomes [14]. Subsequently, the same authors performed a study including approximately 60,000 patients undergoing surgery for hip fracture utilizing a New York State-based registry. While the authors found a modestly decreased length of stay in the regional anesthesia group (5 % CI, −0.8 to −0.4, $P < .001$), a difference in mortality could not be detected [15]. A study by Patorno et al on more than 70,000 patients undergoing surgery for hip fracture yielded similar results in terms of all cause mortality when regional anesthesia was compared to general anesthesia [16]. The authors argue that previously reported advantages of regional anesthesia with regard to this outcome may exist but might have been overestimated. It should be noted that despite the fact that some investigations could not identify a benefit of regional over general anesthesia, virtually no evidence exists that the use of neuraxial anesthesia performs worse.

1.4.5 *Peripheral Nerve Blocks and Outcomes*

Evidence on differential perioperative outcomes using regional anesthetic techniques other than neuraxial anesthesia is still sparse. For shoulder surgery, a population based study compared complication incidence in 17,157 patients undergoing total shoulder arthroplasty. Of those, approximately 20 % received a nerve block in combination with general anesthesia [17]. There were no differences in the incidence of mortality, perioperative pulmonary, cardiac, renal or neurologic complications, as well as in need for mechanical ventilation, ICU admission or blood product transfusion. Moreover, there was no difference in length of stay between groups. These findings of equal safety promote the addition of a nerve block to general anesthesia, particularly in light of the analgesic benefits invoked through this step.

1.5 General Surgery: Complications and Mortality

1.5.1 *Open Colectomy*

The impact of anesthetic technique has been less well studied in the non-orthopedic patient population, but the literature is evolving in light of encouraging results. Outcomes among patients undergoing general surgical procedures were reported in a study by Poeran et al. The authors compared outcomes of patients undergoing open colectomy under general anesthesia with or without addition of neuraxial anesthesia between 2006 and 2012 [18]. The majority of patients underwent surgery with general anesthesia alone (93.9 % vs 6.1 %). While the group receiving epidural anesthesia had significantly lower risk for thromboembolism and cerebrovascular events, the association was non-significant for wound infections, pneumonia and mechanical ventilation. Moreover, the risk for acute myocardial infarction, urinary tract infection, postoperative ileus, blood transfusion and critical care admission was higher when regional anesthesia was utilized. No clear pattern of advantages invoked by addition of neuraxial to general anesthesia could be detected and significant cofounding may have influenced results, including the extent of surgery, which may have been more invasive in patients in whom neuraxial techniques were added for better pain control.

1.5.2 *Lung Resection*

Another population-based study on open lung resection patients published by Özbek et al found lower adjusted odds ratios for blood transfusion (OR 0.82 [0.69–0.98]) and mechanical ventilation (OR 0.81 [0.67–0.98]) in patients receiving combined neuraxial-general anesthesia, compared to those receiving general anesthesia alone.

However, in the same group, higher odds were detected for deep vein thrombosis (OR 1.50 [1.01–2.23]), suggesting higher thromboembolic risk when regional anesthesia was utilized [19]. This might be explained by differences in the use of pharmacologic thromboprophylaxis in those with epidural catheters.

1.6 Outcomes Beyond Complications and Mortality

1.6.1 Functional Outcomes

An important determinant of surgical success is functional outcome, as the primary goal of surgery is preservation or restoration of function. Quality of life can be significantly influenced by how well patients are able perform daily activities. To date, available evidence on how perioperative management influences long-term functional outcome is scarce and controversial. Most of the limited information available stems from the orthopedic literature. In such studies, function is often evaluated by the ability to perform range-of-motion tests, "time up and go"-measures or dedicated scores like the Short Musculoskeletal Function Assessment (SFA) or Disability of the Arm, Shoulder and Hand (DASH) questionnaire [20]. In surgery to the upper extremity, numerous studies found better short-term range of motion and more effective attainment of rehabilitation goals (mostly due to less pain) in patients receiving regional anesthesia, compared to those undergoing their procedures under general anesthesia [21]. However, while higher satisfaction scores among patients receiving regional anesthesia and better analgesic profile clearly favor this modality, long-term results to date fail to show significant differences.

1.6.2 Oncologic Outcomes

Recent publications have suggested that the type of anesthesia may impact on cancer recurrence, however considerable controversy persists and more research is needed [22]. Most studies available were generated from retrospective chart review, rendering causal inference difficult. A meta-analysis by Chen et al synthesized 14 studies analyzing overall survival and recurrence-free survival after oncologic surgery with or without epidural anesthesia [23]. For overall survival, the addition of epidural to general anesthesia was found to confer benefits compared to general anesthesia alone. This was especially the case in patients suffering from colorectal cancer; however, these findings could not be reproduced for recurrence-free survival. Further, synthesis of results available to date is severely limited by inconsistencies in observation time frame, outcomes reported, and type of malignancy studied, as noted in a Cochrane review by Cakmakkaya et al [24]. However, with an increasing number of prospective, randomized studies currently ongoing, more reliable evidence is expected to be available in the near future.

1.6.3 Infectious Outcomes

Regional anesthesia was suggested to blunt immunological responses to surgical insults, and thereby also reduce the risk of systemic and localized infection. Liu et al analyzed the American College of Surgeons National Surgical Quality Improvement Program (NSQIP) database, comparing infectious complication incidences between patients undergoing total knee arthroplasty under general versus neuraxial anesthesia. The authors found lower adjusted odds ratios of pneumonia (OR 0.51 [0.29–0.90]) and composite systemic infections (OR 0.77 [0.64–0.92]) in the latter group [25]. Moreover, while postoperative surgical site infection is of high concern throughout all surgical subspecialties, it represents an even more problematic issue in orthopedic surgery, where implantation of foreign material prone to bacterial colonization is often the norm. Utilizing a health insurance data source from Taiwan, Chang et al were able to report an odds ratio of 2.21 ([1.25–3.90], $P=0.007$) for surgical site infection in patients undergoing total lower extremity joint replacement under general anesthesia, compared to regional anesthesia [26].

1.6.4 Chronic Pain

Similarly, evidence on the impact of regional anesthesia on development of persistent postsurgical pain is sparse and inconsistent. While ample preclinical evidence is available (the concept of "preemptive analgesia" [27]), only few studies were able to prove this benefit in clinical practice. A Cochrane meta-analysis by Andreae and Andreae found significantly lower risk for the development of persistent pain after thoracotomy or breast cancer surgery when epidural or paravertebral anesthesia were used, respectively (OR 0.33 (95 % CI 0.20–0.56) and OR 0.37 (95 % CI 0.14–0.94)) [28]. However, the authors acknowledge that small sample size, performance bias, attrition and inconsistent reporting of outcomes in the studies included weaken their conclusions. Thus, further research is warranted to evaluate the impact of regional anesthesia on this potentially debilitating outcome.

1.7 Blood and Medication Management

1.7.1 Management of Bleeding, Blood and Coagulation

Various perioperative blood management strategies have been described in the literature and are available. Liberal transfusion strategies are increasingly viewed skeptically, as their ability to improve outcome has been disputed [29]. Subsequently, a restrictive transfusion regimen using lower hemoglobin triggers was found to reduce cardiac events, re-bleeding, bacterial infection and total mortality in a recent large-scale meta-analysis [30]. As a consequence, approaches to proactively counteract blood loss are highly sought after. A number of intraoperative interventions controlled by anesthesiologists have been shown to reduce the extent of blood loss and need for allogeneic blood transfusions. In this context, the use of neuraxial anesthesia has been consistently linked to reduced bleeding and need for transfusions [31], likely due to sympathectomy-induced avoidance of blood pressure spikes. A specific technique employing controlled hypotension via this approach has been described and successfully used in many surgical settings [32, 33]. Further, the use of blood salvage equipment and hemodilution techniques remain within the anesthesiologist's armamentarium. Pharmacologic means to reduce bleeding such as antifibrinolytics have gained wide popularity. Tranexamic acid was found to reduce the need for blood transfusion in numerous studies [34]. A population based study examining complication incidence of 872,416 patients undergoing elective total hip or knee arthroplasty found lower rates of allogeneic or autologous transfusion requirement (7.7 % vs 20.1 %), lower incidence of thromboembolic complications (0.6 % vs 0.8 %), acute renal failure (1.2 % vs 1.6 %) and combined complications (1.2 % vs 1.6 %; all $p<0.01$) among those patients that received tranexamic acid, compared to those that did not. These data contribute to the notion that tranexamic acid administration might reduce need for blood transfusions and associated complications, while at the same time not increasing the risk for prothrombotic complications [35].

1.7.2 Cardio-protective Drugs

Anesthesiologists are frequently faced with the decision to administer cardioprotective medications. There has been considerable controversy as to whether the intake of cardioprotective drugs, including beta-blockers, aspirin and statins, should be maintained throughout, or even started in the perioperative period in patients at high risk for cardiovascular complications. One of the largest studies to examine the effect of perioperative beta-blockade is the Perioperative Ischemic Evaluation (POISE) trial [36]. Here, 8,351 patients with or at risk for cardiac atherosclerotic disease were randomized to receive extended-release metoprolol succinate versus placebo before and up to 30 days after surgery. While myocardial infarction was less frequent in the metoprolol group (4.2 % vs 5.7 %, $p=0.0017$), stroke incidence and all-cause mortality was significantly higher (1.0 % vs. 0.5 %, $p=0.0053$ and 3.1 % vs 2.3 %, $p=0.0317$). By contrast, studies in cardiac surgery provided evidence of a significantly lower risk of postoperative supraventricular and ventricular arrhythmias [37]. These advantages must be carefully weighed against the potential complications. Similarly, perioperative statin therapy was shown to reduce cardiac mortality after primary coronary artery bypass graft surgery [38], but evidence is insufficient to prove benefits in non-cardiac surgery [39]. The role of aspirin is even more controversial, with results from the recent POISE2 trial reporting a higher incidence of major bleeding without changes in myocardial ischemia or mortality when the drug was continued or commenced in the perioperative period [40]. Similar lack of benefit for these outcomes was seen with the perioperative use of clonidine [41].

1.8 The Economic and Public Health Impact

According to estimates based on economic data of all 192 member states of the world health organization, more than 230 million surgeries are performed annually worldwide [42]. In light of this vast number, even small reductions in complications and individual expenditure have a large cumulative impact. Given the encouraging data on the positive outcomes associated with regional anesthesia – and the consistently low utilization rates of these techniques across many population based studies – an expansion in use could potentially lead to significant improvements in the health of entire nations.

As outlined above, management choices made by the anesthesiologist – for instance timely detection of high-risk comorbidities, appropriate perioperative drug maintenance, and use of regional anesthesia where appropriate – can contribute to lower complication rates, lower need for advanced treatment modalities, and shorter hospital stay.

Despite the increasing awareness for this interrelation, only very few economic analyses focusing on the direct comparative impact of anesthesiologic and/or perioperative medical interventions have emerged so far. One challenge is likely related to the inability to reliably identify realistic procedure-related cost – including ancillary expenses associated with complications, readmission and other outcomes [43]. Even more difficult is the comparison of such differences across institutions, let alone across different countries and health care systems.

To this date, monetary benefits have been verified most clearly comparing different anesthetic techniques to each other: numerous studies evidence lower cost of hospitalization, shift to less expensive analgesic modalities, shorter hospital stay, outpatient management of comparatively painful procedures, shorter anesthesia-controlled (i.e. non-surgical) operating room time, and lower rates of hospital readmissions when regional anesthesia is used for a subset of surgical procedures [6, 44–46].

The economic impact of perioperative care improvement is a topic of high interest in an environment of value-based payment initiatives. The actual process cycle of undergoing a surgical procedure has previously been fragmented across multiple disciplines, physical locations and individual caretakers. To streamline care, a "triple aim" of better health, improved delivery of healthcare and reduced cost has been targeted and conceptualized as a Perioperative Surgical Home model [47]. Ideally, the patient should be provided with a better experience through seamless continuity of care, and practitioners should benefit from facilitated interaction with their perioperative team peers at numerous boundary points. Most available studies evidence clear advantages of surgical home-like models in terms of quality of care metrics, patient satisfaction, and cost outcomes [48]. However, with the implementation still remaining in an early stage, future studies will have to prove its effectiveness and efficiency on a greater scale. It must be mentioned, that similarly to the medical field, anesthesiology specialty societies have for many years worked on promoting the Perioperative Surgical Home model, with many institutions reporting successes [49].

1.9 Conclusion

Anesthesiologists as members of the perioperative team, and their care choices, are increasingly identified as influential in shaping outcomes by means of choice of the anesthetic approach as well as the perioperative management. Quality of perioperative care is a major determinant of immediate, short-term and longer-term complications, functional outcomes, and quality of life. The literature provides ample evidence regarding the superiority of regional over general anesthesia in terms of mortality or complication risk. Despite some inconclusive study results, virtually no investigations show inferiority of regional techniques compared to the general approach. Most consistent benefits of regional anesthesia can be found among elective orthopedic surgical patients. Taking factors like low anesthetic procedure related complications and improved pain control into consideration, regional anesthesia may represent an effective modality to reduce complications for selected patient populations. Beyond streamlining the perioperative handling through optimization of such management and implementation of perioperative home models (with anesthesiologists at key positions), perioperative medicine is on its way to gaining more recognition as a discrete discipline, allowing for focused and goal-oriented evaluation of its effectiveness. Moreover, with research in the field evolving and generating new insights almost by the day, treatment paradigms and institutional standards of care must be re-evaluated and adapted frequently. Interest in the monetary implications of perioperative care is growing rapidly, in part as a result of the increasing pressure on policymakers to economize health care and to appropriately partition scarce resources. Critical evaluation of patient outcome will become even more central to this discussion as value-based payment paradigms are put into place over the course of the next years. In light of these developments, the duties of the anesthesiologist shift from provision of service to comprehensive stewardship in the perioperative arena.

References

1. Fleischut PM, Mazumdar M, Memtsoudis SG (2013) Perioperative database research: possibilities and pitfalls. Br J Anaesth 111(4):532–534
2. Moskowitz A, McSparron J, Stone DJ, Celi LA (2015) Preparing a new generation of clinicians for the era of big data. Harv Med Stud Rev 2(1):24–27
3. Rodgers A, Walker N, Schug S et al (2000) Reduction of postoperative mortality and morbidity with epidural or spinal anaesthesia: results from overview of randomised trials. BMJ 321(7275):1493
4. Macfarlane AJ, Prasad GA, Chan VW, Brull R (2009) Does regional anesthesia improve outcome after total knee arthroplasty? Clin Orthop Relat Res 467(9):2379–2402
5. Hu S, Zhang ZY, Hua YQ, Li J, Cai ZD (2009) A comparison of regional and general anaesthesia for total replacement of the hip or knee: a meta-analysis. J Bone Joint Surg Br 91(7):935–942
6. Memtsoudis SG, Sun X, Chiu YL et al (2013) Perioperative comparative effectiveness of anesthetic technique in orthopedic patients. Anesthesiology 118(5):1046–1058

7. Basques BA, Toy JO, Bohl DD, Golinvaux NS, Grauer JN (2015) General compared with spinal anesthesia for total hip arthroplasty. J Bone Joint Surg Am 97(6):455–461
8. Pugely AJ, Martin CT, Gao Y, Mendoza-Lattes S, Callaghan JJ (2013) Differences in short-term complications between spinal and general anesthesia for primary total knee arthroplasty. J Bone Joint Surg Am 95(3):193–199
9. Memtsoudis SG, Ma Y, González Della Valle A et al (2009) Perioperative outcomes after unilateral and bilateral total knee arthroplasty. Anesthesiology 111(6):1206–1216
10. Stundner O, Chiu YL, Sun X et al (2012) Comparative perioperative outcomes associated with neuraxial versus general anesthesia for simultaneous bilateral total knee arthroplasty. Reg Anesth Pain Med 37(6):638–644
11. American Society of Anesthesiologists Task Force on Perioperative Management of patients with obstructive sleep apnea (2014) Practice guidelines for the perioperative management of patients with obstructive sleep apnea: an updated report by the American Society of Anesthesiologists Task Force on Perioperative Management of patients with obstructive sleep apnea. Anesthesiology 120(2):268–286
12. Memtsoudis SG, Stundner O, Rasul R et al (2013) Sleep apnea and total joint arthroplasty under various types of anesthesia: a population-based study of perioperative outcomes. Reg Anesth Pain Med 38(4):274–281
13. Memtsoudis SG, Rasul R, Suzuki S et al (2014) Does the impact of the type of anesthesia on outcomes differ by patient age and comorbidity burden? Reg Anesth Pain Med 39(2):112–119
14. Neuman MD, Silber JH, Elkassabany NM, Ludwig JM, Fleisher LA (2012) Comparative effectiveness of regional versus general anesthesia for hip fracture surgery in adults. Anesthesiology 117(1):72–92
15. Neuman MD, Rosenbaum PR, Ludwig JM, Zubizarreta JR, Silber JH (2014) Anesthesia technique, mortality, and length of stay after hip fracture surgery. JAMA 311(24):2508–2517
16. Patorno E, Neuman MD, Schneeweiss S, Mogun H, Bateman BT (2014) Comparative safety of anesthetic type for hip fracture surgery in adults: retrospective cohort study. BMJ 348:g4022
17. Stundner O, Rasul R, Chiu YL et al (2014) Peripheral nerve blocks in shoulder arthroplasty: how do they influence complications and length of stay? Clin Orthop Relat Res 472(5):1482–1488
18. Poeran J, Yeo H, Rasul R, Opperer M, Memtsoudis SG, Mazumdar M (2015) Anesthesia type and perioperative outcome: open colectomies in the United States. J Surg Res 193(2):684–692
19. Ozbek U, Poeran J, Mazumdar M, Memtsoudis SG (2015) Patient safety and comparative effectiveness of anesthetic technique in open lung resections. Chest 148(3):722–730
20. Stundner O, Ortmaier R, Memtsoudis SG (2014) Which outcomes related to regional anesthesia are most important for orthopedic surgery patients? Anesthesiol Clin 32(4):809–821
21. Bernucci F, Carli F (2012) Functional outcome after major orthopedic surgery: the role of regional anesthesia redefined. Curr Opin Anaesthesiol 25(5):621–628
22. Cassinello F, Prieto I, Del Olmo M, Rivas S, Strichartz GR (2015) Cancer surgery: how may anesthesia influence outcome? J Clin Anesth
23. Chen WK, Miao CH (2013) The effect of anesthetic technique on survival in human cancers: a meta-analysis of retrospective and prospective studies. PLoS One 8(2), e56540
24. Cakmakkaya OS, Kolodzie K, Apfel CC, Pace NL (2014) Anaesthetic techniques for risk of malignant tumour recurrence. Cochrane Database Syst Rev (11):CD008877
25. Liu J, Ma C, Elkassabany N, Fleisher LA, Neuman MD (2013) Neuraxial anesthesia decreases postoperative systemic infection risk compared with general anesthesia in knee arthroplasty. Anesth Analg 117(4):1010–1016
26. Chang CC, Lin HC, Lin HW, Lin HC (2010) Anesthetic management and surgical site infections in total hip or knee replacement: a population-based study. Anesthesiology 113(2):279–284
27. Katz J, Clarke H, Seltzer Z (2011) Review article: preventive analgesia: quo vadimus? Anesth Analg 113(5):1242–1253
28. Andreae MH, Andreae DA (2013) Regional anaesthesia to prevent chronic pain after surgery: a Cochrane systematic review and meta-analysis. Br J Anaesth 111(5):711–720
29. Carson JL, Terrin ML, Noveck H et al (2011) Liberal or restrictive transfusion in high-risk patients after hip surgery. N Engl J Med 365(26):2453–2462

30. Salpeter SR, Buckley JS, Chatterjee S (2014) Impact of more restrictive blood transfusion strategies on clinical outcomes: a meta-analysis and systematic review. Am J Med 127(2):124–131.e3
31. Guay J (2006) The effect of neuraxial blocks on surgical blood loss and blood transfusion requirements: a meta-analysis. J Clin Anesth 18(2):124–128
32. Kiss H, Raffl M, Neumann D, Hutter J, Dorn U (2005) Epinephrine-augmented hypotensive epidural anesthesia replaces tourniquet use in total knee replacement. Clin Orthop Relat Res 436:184–189
33. Sharrock NE, Mineo R, Urquhart B, Salvati EA (1993) The effect of two levels of hypotension on intraoperative blood loss during total hip arthroplasty performed under lumbar epidural anesthesia. Anesth Analg 76(3):580–584
34. Perel P, Ker K, Morales Uribe CH, Roberts I (2013) Tranexamic acid for reducing mortality in emergency and urgent surgery. Cochrane Database Syst Rev (1):CD010245
35. Poeran J, Rasul R, Suzuki S et al (2014) Tranexamic acid use and postoperative outcomes in patients undergoing total hip or knee arthroplasty in the United States: retrospective analysis of effectiveness and safety. BMJ 349:g4829
36. Devereaux PJ, Yang H, Yusuf S et al (2008) Effects of extended-release metoprolol succinate in patients undergoing non-cardiac surgery (POISE trial): a randomised controlled trial. Lancet 371(9627):1839–1847
37. Blessberger H, Kammler J, Domanovits H et al. (2014) Perioperative beta-blockers for preventing surgery-related mortality and morbidity. Cochrane Database Syst Rev (9):CD004476
38. Collard CD, Body SC, Shernan SK, Wang S, Mangano DT (2006) Preoperative statin therapy is associated with reduced cardiac mortality after coronary artery bypass graft surgery. J Thorac Cardiovasc Surg 132(2):392–400
39. Sanders RD, Nicholson A, Lewis SR, Smith AF, Alderson P (2013) Perioperative statin therapy for improving outcomes during and after noncardiac vascular surgery. Cochrane Database Syst Rev (7):CD009971
40. Devereaux PJ, Mrkobrada M, Sessler DI et al (2014) Aspirin in patients undergoing noncardiac surgery. N Engl J Med 370(16):1494–1503
41. Devereaux PJ, Sessler DI, Leslie K et al (2014) Clonidine in patients undergoing noncardiac surgery. N Engl J Med 370(16):1504–1513
42. Weiser TG, Regenbogen SE, Thompson KD et al (2008) An estimation of the global volume of surgery: a modelling strategy based on available data. Lancet 372(9633):139–144
43. Martin J, Cheng D (2013) Role of the anesthesiologist in the wider governance of healthcare and health economics. Can J Anaesth 60(9):918–928
44. Duncan CM, Hall Long K, Warner DO, Hebl JR (2009) The economic implications of a multimodal analgesic regimen for patients undergoing major orthopedic surgery: a comparative study of direct costs. Reg Anesth Pain Med 34(4):301–307
45. Macario A, McCoy M (2003) The pharmacy cost of delivering postoperative analgesia to patients undergoing joint replacement surgery. J Pain 4(1):22–28
46. Mariano ER, Chu LF, Peinado CR, Mazzei WJ (2009) Anesthesia-controlled time and turnover time for ambulatory upper extremity surgery performed with regional versus general anesthesia. J Clin Anesth 21(4):253–257
47. Cannesson M, Kain Z (2015) The Perioperative Surgical Home: An Innovative Clinical Care Delivery Model. J Clin Anesth 27(3):185–187
48. Kash BA, Zhang Y, Cline KM, Menser T, Miller TR (2014) The perioperative surgical home (PSH): a comprehensive review of US and non-US studies shows predominantly positive quality and cost outcomes. Milbank Q 92(4):796–821
49. American Society of Anesthesiologists. Perioperative surgical home. https://www.asahq.org/psh. Accessed 05 Apr 2015

Chapter 2
Prehabilitation

James Durrand, Rhiannon Hackett, David Yates, and Gerard Danjoux

Abstract Despite advances in surgical care over recent decades, major surgery still carries a significant burden of morbidity. Perioperative research and improvement strategies have classically focused on the intraoperative and postoperative phase of the surgical journey. Prehabilitation describes the concept of optimizing postoperative outcome through preoperative intervention. There are many potential targets for prehabilitation ranging from the identification and management of abnormal preoperative test results, through optimising functional status using exercise programmes, to psychological interventions addressing unhealthy behaviours known to impact on postoperative outcome.

The preoperative assessment of surgical patients has, to date, focused on 'single organ pathology'- the detection and management of ischaemic heart disease or chronic obstructive pulmonary disease for instance. Whilst the preoperative optimisation of these comorbidities is of the utmost importance, recent evidence has emerged highlighting the detrimental impact of other, previously neglected, factors that are prevalent at significant levels in the surgical population. These factors include a variety of habitual activities such as smoking, hazardous alcohol intake, obesity and sedentary behaviour as well as comorbid states over which an individual may have less control- malnutrition, anaemia, frailty and cognitive dysfunction.

In this chapter the prevalence, relevance and impact of these factors on postoperative outcomes are presented. Potential prehabilitation strategies to address and optimize patient outcomes are described.

J. Durrand • R. Hackett
Northern School of Anaesthesia and Intensive Care Medicine, Newcastle, UK
e-mail: jdurrand@doctors.org.uk; Rhiannon.hackett@doctors.org.uk

D. Yates
Anaesthesia and Critical Care, York Teaching Hospitals NHS Foundation Trust, York, UK
e-mail: david.yates@york.nhs.uk

G. Danjoux (✉)
Department of Anaesthesia, South Tees Hospitals NHS Foundation trust, Middlesbrough, UK
e-mail: Gerard.danjoux@stees.nhs.uk

K. Stuart-Smith (ed.), *Perioperative Medicine – Current Controversies*,
DOI 10.1007/978-3-319-28821-5_2

Keywords Prehabilitation • Preoperative intervention • Exercise • Frailty • Smoking • Nutrition • Anaemia • Alcohol

Abbreviations

AUDIT-C	The Alcohol Use Disorders Identification Test
RCT	Randomised Controlled Trial
NNT	Number Needed to Treat

2.1 Introduction

The term ‘Prehabilitation’ has evolved rapidly in the medical literature over a short period of time. The exact contextual definition appears to be relatively flexible, however there is a common theme of early intervention aimed at subsequent injury prevention. In the setting of preoperative medicine this can have a broad remit. In this chapter we present the concept of prehabilitation as recognition and optimisation of several unhealthy patient behaviours, and lesser-recognised or ‘marginal’ comorbidities, which are often neglected in the preoperative setting. Nonetheless each entity has a developing evidence-base for adverse perioperative outcome without optimisation prior to surgery.

Traditionally in medical practice clinicians often identify and treat organ specific comorbidities in isolation (e.g. cardiac or respiratory disease), whether in the preoperative period or beyond. The concept of prehabilitation facilitates a move away from this, towards more integrated, comprehensive and proactive patient management. Proactive, comprehensive patient assessment and optimisation has a developing evidence-base in high-risk elderly patients perioperatively, through implementation of Comprehensive Geriatric Assessment (CGA) programmes [1]. There are non-medical parallels as to the benefits of such a comprehensive strategy through the ‘Aggregation of Marginal Gains’ notably elite level sport, with the British Olympic Cycling Team representing a contemporary example of marginal gains theory in action. Through implementation of such an approach, the combined magnitude of benefit through addressing several, seemingly insignificant components, is far superior to the sum of the individual parts. Translating this strategy into medicine makes intuitive sense, with the preoperative setting representing an ideal opportunity [2].

Although the prevalence and impact of unhealthy behaviours and comorbidities we have identified could not be described as marginal, the above concept holds true. Early comprehensive patient assessment and management can optimise problems likely to impact on surgical outcome, which are often remediable in a short period of time available prior to surgery, through simple behavioural or medical interventions. Where intervention may not always be possible e.g. cognitive impairment, simple problem identification in a timely fashion strengthens appropriate shared decision-making between healthcare professionals and patient/family.

Managing patients in this way requires a challenging behavioural change for healthcare professionals and undoubted need for closer collaborative working across primary and secondary care. Future implementation of prehabilitation programmes is likely to achieve an unmet need in perioperative patient management, facilitating improved surgical outcomes whilst hopefully embedding healthy patient behaviours in the longer term.

2.2 Unhealthy Behaviours and Preoperative Lifestyle Interventions

2.2.1 Achieving Behavioural Change in Healthcare

Human behaviour can be defined as: 'the product of individual or collective human actions, seen within and influenced by their structural, social and economic context' [3]. It can be difficult to change ingrained behaviour [3], with success reliant on identifying appropriate strategies for the delivery and evaluation of planned interventions. There has been extensive research into the determinants of human behaviour and the factors affecting change, with multiple theories existing. Whilst these are beyond the scope of this chapter, there are perhaps important points to consider for those involved in prehabilitation.

Life events, such as the lead up to major surgery, are recognised as important triggers to facilitate behaviour change. With this in mind, the preoperative period may offer a unique opportunity to address lifestyle changes that are likely to lead to both improved surgical outcomes and longer-term health benefits. The combination of personal health focus and interaction with multiple healthcare professionals creates an appropriate environment, and opportunity, to embed new and healthy behaviours. It is well recognized that multiple unhealthy behaviours tend to co-exist, such as smoking, alcohol excess, lack of physical activity and poor nutritional status. There is evidence that all of these factors can independently influence perioperative outcomes as outlined elsewhere in this chapter. At first glance it may appear a tall order to tackle multiple unhealthy behaviours simultaneously, however there is evidence to the contrary that this strategy may indeed be more effective [4].

In order to change unhealthy behaviours, a number of factors are required. These include knowledge, motivation, the presence of a trigger, available opportunity to address issues, the skills to be able to change, the ability to empower or support the individual through the process and an environment in which to facilitate change [5]. These are all factors that could be available to varying degrees within the preoperative process, requiring close collaboration between patient and healthcare professional.

There remains little consensus as to the optimal approach to achieve this change. Combinations of ideas from coaching methods, psychological concepts and motivational interviewing would possibly provide the platform for healthcare professionals to support patients in achieving change. Goal setting, motivational support and an understanding of an individual patients' social background/environment are cornerstones in achieving sustainable change. Brief encounter and longer duration interactions have both been utilised with varying degrees of success in order to achieve this [6].

The possibility of achieving sustainable positive behavioural changes within the preoperative setting requires significantly more research. Clearly this does represent a 'teachable moment' where patients may be amenable to addressing several unhealthy behaviours simultaneously and a powerful opportunity for healthcare professionals to embed good habits. The combined marginal gains likely to be achieved through such an approach are compelling from such a safe, relatively low cost and easy to deliver intervention. If lifestyle changes for long term health after surgery can be encouraged and embedded, with the large numbers of patients attending for surgery, then the potential impact on population health and its subsequent economic burden would be significant.

2.2.2 Inactivity and Exercise

International consensus statements from the World Health Organisation (WHO) and National Institute for Health and Care Excellence (NICE) recommend 150 min of moderate intensity exercise weekly, for adults >65 years, to maintain a healthy lifestyle [7, 8]. Alternatively, 75 min of more vigorous weekly exercise is recommended. Both should be supplemented by muscle strengthening activities. Behavioural changes associated with modern society have resulted in less than one-third of adults >65 years of age achieving this recommendation [9, 10]. To compound this, background physical activity levels have dropped considerably over the last five decades with approximately 50 % of adults now spending >5 h per day sedentary e.g. desk work, watching television [9, 11]. There is clear evidence that sedentary time is associated with reduced aerobic fitness and a two to threefold increased risk of cardiovascular disease and death, diabetes and all cause mortality [12]. Inactivity is therefore associated with reduced aerobic fitness levels and increased likelihood of significant comorbid disease.

Modifying patient and healthcare professionals' behaviour in a bid to reverse this trend is challenging. This is despite compelling evidence that undertaking regular exercise can reduce the risk of developing dementia, type 2 diabetes, some cancers, heart disease and all-cause mortality by at least 30 % [5]. For a wide spectrum of chronic diseases regular exercise has also demonstrated significant improvements, often superior to drugs and with a lower side-effect profile. Such conditions include; coronary disease and heart failure, peripheral vascular disease, hypertension, Chronic Obstructive Pulmonary Disease (COPD), cancer and diabetes. It is therefore little wonder that exercise has recently been referred to as 'the miracle cure' in a publication by the Academy of Medical Royal Colleges [5].

2.2.2.1 Perioperative Implications

The introduction of objective exercise testing preoperatively over recent years demonstrates a clear association between aerobic fitness levels and outcome following surgery, with unfit patients at increased risk [13]. Several recent publications, across differing surgical specialties, have demonstrated an association between physical fitness and surgical outcome in the short to medium-term [13–15]. Identification of patients with poor aerobic fitness levels, who may be at greater risk, remains challenging. Limitations in subjective patient self-reported exercise capacity [16, 17] have prompted many units to employ more formal objective assessments of aerobic fitness preoperatively. Several testing modalities have been employed in this regard including; timed stair climbing, 6-min walk test, shuttle-walk test and cardiopulmonary exercise testing (CPET). Further discussion of the evidence base in this area is beyond the scope of this chapter.

The recognition of the association between objectively evaluated fitness and surgical outcome supports the compelling idea that a structured and active improvement of fitness levels, in targeted clinical populations prior to surgery, may yield substantial benefits to perioperative outcome. This growing interest in preoperative exercise training is not yet matched by sufficient prospective RCTs in surgical populations, however a signal is developing.

2.2.2.2 Prehabilitation Strategies

Aerobic Training

Two recent systematic reviews [18, 19] have highlighted the dearth of published studies demonstrating improved perioperative outcome secondary to preoperative aerobic exercise interventions. Only 1 study was identified in this respect, which utilised an 8-week moderate intensity exercise programme prior to low-risk cardiac bypass surgery. The intervention group demonstrated a significant reduction in overall hospital length of stay but no obvious survival benefit [20].

Other published research has concentrated on the concept that fitness improvements are achievable in preoperative patient populations. A study published in 2009 [21] demonstrated that patient fitness levels could be significantly improved with a 6-week supervised, cycle-based, moderate-intensity training programme (NNT = 5). There were a number of non-responders (participants in whom there were no improvements in fitness) in the study cohort, which assessed patients with Abdominal Aortic Aneurysm disease under surveillance. The authors concluded that higher volumes of exercise could have achieved greater fitness improvements. The concept of responders and non-responders to exercise is well recognised, however there is evidence that this problem can be partially offset in clinical populations through increasing exercise volume [22].

Focus has more recently centred on preoperative exercise initiatives in patients with colorectal cancer. A study published in 2015 [23] demonstrated a 16–19 %

drop in patient fitness levels (measured through CPET) following preoperative chemoradiotherapy (CRT). This fitness decrement was subsequently reversed in patients allocated to receive a supervised, 6-week cycle-based, high intensity interval training (HIIT) programme (3 times per week). At the time of surgery, 8 weeks post-exercise, the intervention group had maintained this level of fitness, however fitness levels in the control group had dropped further, to 34 % below initial baseline levels (pre-CRT). Whilst this study was not powered to detect differences in surgical outcome, these compelling results demonstrate that the negative impact on patient fitness levels induced through CRT can be completely offset through appropriately constructed exercise training, in the often short time available prior to surgery.

To draw further conclusions about the feasibility and benefits of preoperative exercise it is pertinent to evaluate evidence from other settings. Appropriately constructed exercise programmes can create significant fitness improvements within a minimum timeframe of 4 weeks with similar trajectory improvements for young and old patients [24]. Despite this, it is important to note that greater fitness benefits are accrued through longer training programmes [22]. HIIT has demonstrated superior fitness improvements in high-risk heart failure patients when compared to a moderate continuous programme with associated higher levels of reported patient enjoyment [25]. The absolute risk of harm to high-risk cardiac patients through HIIT and moderate intensity exercise appears to be low. A recent publication has put this risk at 1 cardiac event per 23,000 and 130,000 patient exercise hours for HIIT and moderate intensity training respectively [26]. This risk may be increased substantially in previously sedentary individuals, however the absolute risk of harm still remains low [27]. This evidence reinforces that significant fitness improvements can be made in all age groups, safely, within the short timeframes often available prior to surgery.

At the time of writing the authors are aware of 2 ongoing feasibility studies assessing the benefits of preoperative HIIT prior to Abdominal Aortic Aneurysm repair, and surgery for colorectal cancer [28, 29]. It is hoped that this critical preliminary work will form the platform for future definitive studies required to answer the ultimate question as to whether preoperative exercise can lead to significant improvements in survival following major surgery.

Musculoskeletal and Strength Training

A systematic review published in 2011 identified seven studies that had employed musculoskeletal or strength training prior to lower limb joint replacement surgery [18]. Studies employed a range of preoperative interventions, however outcome measures were generally recorded as hospital length of stay (LOS) and complications. The authors concluded that patients receiving preoperative training, versus usual care, had a non-significant reduction in postoperative complications but no benefit with regards LOS. It makes intuitive sense that undertaking preoperative strength training should translate to improved outcome for certain patient groups. This may include patients identified with frailty and/or sarcopaenia. Further studies are awaited in this respect.

2.2.2.3 Patient Attitudes and Clinical Implementation

There appears to be a generally positive attitude amongst patients to the concept of preoperative exercise training, with health related quality of life generally improved amongst participants [19]. Compliance and enjoyment appear superior with supervised, group exercise [30] employing higher intensity programmes [25].

The cornerstone of clinical implementation is to increase general population activity levels and recommend exercise within the primary care setting. This would require a major population behavioural change, through mass education, to reverse relentless trends that have developed over five decades. Ultimately this may not prove to be a feasible option. As such, it may be more appropriate to concentrate efforts and resources on specific patient groups presenting for intermediate to high-risk surgery, who are likely to achieve greatest benefit. In this regard, previously sedentary individuals appear to achieve greatest fitness benefits in the short term by moving to low and moderate levels of fitness. Additionally, it would appear that patient groups undergoing preoperative CRT are also likely to achieve significant benefit with exercise prior to surgery, through restoring baseline fitness levels.

Implementation of exercise initiatives at local level will vary and depend upon several factors including infrastructure and resources. Supervised group sessions conducted in primary or secondary care are likely to achieve greatest benefit, enjoyment and patient compliance. Implementation could be alongside existing initiatives such as cardiac or pulmonary rehabilitation programmes. Where such opportunities are not presently available, a more pragmatic approach may be required to allow the maximum number of patients the opportunity to improve fitness before surgery. In our institution we are trialing exercise advice given to patients within the preoperative assessment setting through an information leaflet reinforced with verbal advice. We advocate the use of moderate intensity exercise (or activity) aiming to achieve at least 3 30-min sessions per week or ideally 5 sessions per week as per international guidance [7, 8]. Figure 2.1 demonstrates this guidance.

2.2.2.4 General Recommendations

- Sedentary individuals should be identified well in advance of surgery with increased activity levels and regular exercise encouraged
- Moderate intensity exercise for a minimum of 4 weeks (consistent with WHO recommendation) represents an achievable, safe and likely beneficial intervention prior to surgery
- At a minimum patients should be provided with written exercise information preoperatively, reinforced with verbal encouragement
- Where exercise on prescription or group classes are available in primary or secondary care, patients should be encouraged to partake due to greater enjoyment and compliance compared with undertaking exercise in isolation

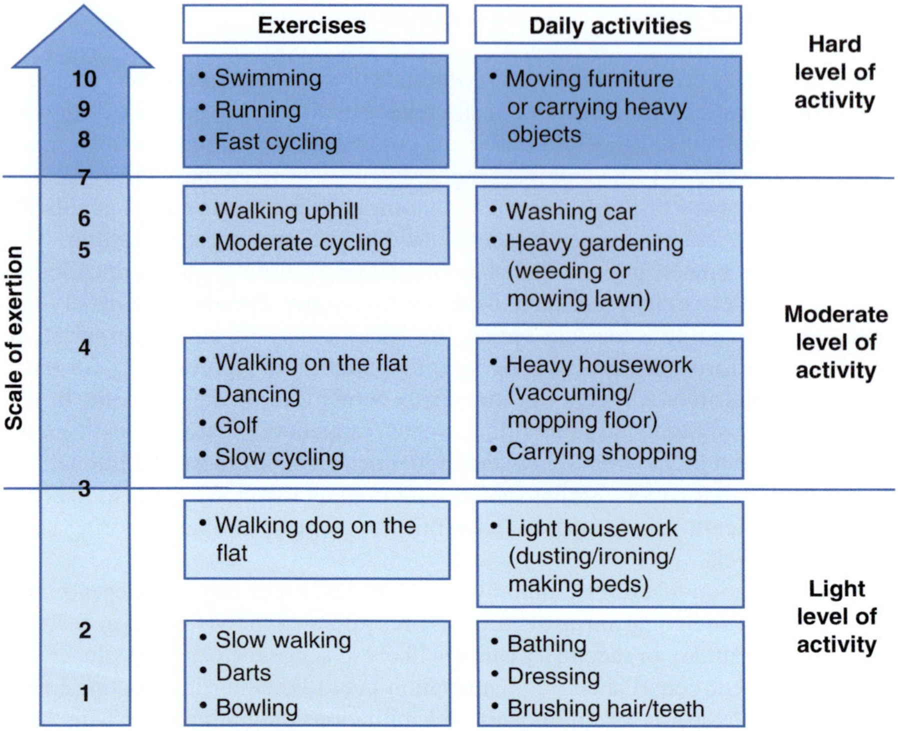

Fig. 2.1 Excerpt from patient information leaflet on exercise prior to surgery. Patients are encouraged to exercise between levels 3–7

2.2.3 Alcohol

In a 2011 UK health survey 23 % of men and 18 % of women reported a weekly alcohol consumption in excess of government recommended levels: 21–28 units per week for men and 14–21 units per week for women. A quarter of those drinking excessively were reported to be consuming alcohol at a very high risk level; more than 50 units per week for men and 35 units per week for women [31]. Whilst the common perception is that it is the younger generation who are consuming more and more alcohol, another study documented that it was the older age group, the over 65's who were more likely to be drinking alcohol every day of the week (17 % of men and 9 % of women over 65) rather than the 16–24 age group (1 % of both men and women) [32]. Managing excessive alcohol consumption presents a genuine challenge to perioperative teams aiming to effectively prehabilitate patients in the lead up to major surgery. Whilst chronic alcoholism remains a leading international public health concern, lesser levels of consumption also appear to confer an increased risk of poor postoperative outcome. People consuming hazardous levels of alcohol are also more likely to indulge in other unhealthy behaviours e.g. smoking [4].

2.2.3.1 Perioperative Implications

The association between established liver cirrhosis and postoperative morbidity and mortality is stark [33]. Whilst chronically elevated alcohol intake resulting in end organ damage will clearly increase surgical risk, heavy drinking has deleterious effects beyond the liver, pancreas and nervous system. These include an impairment of cardiac and immune function and an exaggerated metabolic response to surgical stress. Latent dysfunction of these systems may lead to significant perioperative derangement when subjected to the stress response of major surgery (Table 2.1). In addition, patients with chronic alcohol excess approaching surgery are at risk of the alcohol withdrawal syndrome perioperatively [34].

However, consumption well below the levels traditionally associated with alcohol dependence has implications perioperatively. Evidence suggests that a clear dose–response relationship exists regarding perioperative risk. Postoperative complication rates are 50 % higher with the consumption of 3–4 units daily compared to 0–2 units. When consumption exceeds 5 units daily, rates of complication may be three to five times greater. Relatively low perceived levels of consumption may therefore lead to a significantly increased risk of perioperative complications. [34, 35]

2.2.3.2 Patient Assessment

Accurate quantification of alcohol intake is useful preoperatively and may provide insight into a given patient's associated risk. Traditional screening tools such as the widely used CAGE and AUDIT-C questionnaires are designed to identify severe excess consumption, and in particular, search for features of dependence. Patients can conceivably present for preoperative assessment and intervention, unknowingly consuming excess alcohol at 'subclinical' levels sufficient to substantially elevate their surgical risk and without an established 'label' of alcohol excess or a manifest associated pathology. This so-called 'hazardous drinking' is typically classified as >3 units daily [36]. In the absence of a dedicated perioperative screening tool, we recommend that obtaining an accurate and honest assessment of intake (quantified in alcohol units) at the time of referral may be useful in the surgical setting, with the

Table 2.1 Pathological effects of elevated alcohol intake perioperatively

System	Effect of elevated alcohol intake
Immunological	Impaired wound healing Suppressed delayed-type hypersensitivity
Cardiovascular	Subclinical cardiac insufficiency Increased risk of perioperative arrhythmias
Haematological	Coagulopathy and increased perioperative haemorrhage
Gastrointestinal	Disordered hepatic handling of drugs
Endocrine	Accentuated endocrine response to surgery with higher circulating cortisol and catecholamines

understanding that patient self-reported consumption is often underestimated but very rarely overestimated. It may be possible to identify patients preoperatively who are drinking at unappreciated 'hazardous' levels that may benefit from more informal advice surrounding their consumption in the approach to surgery.

2.2.3.3 Prehabilitation Strategies

The lead-time required to derive maximum benefit for the recovery of organ systems following preoperative alcohol cessation appears to be greater than for cigarette smoking. Whilst a normalization of immune functioning and the endocrine stress response to surgery may become apparent within 2 weeks of complete cessation, full recovery of coagulation, homeostatic and cardiac functioning may take more than 2 months depending on the initial severity of derangement. Thus effective management is reliant on early preoperative identification of hazardous drinking and prompt intervention [34].

Evidence supporting formal preoperative cessation is sparser than for other lifestyle factors. Nonetheless, a recent Cochrane review evaluated intensive intervention programmes, namely, 4–8 week initiatives aiming to achieve complete cessation through a combination of pharmacological and psychosocial methods [36]. The review identified two randomized controlled trials of pharmacological intervention (disulfiram) in major colorectal surgery and combined pharmacological intervention (disulfiram alongside chlordiazepoxide for withdrawal symptoms) with motivational counseling prior to hip arthroplasty. The authors concluded that an intensive cessation intervention appears to reduce complication rates substantially, without influencing 30-day mortality or hospital length of stay. Questions remain surrounding impact on longer-term alcohol consumption. It should be noted that such interventions are certainly at the 'extreme' end of available options to address alcohol intake prior to surgery. Simpler initiatives, including short motivational interviewing are available with an evidence base for reducing intake in both primary care patients and general hospital populations. These techniques may be more applicable to the previously under-recognised group of 'hazardous drinkers' presenting for surgery, although they are yet to be scrutinized in the perioperative setting [36].

Excessive alcohol consumption is an important risk factor for postoperative morbidity. Careful questioning and screening in the preoperative assessment clinic to detect those individuals consuming more than the recommended weekly amount is crucial. Research to date only really supports intensive interventions lasting several weeks preoperatively, but the evidence base is certainly sparse. In keeping with the concept of 'marginal gains' theory the authors advocate appropriate counselling and advice for those individuals at risk, with complete cessation for a period up to 8 weeks preoperatively being the aim. Where high levels of consumption are identified, investigations seeking evidence of end organ damage are warranted and a treatment plan should be put in place to manage the potential for alcohol withdrawal.

2.2.3.4 General Recommendations

- Alcohol consumption should be documented as soon as possible prior to surgery. Chronic, low-level intake is extremely common in the elderly
- Patients identified as 'hazardous' drinkers (>3 units daily) should be counselled regarding the increased perioperative risk that this confers. Abstinence prior to surgery should be advised.
- Where excessive alcohol consumption is identified, consideration should be given to more formal abstinence regimes involving specialist healthcare professionals
- Where excessive drinking is identified, investigations seeking end-organ damage should be performed. Treatment plans should also be put in place for the management of alcohol withdrawal syndromes perioperatively

2.2.4 *Smoking*

A long recognized and independent surgical risk factor, smoking remains the leading preventable cause of morbidity and mortality in the developed world. Persistently increasing rates across developing nations present a genuinely global threat to public health, reflected in the enormous directly attributable medical and social costs internationally. One hundred million deaths worldwide were attributed to smoking during the twentieth century, with over five million deaths annually considered to be smoking related. The prevalence of active cigarette smoking in adults presenting for surgery is 20–30 % [37].

2.2.4.1 Perioperative Implications

The deleterious effects of cigarette smoke on the surgical patient develop over a wide time course, from the hyperacute effects of the most recent cigarette to more chronic manifestations such as COPD and a wide range of malignancies. Therefore, even younger smokers without demonstrable features of an established smoking related comorbidity are still at increased risk in the perioperative period [38].

Regular cigarette smoke exposure, including secondhand smoke, drives pathological change across a range of organ systems. These effects are mediated by direct cellular injury, disordered coagulation and free radical release in response to toxic smoke contents. The cardiovascular and respiratory systems alongside wound healing are most extensively affected (Table 2.2) leading to an increased risk of multiple perioperative complications and worse surgical outcomes, reflected in a 1.38 fold increase in surgical mortality [37].

Table 2.2 Pathological effects and systemic consequences of tobacco smoke exposure

System	Effects of cigarette smoke exposure	Associated complications	Odds ratio (95 % confidence interval)
Respiratory	• Dysfunction of the mucociliary escalator, goblet cell hyperplasia and poor clearance of pathogens • Macrophage dysfunction • Alveolar destruction and impaired gas exchange • Bronchial hyper-reactivity	Unplanned intubation	1.87 (1.58–2.21)
		Post-op mechanical ventilation	1.53 (1.31–1.79
		Pneumonia	2.09 (1.80–2.43)
Cardiovascular	• Direct endothelial injury, platelet activation and thrombogenesis • Stimulated erythropoiesis and increased blood viscosity • Formation of carboxyhaemoglobin • Sympathetic response to nicotine	Myocardial infarction	1.80 (1.11–1.92)
		Postoperative cardiac arrest	1.57 (1.10–2.25)
		Stroke	1.73 (1.18–2.53)
Wound healing/ infection	• Free radical release and direct cellular injury • Local tissue vasoconstriction • Impaired collagen synthesis	Wound infection	
		Superficial Deep	1.30 (1.20–1.42) 1.42 (1.21–1.68)
		Sepsis	1.30 (1.15–1.46)
		Septic Shock	1.55 (1.29–1.87)

2.2.4.2 Patient Assessment

Assessment of chronicity and severity of cigarette smoke exposure, often quantified in pack years, has classically been based upon subjective patient reporting, with the associated risk of significant under-reporting of consumption particularly in the heaviest smokers. Leading intervention programmes utilise the Fagerstrom test (Table 2.3) to objectively evaluate nicotine dependence and guide replacement therapies [39, 40].

2.2.4.3 Prehabilitation Strategies

As with other unhealthy patient behaviours, the preoperative period represents an ideal 'teachable moment' for clinicians to provide smoking cessation advice. Clinical concerns regarding the potential rebound deterioration in respiratory function, if an inadequate window of cessation is not permitted prior to surgery, have been the subject of a recent systematic review and meta-analysis. These concerns appear to be entirely unfounded and probably originated from small, observational data from nearly 30 years ago. Whilst there is possibly a greater benefit from longer periods of abstinence, smoking cessation even for short durations prior to surgery may be beneficial and there is certainly no evidence of any significant quality that it is detrimental [41].

Table 2.3 Fagerstrom score for nicotine dependence and guidance of nicotine replacement therapy

Questions		Score
1. How soon after waking do you smoke your first cigarette?	Within 5 min 5–30 min 31–60 min	3 points 2 points 1 point
2. Do you find it difficult to refrain from smoking when required?	Yes No	1 point 0 points
3. Which cigarette would you hate to give up?	First in the morning Any other	1 point 0 points
4. How many cigarettes do you smoke daily?	31 or more 21–30 11–20 10 or less	3 points 2 points 1 point 0 points
5. Do you smoke more frequently in the morning?	Yes No	1 point 0 points
6. Do you smoke even if you are sick in bed most of the day?	Yes No	1 point 0 points

Total score	Result	Suggested Nicotine replacement therapy
1–2	Low dependence	May not need nicotine replacement therapy monitor for withdrawal symptoms **Patches**: 7 mg/24 h patch or 5 mg/16 h **Lozenge**: 2 mg **Gum**: 2 mg
3–4	Low/moderate dependence	**Patches**: 14 mg/24 h patch or 10 mg/16 h **Inhaler**: 6–12 cartridges per day **Lozenge**: 2 mg **Gum**: 2 mg **Combination**: Patches + gum/lozenge
5–7	Moderate dependence	**Patches**: 21 mg/24 h or 15 mg/16 h **Inhaler**: 6–12 cartridges per day **Lozenge**: 4 mg **Gum**: 4 mg **Combination**: Patches + gum/lozenge
≥8	High dependence	**Patches**: 21 mg/24 h or 15 mg/16 h **Inhaler**: 6–12 cartridges per day **Lozenge**: 4 mg **Gum**: 4 mg **Combination**: Patches + gum/lozenge

Adapted with permission from Vendome Healthcare Media [39, 40]

In the same fashion that cigarette smoking can exert detrimental physiological effects within minutes to hours, measurable recovery in cardiorespiratory functioning begins equally rapidly following cessation. The short half-lives of nicotine and carbon monoxide allow some detrimental effects to resolve in less than 12 h. However, in terms of an actual measurable effect on complication rates and postoperative outcome, debate continues surrounding the benefit derived from interventions commenced less than 4 weeks before surgery. Systematic reviews have

Table 2.4 Key elements of gold standard programme (GSP) intervention for smoking cessation

Elements of the Gold standard programme (GSP) for smoking cessation
Introductory meeting incorporating motivational conversation.
5 meetings over 6–8 weeks delivering a structured patient education programme.
Nicotine replacement therapy guided by the Fagerstrom score.
Monitoring of smoking status using expired carbon monoxide monitoring
Support-line available during daytime hours.
A dedicated follow-up programme delivered over the following 6–8 weeks for up to 1 year.

supported these ‘shorter term’ interventions [42], which appear particularly beneficial for wound healing. Evidence suggests that the estimated times from abstinence for measurable recovery of pulmonary capacity may be longer at 4–6 weeks and the ‘ideal’ lead time required to achieve improvements in cardiovascular risk is the least well defined.

In particular, two randomized trials demonstrated a halving of complication rates when the ‘gold standard programme’ (GSP), was commenced at 6–8 weeks and 4 weeks prior to surgery (Table 2.4), in the context of elective orthopaedic and general surgery. The GSP is an entirely outpatient delivered intervention and has been demonstrated to be cost-effective in reducing the economic burden of perioperative complications. Whilst less intensive interventions such as a single short 3-min counselling session from surgeons have been shown to assist in achieving abstinence, best results appear to be obtained in the context of a structured intensive intervention such as the GSP. This approach relies on individual counselling in combination with tailored nicotine replacement therapy. These studies informed a recent Cochrane review concluding that based on available evidence, interventions beginning 4–8 weeks prior to surgery combining nicotine replacement with regular counselling appear to derive the greatest postoperative benefit in terms of overall complications and long term cessation [43]. Effective delivery of programmes such as GSP is reliant upon specifically trained and competent staff. Interestingly, an established barrier to delivery of such programmes is reluctance among staff who themselves smoke to inform patients about the associated risks, a consideration when planning service provision.

Thus, whilst avoiding even a single extra cigarette preoperatively is likely to be beneficial to a small degree, the sooner abstinence from cigarette smoking can be achieved prior to surgery the better [44, 45]. The benefits of achieving preoperative cessation may project well beyond the perioperative period through triggering a permanent lifestyle change with wider benefits for public health. Good quit rates at 1 year postoperatively have been reported following preoperative intervention.

2.2.4.4 General Recommendations

- Smoking cessation prior to surgery should be encouraged no matter how short the window is prior to surgery
- Surgery is increasingly recognised as a teachable moment for smoking cessation and the opportunity to try to influence long-term behaviour should be taken
- Preoperative assessment clinics should be able to provide written advice on how patients can access nicotine replacement therapy and smoking cessation services

2.3 'Marginal' Comorbidities

2.3.1 Obesity and Malnutrition

Formal evaluation and optimisation of nutritional status remains an often-neglected element of modern preoperative assessment [46]. This is despite a wealth of data linking malnutrition with poorer postoperative outcome. Surgical patients are at risk of malnutrition in the perioperative period for a range of reasons including: poor access to adequate nutrition whilst hospitalized, alimentary tract dysfunction, unmet calorific demand due to chronic neoplastic and inflammatory processes and disordered handling of nutrients from metabolic disturbance. The surgical insult itself constitutes a significant metabolic stress with documented increases in catabolism, insulin resistance and loss of muscle mass. Hence the optimisation of nutritional status prior to surgery lends itself as a potential means of improving perioperative outcomes [47, 48]. At the other extreme to this 'undernourished malnourishment', a rapidly increasing number of patients in the western world are presenting for surgical intervention in the context of obesity.

2.3.1.1 Perioperative Implications

Excess catabolism in the perioperative period leads to a breakdown in lean muscle mass resulting in delayed recovery and an increased incidence of complications. An adequate nutritional status provides the bedrock to respond to the surgical insult and is intrinsically linked to the concepts of aerobic capacity and physical frailty. Beyond this fundamental requirement, nutritional strategies have expanded to address glucose control, modulation of the immune system and attenuation of the inflammatory process postoperatively.

2.3.1.2 Patient Assessment

Defining the malnourished patient can be challenging. Established measures such as body mass index (BMI) remain useful with malnourished patients presenting at both extremes of the BMI scale. Whilst we might associate obese patients with an excessive intake of macronutrients, they are often found deficient in at least one micronutrient. In addition, lack of lean muscle mass is recognized as an independent marker of risk with patients at both the upper and lower extremes of BMI also likely to be sarcopaenic. Of note, the picture of 'sarcopaenic obesity' is particularly associated with prolonged ventilation and Critical Care admission alongside infectious complications. This has driven increased efforts to utilize cross sectional radiology to accurately assess the ratio of adipose to lean body tissue. The most substantial increase in risk is thought to emerge when BMI exceeds 30 kg/m^2. The picture is complicated, however, by the emergence of an 'obesity paradox' whereby a BMI of 25–34 kg/m^2 is seen to confer a degree of protection against perioperative complications [49]. Whilst this apparent protection has been documented across a wide variety of surgical (and indeed non-surgical) conditions, it is important to note that when the metabolic syndrome is also present, the risk profile is significantly elevated.

A leading risk scoring system validated in surgical populations for identification of 'under-nourishment' is the Nutritional Risk Screening Tool (NRS-2002) (Table 2.5) [50]. In one observational cohort study in which it was used to identify patients at high nutritional risk, preoperative nutritional supplementation of those patients, was associated with a 50 % drop in major postoperative morbidity.

Obesity

Whilst an abnormally low BMI is associated with the greatest surgical risk, the obese surgical patient faces a different but equally serious range of perioperative complications [51]. Rather than the actual excess weight itself, these stem from the extensive range of comorbidities known to accompany an elevated BMI.

Obstructive sleep apnoea syndrome (OSAS) has a reported prevalence of up to 40 % in obese individuals and approximately 70 % of OSAS patients are obese. OSAS increases the risk of postoperative respiratory complications and the need for unplanned ventilation. This is in the context of a group of patients who have a significant reduction in longer-term survival mainly due to increased cardiovascular events, irrespective of whether they undergo surgery or not.

There is a sevenfold increase in the incidence of Type 2 Diabetes in the obese patient and over 90 % of patients with Type 2 diabetes are obese. The condition is associated with a wide range of postoperative complications, ranging from disturbances of the actual blood glucose concentration resulting in serious metabolic compromise to a significant increase in infectious complications.

Coronary heart disease and cardiomyopathy resulting in left ventricular systolic dysfunction are also significantly more prevalent in this population.

Table 2.5 NRS-2002 for preoperative nutritional screening in surgical patients

Initial screening			
		Yes	**No**
1.	Is BMI <20.5?		
2.	Has the patient lost weight in the last 3 months		
3.	Has the patient had a reduced dietary intake in the last week?		
4.	Is the patient severely ill (e.g. requiring ICU)		

Answering 'yes' to any question should prompt further screening.
Patients without any risk factors require regular re-screening and may still require a dedicated nutrition plan approaching major surgery if high risk.

Further screening			
1. Assess nutritional status		**2. Assess disease severity (increased requirements)**	
	Score		**Score**
Normal nutritional status	0	No increased requirement	0
Weight loss >5 % in 3 months **or** Reduction in food intake 50–75 % in last week	1	Hip fracture Co-morbidities: Cirrhosis, COPD, haemodialysis, diabetes, malignancy	1
Weight loss >5 % in 2 months **or** BMI 18.5–20.5 **and** general impaired condition **or** food intake 25–60 % of normal in last week	2	Major abdominal surgery Co-morbidities: Severe pneumonia, stroke, haematological malignancy	2
Weight loss >5 % in 1 month **or** BMI <18.5 **and** general impaired condition **or** food intake 0–25 % of normal in last week	3	Head injury Bone marrow transplant ICU patients (APACHE >10)	3
Nutritional status total	=	**Disease severity total**	=
Total score Nutritional status total + disease severity total + 1 if age ≥70 (age-adjustment)	=		

Total score ≥3: Patient is nutritionally at risk and requires dedicated nutritional planning.
Total score <3: Patient may require regular re-screening and may still require a dedicated nutrition plan approaching major surgery if high risk.

Adapted with permission from Elsevier [50]

The metabolic syndrome describes the combination of insulin resistance resulting in low glucose tolerance, dyslipidaemia, hypertension and central obesity. It was first described as a syndrome in 1988 and has been associated with poorer outcomes in cardiac surgery with a reported prevalence of up to 50 % in this population [52].

Malnutrition

In 1936 Hiram Studley was the first to report the relationship between poor nutritional status and adverse perioperative outcomes in a cohort of patients undergoing surgery for chronic peptic ulcer disease [53]. Since then, many observational studies have documented a clear association between undernourishment and poorer surgical outcomes. Malnutrition seems to confer a particular increase in the risk of infectious complications but it has also been, perhaps unsurprisingly, associated with increased mortality and hospital length of stay, increased Critical Care utilization, a decrease in long-term survival and delayed wound healing. Systematic incorporation of a comprehensive nutritional assessment prior to surgery is still disappointingly rare [54, 55].

2.3.1.3 Prehabilitation Strategies

An ideal comprehensive prehabilitation strategy would not only address the nutritional requirements of an individual patient - be that calorie restriction or supplementation- but would incorporate that into a package to counter poor aerobic capacity and physical frailty as well. The potential for surgical patients to present with complex and nuanced malnutrition necessitates multidisciplinary input led by a dietician.

Enteral and Parenteral Supplementation

Research studies investigating the impact of preoperative caloric and nitrogenous supplementation have demonstrated mixed results in terms of perioperative morbidity, lengths of hospital and Critical Care stays and infection rates. It would appear that it is mainly those patients with more advanced malnutrition, demonstrated by a Nutritional Risk Score 2002 ≥ 5, who most benefit from preoperative nutritional support. Several studies have investigated the routes via which feeding may be established. In the setting of inadequate calorific and nitrogenous intake, supplementation by the oral route is the preferred method of correction. This promotes an anabolic state and acts synergistically to improve aerobic capacity and increase lean muscle mass. For reasons that remain unclear, administration of supplements orally appears more effective than via nasogastric or nasojejunal routes however, evidence also suggests that in patient groups in which there is no other choice, such as head, neck and oesophageal cancers, supplementation via tube feeding still confers benefit. Indeed, it is this group, whose caloric intake may have been significantly reduced for a long period of time, who benefit most.

Whilst use of the enteral route is preferred whenever possible, a subgroup of patients will inevitably struggle to achieve nutritional targets via oral or tube feeding. These patients may present with severe malnourishment meaning consideration must be given to parenteral feeding. This approach is supported by evidence of reduced overall complications and postoperative infection but must be balanced against the risks associated with long-term parenteral nutrition such as electrolyte disturbance, re-feeding syndrome and complications of central venous access. Once again, it appears to be only the most severely affected patients who benefit [54, 55].

Metabolic and Immune Modulating Strategies

The supplementation of specific nutrients thought to improve metabolic and immune responses in the perioperative period is of increasing interest. Typical supplements include omega-3-fatty acids, arginine, nucleic acids and antioxidants. These agents are thought to mediate alterations in gut mucosal integrity, enhance cell-mediated immunity, improve blood flow to ischaemic tissues and attenuate the inflammatory response. Whilst the majority of research to date has focused on the potential benefits of postoperative supplementation with these agents, support is growing for the strategy of preoperative optimization to prime the patient's response to the surgical insult. The evidence base for this approach is far from concrete, although there does appear to be emerging evidence to support its use in major abdominal surgery. A recent Cochrane analysis of randomised trials in patients undergoing a variety of surgeries concluded that preoperative supplementation reduced both total and infectious complication rates, however recent meta-analyses that have applied more stringent methodologies have not demonstrated a clear benefit in comparison to standard oral supplementation [54, 55].

Weight Reduction Strategies

Achieving any meaningful weight reduction preoperatively is challenging. Conservative measures, such as dietary modification and exercise programmes may dovetail with attempts to improve aerobic capacity and sarcopaenia. These are effective and may be sufficient in the lead-up to a major surgical procedure, however longer term weight reduction often proves elusive. Evidence for pharmacological appetite suppressants and modifiers of lipid absorption such as sibutramine, rimonabant and orlistat, have failed to demonstrate sustained weight reduction with questions raised over long-term safety, particularly when deployed in combination. Consequently, the field of bariatric surgery has rapidly expanded to provide the only proven method to achieve long-term weight control with demonstrable improvements in quality of life and associated co-morbidities. The benefit of undertaking bariatric surgery aiming to modify subsequent perioperative risk for another procedure is, however, entirely unproven.

Glycaemic Control

The importance of glucose control in the perioperative period is well recognized. Longer term diabetic control also appears to be important and achieving a preoperative HbA1c of <7 % has been shown to reduce postoperative wound infection.

Indeed, glycaemic control in the weeks preceding surgery appears to be more influential on postoperative outcome than individual blood glucose measurements on the day of surgery. Some authorities even go as far as suggesting delaying surgery, where practical, if the HbA1c is >8.5 %. An additional reported benefit of the advanced identification and correction of poor glycaemic control is a reduced rate of day-of-surgery cancellation due to hyperglycaemia although again, more recent literature suggests that less heed be paid to elevated blood sugars on the day of surgery if the HbA1c value is acceptable [55].

2.3.1.4 General Recommendations

- Nutritional risk should be assessed in all patients presenting for surgery
- Caloric and nitrogenous supplementation by any route should only be used to treat severely undernourished patients (NRS-2002 $\geq$ 5) and only then under the direct supervision of a dietician
- Preoperative immunonutrition should be considered in patients presenting for major abdominal surgery
- HbA1c should be measured as early as possible in the diabetic patient's journey through to surgery and hypoglycaemic agents optimised to achieve a value <7 %. Delaying elective surgery should be considered where HbA1c > 8.5 %

2.3.2 Frailty

Life expectancy continues to increase in developed countries, with an average increase in the UK of 8 years for women and 10 years for men since 1960. The number of centenarians is projected to rise from 14,000 in 2013 to 111,000 by 2037 [56]. The number of older people undergoing surgery has actually outweighed this growth in population ageing, perhaps due to increased patient expectations and advances in anaesthesia and surgical techniques [57]. It has been estimated that 25–50 % of patients >85 years have established frailty [58].

Clegg and colleagues [58] describe frailty as occurring secondary to an age-related decline in multiple physiological systems rendering affected individuals susceptible to sudden health deterioration often precipitated by minor stressor events, unlikely to affect non-frail individuals. This increased state of vulnerability to stressor events, such as surgery, characterizes frailty as a specific medical condition and leads to affected patients being unable to return to their previous level of function e.g. activities of daily living. This is represented in Fig. 2.2, where a stressor event results in a disproportionate drop in functional ability and health status with prolonged recovery, rendering previously independent patients dependent, mobile patients immobile or lucid patients cognitively impaired.

The underlying pathophysiology has yet to be established, however there is increasing evidence linking the frailty syndrome with underlying inflammation and

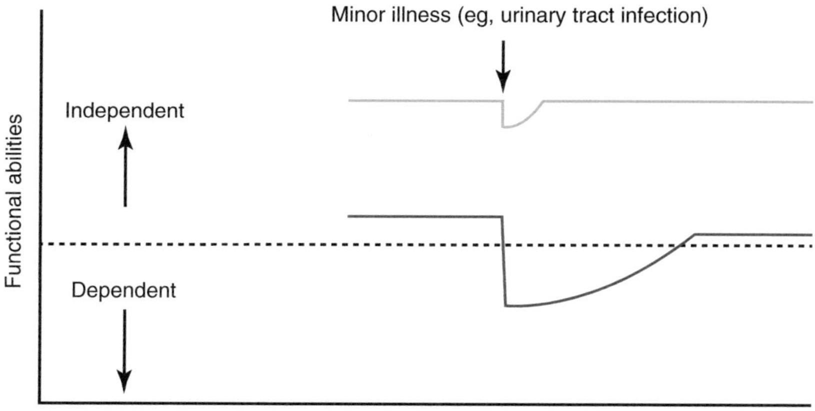

Fig. 2.2 The vulnerability of the frail patient to a change in health following a precipitating event. The green line represents the non-frail patient who, after a minor illness experiences a small drop in their level of function but returns to their baseline after a short recovery period. The red line represents the frail patient who, after the same precipitating event, experiences a more marked drop in their level of function, which may result in full dependency. They fail to return to their previous level of function and recovery is significantly prolonged. The dashed line represents the cut off between independent and dependent living (Reproduced with permission from Elsevier [58])

dysregulation within the adaptive and innate immune systems of the frail older adult [59, 60]. In addition the development of frailty appears to be linked with genetic susceptibility combined with extrinsic factors such as smoking, nutritional status, and lack of physical exercise [58], which is of increasing concern given the substantial prevalence of sedentary and inactive lifestyles previously described.

2.3.2.1 Perioperative Implications

The prevalence of frailty in the perioperative setting has been found to be as high as 28 % in patients >65 years, with affected patients having a threefold increased risk of death within 90 days of surgery [61]. Frail patients also have an increased risk of non-fatal complications, prolonged hospital stay and long term institutionalized care following discharge from hospital [61–63]. Several factors including comorbid disease and reduced physiological reserve appear to be important in this respect.

Preoperative recognition of frailty, and an appreciation of its profound adverse effect following surgery, are therefore crucial to facilitate appropriate patient selection for surgery and shared decision-making between patients and healthcare professionals. Where surgery is deemed appropriate, frailty recognition facilitates optimisation of associated modifiable risk factors and a comprehensive perioperative management plan focused towards the individual patient. The preoperative period therefore creates the ideal platform for identifying at risk patients whose care can then be tailored appropriately between primary and secondary care.

2.3.2.2 Patient Assessment

Given the perioperative implications described, the authors recommend formal assessment for identification of frailty to be a useful component of general preoperative assessment. There are multiple assessment tools for frailty described within the literature [64], the majority of which require training, space and time to administer. This makes assessment and recognition of frailty challenging in the preoperative setting. It is also yet to be established as to whether this task falls within the remit of the preoperative assessment clinic, or should be undertaken as part of a more formal CGA model. A pragmatic model would allow 'at risk' patients to be identified through established preoperative assessment pathways with referral for more comprehensive specialist assessment, through Geriatricians, when required.

The two main models of frailty described within the literature are the 'phenotype' and 'cumulative deficit models'. The phenotype model described in a landmark study by Fried and colleagues [65], consists of five indicators:

1. Self-reported weight loss (>4.5 kg)
2. Self reported exhaustion (3–4 days per week)
3. Low energy expenditure (males <383 kcal/week, females <270 kcal/week)
4. Slow gait speed
5. Weak grip strength

People with ≥3 indicators are judged to be frail, 1–2 indicators as pre-frail and those with no factors, as not frail. However this model predominantly looks at changes in functional ability and muscle strength and does not account for any changes in cognition or inability to demonstrate gait speed secondary to surgical pathology.

The cumulative deficit model, which was developed as part of the Canadian Study of Health and Ageing [66] produced a frailty index of 70 parameters including symptoms, signs, abnormal blood values, disease states and disabilities to define the presence of frailty. An abridged 36 variable version of the index has been subsequently validated to predict risk of mortality in elderly individuals [67]. Whilst there are obvious practical difficulties in implementing such lengthy assessment of frailty clinically, a much shorter 7-point frailty scale [68] has also been proposed and validated against the original index presenting a more readily useful tool in daily practice. An updated revised 9-point scale is shown if Fig. 2.3.

No single tool has yet been robustly validated to capture frailty in a time-efficient manner in the preoperative assessment clinic. The Frailty Scale outlined above, and 2 other models demonstrate promise in identification of 'at risk' individuals preoperatively. At the time of writing the Edmonton Frail Scale is probably the most widely utilised assessment for this purpose, being validated for use by non-geriatricians [69]. This comprises a 17-point scale that can be completed within 5 min. It includes tests of cognition, and the 'get-up and go' test which has been shown to independently predict mortality and morbidity within surgical patients. An alternative is the time taken to walk 4 m, which has also demonstrated good discriminatory ability in frailty detection [70], with a person taking >5 s to complete this task at increased risk.

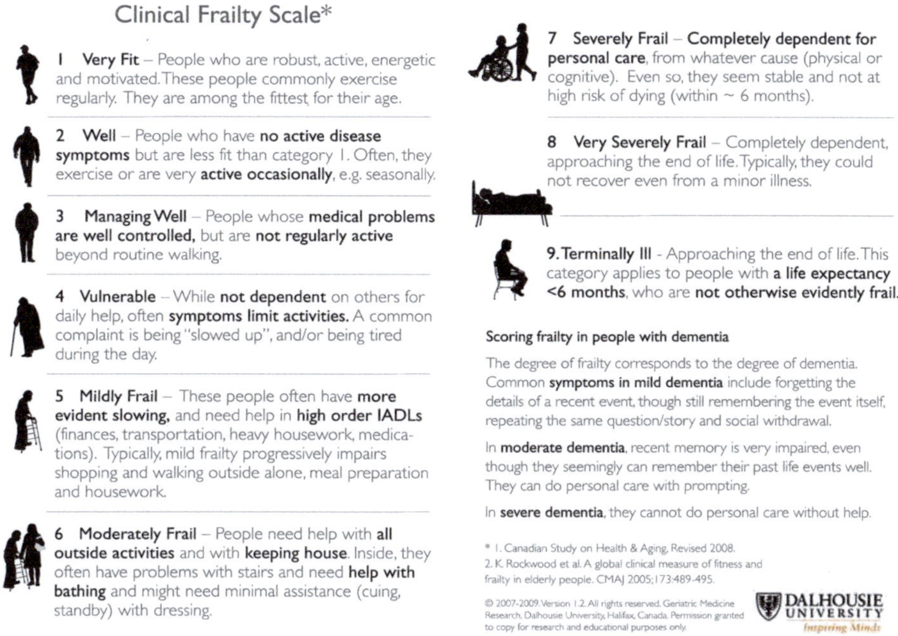

Fig. 2.3 The CSHA study 9-point frailty scale [68]

2.3.2.3 Prehabilitation Strategies

The natural overlap in assessment and management of preoperative frailty between perioperative clinical teams and clinicians caring for the elderly, dictates that establishing multi-professional links between these groups alongside primary care must form the foundation of any frailty prehabilitation strategy. This facilitates more holistic management of this high-risk patient group. Comprehensive Geriatric Assessment is the internationally established method to clinically assess older people, consisting of a multidisciplinary process to determine medical, psychological and functional capacity. It has been shown that when combined with proactive interventions in the elective surgical population, CGA can have a significant impact on the postoperative outcome of frail patients [1, 71] reducing length of stay and morbidity.

Whilst close liaison with geriatric specialists may overcome some of the challenges of preoperative patient assessment, there are currently no evidence-based strategies to actively reverse frailty once diagnosed. Exercise initiatives are of particular interest given the likely benefits in improving several deficit accumulations including; skeletal muscle, aerobic fitness, balance, cognitive decline, the endocrine and immune systems [58]. Exercise programmes are well tolerated, feasible and safe even within the very frail [72], with highest levels of compliance achieved when conducted in a supervised environment. It is perhaps unclear as to whether exercise initiatives may have most impact within the pre-frail or established frail populations, however potentially small gains in strength may have significant functional impact [58]. Current evidence suggests a combination of strength and aerobic training, to address sarcopaenia and lack of aerobic fitness, as the optimal package.

2.3.2.4 General Recommendations

- Implementation of simple screening assessment tools e.g. Edmonton Frail Scale or 4-m walk test should be considered for older patients during preoperative assessment allowing identification of high-risk patients and referral for specialist CGA
- Close collaboration between preoperative clinics, Geriatricians and Surgeons in patients with frailty to facilitate shared decision-making with regards to appropriateness of surgery
- Where surgery is deemed appropriate, proactive optimisation and management of frail patients should be undertaken, ideally through Geriatricians, to facilitate improved outcomes and reduced length of stay. This should include discharge planning in collaboration with family members
- Where balance of perioperative risks is considered excessive, consideration should be given to non-operative management of frail patients

2.3.3 Cognitive Impairment

Normal ageing is associated with physiological and structural changes within the brain. Altered synaptic function, protein transport and mitochondrial activity contribute to loss of neuronal function [58] and cognitive decline. Mounting evidence shows that frailty is independently associated with an increased risk of developing cognitive impairment, with increasing frailty linked to a faster rate of decline [73–76]. Despite this strong association it is important to be aware that mild cognitive impairment does occur in the absence of established frailty.

2.3.3.1 Perioperative Implications

There is emerging evidence that preoperative cognitive dysfunction is under diagnosed and has a significant adverse impact on perioperative outcome. An observational study of 114 patients aged 60 years and over, presenting for vascular surgery, showed a prevalence of 60 % of undiagnosed cognitive impairment [73]. Although this represents a homogenous high-risk group, it is likely that a similar picture exists in other older surgical patient groups. This is particularly pertinent given the developing recognition that patients with cognitive decline are at increased risk in the postoperative period [77]. Preoperative cognitive impairment has emerged as the strongest predictor of the development of postoperative delirium amongst multiple risk factors (ASA classification, advancing age, poor functional status, medical co-morbidities and depression) [79]. This is of significance as delirium is a common complication affecting up to 70 % of patients aged 60 years or older who undergo major surgery [80] and its occurrence is strongly associated with prolonged length of stay, the need for institutionalized care, persistent cognitive decline [81] and all-cause mortality [82]. Recent work has reinforced this risk for the wider surgical population, showing that any neurological postoperative morbidity is associated with a reduction in longer-term survival [82].

2.3.3.2 Patient Assessment

Given the evolving evidence base, it would seem appropriate to assess a patient's cognitive function, in addition to their degree of frailty, within the preoperative assessment period in order to identify those most at risk. In general, screening tests for cognitive impairment, such as the Mini-Mental State Examination [83] are time consuming and therefore not suited to the preoperative clinic setting. A review of practical screening tests by Long and colleagues in 2012 found only one test validated for use within the preoperative environment, the Mini-Cog, a 3 item recall and clock drawing test [78]. The Mini-Cog can be delivered in <5 min, identifies pre-existing cognitive impairment and therefore represents a potentially useful tool for wider screening in the preoperative clinic. Any tool delivered in the preoperative clinic setting may be best suited to screening for 'at risk' individuals with more formal specialist assessment subsequently indicated.

2.3.3.3 Prehabilitation Strategies

Given the implications of cognitive dysfunction in the perioperative period a coordinated multidisciplinary approach may well represent the best model of care. Closer working relationships between general practitioners, anaesthetists, surgeons and geriatricians would facilitate more comprehensive assessment and management. At the time of writing, CGA represents the optimal model to facilitate improved outcomes [1]. However it is clear that this is an area that requires further research and development of healthcare delivery to reduce the potentially significant burden of postoperative morbidity related to cognitive impairment.

2.3.3.4 General Recommendations

- Assessment of cognitive impairment should be considered in older high-risk patients during the preoperative assessment process
- Due to the high degree of overlap with frailty, further recommendations for management of this patient group are as outlined above in the frailty section

2.3.4 *Anaemia*

The WHO definition of anaemia is a haemoglobin level of less than 130 g/L for men and 120 g/L for women. The most recent European prevalence data available demonstrated that 31 % of men and 27 % of women were anaemic prior to major non-cardiac surgery [84]. This figure of 30 % has been reported repeatedly in the preoperative setting. It increases with advancing age and the high prevalence is remarkable given both the detrimental impact of anaemia on postoperative outcomes and the potential ease of identification [85, 86]. For many patients, anaemia may be present as a result of underlying surgical pathology such as bowel cancer but is also frequently due to medical conditions such as chronic renal failure. It is however important to consider that approximately 10 % of anaemic patients presenting for surgery on a non-malignant condition, will have an underlying occult malignancy.

The identification and management of anaemia preoperatively is therefore of paramount importance and indeed it represents the first of the 'three pillars of Patient Blood Management' [87].

2.3.4.1 Perioperative Implications

Reduced circulating haemoglobin has well-understood implications on blood oxygen carriage and by extension on tissue oxygen delivery. In combination with poor aerobic capacity, preoperative anaemia has arguably the greatest ability to hamper the capacity of a patient to meet the increased oxygen demand placed upon him/her in the perioperative period. It is therefore unsurprising that the presence of anaemia has been repeatedly documented as strongly associated with adverse outcome across most surgical specialties and age groups for a wide range of perioperative measures of interest. It is associated with a 1.5–3 fold increase in 30-day mortality, a significantly increased length of hospital stay, a 30–50 % increase in cardio-respiratory morbidity and a 20–30 % increase in wound infections or septic complications [84, 85]. In addition, the anaemic patient is more likely to require perioperative blood transfusion, an additional independent risk factor for morbidity and mortality. There are few preoperative factors more readily identifiable and evidently threatening to the surgical patient [88].

2.3.4.2 Patient Assessment

The measurement of a full blood count (FBC) immediately identifies reduced circulating haemoglobin. Frequently this is a routine screening investigation but may have precipitated the surgical referral or been triggered by suggestive symptoms such as fatigue, shortness of breath or an unexplained tachycardia. An unexpected finding of anaemia preoperatively should trigger appropriate investigations to determine the underlying cause. These investigations may include a combination of further haematological and ancillary tests guided by the initial FBC findings in the context of the patient's past medical history. Rational testing can be supported by an algorithm-based approach such as that shown in Fig. 2.4 [89]. The cause of anaemia may be directly related to the proposed surgical pathology and procedure, an established co-morbidity, or may have been identified unexpectedly prior to surgical referral. The most likely causes encountered in the surgical patient preoperatively are iron-deficiency anaemia from chronic blood loss, nutritional deficiency and anaemia of chronic disease in the context of renal or hepatic disease. In reality, anaemia of chronic disease is often iron deficiency anaemia in another guise. Rather than the traditional absolute depletion of body iron stores secondary to excessive loss or inadequate intake, chronic inflammatory processes cause an upregulation of the glycoprotein hormone hepcidin. Hepcidin has recently been identified as the key regulatory hormone responsible for iron absorption and homeostasis. Increased production of hepcidin from the liver in response to increased iron stores prevents the absorption of iron from the duodenum and also inhibits the release of stored iron from hepatocytes and macrophages. This results in decreased transport of iron to the bone marrow where it is required for erythropoiesis generating a state of 'functional iron deficiency' [90]. In this situation plasma ferritin levels may not be as low as classically required for the diagnosis of 'true' iron deficiency anaemia and some authorities

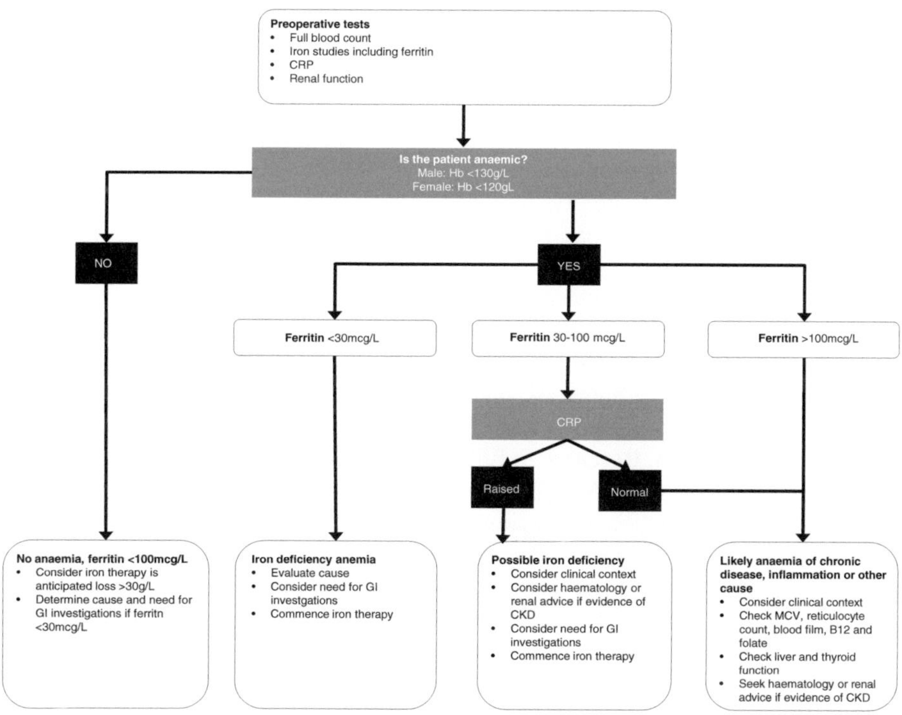

Fig. 2.4 An example algorithm for assessment of preoperative anaemia (Adapted with permission from the Australian National Blood Authority [89])

recommend using a serum ferritin <100 μg/L with a transferrin saturation (TSAT) <20 % in the presence of anaemia as a diagnostic threshold [90]. The main perioperative implication of diagnosing functional iron deficiency correctly is the lack of efficacy of oral iron supplementation - intravenous preparations are often required.

In all cases a diagnosis of iron deficiency anaemia should prompt the active exclusion of ongoing occult blood loss if not otherwise explained.

2.3.4.3 Prehabilitation Strategies

Anaemia treatment should logically be focused on addressing the underlying cause, with the acceptance that an ultimate solution could be the surgical procedure itself. Alongside addressing any ongoing occult haemorrhage, iron deficiency anaemia can be managed by iron supplementation by either the oral or intravenous route. In chronic disease states oral iron may have limited efficacy due to decreased absorption and increased sequestration in macrophages (under the influence of upregulated hepcidin) [90]. Iron is absorbed via the enteral route at a maximum rate of 25 mg per day. Reticulocytosis is not seen until 7–10 days after the commencement of oral iron. A subsequent increase in haemoglobin levels of 7–10 g/L per week is then typically seen meaning many weeks may be required, as long as there is no significant ongoing blood loss or functional resistance, before normal values are attained. Intravenous iron bypasses the difficulties of patient compliance and bioavailability associated with the oral route and is often a faster solution. Typical doses of 1000 mg of elemental iron are administered in one infusion meaning circulating haemoglobin levels rise much faster than with the oral route. It still takes several days for mature erythrocytes to be formed and it is advised that intravenous iron be administered at least 7–10 days prior to surgery. In a systematic review of 6 trials comparing the routes of iron administration, haemoglobin levels increased by a significantly greater amount when iron was given intravenously than with oral supplementation (mean difference +50 g/L, 95 % CI 27–73 g/L) [90]. However, this improved efficacy is counterbalanced by an established risk of anaphylaxis although this is largely mitigated by modern preparations. Cost is often considered a drawback of intravenous replacement but proponents suggest this is likely to be offset by savings achieved in reduced transfusion rates, hospital length of stay and complications attributable to anaemia. Research to fully explore cost-effectiveness is ongoing.

Nutritional deficiencies, typically in vitamin B12 and folate are readily managed by dietary supplementation and anaemias of chronic disease can be addressed both by optimisation of the associated condition and specific therapies such as erythropoietin supplementation in chronic kidney disease. It is advisable that this occurs in conjunction with nephrology and haematology input, particularly in view of concerns surrounding hypertension, thrombosis and ischaemic events associated with erythropoietin supplementation [88]. Regardless of the underlying cause, a modern Patient Blood Management program should include blood transfusion preoperatively only as a last resort, once a clear cause is determined and the above strategies prove inadequate [87].

2.3.4.4 General Recommendations

- Haemoglobin levels should be measured as early as possible before major surgery to allow sufficient time for correction of anaemia
- Patients found to be anaemic should also have serum ferritin, CRP, transferrin saturation and creatinine measured
- Unexplained preoperative iron deficiency anaemia requires investigation
- Patients should be managed according to a locally agreed protocol for both investigation and treatment of their preoperative anaemia

References

1. Partridge JSL, Harari D, Martin FC et al (2014) The impact of pre-operative comprehensive geriatric assessment on postoperative outcomes in older patients undergoing scheduled surgery: a systematic review. Anaesthesia 69(1):8–16
2. Durrand JW, Batterham AM, Danjoux GR (2014) Pre-habilitation (i): aggregation of marginal gains. Anaesthesia 69:403–406
3. National Institute for Health and Care Excellence (2007) [PH6] Behaviour change: the principles for effective interventions. http://www.nice.org.uk/guidance/ph6. Accessed Feb 2015
4. Buck D, Frosini F (2012) Clustering of unhealthy behaviours over time. Implications for policy and practice. Kings Fund, London, http://www.kingsfund.org.uk/sites/files/kf/field/field_publication_file/clustering-of-unhealthy-behaviours-over-time-aug-2012.pdf. Accessed Feb 2015
5. Academy of Medical Royal Colleges (2015) Exercise: the miracle cure and the role of the doctor in promoting it. http://www.aomrc.org.uk. Accessed Feb 2015
6. Rubak S, Sandbæk A, Lauritzen T et al (2005) Motivational interviewing: a systematic review and meta-analysis. Br J Gen Pract 55(513):305–312
7. World Health Organization (2010) Global recommendations on physical activity for health. http://www.who.int/dietphysicalactivity/publications/9789241599979/en/. Accessed Feb 2015
8. National Institute for Health and Care Excellence (2013) Physical activity: brief advice for adults in primary care. http://www.guidance.nice.org.uk/ph44. Accessed Feb 2015
9. Health Survey Trends (2008) Health survey for England. http://www.hscic.gov.uk/catalogue/PUB00430/heal-surv-phys-acti-fitn-eng-2008-rep-v1.pdf. Accessed Feb 2015
10. Department for Culture, Media and Sport (2012) Taking Part 2011/12 quarter 4: Statistical Release Quarter 4. http://www.culture.gov.uk/publications/9135.aspx. Accessed Feb 2015
11. Scottish Government (2014) Health of Scotland's population – physical activity. http://www.scotland.gov.uk/Topics/Statistics/Browse/Health/TrendPhysicalActivity. Accessed Feb 2015
12. Wilmot EG, Edwardson CL, Achana FA et al (2012) Sedentary time in adults and the association with diabetes, cardiovascular disease and death: systematic review and meta-analysis. Diabetologia 55(11):2895–2905
13. Hennis PJ, Meale PM, Grocott M (2011) Cardiopulmonary exercise testing for the evaluation of perioperative risk in non-cardiopulmonary surgery. Postgrad Med J 87:550–557
14. Snowden C, Prentis J, Jacques B et al (2013) Cardiorespiratory fitness predicts mortality and hospital length of stay after major elective surgery in older people. Ann Surg 257:999–1004
15. Carlisle J, Swart M (2007) Midterm survival after abdominal aortic aneurysm surgery predicted by cardiopulmonary exercise testing. Br J Surg 94:966–969
16. Sims J, Smith F, Duffy A et al (1999) The vagaries of self-reports of physical activity: a problem revisited and addressed in a study of exercise promotion in the over 65's in general practice. Fam Pract 16:152–157

17. Lindholm E, Brevinge H, Bergh C-H et al (2003) Relationships between self-reported health related quality of life and measures of standardized exercise capacity and metabolic efficiency in a middle-aged and aged healthy population. Qual Life Res 12(5):575–582
18. Valkenet K, van de Port IGL, Dronkers JJ et al (2011) The effects of preoperative exercise therapy on postoperative outcome: a systemic review. Clin Rehabil 25:99–111
19. O'Doherty AF, West M, Jack S et al (2013) Preoperative aerobic exercise training in elective intra-cavity surgery: a systematic review. Br J Anaesth 110:679–689
20. Arthur HM, Daniels C, McKelvie R et al (2000) Effect of a preoperative intervention on pre-operative and postoperative outcomes in low-risk patients awaiting elective coronary artery bypass graft surgery. Ann Int Med 133:253–262
21. Kothmann E, Batterham A, Owen SJ et al (2009) Effect of short-term exercise training on aerobic fitness in patients with abdominal aortic aneurysms: a pilot study. Br J Anaesth 103(4):505–510
22. Sisson S, Katzmarzyk PT, Earnest CP et al (2009) Volume of exercise and fitness non-response in sedentary, post-menopausal women. Med Sci Sports Exerc 41(3):39–45
23. West MA, Loughney L, Lythgoe D et al (2015) Effect of prehabilitation on objectively measured physical fitness after neoadjuvant treatment in preoperative rectal cancer patients: a blinded interventional pilot study. Br J Anaesth 114(2):244–251
24. Murias M, Kowalchuk JM, Paterson DH (2010) Time course and mechanisms of adaptation in cardiorespiratory fitness with endurance training in older and young men. J Appl Physiol 108:621–627
25. Wisløff U, Støylen A, Loennechen JP et al (2007) Superior cardiovascular effect of aerobic interval training versus moderate continuous training in heart failure patients. Circulation 115:3086–3094
26. Rogmo Ø, Moholdt T, Bakken H et al (2012) Cardiovascular risk of high- versus moderate-intensity aerobic exercise in coronary heart disease patients. Circulation 126: 1436–1440
27. American Heart Association Council (2007) Exercise and acute cardiovascular events: placing the risks into perspective: a scientific statement from the american heart association council on nutrition, physical activity, and metabolism and the council on clinical cardiology. Circulation 115:2358–2368
28. Tew G, Weston M, Kothmann E et al (2013) High-Intensity interval exercise training before abdominal aortic aneurysm repair (HIT-AAA): protocol paper for a randomised controlled feasibility trial. BMJ Open. doi:10.1136/bmjopen-2013-004094
29. Jack S (2013) The Effects of a 9 -Week Exercise Programme on Fitness and Quality of Life in Rectal Cancer Patients After Chemoradiotherapy and Before Surgery (SRETP). https://clinicaltrials.gov/ct2/show/NCT01914068?term=preoperative+exercise+colorectal+cancer&rank=13. Accessed Feb 2015
30. Burke SM, Carron AV, Eys MA et al (2006) Group versus individual approach? A meta-analysis of the effectiveness of interventions to promote physical activity. Sport Exer Psychol Rev 2(1):19–35
31. Craig R, Mindell J (2012) Health Survey for England 2011: Health, social care and lifestyles. http://www.hscic.gov.uk/catalogue/PUB09300/HSE2011-All-Chapters.pdf. Accessed April 2015
32. Office for national statistics (2013) General lifestyle survey 2011- Chapter 2- Drinking. http://www.ons.gov.uk/ons/dcp171776_302636.pdf. Accessed Apr 2015
33. Pearse RM, Moreno RP, Bauer P et al (2012) Mortality after surgery in Europe: a 7 day cohort study. Lancet 380:1059–1065
34. Tønnesen H, Nielsen PR, Lauritzen JB et al (2009) Smoking and alcohol intervention before surgery: evidence for best practice. Br J Anaesth 102(3):297–306
35. Tønnesen H (2003) Alcohol abuse and postoperative morbidity. Dan Med Bull 50(2):139–160
36. Oppedal K, Møller AM, Pedersen B et al (2012) Preoperative alcohol cessation prior to elective surgery. Cochrane Database Syst Rev (7):CD008343. doi:10.1002/14651858.CD008343.pub2. Accessed Mar 2015

37. Turan A et al (2011) Smoking and perioperative outcomes. Anesthesiology 114(4):837–846
38. Hans P, Marechal H, Bonhomme V (2008) Effect of propofol and sevoflurane on coughing in smokers and non-smokers awakening from general anaesthesia at the end of a cervical spine surgery. Br J Anaesth 101:731–737
39. Fagerstrom KO, Heatherton TF, Kozlowski LT (1990) Nicotine addiction and its assessment. Ear Nose Throat J 69(11):763–765
40. Curtin University National Drug Research Institute (NDRI) The brief tobacco intervention training program. http://ndri.curtin.edu.au/btitp/. Accessed Mar 2015
41. Myers K, Hajek P, Hinds P et al (2011) Stopping smoking shortly before surgery and postoperative complications: a systematic review and meta-analysis. Arch Inter Med 171(11): 983–989
42. Wong JW, Lam JP, Chung F et al (2012) Short-term preoperative smoking cessation and postoperative complications: a systematic review and meta-analysis. Can J Anesth 59:268–279
43. Thomsen T, Villebro N, Møller AM (2014) Interventions for preoperative smoking cessation (review). Cochrane Database Syst Rev. doi:10.1002/14651858.CD002294.pub4
44. Theadom A, Cropley M (2006) Effects of preoperative smoking cessation on the incidence and risk of intraoperative and postoperative complications in adult smokers: a systematic review. Tob Control 15:352–358
45. Mills E, Eyawo O, Lockhart I et al (2011) Smoking cessation reduces postoperative complications: a systematic review and meta-analysis. Am J Med 124:144–154
46. Grass F, Cerantola Y, Schafer M et al (2011) Perioperative nutrition is still a surgical orphan: results of a Swiss- Austrian survey. Eur J Clin Nutr 65(5):642–647
47. Schiesser M, Kirchhoff P, Muller MK et al (2009) The correlation of nutrition risk index, nutrition risk score, and bioimpedance analysis with postoperative complications in patients undergoing gastrointestinal surgery. Surgery 145:519–526
48. Kassin MT, Owen RM, Perez SD et al (2012) Risk factors for 30-day hospital readmission among general surgery patients. J Am Coll Surg 215(3):322–330
49. Mullen JT, Moorman DW, Davenport DL (2009) The obesity paradox: body mass index and outcomes in patients undergoing nonbariatric general surgery. Ann Surg 250(1):166–172
50. Kondrup J, Rasmussen HH, Hamberg O (2003) Nutritional risk screening (NRS 2002): a new method based on an analysis of controlled clinical trials. Clin Nutr 22(3):321–336
51. Doyle SL, Lysaght J, Reynolds JV (2010) Obesity and post-operative complications in patients undergoing non-bariatric surgery. Obes Rev 11:875–886
52. Echahidi N, Pibarot P, Despres J-P et al (2007) Metabolic syndrome increases operative mortality in patients undergoing coronary artery bypass grafting surgery. J Am Coll Cardiol 50:843–851
53. Studley H (1936) Percentage of weight loss: basic indicator of surgical risk with chronic peptic ulcer. J Am Med Assoc 106(6):458–460
54. Martindale RG, McClave SA, Taylor B et al (2013) Perioperative nutrition: what is the current landscape? J Parenter Enteral Nutr. doi:10.1177/0148607113496821
55. Evans DC, Martindale RG, Kiraly LN et al (2014) Nutrition optimization prior to surgery. Nutr Clin Pract. doi:10.1177/0884533613517006
56. Office for national statistics (ONS). http://ons.gov.uk. Accessed Feb 2015
57. Partridge JSL, Harari D, Dhesi JK (2012) Frailty in the older surgical patient: a review. Age Ageing 41(2):142–147
58. Clegg A, Young J, Olde Rikkert M et al (2013) Frailty in elderly people. Lancet 381:752–762
59. Hubbard RE, Story DA (2014) Patient frailty: the elephant in the operating room. Anaesthesia 69(1):26–34
60. Huifen L, Manwani B, Leng S (2011) Frailty, inflammation and immunity. Ageing Dis 2(6):466–473
61. Hewitt J, Moug SJ, Middleton M et al (2014) Prevalence of frailty and its association with mortality in general surgery. Am J Surg 209(2):254–259
62. Hamel MB, Henderson WG, Khuri SF et al (2005) Surgical outcomes for patients aged 80 and older: morbidity and mortality from major noncardiac surgery. J Am Geriatr Soc 53:424–429

63. Beggs T, Sepehri A, Szwajcer A et al (2015) Frailty and perioperative outcomes: a narrative review. Can J Anesth 62:143–157
64. Afilalo J, Alexander KP, Mack MJ et al (2014) Frailty assessment in the cardiovascular care of older adults. J Am Coll Cardiol 63(8):747–762
65. Fried LP, Tangen CM, Walston J et al (2001) Frailty in older adults: evidence for a phenotype. J Gerontol Series A: Biol Sci Med Sci 56(3):146–156
66. Rockwood K, Howlett SE, MacKnight C et al (2004) Prevalence, attributes, and outcomes of fitness and frailty in community-dwelling older adults: report from the Canadian study of health and aging. J Gerontol 59(12):1310–1317
67. Song X, Mitnitski A, Rockwood K (2010) Prevalence and 10 year outcome of frailty in older adults in relation to deficit accumulation. J Am Geriatr Soc 58(4):681–687
68. Rockwood K, Song X, MacKnight C et al (2005) A Global clinical measure of fitness and frailty in elderly people. Can Med Assoc J 173(5):489–495
69. Rolfson DB, Majumdar SR, Tsuyuki RT et al (2006) Validity and reliability of the Edmonton frail scale. Age Ageing 35:526–529
70. Van Kan GA, Rolland Y, Andrieu S et al (2009) Gait speed at usual pace as a predictor of adverse outcomes in community dwelling older people. An international academy on nutrition and ageing (IANA) task force. J Nutr Health Ageing 13(10):881–889
71. Harari D, Hopper A, Dhesi J et al (2007) Proactive care of older people undergoing surgery ('POPS'): designing, embedding, evaluating and funding a comprehensive geriatric service for older elective surgical patients. Age Ageing 36:190–196
72. Oosting E, Jans MP, Dronkers JJ et al (2012) Preoperative home based physical therapy versus usual care to improve functional health of frail older adults scheduled for elective total hip arthroplasty: a pilot randomized controlled trial. Arch Phys Med Rehabil 93(4):610–616
73. Partridge J, Dhesi JK, Cross JD et al (2014) The prevalence and impact of undiagnosed cognitive impairment in older vascular surgical patients. J Vasc Surg 60(4):1002–1011
74. Boyle PA, Buchman AS, Wilson RS et al (2010) Physical frailty is associated with incident mild cognitive impairment in community based older persons. J Am Geriatric Soc 58(2):248–255
75. Buchman AS, Boyle PA, Wilson RS et al (2007) Frailty is associated with incident Alzheimer's disease and cognitive decline in the elderly. Psychosom Med 69(5):483–489
76. Song X, Mitnitski A, Rockwood K (2011) Nontraditional risk factors combine to predict Alzheimer disease and dementia. Neurology 77(3):227–234
77. Nadelson MR, Sanders RD, Avidan MS (2014) Perioperative cognitive trajectory in adults. Br J Anaesth 112(3):440–451
78. Long LS, Shapiro W, Leung JM (2012) A brief review of practical preoperative cognitive screening tools. Can J Anaesth 59(8):798–804
79. Mashour GA, Woodrum DT, Avidan MS (2015) Neurological complications of surgery and anaesthesia. Br J Anaesth 114(2):194–203
80. Neufeld KJ, Leoutsakos JMS, Sieber FE et al (2013) Outcomes of early delirium diagnosis after general anaesthesia in the elderly. Anesth Analg 117:471–478
81. Kiely DK, Marcantonio ER, Quach L et al (2009) Persistent delirium predicts greater mortality. J Am Geriatric Soc 57:55–61
82. Moonesinghe SR, Harris S, Mythen MG et al (2014) Survival after postoperative morbidity: a longitudinal observational cohort study. Br J Anaesth 113(6):977–984
83. Folstein MF, Folstein SE, McHugh PR (1975) 'Mini-Mental State': a practical method for grading the cognitive state of patients for the clinician. J Psychiatr Res 12:189–198
84. Baron DM, Hochrieser H, Posch M et al (2014) Preoperative anaemia is associated with poor clinical outcome in non-cardiac surgery patients. Br J Anaesth 113:416–423
85. Musallam KM, Tamim HM, Richards T et al (2011) Preoperative anaemia and postoperative outcomes in non-cardiac surgery: a retrospective cohort study. Lancet 378:1396–1407
86. Spahn DR (2010) Anemia and patient blood management in hip and knee surgery: a systematic review of the literature. Anesthesiology 111:485–492

87. Ibister JP (2013) The three pillar matrix of patient blood management- an overview. Best Pract Res Clin Anaesthesiol 27:69–84
88. Hans GA, Jones N (2013) Preoperative anaemia. Br J Anaesth CEACCP 13(3):71–74
89. National Blood Authority (NBA) (2012) Patient blood management guidelines: module 2 – perioperative. NBA, Canberra, http://www.blood.gov.au. Accessed Mar 2015
90. Clevenger B, Richards T (2015) Pre-operative anaemia. Anaesthesia 70:20–28

Chapter 3
Impact of Co-morbidites, Physiological Status and Age on Survival

J. Carlisle

Abstract This chapter describes the preoperative calculation of survival, illustrating how the historical failure to generate appropriate postoperative survival curves has led to bad practice for scheduled repair of abdominal aortic aneurysm. Survival in the general population, including patients who have had surgery, can be modelled using published mortality rates specific for year, age and sex, adjusted as necessary for morbidity, fitness and the transient effects of surgery. The mortality hazard is a proxy for quality of life – people more likely to die are less likely to have a good quality of life: it therefore serves as an indicator of life as well as death, until we have more information about how patients live after surgery.

Keywords Survival • Kaplan-Meier • Abdominal aortic aneurysm

3.1 Introduction

Death is an attractive outcome, to count, not to experience. You and I are very likely to agree whether someone is dead or not dead. Life, on the other hand, is distinctly heterogeneous, which is why I chose to write 'not dead' rather than 'alive' in the preceding sentence. We may disagree about how alive someone is, which makes life harder to measure than death. The two problems with using death as a postoperative outcome are: it is a weak measure because most people don't die during traditional observation periods; it is uninformative for the majority who survive, who want to know how *alive* they might be, how happy or sad, how much pain, how much pleasure. To paraphrase Florence Nightingale, death is an outcome for killers; doctors should be measuring life [1].

Why, therefore, am I going to concentrate on survival in this chapter? The utilitarian answer is that survival is the outcome most commonly measured and it can be pooled across studies without fear of pooling different things. In addition one's median life expectancy – how long until you have a 50:50 probability of being

J. Carlisle
South Devon and Torbay NHS Foundation Trust, Lowes Bridge, Torquay, Devon TQ2 7AA, UK
e-mail: John.carlisle@nhs.net

K. Stuart-Smith (ed.), *Perioperative Medicine – Current Controversies*, DOI 10.1007/978-3-319-28821-5_3

alive – is intrinsically linked with quality of life. The closer you are to death the worse your quality of life is likely to be. The probability of dying in one's sleep in fine fettle is fortunately low; unfortunately that means that you are likely to die ill, having diminished over the preceding months and years. We can therefore use survival to measure one dimension of life's matrix and median life expectancy as a proxy for quality of life, which is life's other dimension.

This chapter is not a recipe for anaesthetists. It will benefit you most if you read it as a 'peri-operative physician', with a role that extends to counselling patients on the merits of surgery, or perhaps just as a human being interested in the course of life. Deaths directly attributable to anaesthesia are very rare. Deaths in the first month or so after scheduled surgery are also fairly infrequent, but more frequent than those observed in an equivalent population that has not had surgery. Postoperative mortality rates are also not representative of the mortality rates observed months after scheduled surgery: the initial postoperative surge in mortality wanes, occasionally being replaced by a mortality rate less than that exhibited by an equivalent cohort that has not had surgery (see later). Death in the first month after surgery measures a number of things, but it does not measure the value and wisdom of an operation; it is too brief a time to determine whether surgery has helped, in fact it is so brief that all one month's worth of death can prove is that scheduled surgery is harmful. I will therefore not be discussing deaths within 1 month of surgery, but the patient variables that affect longer term survival also affect survival in hospital, with the addition of hospital-specific factors, such as surgery, infections, wrong drugs and food quality that historically has been patchy.

Factors that predispose to death also predispose to ill health, so I hope that by writing about death I will also be writing about life.

3.2 Principles of Survival

Survival is usually depicted as a graph, with time on the horizontal axis and the number (or proportion) alive on the vertical axis. As time increases (left to right) the number alive decreases (top to bottom): survival is therefore a line from top-left to bottom-right.

Figure 3.1a shows a representative survival curve (solid red line) for a population that has a mortality rate of 5 % in the first year i.e. in the first year the proportion alive decreases from 1 to 0.95 [2]. I have also plotted lines for a constant number of deaths (black line) and for a constant proportion of deaths (blue line). Their respective annual mortality rates are shown by the dotted lines. I have drawn all three lines to have the same mortality rate in the first year (0.05 or 5 %). The blue exponential line plots the decreasing cohort survival as 5 % of those left alive die each year. The black line plots a greater loss of life as a constant number in the cohort dies each year: for instance, if 500/10,000 (5.0 %) died in the first year then 500/9500 (5.3 %) would die in the second year and 500/9000 (5.6 %) would die in the third year, and so on. You should note that a straight line on a survival graph is

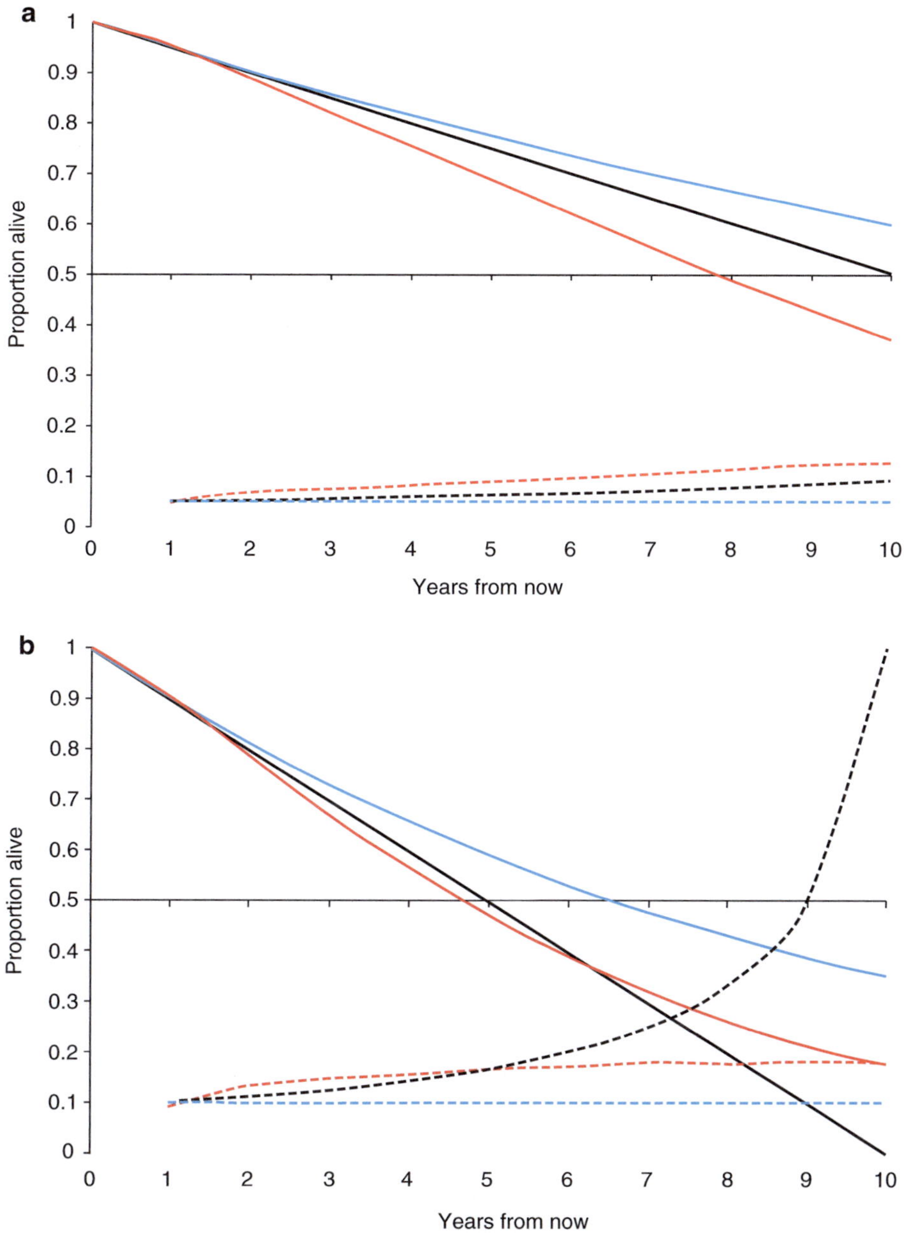

Fig. 3.1 Representative Survival curves (——) and their annual mortality rates (- - -) of populations for which in the first year: (**a**), 5 % die; (**b**), 10 % die; (**c**), 15 % die. For comparison, survival and mortality rates when a constant number of people die (—— and - - -) and a constant proportion of people die (—— and - - -)

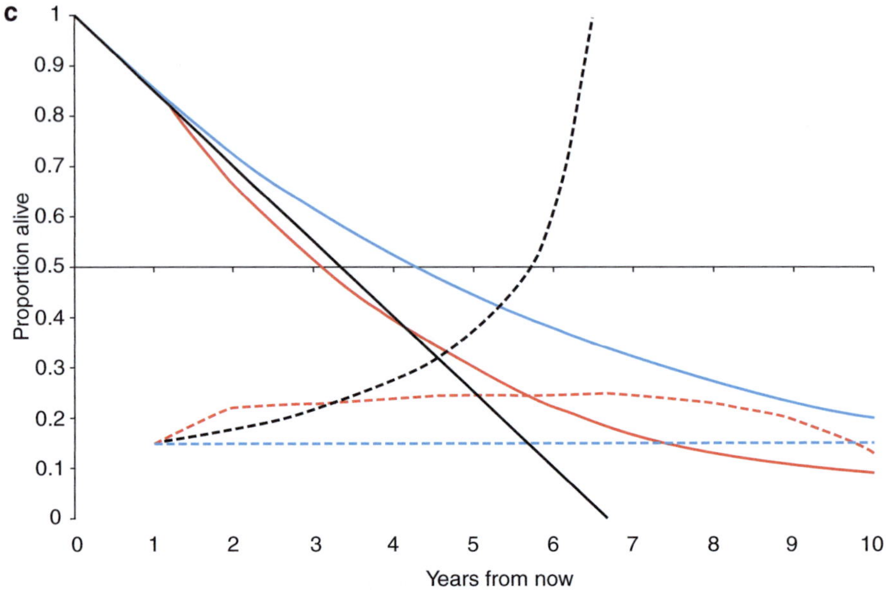

Fig. 3.1 (continued)

not a constant mortality rate: the rate of death increases left-to-right along a straight line. The blue curved line plots a stable mortality rate: 500/10,000 (5 %) die in the first year, 475/9500 (5 %) die in the second year, 451/9025 (5 %) die in the third year, and so on. The red line, which plots what happens in a cohort, is more complicated in shape and after the first year the mortality rate accelerates faster than for either of the other lines.

Figure 3.1b shows a cohort with a mortality rate of 0.10 (10 %) in the first year, that is double the rate of the cohort in Fig. 3.1a. Unsurprisingly the median life expectancy, when the population has halved, is shorter in Fig. 3.1b than in Fig. 3.1a. In the first cohort the median life expectancy is 8 years and in the second cohort it is 4.6 years: the number dying in Fig. 3.1b does not remain double that in Fig. 3.1a, if it did, one would expect the median life expectancy to be halved (as shown by the black lines). The dashed black line in Fig. 3.1b illustrates clearly how a constant number of deaths results in an increasing proportion of deaths, also shown in Fig. 3.1c, in which the first year mortality rate is 0.15 (15 %). These two figures also show how the mortality rate in the population, which initially accelerated after the first year, levels off and might even decrease (Fig. 3.1c), although one should be circumspect that mortality rates truly decrease as the data on survivors are limited for this part of the population.

Everybody in a cohort dies, if you wait long enough, which would allow you to draw lines that terminated on the horizontal axis. However, even researchers who are prepared to observe longer than one lunar cycle want to publish their research without having to wait for everyone to die. Cohorts are usually recruited over a period of time, some people might be observed for nine postoperative years and others for nine postoperative days. You could draw a simple survival curve for the whole cohort, but it would only extend for nine postoperative days, so you would waste the 3278 additional days that you have observed the patients recruited first, 9 years ago. Alternatively, you could generate a survival curve up to 9 years, but how should you treat patients you've observed for less than 9 years who were still alive when you last checked? You don't know when patients will die, but if you have data for other patients who've died during a longer observation period you can hazard a guess i.e. you can estimate the survival curve that is the composite of patients who survived for the whole of the longest observation period, patients who died, and patients who survived but who have not been observed for the whole observation period ('censored' patients). The most commonly-used estimate of survival data is that described by Edward Kaplan and Paul Meier in 1958, i.e. the Kaplan-Meier estimator [3].

3.3 The Kaplan-Meier Estimator

In this chapter we're looking at survival; time to death. 'Survival' or time-to-event curves can be constructed for any binary state – it needn't be 'alive or dead'. All such curves are constructed in the same way.

Kaplan-Meier estimated curves are constructed by integrating or 'sticking together' mortality rates during brief intervals. The start time, time zero, is the time you start observing each patient in the cohort, irrespective of the particular date from which each patient was observed. An interval is the time to the next event, or for our purposes, the time to the next death, i.e. intervals vary in duration, determined by how long the next death in the cohort is from the preceding death. The mortality rate during the interval is the single death in the interval divided by the number observed for the whole period (including the patient who died). Censored patients whose observation period completed during the interval do not contribute to the calculation for that particular period. For instance, we might start an interval with 100 patients. Before the next death 10 patients are censored. The mortality rate for the interval is 1/90 and the survival rate is 89/90 (0.989). In the next interval 5 patients are censored. The mortality rate for the interval is 1/84 and the survival rate is 83/84 (0.988). The estimated cumulative survival at the end of the two intervals is 0.989×0.988, which is 0.977.

A table of the number of patients 'at risk' may be displayed below the Kaplan-Meier graph. Figure 3.2a is the Kaplan-Meier estimated survival after scheduled repair of abdominal aortic aneurysms in over 1000 patients [4]. By four postoperative years only 324 patients remain 'at risk' (33 % of the original cohort), but the estimated survival is 80 % i.e. estimated mortality is 20 % at 4 years. Shouldn't the mortality be 67 %? What happened to the 'unaccounted' 47 % of the original cohort? You can see that there are 'ticks' attached to the survival line. These ticks mark the missing patients, indicating that they were censored i.e. at the marked

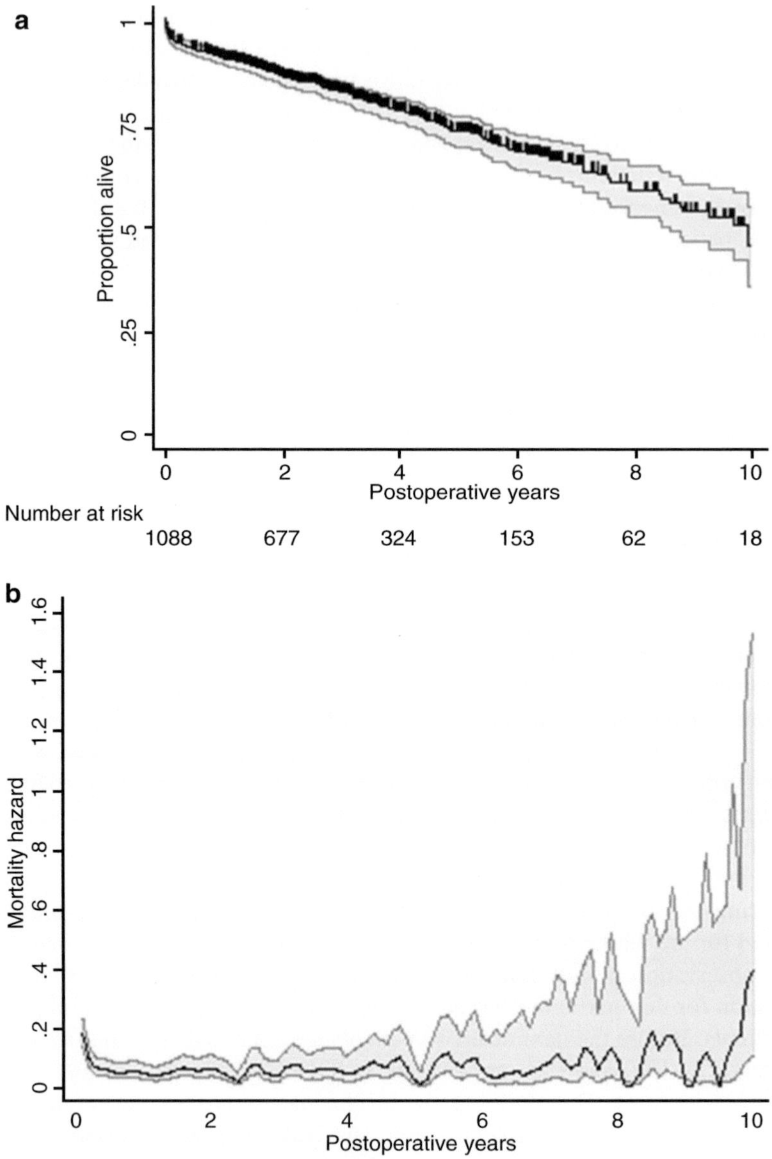

Fig. 3.2 Kaplan-Meier estimated curves (95 % CI) after scheduled abdominal aortic aneurysm repair in 1088 patients: (**a**), survival; (**b**), smoothed mortality hazard [4]

times they were alive, after which they were unobserved, to live for 'who knows' how much longer. The estimated survival line is also bracketed by a shaded 95 % confidence interval. The corresponding mortality hazard estimate is illustrated in Fig. 3.2b, again bracketed by a 95 % confidence interval. You might note that the confidence interval is much wider around the hazard estimate than the survival estimate: this is because the survival estimate is cumulative whilst the smoothed hazard estimate in this particular example has been set to sample from single sequential months.

Open abdominal aortic aneurysm repair temporarily increases mortality tenfold and endovascular repair increases mortality fourfold. Approximately half the cohort in Fig. 3.2 had each type of aneurysm repair. The heightened mortality rate caused by surgery is evident in the initial steep survival curve and the steep decreasing mortality hazard curve. Mortality hazards generally increase with time: a decreasing mortality hazard demonstrates that the cohort has been exposed to an unusual hazard. One can envisage two general effects of major surgery on the normal population survival curve. The survivors might be left in a weakened state, so that the mortality hazard remains higher than expected (for an equivalent non-surgical population). Alternatively, surgery may have preferentially killed the weak, leaving the stronger, albeit numerically depleted cohort, which could have a mortality hazard less than expected.

3.4 Analysing Survival

Death within 30 days of surgery is usually analysed as a homogenous outcome: you're alive or dead at 30 days, and if you're dead it doesn't matter whether you died on day 1 or day 30. The ease of such an analysis does not outweigh the disadvantage that the postoperative period is not representative of subsequent survival, in addition to which few deaths occur during such a short interval. It does matter when you die if mortality and survival are analysed over months or years.

Kaplan-Meier survival curves are analysed for two purposes: to determine whether patient characteristics are associated with survival; to determine whether an intervention changes survival. The most common statistical tests for both purposes assume a constant ratio of mortality hazards between two groups, for instance one with a characteristic and one without, or one exposed to an intervention and an unexposed group. This assumption holds perfectly true between two exponential survival curves, as illustrated by the blue lines in Fig. 3.3, for which the mortality hazard ratio remains two (dotted blue line). General population survival curves (red lines) maintain a fairly steady hazard ratio, but constant numbers of deaths in each year (black lines) exhibit a hazard ratio that rapidly increases with time (approaching infinity when survivors near zero in one cohort).

The log-rank test calculates the probability that two survival curves are the same i.e. derived from the same population. Cox regression analyses can adjust the comparison for prognostic variables that might have been distributed differently in the two groups. Both types of analysis require fairly stable hazard ratios throughout the observation period. Hazard ratios that cross (with or without crossing of the survival curves) limit the power of these tests. There are alternative methods to analyse hazards ratios that cross, which can also allow one to weight early or late events: for instance, one might use survival analyses for length of postoperative stay and one might want to emphasise lengths of stay longer than the median as indicating

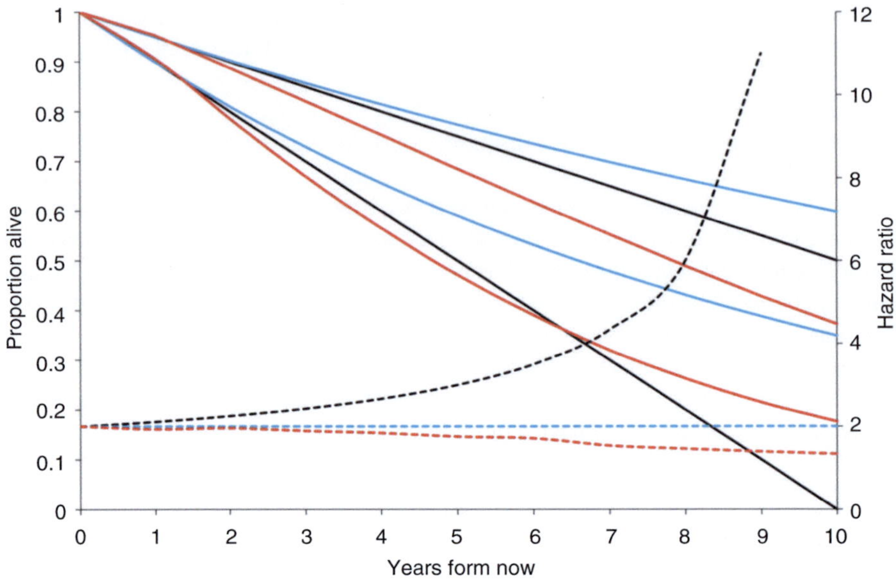

Fig. 3.3 Survival curves and mortality hazard ratios for three pairs of data: general population cohorts (—— and - - -); constant numbers of deaths per year (—— and - - -); and constant proportions of deaths per year (—— and - - -)

a burden of morbidity. Standard log-rank tests do not permit this more nuanced exploration of the data. Researchers should familiarise themselves with Fleming-Harrington tests and Renyi supremum tests, for example, as disproportionate changes in hazards are likely in two groups if , for instance, one has had surgery and the other has not.

3.5 Principles for Picking Prognostic Factors

Firstly, we have to identify factors associated with survival. A logical but expensive method is to measure everything about the patient and their environment and then test for an association with survival. Another disadvantage of such a method is that one would falsely identify associations that are a consequence of data trawling.

Orthodox survival research might be characterised as some experiments testing whether a single new variable, for instance a genetic mutation, is a survival factor, whilst other experiments analyse known factors for associations with survival in a particular patient cohort.

We should make the peri-operative period part of the patient's life. Most published papers in surgical and anaesthetic journals treat an operation in isolation [5–8]. Fragmentation of a patient's life into time silos limits the power of each to identify survival factors, the importance of which might only be fully appreciated over the duration of a patient's life, an example of which is age. Underpowered periods of observation not only miss important survival factors, they also generate spurious highly-significant associations. Spurious associations fragment the evidence base as they are interpreted as being real but context-sensitive i.e. specific to particular cohorts, whether due to patient factors, surgical factors or other factors. An approach to limit spurious associations is to include into all survival models those factors that have been proven over long observation periods i.e. in primary care. It is unnecessary and counterproductive to demonstrate statistical association of these default factors with survival in each cohort: they should be forced in to the model so that the associations of other variables with survival can be assessed in the context of known general survival factors. The association of age with survival, for instance, is not suspended by walking through the hospital doors.

We want to estimate survival as accurately as possible, without underestimation or overestimation. We therefore want to take into account all the prognostic associations that are independent of all the other prognostic associations: we don't want to miss any prognostic factors, but we don't want to double-count associations that are shared with other factors. Researchers usually test survival association with individual candidate variables, one at a time. It would be more logical to test candidate variables with the default survival variables, rather than in isolation. The threshold p value for choosing candidate variables for the multivariable analyses is a matter of choice, but is often set at 0.10 rather than 0.05. One could argue that smaller p values would encourage researchers to observe survival in larger cohorts over longer periods of time, which would reduce spurious associations. Interactions between candidate factors should also be tested, with interactions being included or excluded independently of the parent variables.

The association of each selected variable that is independent of other variables is determined in a multivariable model. The final multivariable model is selected from candidate models, each of which contains different combinations of the candidate factors. Stepwise selection is usual, by sequentially including or excluding candidate factors. There are a number of metrics that can be used to determine whether the survival model is improved by the inclusion of a candidate factor (or interaction term), including the Akaike information criterion and the false discovery rate. The problem encountered by all studies is that the selected model cannot be properly applied to the cohort used to generate the model: the performance of the model can only be properly assessed on an independent cohort. The performance of a model within the cohort from which it was derived overestimates its performance: it performs less well when applied to other cohorts – it is 'over-fitted'.

3.6 Year, Age and Sex

These three variables by default should contribute to any survival model because they strongly associate with general survival [1].

Mortality rates have fallen by about 2 % each year for at least five decades, with a corresponding extension to survival (Fig. 3.4). Age is the most important survival variable (Fig. 3.5). The difference in survival between men and women has reduced

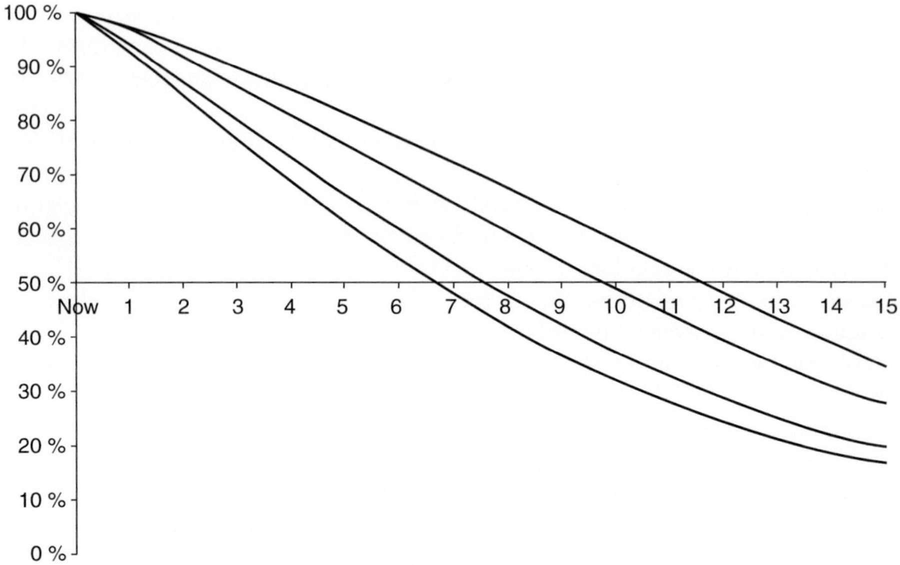

Fig. 3.4 Survival for an average 75 year old man in 1983, 1993, 2003 and 2013 (*left to right*). The median life expectancy increases from 6.7 years, to 7.5 years, 9.7 years and 11.5 years

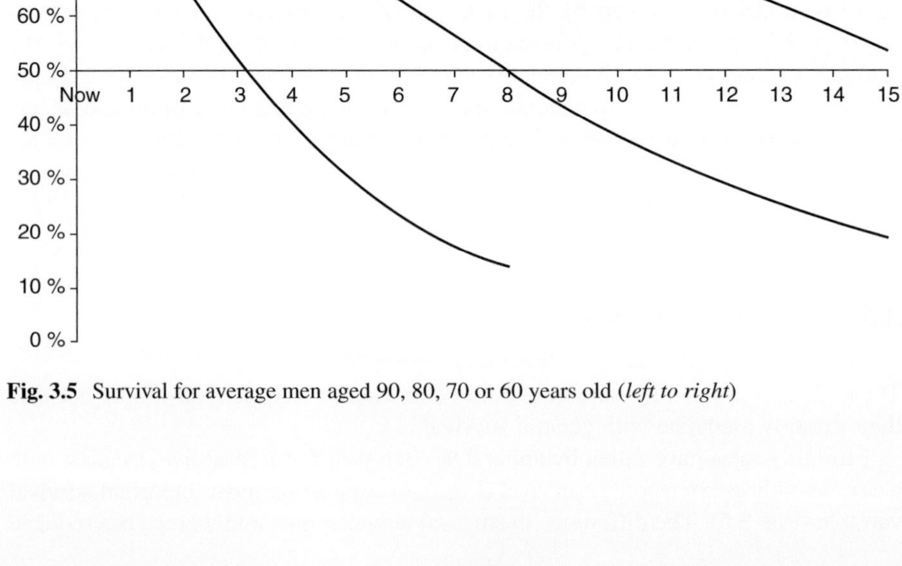

Fig. 3.5 Survival for average men aged 90, 80, 70 or 60 years old (*left to right*)

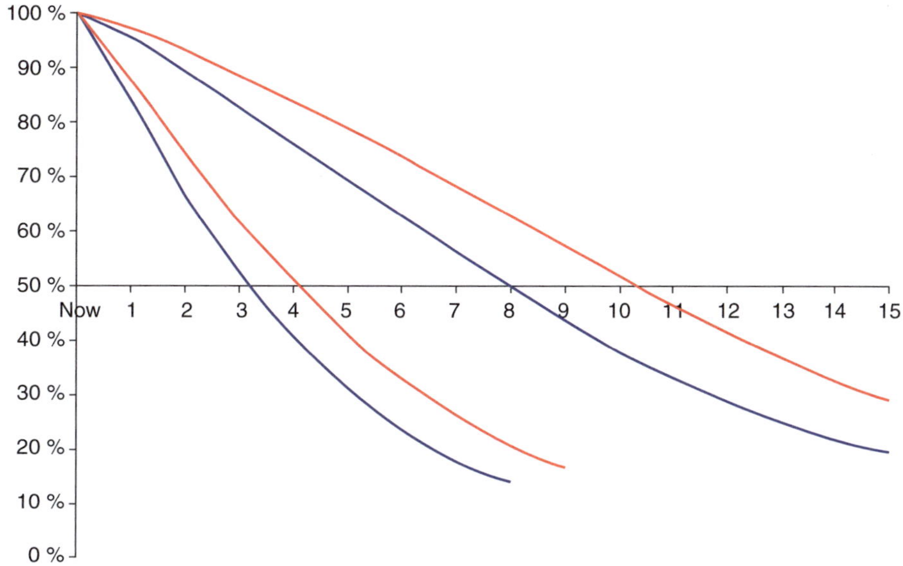

Fig. 3.6 Survival for men (——) and women (——) aged 90 years and 80 years (*left to right*)

over the past five decades. The current lifetime difference in median survival is around 4.5 years now (2015), with the difference in median life expectancies at 90 years and 80 years of age being 1 year and 2 years, respectively (Fig. 3.6).

3.7 Organ Ischaemia and Failure

Acute coronary syndrome, stroke and peripheral arterial disease independently decrease long-term survival. Heart failure and cognitive impairment also decrease survival. A broad-brush estimate is that each of these diagnoses independently increases mortality hazard 1.5–2.0 times [9–22]. Functional tests, including physical fitness, increase the precision of this estimate (see below). The cumulative relative mortality hazard associated with multiple diagnoses is their product: for instance, two of these diagnoses would increase the mortality hazard by 2.25 times compared with someone without either diagnosis, if one assumes each increases mortality by 1.5 times. If a patient has not had an acute coronary syndrome or stroke their mortality hazard is increased by about 1.2 times for angina and TIAs.

Renal dysfunction increases the mortality hazard [23]. The estimated glomerular filtration rate permits one to use a 'sliding scale' of risk associated with the severity of the dysfunction. The expected glomerular filtration rate (ml.min^{-1}.1.73 m^{-2}) is (0.003 × age × age) – (1.017 × age) + 139.4. The Cockcroft-Gault estimated filtration rate is ((140 – age) × mass × number) ÷ creatinine (μmol.l^{-1}), where 'number is 1.04 for women and 1.23 for men. The mortality hazard increases by 1.0085 times per ml.min^{-1}.1.73 m^{-2} that the estimated filtration rate is below the expected rate.

3.8 Whole Body Fitness

Physical fitness is an independent survival factor [24, 25]. Its measurement also serves to gauge the functional impact of other survival factors, for instance a stroke, heart failure or angina. It is unclear whether a single functional test can capture the prognostic information associated with physical function. There is certainly substantial overlap in the prognostic value of functional tests. The multiplication of the mortality hazards associated with two measures of function is likely to severely overestimate the hazard of physical dysfunction.

A pertinent example for the peri-operative physician of the problems of measuring different but dependent variables is the cardiopulmonary exercise test (CPET). The CPET measures five main variables during incrementally harder exercise, usually by increasing the braking on a cycle ergometer or the speed and slope of a treadmill: power; oxygen consumption; carbon dioxide excretion; minute ventilation; and heart rate. Particular features of these measurements are extracted for prognostic testing, which include oxygen consumption, both peak and at the anaerobic threshold, and the ratios of minute ventilation to oxygen consumption and particularly to carbon dioxide excretion (often at the anaerobic threshold).

These measurements are different. However, their deviation from normal values results from abnormal function. The dysfunction is of the whole body, so the prognostic overlap of these variables is substantial. Table 3.1 is taken from reference [4], with permission. Table 3.1a lists the univariate associations of different variables with survival after scheduled abdominal aortic aneurysm repair. The column titled 'z' indicates how much the variables were associated with survival, the effect increasing with the difference from zero. The column titled 'p value' indicates the precision of that effect. In isolation the peak power was the factor most associated with postoperative survival, followed by age. The other CPET variables were also strongly associated with survival. You might like to note that serum creatinine concentration was only weakly associated with survival, whereas the estimated glomerular filtration rate was strongly associated with survival. When the prognostic overlap between variables was accounted for in multivariable analysis (Table 3.1b), only four variables contributed statistically independent prognostic information (sex and date were retained for the reasons explained previously): age; renal function; peak oxygen consumption; and the ventilatory equivalent for carbon dioxide excretion. Although in this particular dataset peak oxygen consumption and ventilatory equivalents for carbon dioxide carried the available CPET prognostic information, another dataset might be characterised by a different combination of CPET variables. In any dataset a single CPET variable will be dominant, perhaps largely by chance, and will be incorporated in the multivariable model. In all cases the prognostic value of the other CPET variables will be largely exhausted, due to the prognostic overlap with the dominant CPET variable.

Table 3.1 Univariate (a) and multivariate (b) analyses of preoperative variables associated with postoperative survival in 1096 patients who had scheduled repair of abdominal aortic aneurysms

(a)	Number of patients	Hazard ratio	Standard error	z	p value
Age; years	1096	1.075 (1.054–1.098)	0.011	7.19	3.3×10^{-11}
Weight; kg	943	0.978 (0.968–0.988)	0.005	−4.40	6.4×10^{-06}
Height; cm	943	0.965 (0.948–0.982)	0.009	−3.98	8.0×10^{-05}
BMI; $kg.m^{-2}$	943	0.947 (0.915–0.981)	0.017	−3.07	0.0015
Creatinine; $\mu mol.kg^{-1}$	943	1.003 (1.001–1.004)	0.001	2.03	0.043
eGFR; $ml.min^{-1}.1.73\ m^{-2}$	943	0.979 (0.973–0.985)	0.003	−6.63	4.3×10^{-10}
Expected eGFR; $ml.min^{-1}.1.73\ m^{-2}$	1080	0.965 (0.951–0.978)	0.007	−5.42	6.0×10^{-08}
Haemoglobin; $g.dl^{-1}$	576	0.72 (0.645–0.805)	0.041	−5.8	2.4×10^{-08}
CPET					
Peak $\dot{V}_{O2}$; $ml.kg^{-0.83}.min^{-1}$	943	0.950 (0.933–0.967)	0.009	−6.61	3.9×10^{-11}
Anaerobic threshold; ml $O_2.kg^{-1}.min^{-1}$	1096	0.880 (0.835–0.927)	0.024	−4.77	2.5×10^{-06}
$\dot{V}_E / \dot{V}_{CO2}$ at the anaerobic threshold	1096	1.058 (1.039–1.077)	0.009	6.31	2.8×10^{-09}
Peak power; W	932	0.983 (0.978–0.987)	0.003	−7.67	1.7×10^{-14}

(b)	Hazard ratio	Standard error	z	p value
Age; years	1.049 (1.016–1.066)	0.012	2.99	0.0028
Female	0.939 (0.616–1.430)	0.202	−0.47	0.64
Date of surgery	1.000 (0.999–1.000)	0.000	−0.04	0.96
eGFR; $ml.min-1.1.73\ m^{-2}$	0.988 (0.981–0.996)	0.004	−2.71	0.0067
Peak $\dot{V}_{O2}$; $ml.kg-0.83.min^{-1}$	0.961 (0.942–0.982)	0.010	−4.88	0.0000011
$\dot{V}_E / \dot{V}_{CO2}$ at the anaerobic threshold	1.033 (1.010–1.056)	0.012	2. 70	0.0069

Reproduced from Ref. [4] with permission

3.9 Missing Survival Factors

You can download the survival calculator that I've developed at https://www.dropbox.com/sh/i57xhrtyop0pq31/AACEvMTOPRzdMEQnhLbCL2G8a?dl=0#. In addition references [4, 26] describe in more detail its derivation and use. My intention is to continuously update and improve the calculation as its performance is tested against more cohorts and as additional independent survival factors are discovered. The calculator is intended to estimate survival for the majority of people in non-surgical and surgical cohorts. At the moment it does not aspire to calculating survival in patients with severe but rare disease.

You might wonder why the calculator does not incorporate some common survival factors, such as smoking, hypercholesterolaemia, hypertension and diabetes. These factors have substantial prognostic overlap (see above), but one could still estimate the independent contribution of each to survival. However, many of the studies of survival that have measured physical function have found that the independent prognostic value of these variables is diminished, to become inconsistent or entirely lost. I have therefore chosen to exclude them when the calculator incorporates measured or estimated physical fitness (as described above and in the references).

I might have incorporated additional variables between the time I'm writing (June 2015) and the time you read this chapter. Both anaemia and cognitive impairment have been independently associated with survival. Serum concentrations of albumin, brain natriuretic peptide and other substances might also contribute independent survival associations to the calculator, once dose-effect relationships have been better characterised.

3.10 A Study of Errors: Abdominal Aortic Aneurysm Scheduled Repair and Screening

The current practice is to repair male abdominal aortic aneurysms in excess of 54 mm diameter and female aneurysms in excess of 49 mm diameter. Smaller aneurysms are repeatedly scanned by ultrasound until they approach these sizes. Endovascular repair is preferred to open repair as it is followed by a lower mortality rate in the first postoperative month.

Current practice is wrong and it doesn't take much thought to realise that must be the case, even if one cannot determine what the correct practice should be. Current practice is wrong because it only quantifies aneurysm diameter to determine whether a patient's survival will be prolonged. Current practice is wrong because it only uses a single aneurysm diameter to determine whether repair is appropriate. Current practice is wrong because it presents aneurysm repair as either being good or bad: there is usually no attempt to quantify how much survival might be increased or decreased. I will also suggest that waiting for aneurysm growth will be wrong for most patients: either the aneurysm is worth repairing now or it will never be worth repairing.

These errors arise because the survival variables discussed in this chapter have not been used to determine how much survival scheduled abdominal aortic aneurysm repair might give (or take away). This is a bizarre situation, given that the sole purpose of reinforcing an asymptomatic dilatation of the aorta is to prolong survival. An appreciation of how factors affect individual survival trajectories should have led to the obvious conclusion that we have to incorporate patient variables into a calculation to determine the wisdom of scheduled aneurysm repair. I don't know why authors of randomised controlled trials of aneurysm repair, or subsequent researchers, haven't attempted to generate individual predictions. Perhaps they believed that they could only use the survival of the patients in trials of aneurysm repair to model survival. Trials of aneurysm repair don't have enough patients to generate reliable survival models. In isolation those trials, conducted 22 years ago, couldn't determine how applicable or inapplicable their results are to today's population (but researchers might have been unaware of this point). Trials of early surgery versus the combination of surveillance and later surgery also could not provide much information about the growth and rupture rate of aneurysms. However, the whole enterprise of scheduled repair of abdominal aortic aneurysm repair is based upon estimation of rates of aneurysm dilatation and rupture: it is beholden on researchers and clinicians to make explicit the assumptions on which they base their decision-making. Rather than embark on survival estimation researchers and clinicians resorted to aneurysm size. The results of the UK small aneurysm trials and the US aneurysm detection and management trial have been misinterpreted: they are both consistent with repair of aneurysms less than 55 mm in diameter, but to see that we need to examine the 'anatomy' of the studies.

Both the UK SAT [27] and ADAM [28] were randomised controlled trials of surgical management of 1090 and 1136 patients, respectively, with abdominal aortic aneurysms with maximum diameters less than 55 mm on ultrasound (mean diameter 46 mm, UK SAT) or CT (mean diameter 47 mm, ADAM). One group in each trial was allocated to 'early' open repair whilst the other group was allocated to 'later' repair: these two trials are often incorrectly viewed as 'immediate repair' vs 'repair at 55 mm'. Within 5 years of allocation 92 % of early surgery patients had had aneurysm repair, compared with 62 % of later surgery patients. It took 6 months for the aneurysms in 80 % of the early group to be repaired. The delay in repairing the aneurysms of participants allocated to early surgery, which 8 % never had, is understandable and might appear trivial. However, the power of both studies to detect an effect of operating on smaller aneurysms was very sensitive to incomplete and delayed repair. Many patients in the surveillance group had repair before their aneurysms reached a diameter of 55 mm: the best available models suggest that it would take 5 years for an abdominal aortic aneurysm 46 mm diameter to expand to 55 mm. One would therefore expect half of the surveillance cohorts to have not had surgery at 5 years, at which time only 30 and 35 % of the UK SAT and ADAM groups had not had aneurysm repair. The power of the two trials to detect an effect

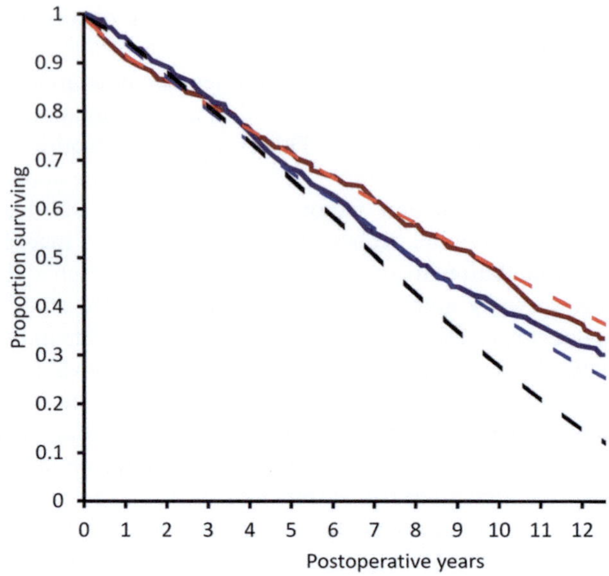

Fig. 3.7 The reported and simulated results of the UK Small Aneurysm Trial [27] that allocated participants with a mean abdominal aortic diameter of 46 mm to early surgery (reported ——, simulated – – – .) or surveillance up to a diameter of 55 mm (reported ——, simulated – – – .). The mean follow-up was 12 years, when the mean diameter of aneurysms without surgery would have been about 80 mm. The trial as conducted was limited to detecting a difference in median survival of 6–12 months if follow-up was complete at 10 years: early surgery was not immediate and was incomplete (94 %), whilst more than a third of the 76 % of repairs in the surveillance cohort were on aneurysms smaller than 55 mm. Reproduced with permission [31]

of timing of aneurysm repair on survival was diminished by the treatment that the two groups received being more similar than intended.

Simulations of both trials, using the calculator referenced previously, generated the same survival curves as observed (Fig. 3.7, reproduced with permission). This demonstrates that we can explain the results of UK SAT and ADAM through simulations using the survival calculator that I developed and published estimates of aneurysm rates of expansion and rates of rupture [29, 30].

I therefore used the same method to simulate additional scenarios for scheduled repair of abdominal aortic aneurysm and screening [31]. My conclusions were:

- It is a patient's characteristics, not aneurysm diameter, which predominantly determine the benefit or harm caused by scheduled aneurysm repair.
- Scheduled repair of abdominal aortic aneurysms as small as 40 mm diameter can prolong survival in some patients, therefore a threshold of 55 mm for intervention will fail to prevent avoidable deaths.
- Scheduled repair of abdominal aortic aneurysms larger than 55 mm, or 65 mm, or 75 mm, can shorten survival in some patients, therefore an intervention threshold of 55 mm will cause avoidable deaths.

- Surveillance is predominantly futile – the absolute increase in median survival caused (in appropriate patients) by scheduled aneurysm repair is not increased by waiting: the increased probability of a larger aneurysm's rupturing is counterbalanced by the patient's diminished survival from other causes. In addition, the absolute benefit that might be realised from operating on a cohort early is partly lost by waiting (see point 2).
- The results of the UK Small Aneurysm Trial and, to a lesser extent, the US Aneurysm Detection and Management trial, can be reproduced in simulation. Simulation suggests that their results have been misinterpreted. Scheduled repair at 46 mm and 47 mm, respectively, would have prolonged median survival by about 3 years compared with repair at 55 mm. Repairs of aneurysms smaller than 55 mm in the surveillance group, coupled with incomplete and delayed repair in the early surgery group, reduced the power of both trials to detect a difference in survival.
- The results, reported or simulated, of the UK SAT and US ADAM trial cannot be used today without adjustment. The UK SAT would have the power to determine a difference in median survival if it were conducted in exactly the same way today (2015), as early surgery would extend median survival by about 4.5 years, or by 6 years with 100 % adherence to the protocol. Historical results become more inaccurate with time: the value of 4.5 years will underestimate the effect of early surgery in the future if the year-on-year reduction in general population mortality continues.
- Endovascular repair extends median survival by 2–3 months for most patients compared with open repair, if one ignores the annual 1 % fatalities due to rupture after endovascular repair, which might negate the small extension to survival afforded by EVAR. Survival is dominated by patients' variables and whether an aneurysm is repaired or not; it is hardly affected by how the aneurysm is repaired.
- Screening should be based upon an individual's calculated monthly mortality coupled with the probability that he/she has an aneurysm sufficiently large that treatment would prolong survival by a meaningful period. A more effective screening programme might be to identify and treat aneurysms 40 to 55 mm diameter in people with low mortality hazards, which would involve repairing many more abdominal aortic aneurysms than currently undertaken. The harm or benefit caused by screening is dependent upon treating the right people and also screening the right people, which cannot be determined by age and sex alone.

The dynamic of harm and benefit for patients with abdominal aortic aneurysms contemplating scheduled surgery is summarised in Fig. 3.8. All the survival factors contribute to determining the patient's monthly mortality, with its associated survival curve, plotted on the vertical axis, whilst the aneurysm diameter determines a patient's coordinate on the horizontal axis.

The fitted polynomial trend lines ($r^2 > 0.995$ for all) relate monthly mortality hazard (%) to aneurysm diameter ('mm') with the following equations (top to bottom

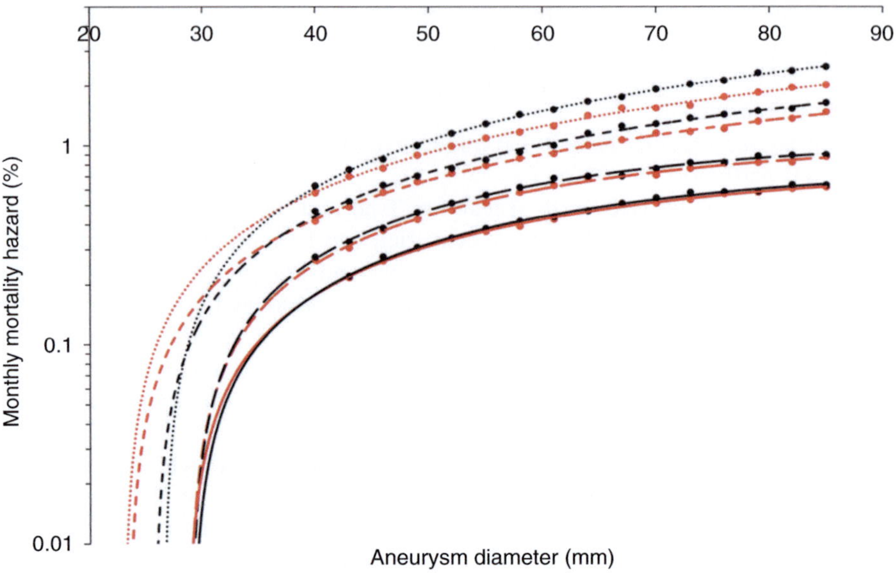

Fig. 3.8 Isosurvival curves of median life extension after open (*black*) and endovascular (*red*) abdominal aortic aneurysm repair by: 1 year (……… and ………); 2 years (– – – – and – – – –); 4 years (— — and — —); or 6 years (——— and ———). For example, a patient with a monthly mortality hazard of 0.5 % would expect surgery to extend their survival by: 2 years (aneurysm 41 mm); 4 years (aneurysm 52 mm); or 6 years (aneurysm 66 mm). Combinations of aneurysm size (horizontal axis) and patient monthly mortality hazard (vertical logarithmic scale) that intersect below the lines indicate larger increases in median survival, intersections above the line indicate smaller increases in median survival (Reproduced with permission [31])

lines), starting with 1 year extension for open repair (black) and finishing with 6 year extension for endovascular repair (red). Monthly mortality hazard = (−0.00006 × mm × mm + 0.0499 × mm − 1.2824), (−0.00004 × mm × mm + 0.0369 × mm − 0.8297), (−0.00008 × mm × mm + 0.036 × mm − 0.8754), (−0.00005 × mm × mm + 0.0291 × mm − 0.6529), (−0.0002 × mm × mm + 0.037 × mm − 0.9173), (−0.0002 × mm × mm + 0.0337 × mm − 0.8357), (−0.0001 × mm × mm + 0.0241 × mm − 0.6075), (−0.0001 × mm × mm + 0.0226 × mm − 0.5618).

3.11 Conclusions

The patient comes first. Operations happen within the context of a patient's life; we will unnecessarily harm some patients and fail to help others if we do not embed the immediate effects of surgery within the context of their overall survival and quality of life. We can use survival data from non-surgical populations and adapt them for this purpose. The peri-operative physician ignores the sum of a patient's life to their detriment. We must broaden our horizons.

References

1. Nightingale F (1859) Notes on hospitals, 1st edn. Longman, Green, Longman, Roberts and Green, London
2. Office for National Statistics. Statistical bulletin: Interim life tables, England and Wales, 2010–12. http://www.ons.gov.uk/ons/rel/lifetables/interim-life-tables/2010-2012/stbilt2012.html. Accessed 12/06/2015
3. Kaplan EL, Meier P (1958) Nonparametric estimation from incomplete observations. J Am Stat Assoc 53:457–481
4. Carlisle JB, Danjoux G, Kerr K, Snowden C, Swart M (2015) Validation of long-term survival prediction for scheduled abdominal aortic aneurysm repair with an independent calculator using only pre-operative variables. Anaesthesia 70:654–665
5. Moonesinghe SR, Mythen MG, Das P, Rowan KM, Grocott MPW (2013) Risk stratification tools for predicting morbidity and mortality in adult patients undergoing major surgery. Anesthesiology 119:959–981
6. Grant SW, Hickey GL, Carlson ED, McCollum CN (2014) Comparison of three contemporary risk scores for mortality following elective abdominal aortic aneurysm repair. Eur J Vasc Endovasc Surg 48:38–44
7. Lee TH, Marcantonio ER, Mangione CM et al (1999) Derivation and prospective validation of a simple index for prediction of cardiac risk of major noncardiac surgery. Circulation 100:1043–1049
8. Protopapa KL, Simpson JC, Smith NCE, Moonesinghe SR (2014) Development and validation of the surgical outcome risk tool (SORT). Br J Surg 101:1774–1783
9. Fox KAA, Carruthers KF, Dunbar DR et al (2010) Underestimated and under-recognized: the late consequences of acute coronary syndrome (GRACE UK-Belgian Study). Eur Heart J 31:2755–2764
10. Brindle P, Beswick A, Fahey T et al (2006) Accuracy and impact of risk assessment in the primary prevention of cardiovascular disease: a systematic review. Heart 92:1752–1759
11. Wannamethee SG, Sharper AG, Lennon L (2004) Cardiovascular disease incidence and mortality in older men with diabetes and in men with coronary heart disease. Heart 90:1398–1403
12. Empana JP, Ducimetiere P, Arveiler D (2004) Are the Framingham and PROCAM coronary heart disease risk functions applicable to different European populations? The PRIME study. Eur Heart J 24:1903–1911
13. Brindle P, Emberson J, Lampe F et al (2003) Predictive accuracy of the Framingham coronary risk score in British men: prospective cohort study. Br Med J 327:1267–1273
14. Conroy RM, Pyörälä K, Fitzgerald AP et al (2003) Estimation of ten-year risk of fatal cardiovascular disease in Europe: the SCORE project. Eur Heart J 24:987–1003
15. Lampe FC, Whincup PH, Wannamethee SG et al (2001) The natural history of prevalent ischaemic heart disease in middle-aged men. Eur Heart J 21:1052–1062
16. Clayton TC, Lubsen J, Pocock SJ et al (2005) Risk score for predicting death, myocardial infarction, and stroke in patients with stable angina, based on a large randomised trial cohort of patients. Br Med J 331:869–874
17. Hemingway H, McCallum A, Shipley M et al (2006) Incidence and prognostic implications of stable angina pectoris among women and men. JAMA 295:1404–1411
18. Hippisley-Cox J, Coupland C, Vinogradova Y et al (2007) Derivation and validation of QRISK, a new cardiovascular disease risk score for the United Kingdom: prospective open cohort study. Br Med J 335:136
19. D'Agostino RB, Vasan RS, Pencina MJ et al (2008) General cardiovascular risk profile for use in primary care. Circulation 117:743–753
20. Steg PG, Bhatt DL, Wilson PWF et al (2007) One-year cardiovascular event rates in outpatients with atherothrombosis. JAMA 297:1197–1206
21. Perna L, Wahl H-W, Mons U, Saum K-U, Holleczek B (2015) Cognitive impairment, all-cause and cause-specific mortality among non-demented older adults. Age Ageing 44:445–451

22. Sachs GA, Carter R, Holtz LR et al (2011) Cognitive impairment: an independent predictor of excess mortality. A cohort study. Ann Intern Med 155:300–308
23. Coresh J, Turin TC, Matsushita K, et al. for the CKD Prognosis Consortium (2014) Decline in estimated glomerular filtration rate and subsequent risk of end-stage renal disease and mortality. J Am Med Assoc 311:2518–2531
24. Myers J, Prakash M, Froelicher V (2002) Exercise capacity and mortality among men referred for exercise testing. New Engl J Med 346:793–801
25. Gulati M, Black HR, Shaw LJ et al (2005) The prognostic value of a nomogram for exercise capacity in women. New Engl J Med 353:468–475
26. Carlisle JB (2014) Pre-operative co-morbidity and postoperative survival in the elderly: beyond one lunar orbit. Anaesthesia 69(Suppl 1):17–25
27. Thompson SG, Ashton HA, Gao L et al (2012) Final follow-up of the Multicentre Aneurysm Screening Study (MASS) randomized trial of abdominal aortic aneurysm screening. Br J Surg 99:1649–1656
28. Lederle FA, Wilson SE, Johnson GR et al (2002) Immediate repair compared with surveillance of small abdominal aortic aneurysms. New Engl J Med 346:1437–1444
29. Thompson SG, Brown LC, Sweeting MJ et al (2013) Systematic review and meta-analysis of the growth and rupture rates of small abdominal aortic aneurysms: implications for surveillance intervals and their cost-effectiveness. Health Tech Assess 17:1–118
30. Grant SW, Sperrin M, Carlson E et al (2015) Calculating when elective abdominal aortic aneurysm repair improves survival for individual patients: development of the Aneurysm Repair Decision Aid and economic evaluation. Health Technol Assess 19:1–154
31. Carlisle JB (2015) Simulations of the effects of scheduled abdominal aortic aneurysm repair on survival. Anaesthesia 70:666–678

Chapter 4
Transthoracic Echocardiography in the Preoperative Clinic

David J. Canty and Colin F. Royse

Abstract Whilst conventional ('quantitative') transthoracic echocardiography (TTE) performed by an echocardiography laboratory remains a useful diagnostic tool for non-invasive cardiac assessment, limited availability, cost, and a time-delay between request and result restrict its use. However, increased availability, portability and imaging capability of TTE equipment has enabled treating physicians to perform their own 'focused TTE' as an aid to initial clinical assessment at the 'point-of-care' when required. Initially embraced by critical care and emergency physicians for rapid assessment of shock, anaesthetists are increasingly using focused TTE for perioperative cardiac assessment. An abbreviated 'qualitative' study enables convenient separation of significant from non-significant cardiac pathology and helps identify the causes for persistent hypotension, such altered preload, afterload, pump or valve failure, tamponade or pulmonary hypertension. When incorporated into routine initial patient assessment of patients at increased cardiac risk presenting to the preoperative clinic before non-cardiac surgery, focused TTE changes the diagnosis and management in approximately 50 % of patients. Identification of previously unknown cardiac disease alerts the anaesthetist to increased perioperative risk, prompting an escalation in preoperative work-up, perioperative haemodynamic monitoring and management and increased use of postoperative intensive or high dependency care. Conversely, reassuring TTE enables a step-down in care, preventing unnecessary surgical delay for investigations or referral, and circumventing the need for invasive monitoring during or after surgery. Focused TTE may assist anaesthetists assess and manage the increasing number of patients presenting to surgery with increased age and increasingly complex co-morbidities admitted on the day of surgery.

Keywords Echocardiography • Transthoracic echo • Point-of-care systems • Cardiac pathology • Preoperative assessment

---

D.J. Canty, MBBS(Hons), PhD, FANZCA, PGDipEcho (✉)
C.F. Royse, MBBS, MD, FANZCA
Ultrasound Education Group, Department of Surgery, University of Melbourne, Parkville, VIC 3070, Australia
e-mail: dj_canty@hotmail.com

K. Stuart-Smith (ed.), *Perioperative Medicine – Current Controversies*,
DOI 10.1007/978-3-319-28821-5_4

4.1 Introduction

4.1.1 The Problem of Preoperative Evaluation

There were 2.4 million non-cardiac surgical procedures performed in Australia alone in 2011, a 5-year increase in volume of 4 %, and a 5-year increase in waiting time of 12.5 % [1]. Despite a very low rate of anaesthesia related mortality (1 in 55,490) after non cardiac surgery [3–6], cardiac complications occur in 2.0–3.5 % of patients, accounting for a third of postoperative deaths (0.5–1.5 %), rising to complication rate of 20 % and mortality rate of 5 % in patients aged over 70 years [7]. Patient comorbidity has a stronger association with poor outcome than type of surgery, which may be contributed to by sub-optimal preoperative assessment and postoperative care [3].

An ageing population (Fig. 4.1) with a higher prevalence of comorbid disease and increased financial pressure to minimise unnecessary preoperative investigations, have increased demands on anaesthetists to assess and manage patients rapidly pre-operatively, to improve hospital efficiency without exposing patients to increased risk.

Preoperative clinics were introduced to reduce the incidence of late surgical cancellation from inadequate patient preparation and to reduce the requirement of day-before admission, thus improving perioperative efficiency. The American Heart Association recommends that the goals of pre-operative cardiac evaluation include provision of medical clearance for surgery, accurate assessment and management of cardiac disease and collaborating with the perioperative anaesthesia team in management that might improve cardiac outcome [8]. Traditional clinical assessment

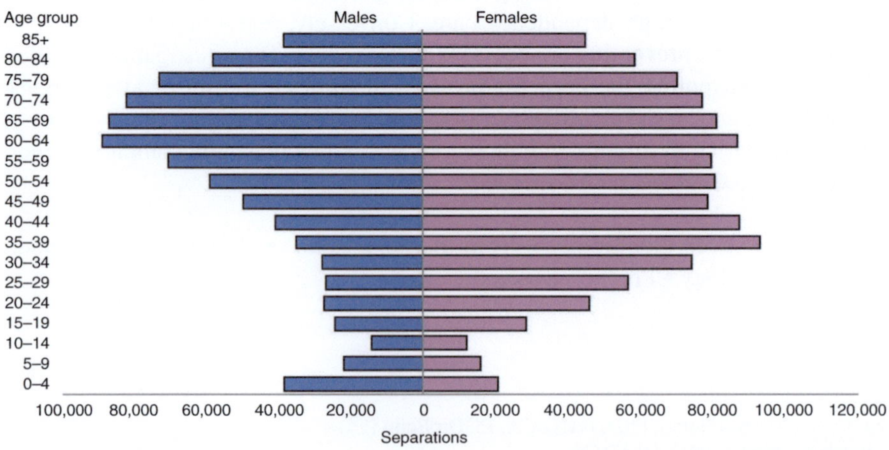

Fig. 4.1 Summary of patients undergoing elective non-cardiac surgery in Australian and New Zealand Hospitals in 2010–11 [1]. Patients aged over 65 years are over-represented compared to other age groups and reflects an ageing population (Reproduced with permission from Australian Bureau of Statistics)

includes history taking, physical examination (clinical assessment), and ordering of indicated tests. Principal cardiac risk factors of adverse postoperative cardiac outcome include left ventricular dysfunction [9–11], ischaemic heart disease, aortic stenosis [12, 13] and pulmonary hypertension [14]. Clinical assessment for these conditions is unreliable, especially if symptoms are masked by exercise limitations from non-cardiac disease (such as painful joints or peripheral vascular disease), unreliable patient history, or incomplete medical records. Detection of a murmur with auscultation is common, but diagnosis or exclusion of aortic stenosis and pulmonary hypertension is unreliable without transthoracic echocardiography [15–17]. Investigation for ischaemic heart disease also requires preoperative testing that may not be available before surgery [8]. As a result, anaesthetists are frequently faced with the dilemma of deferring surgery to obtain pre-operative tests and cardiology assessment in patients with suspected cardiac disease, or allowing non-cardiac surgery to proceed without adequate investigation.

4.2 Transthoracic Echocardiography

4.2.1 Comprehensive (Traditional) Transthoracic Echocardiography

The ability to non-invasively assess left and right systolic and diastolic ventricular function, valvular disease and pulmonary hypertension makes transthoracic echocardiography (TTE) a commonly performed preoperative investigation. Exclusion of left ventricular failure or severe aortic stenosis may assist in clearance for surgery and, on the other hand, if diagnosed, then may prompt referral to a cardiologist for an opinion on perioperative risk, or to reduce risk with either medical or surgical treatment prior to surgery. The current British, European and American cardiology societal recommendations for TTE before elective non-cardiac surgery are listed in Table 4.2 [18–20]. Although TTE may identify structural evidence of ischaemic heart disease, such as regional wall motion abnormality and scar, it has a low sensitivity and specificity for ischaemic heart disease, for which exercise testing is recommended [8]. Guidelines also discourage repeat TTE with no intervening change in clinical status within 12 months [18].

Despite frequent use, the current evidence for the use of preoperative TTE for risk estimation is not strong (Table 4.2) and improvements in outcome after elective non-cardiac surgery has not yet been subjected to testing with an adequately powered randomised controlled trial. However, in a large retrospective cohort study of 264,823 patients aged over 40 years presenting for intermediate to high-risk non-cardiac surgery, preoperative echocardiography was not associated with improved survival or shorter hospital stay [21]. The only other report is a small retrospective cohort study that demonstrated a twofold 12-month survival advantage in elderly patients undergoing hip fracture surgery who received preoperative TTE compared with controls [22].

Table 4.1 Current societal recommendations for indications of transthoracic echocardiography before non-cardiac surgery

Society	Indication
British Society of Echocardiography [18]	Documented IHD with reduced functional capacity (<4 metabolic equivalents)
	Unexplained shortness of breath in the absence of clinical signs of heart failure
	Abnormal electrocardiogram (ECG) and/or chest X-ray
	Murmur in the presence of cardiac or respiratory symptoms
	Murmur in an asymptomatic individual in whom clinical features or other investigation suggest severe structural heart disease
European Society of Cardiology and Anaesthesiology [20]	Established or suspected heart failure in patients scheduled for intermediate or high-risk surgery
	Established or suspected valvular disease in patients scheduled for intermediate or high-risk surgery
American Society of Echocardiography [19]	Stress TTE to investigate for suspected IHD in patients with reduced functional capacity presenting for high-risk vascular procedures

IHD ischaemic heart disease, *ECG* electrocardiogram

Table 4.2 Cardiac pathology identified with transthoracic echocardiography associated with cardiac complications after non-cardiac surgery

Author	Cardiac pathology	Postoperative cardiac risk		
Rohde [9]		Myocardial infarction Odds ratio (95 % CI)	Cardiogenic pulmonary oedema Odds ratio (95 % CI)	Major cardiac complications Odds ratio (95 % CI)
	Reduced LV systolic function	2.8 (1.1–7.0)	3.2 (1.4–7.0)	2.4 (1.3–4.5)
	LV hypertrophy	3.1 (1.1–8.3)	1.5 (0.6–3.8)	2.2 (1.1–4.3)
	Mitral regurgitation	1 (0.4–4.2)	2.4 (1.0–5.6)	2.3 (1.2–4.6)
	Aortic stenosis	3.8 (0.5–32)	6.4 (1.3–32)	6.3 (1.5–26)
Halm [11]		Ischaemic outcomes	Congestive heart failure	All cardiac outcomes
	Ejection fraction <40 %	2.7 (0.7–11)	3.0 (1.2–7.4)	3.5 (1.8–6.7)
	LV wall motion abnormality	1.3 (0.5–3.3)	1.3 (0.5–3.3)	1.6 (1.3–2.1)
	LV hypertrophy	0.5 (0.1–3.7)	1.2 (0.5–3.3)	1.2 (0.7–2.0)

Table 4.2 (continued)

Author	Cardiac pathology	Postoperative cardiac risk			
Park [10]		Myocardial infarction	Congestive heart failure	Cardiovascular death	
	Ejection fraction <50 %	1.6 (0.9–2.8)	2.3 (1.7–3.1)	8.7 (2.6–26.6)	
	LV wall motion abnormality	1.5 (0.9–2.4)	1.7 (1.2–2.2)	1.7 (1.3–2.2)	
Flu [25]		30-day cardiovascular morbidity Odds ratio (95 % CI)		Long-term cardiovascular morbidity Odds ratio (95 % CI)	
	Reduced LV systolic function	2.3 (1.4–3.6)		4.6 (2.4–8.5)	
	Reduced LV diastolic function	1.8 (1.1–2.9)		3.0 (1.5–6.0)	
Kertai [12]		Perioperative/postoperative cardiac events/cardiac deaths			
O'Keefe [13]	Aortic stenosis	Aortic stenosis 14 % compared with controls 2 %, relative risk 5.2 (1.6–17)			
	Aortic stenosis	Aortic stenosis 14 % compared with controls 0 %			
Cowie [26]		Adverse cardiac events %	Positive predictive value % (95 % CI)	Sensitivity	Specificity
	LV/RV dysfunction	58	58 (39–75)	48	92
	Aortic stenosis	88	14 (3–35)	8	89
	Pulmonary hypertension	63	64 (47–79)	64	92
Ramakrishna [14]	Pulmonary hypertension	Short-term morbid event occurred in 42 %, early death in 7 %			

Access to TTE, traditionally performed by cardiology or radiology departments, is variable and the waiting times in many centres may be prolonged due to high demand and reduced availability [23], potentially interfering with surgical work flow [24]. The expansion in use of TTE coupled with persistent economic pressure has lead to a reduction in availability of TTE for preoperative testing in the USA.

4.2.2 Focused Transthoracic Echocardiography

Over the last 10 years there has been a rapid uptake in the clinical use of TTE by emergency and critical care physicians and anaesthetists to guide decision-making in real time, or at the 'point-of-care' [27]. This has occurred due to improved portability, availability and affordability of equipment as well as the realisation that accuracy of clinical assessment can be substantially improved with only limited training and views [28]. Haemodynamic state and cardiac pathology can be diagnosed very quickly, even during a cardiac arrest [29], guiding important decision-making during initial assessment, rather than waiting for a third party to perform the investigation, essentially representing 'ultrasound assisted examination' [2].

The use of focused TTE has been demonstrated to improve clinical assessment and alter important treatment decisions in the order of 50 % of patients either experiencing or at risk of persistent haemodynamic instability in intensive care [30], emergency department [31] and perioperative settings [17, 26], including the preoperative clinic [2] (Fig. 4.2). In 2012, 'point-of-care' physician-performed TTE was defined as focused cardiac ultrasound (FCU) and endorsed in a position statement

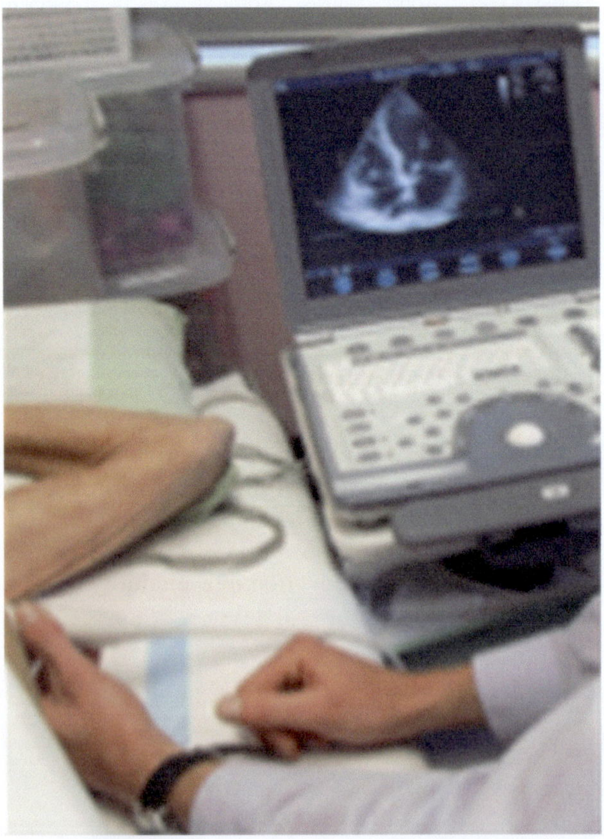

Fig. 4.2 Focused transthoracic echocardiography may be performed at the initial anaesthetist's assessment of patients presenting for noncardiac surgery, enabling convenient, faster and more accurate diagnosis of cardiac disease

from the American Society of Echocardiography [32] and in another position statement on preoperative assessment before non-cardiac surgery from the European Society of Cardiology and Anaesthesiology [20].

In a prospective observational study of 100 patients presenting for assessment by the anaesthetist in the preoperative clinic for non-minor elective noncardiac surgery, the treating anaesthetist's plan was recorded before and after focused TTE was performed by an independent operator. The treating anaesthetists deemed 80 patients to have suspected cardiac disease (Table 4.3) and 20 patients who were not at increased cardiac risk, other than being aged over 65 years. Focused TTE identified at least moderately severe cardiac ventricular or valvular dysfunction in 36 % of the high-risk patients (where TTE was indicated as shown in Table 4.3) and 10 % of the low risk patients (overall 32 %), resulting in a change in management decisions in 54 % (Fig. 4.3). Not only did the TTE result in changes in decision to proceed or postpone surgery without subjecting the patient to postponement of surgery for a repeat appointment, but also altered the anaesthetic technique (regional and general anaesthetic), invasive haemodynamic monitoring and treatment and use of postoperative high dependency care. When a clinically important haemodynamic or valvular abnormality (or worsening of known abnormality) was detected on TTE (e.g. the 11 patients with aortic stenosis), the anaesthetist was alerted to increased cardiac risk, who stepped up their level of care (in 24 %). On the other hand, a clinically unimportant finding or normal TTE had the effect of reassuring the anaesthetist, leading to a reduced requirement for further investigation, invasive haemodynamic monitoring and high dependency postoperative care (step down in care in 30 %). Furthermore, the overall impact of the TTE findings was to improve hospital efficiency by reducing delays to surgery and referrals for further medical opinions, suggesting a resulting cost-savings to the hospital [33].

Another finding was that focused TTE led to a significant change in management in patients in whom the only identified risk factor was age greater than 65 years, raising the question whether focused TTE should be routine in this population. This was also found in a similar study performed in patients undergoing emergency non-cardiac surgery [17].

Table 4.3 Indications for preoperative focused TTE in 100 patients who received focused transthoracic echocardiography in the preoperative clinic

Indication	Percentage (%)
Not indicated – Age >65 years, no cardiac disease suspected	20
Indicated – cardiac disease suspected	80
Signs and/or symptoms of cardiac disease	26
Known cardiac disease but suspected deterioration	15
Murmur on auscultation (not previously documented)	15
Abnormal cardiac investigation (not echo)	8
Unable to exercise and cardiac disease suspected	9
Pulmonary hypertension and/or right ventricular failure suspected	5
Previous peri-operative cardiac complication(s)	2

Reproduced with permission from Wiley, from Canty et al. [2]

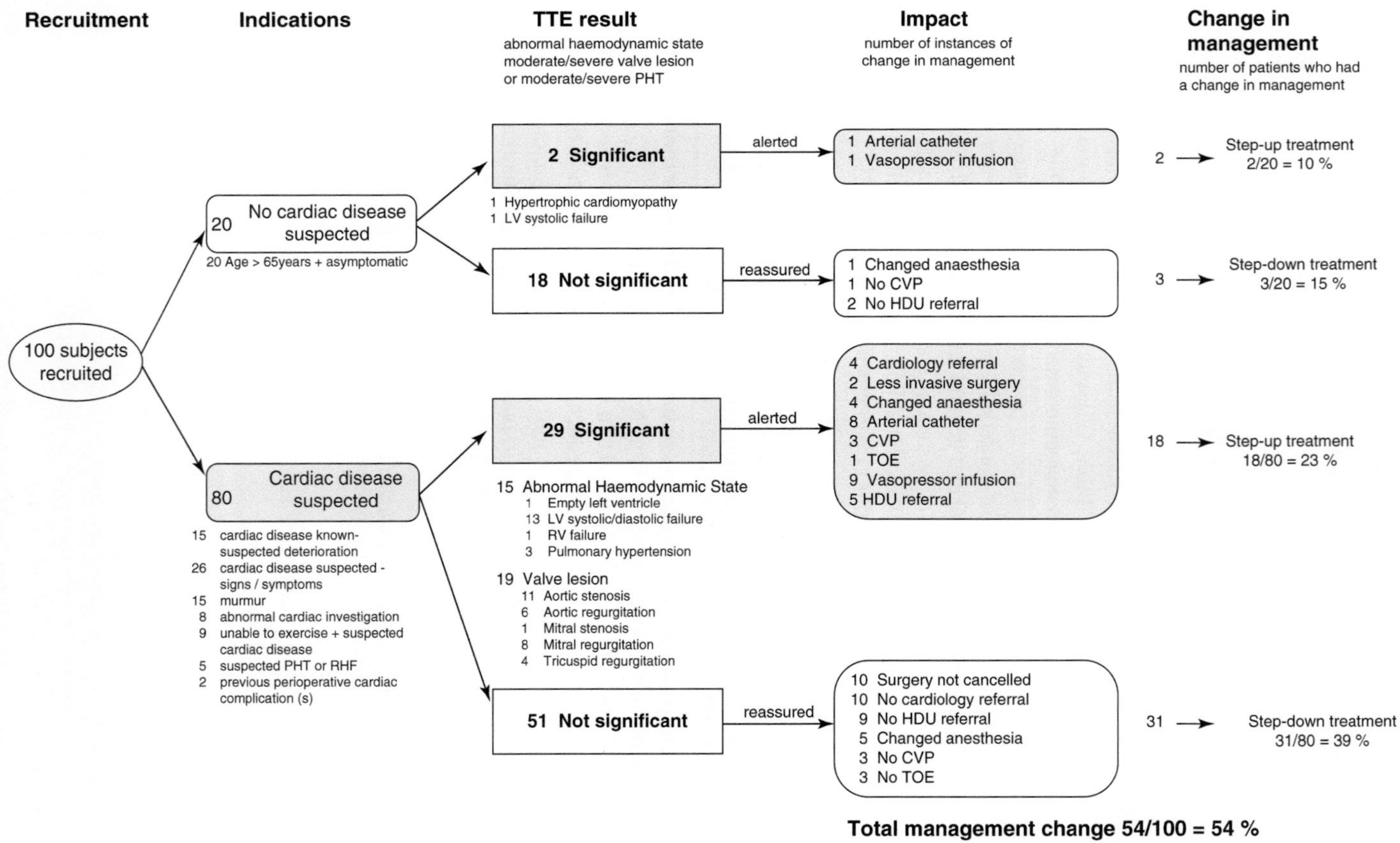

Fig. 4.3 The influence on anaesthetist decision-making on preoperative evaluation after the addition of focused transthoracic echocardiography to standard clinical assessment of patients aged over 65 years or with suspected cardiac disease (Reproduced with permission from Wiley, from Canty et al. [2]) *CVP* central venous pressure monitoring, *HDU* high dependency unit, *PHT* pulmonary hypertension, *RHF* right heart failure, *TOE* transoesophageal echocardiography

The findings from this study not only confirmed that routine anaesthetist-performed TTE was feasible and could be used to assist in decisions for suitability of patients for surgery, but it also influenced the anaesthesia and postoperative medical care in an attempt to reduce poor postoperative outcome, addressing the suggested criteria set by the American Heart Association in preoperative evaluation. However the study was not designed to assess the influence of TTE on postoperative outcome and this question remains unanswered. Another important limitation is lack of consensus and resultant variability in management of specific cardiac disorders and anaesthesia techniques. Despite this, the overall effect of routine focused TTE is earlier detection of patients at increased cardiac risk allowing time for appropriate management decisions. In this study, focused TTE identified clinically significant cardiac pathology in 36 % of patients with suspected cardiac disease (Table 4.3), a similar finding by Faris et al. [34] who reported 27.6 % of patients using comprehensive TTE when used as per published recommendations from the American Society of Echocardiography [35].

4.2.3 Specific Cardiac Pathology Identified with Echocardiography

4.2.3.1 Left Ventricular Function

Patients with a history of congestive heart failure (CHF) have a higher incidence of postoperative CHF and death (Table 4.2), and preoperative TTE is indicated in patients who have suspected CHF or worsening of known CHF (Table 4.1). This is also the case with asymptomatic left ventricular diastolic failure, found in 21 % of patients presenting for vascular surgery [25]. However, ischaemic complications are not clearly correlated with preoperative CHF, for which stress testing is recommended [8]. Preoperative identification of CHF enables optimisation with medical therapy, such as with diuretics and vasodilators, as well as guides perioperative care, including minimising haemodynamic stress from less invasive surgery and anaesthesia and tighter perioperative fluid balance and monitoring for cardiac complications.

4.2.3.2 Aortic Stenosis

Aortic stenosis is the most common valvular heart disease affecting 2–9 % of adults over 65 years of age [36, 37] and is a strong risk factor for cardiac perioperative mortality and morbidity [12, 13]. Aortic stenosis restricts left ventricular stroke volume reducing the ability to cope with haemodynamic demands from anaesthesia and surgery. If not appropriately managed, cardiac function may become irreversibly depressed resulting in cardiac arrest, which is difficult to treat as chest compressions may not be sufficient to provide coronary blood flow. An example is spinal anaesthesia, commonly used in lower limb surgery, which may cause an uncontrollable decrease in systemic vascular resistance. Severe aortic stenosis may be asymptomatic, and therefore may remain undetected until it is too late. Even if detected by auscultation, diagnosis requires echocardiography as clinical assessment is unreliable [15–17].

Preoperative interventions made possible by diagnosis with TTE include aortic valve surgery or valvotomy, and changes in perioperative anaesthetic and medical management. Symptomatic patients should be considered for aortic valve replacement, balloon valvuloplasty or transcatheter aortic valve implantation [38]. If intervention is precluded due to associated comorbidities or patient refusal, then noncardiac surgery should only be performed if essential [38]. In asymptomatic patients, absence of symptoms should be confirmed with exercise testing if possible [38], in which case low to intermediate risk surgery can be performed safely [39].

If the decision is to proceed with surgery without surgery or valvotomy, recommended anaesthetic management includes increased use of invasive haemodynamic monitoring allowing pre-emptive treatment of hypotension (usually with vasopressors), optimisation of volume status, maintenance of sinus rhythm and close attention to effective postoperative analgesia. There is an association of von Willebrand syndrome with aortic stenosis, which can increase bleeding risk [40].

4.2.3.3 Mitral Stenosis

Although less common than aortic stenosis, patients with mitral stenosis are prone to reduced cardiac output and acute pulmonary oedema. However there is little evidence for benefit of valve replacement or valvotomy prior to noncardiac surgery, unless symptomatic or associated with systolic pulmonary pressure is >50 mmHg [8, 38]. In addition to the perioperative anaesthetic management principals for aortic stenosis, avoidance of tachycardia (and atrial fibrillation) and excessive fluid administration may prevent precipitation of acute pulmonary oedema. These patients are also at increased risk of thromboembolism and anticoagulation should be considered [38].

4.2.3.4 Pulmonary Hypertension and Right Ventricular Function

Pulmonary hypertension is a haemodynamic and pathophysiological condition, defined as an increased in mean pulmonary arterial pressure (>25 mmHg) at rest, as assessed by right heart catheterisation, however may be estimated in most patients with TTE by measuring the peak regurgitant velocity of a tricuspid regurgitant jet. Pulmonary hypertension can be either idiopathic or a sign of severity of valvular or ventricular pathology, congenital heart disease, chronic lung disease and obstructive sleep apnoea. Pulmonary hypertension is a strong risk factor for poor outcome after both cardiac and noncardiac surgery. Proposed mechanisms include increases in pulmonary vascular resistance from mechanical ventilation and hypoxia from anaesthesia, exacerbating right ventricular dysfunction and failure. Postoperative respiratory failure is the most common complication [14]. Recommendations for management of patients in whom significant pulmonary hypertension is diagnosed before noncardiac surgery include [20]:

- Avoidance of surgery unless absolutely necessary
- Management in a centre with appropriate expertise with a multidisciplinary team
- Optimised treatment of pulmonary hypertension before any non-emergency surgery, which should be continued before, during and after surgery without interruption
- Close monitoring of patients for at least 24 h in the postoperative period

4.2.3.5 Valvular Regurgitation

Isolated regurgitant valvular disease are not associated with increased perioperative risk of noncardiac surgery [20], however may be associated with other cardiac pathology such as left ventricular systolic or diastolic impairment or ischaemic heart disease (mitral regurgitation), and right ventricular dysfunction and pulmonary hypertension (tricuspid regurgitation).

4.2.4 How to Implement Focused Transthoracic Echocardiography in the Preoperative Clinic

When introducing focused TTE into the preoperative clinic it is essential to address three important factors: the model of care, (i.e. who will perform the TTE and when), training and competency of staff performing TTE, and equipment.

Important common threads include a clear policy and procedure to be followed by the staff in the preoperative clinic and a quality assurance program, including a mechanism for audit and review. Collaboration and support by local cardiology department is advisable in all cases, for assistance in interpretation and quality assurance activity.

4.2.4.1 Models of Care

The model of care depends on number of anaesthetists trained in TTE and the availability of preoperative TTE from existing echocardiography services (cardiology or radiology). This varies considerably from access to TTE at short notice to a waiting list of many months or non-existent. Example models of preoperative focused TTE include the on-call anaesthetist, anaesthetist clinic and the cardiac sonographer clinic.

The On-Call Anaesthetist

Perhaps the simplest model involves having a group of consultant anaesthetists who are highly trained in TTE who can be called upon when required by the anaesthetist who is managing patients in the clinic. This model might suit a larger department containing more than 5 anaesthetists competent in TTE, who are likely to be able to cover the preoperative clinic opening hours, perhaps in a roster or simply a list of phone numbers to ring. Advantages of this system include requirement to train only a small number of staff that perform all of the TTE, making it easier for them to maintain competency. Disadvantages include requirement to relieve the anaesthetist from their clinical duty to perform the TTE. Even if the actual scanning time is brief, the time taken is significant (30–60 min) as there is significant time required to get to the patient, set up the equipment, write the report and archive the images.

Anaesthetist Clinic

A more structured approach is to have a periodic clinic dedicated for preoperative TTE, occurring once or twice per week. If a patient is identified to require a preoperative TTE then returning in several days for a TTE may still avoid cancellation of planned surgery if the result of the TTE is reassuring. Anaesthetists competent in TTE may then be rostered on for the clinic, which may also be used as a teaching session for those learning the skill. This would reduce the delay and time required to find an anaesthetist, and set up the equipment that is required with the on-call anaesthetist model. If given notice, echocardiography laboratory staff may be able to assist in image acquisition or interpretation in difficult cases. This model may suit a hospital with only a few anaesthetists available who are competent in TTE.

Cardiac Sonographer Clinic

A cardiac sonographer stationed in the preoperative clinic trained in focused TTE is another model that has been used successfully. Anaesthetists appropriately credentialed in TTE may review the generated images and report and act on their findings without having to be relieved of their clinical duties as in the on-call model or require being rostered for a whole session to perform TTE as in the anaesthetist clinic model. This model has the other advantage of a potentially larger pool of sonographers that are already trained and can maintain their training in TTE in the echocardiography laboratory.

4.2.4.2 Training and Competency

A common concern is that anaesthetists with limited training in TTE who are performing a focused or abbreviated TTE examination may miss important pathology. One view is that misdiagnosis by TTE could lead to patient harm, and the best way to avoid that is to ensure that only practitioners who are expert in echocardiography perform echocardiography [41]. This provides high quality TTE but restricted to only a few patients, resulting in many patients with time-critical illnesses who may miss out. The alternative view is that clinical diagnosis is frequently incorrect compared with echocardiography (approximately 50 % of the time), and even accepting that practitioners with limited training and experience will miss or misdiagnose pathology on occasions, that ultrasound-assisted examination will reduce the incidence of incorrect diagnosis compared with clinical examination alone, and that there could be a clinical consequence to not performing echocardiography. Practice is likely to be dependent on availability of expert and focused TTE and probably lies between these two views and should be directed by evidence. The literature supports the accuracy of anaesthetist-performed TTE, which is consistently reported at 90–95 % compared with board-certified cardiologist [2, 17, 26]. Novices tend to slightly overestimate severity but importantly tend not to miss significant cardiac pathology. This degree of accuracy is on a par with inter-observer variability between cardiologists [42].

The concept of an expertise pyramid is shown in Fig. 4.4 [41]. This concept follows the alternative view, in that rather than restricting performance of echocardiography to the fully-trained practitioner, novices in training are allowed to perform limited scans but under qualified supervision. This not only allows feasibility of more routine TTE, but also serves as a mechanism of training and quality assurance. At the top of the pyramid are the highly trained and qualified experts in ultrasound, with expertise restricted to one specific area, such as echocardiography (TTE or

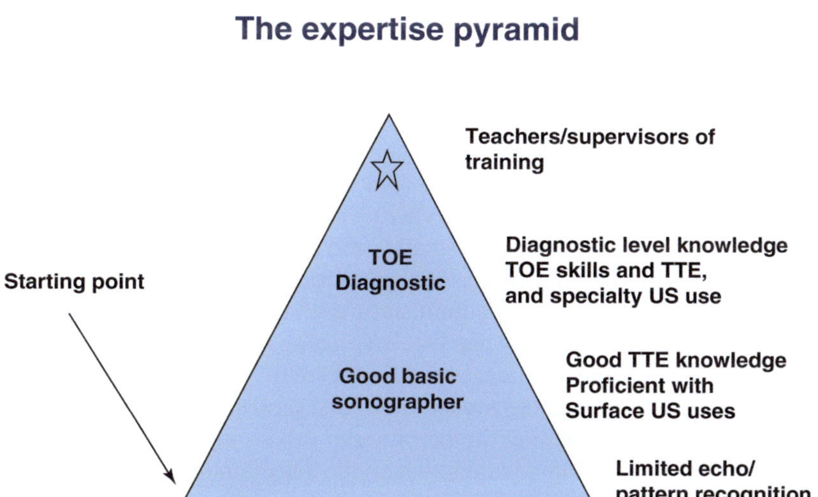

Fig. 4.4 The "expertise pyramid". Four levels of expertise are shown, starting with ultrasound-assisted examination, suitable for all physicians, and the "*star*" at the top of the pyramid representing teachers, supervisors of training, or heads of echocardiography laboratories (From Royse et al. [41]) *TOE* transoesophageal echocardiography, *TTE* transthoracic echocardiography, *UGRA* ultrasound guided regional anaesthesia

TOE), abdominal ultrasound, or lung ultrasound, including specialists from acute care disciplines as well as radiologists and cardiologists.

Recommendations and requirements for training and competency (credentialing), including supervision, are important in ensuring a minimum 'safe' standard. Recommendations for focused echocardiography [32, 43, 44] have lagged behind comprehensive echocardiography, which has only become widely used in the past 5–10 years. As a result, credentialing in focused TTE for the non-cardiology specialties currently rests with the employer (hospital), rather than with the specialty colleges. The three components of competency may be separated into image interpretation (knowledge-base), image acquisition (practical skill) and maintenance of both skills (such as a log book with continuing medical education).

1. *Image interpretation* is assessed either as an exam, such as in North America, United Kingdom and Europe, or a postgraduate university course, such as the diploma of diagnostic ultrasound by the Australian Institute of Ultrasound or the University of Melbourne graduate courses (certificate, diploma and masters).
2. *Image acquisition* may be initially learned during an extended placement, such as at an echocardiography laboratory (TTE) or cardiac anaesthesia rotation/fellowship (TOE), or from a variety of short courses or workshops.
3. *Maintenance of image interpretation and acquisition skills* is usually assessed by means of a logbook of cases, with some specialties also requiring exposure to a variety of specified pathology. Most competency pathways include a minimum number of cases for both the initial training period as well as subsequent maintenance of competency. Some systems require re-sitting the entrance exam (e.g. National Board of Echocardiography, North America) and periodic attendance at courses accredited by the specialty college for Continuous Medical Education.

4.2.4.3 Equipment Considerations

Important factors in deciding on equipment for a preoperative TTE service are financial, quality and portability, and archiving of images for review. As a generalisation, as TTE equipment becomes smaller and more portable, so does the cost, quality and ease of archiving and reviewing images. However as focused TTE is designed to distinguish clinically significant from insignificant cardiac pathology, or qualitative assessment, rather than precise grading of cardiac pathology, known as quantitative assessment, the image quality and archiving capabilities become less important. Where distinguishing moderate from severe aortic stenosis is important for a cardiologist in determining timing of aortic valve surgery, it is less important to the anaesthetist managing a patient, where the detection of previously unknown aortic stenosis is likely to prompt minimising the haemodynamic effects of anaesthesia and use of invasive monitoring and higher level postoperative care, potentially reducing the risk of postoperative haemodynamic instability and adverse outcome. Whilst most focused TTE protocols assess for left and right ventricular function and loading conditions (hypovolaemia, hypervolaemia and vasodilation),

Table 4.4 Features of ultrasound machines capable of transthoracic echocardiography

Size	Cost	Modalities	Advantages	Disadvantages
Full-sized cart	AUD$2-300,000	All, including 3-D	Superior 2-D imaging Good review and archiving	Least portable
Small-sized cart	AUD$100,000	2-D, CFD, MM, SD, TD, DICOM output	Large screen, Good review and archiving (DICOM)	Portability
Lap-top on trolley – cardiology standard	AUD$80,000	2-D, CFD, MM, SD, TD, DICOM output	Portability, acceptable review and archiving (DICOM)	Small screen
Lap-top on trolley – critical care standard	AUD$50,000	2-D, CFD, MM, SD, DICOM output	Portability, durability	Poor review and archiving (DICOM)
Hand-held	AUD$10,000	2-D (including measurements), CFD	Most portable, quick boot-up speed	reduced CFD quality, no spectral Doppler, poor review and archiving (no DICOM)

2-D two dimensional, *CFD* colour flow Doppler, *SD* spectral Doppler, *TD* tissue Doppler

the iHeartScan™ protocol (University of Melbourne) is the only currently published protocol that assesses valvular function. This is accomplished with 2-D assessment of valve separation for stenosis, validated by Godley [45] and Chang [46], and colour flow Doppler for regurgitation, not requiring measurement of valve gradients, which requires continuous wave Doppler as used in comprehensive echocardiography. The specifications, imaging capability, advantages and disadvantages of commonly used TTE equipment are shown in Table 4.4.

4.2.5 Future Directions

As hand carried ultrasound (HCU) devices can be used to assess the lungs [47], abdomen [48], for deep venous thrombosis and carotid atheroma, this global movement has potential to do more than just replace the stethoscope, and has resulted in a paradigm shift in clinical medicine that some have claimed as a 'revolution' [49]. As ultrasound becomes increasingly incorporated into the undergraduate medical curriculum, it is likely that use of hand-held ultrasound devices will be used routinely by resident doctors in initial assessment of all patients, whether in-patient or out-patient. If abnormalities are found then more detailed ultrasound may be quickly obtained with more advanced equipment by a more highly trained operator or supervisor of ultrasound training when required. Ultrasound skills may also be adopted by nursing and para-medical staff, who may use ultrasound assessment as an intermittent monitor, for example tracking cardiac volume status or size of pleural effusion, as is done in some intensive care units with the small sized transoesophageal echocardiography probes (hTEE, Imacor).

4.3 Summary

Focused transthoracic echocardiography is a useful tool that may be used by anaesthetists to non-invasively guide important medical and surgical decisions on patients presenting to the preoperative clinic.

References

1. Australian Government of Health and Welfare. Surgery in Australian Hospitals 2010-11. Authoritative information and statistics to promote better health and wellbeing. http://www.aihw.gov.au/hospitals/surgery-2010-11/. Last accessed 12 June 2015
2. Canty D, Royse C, Kilpatrick D et al (2012) The impact of focused transthoracic echocardiography in the pre-operative clinic. Anaesthesia 67(6):618–625
3. Gibbs NM (2012) Anaesth Intensive Care 40(1):14–6
4. Lee TH, Marcantonio ER, Mangione CM et al (1999) Derivation and prospective validation of a simple index for prediction of cardiac risk of major noncardiac surgery. Circulation 100(10):1043–1049
5. Task Force for Preoperative Cardiac Risk A, Perioperative Cardiac Management in Non-cardiac S, European Society of C et al (2009) Guidelines for pre-operative cardiac risk assessment and perioperative cardiac management in non-cardiac surgery. Eur Heart J 30(22):2769–2812
6. Group PS, Devereaux PJ, Yang H et al (2008) Effects of extended-release metoprolol succinate in patients undergoing non-cardiac surgery (POISE trial): a randomised controlled trial. Lancet 371(9627):1839–1847
7. Story DA (2013) Postoperative complications in Australia and New Zealand (the REASON study). Periop Med 2(1):16
8. Fleisher LA, American College of Cardiology/American Heart A (2009) Cardiac risk stratification for noncardiac surgery: update from the American College of Cardiology/American Heart Association 2007 guidelines. Cleveland Clin J Med 76(Suppl 4):S9–15
9. Rohde LE, Polanczyk CA, Goldman L et al (2001) Usefulness of transthoracic echocardiography as a tool for risk stratification of patients undergoing major noncardiac surgery. Am J Cardiol 87(5):505–509
10. Park SJ, Choi JH, Cho SJ et al (2011) Comparison of transthoracic echocardiography with N-terminal pro-brain natriuretic Peptide as a tool for risk stratification of patients undergoing major noncardiac surgery. Korean Circ J 41(9):505–511
11. Halm EA, Browner WS, Tubau JF et al (1996) Echocardiography for assessing cardiac risk in patients having noncardiac surgery. Study of Perioperative Ischemia Research Group. Ann Intern Med 125(6):433–441
12. Kertai MD, Bountioukos M, Boersma E et al (2004) Aortic stenosis: an underestimated risk factor for perioperative complications in patients undergoing noncardiac surgery. Am J Med 116(1):8–13
13. O'Keefe JH Jr, Shub C, Rettke SR (1989) Risk of noncardiac surgical procedures in patients with aortic stenosis. Mayo Clin Proc 64(4):400–405

14. Ramakrishna G, Sprung J, Ravi BS et al (2005) Impact of pulmonary hypertension on the outcomes of noncardiac surgery: predictors of perioperative morbidity and mortality. J Am Coll Cardiol 45(10):1691–1699
15. Das P, Pocock C, Chambers J (2000) The patient with a systolic murmur: severe aortic stenosis may be missed during cardiovascular examination. QJM 93(10):685–688
16. van Klei WA, Kalkman CJ, Tolsma M et al (2006) Pre-operative detection of valvular heart disease by anaesthetists. Anaesthesia 61(2):127–132
17. Canty D, Royse C, Kilpatrick D et al (2012) The impact of preoperative focused transthoracic echocardiography in emergency non-cardiac surgery patients with known or risk of cardiac disease. Anaesthesia 67(7):714–720
18. British Society of Echocardiography. Clinical indications for echocardiography. Available from: http://www.cdn1.cache.twofourdigital.net/u/bsecho/media/10465/clinical_indications_forechocardiography.pdf. Last accessed on June 2015
19. American College of Cardiology Foundation Appropriate Use Criteria Task Force (2011) ACCF/ASE/AHA/ASNC/HFSA/HRS/SCAI/SCCM/SCCT/SCMR 2011 Appropriate Use Criteria for Echocardiography. J Am Coll Cardiol 57(9):1126–1166
20. Kristensen SD, Knuuti J, Saraste A et al (2014) 2014 ESC/ESA Guidelines on non-cardiac surgery: cardiovascular assessment and management: The Joint Task Force on non-cardiac surgery: cardiovascular assessment and management of the European Society of Cardiology (ESC) and the European Society of Anaesthesiology (ESA). Eur Heart J 35(35):2383–2431
21. Wijeysundera DN, Beattie WS, Karkouti K et al (2011) Association of echocardiography before major elective non-cardiac surgery with postoperative survival and length of hospital stay: population based cohort study. BMJ 342:d3695
22. Canty D, Royse C, Kilpatrick D et al (2012) The impact of preoperative focused transthoracic echocardiography in fractured neck of femur: a retrospective cohort study in mortality. Anaesthesia 67(11):1202–1209
23. Macdonald MR, Hawkins NM, Balmain S et al (2008) Transthoracic echocardiography: a survey of current practice in the UK. QJM 101(5):345–349
24. Badano LP, Nucifora G, Stacul S et al (2009) Improved workflow, sonographer productivity, and cost-effectiveness of echocardiographic service for inpatients by using miniaturized systems. Eur J Echocardiogr 10(4):537–542
25. Flu WJ, van Kuijk JP, Hoeks SE et al (2010) Prognostic implications of asymptomatic left ventricular dysfunction in patients undergoing vascular surgery. Anesthesiology 112(6):1316–1324
26. Cowie B (2011) Three years' experience of focused cardiovascular ultrasound in the perioperative period. Anaesthesia 66(4):268–273
27. Moore CL, Copel JA (2011) Point-of-care ultrasonography. N Engl J Med 364(8):749–757
28. Royse CF, Seah JL, Donelan L et al (2006) Point of care ultrasound for basic haemodynamic assessment: novice compared with an expert operator. Anaesthesia 61(9):849–855
29. Price S, Uddin S, Quinn T (2010) Echocardiography in cardiac arrest. Curr Opin Crit Care 16(3):211–215
30. Vignon P, Mentec H, Terre S et al (1994) Diagnostic accuracy and therapeutic impact of transthoracic and transesophageal echocardiography in mechanically ventilated patients in the ICU. Chest 106(6):1829–1834
31. Labovitz AJ, Noble VE, Bierig M et al (2010) Focused cardiac ultrasound in the emergent setting: a consensus statement of the American Society of Echocardiography and American College of Emergency Physicians. J Am Soc Echocardiogr: Off Publ Am Soc Echocardiogr 23(12):1225–1230
32. Spencer KT, Kimura BJ, Korcarz CE et al (2013) Focused cardiac ultrasound: recommendations from the American Society of Echocardiography. J Am Soc Echocardiogr: Off Publ Am Soc Echocardiogr 26(6):567–581
33. Neale D, Williams D, Canty D (2015) Cost-benefit analysis of focused pre-opewrative transthoracic echocardiography in the pre-operative clinic in patients at increased cardiac risk presenting for non-cardiac surgery – A pilot study. Int J Anesthetics Anesthesiol 2(1):1–5
34. Faris JG, Hartley K, Fuller CM et al (2012) Audit of cardiac pathology detection using a criteria-based perioperative echocardiography service. Anaesth Intensive Care 40(4):702–709

35. Douglas PS, Khandheria B, Stainback RF, et al (2007) ACCF/ASE/ACEP/ASNC/SCAI/SCCT/SCMR 2007 appropriateness criteria for transthoracic and transesophageal echocardiography: a report of the American College of Cardiology Foundation Quality Strategic Directions Committee Appropriateness Criteria Working Group, American Society of Echocardiography, American College of Emergency Physicians, American Society of Nuclear Cardiology, Society for Cardiovascular Angiography and Interventions, Society of Cardiovascular Computed Tomography, and the Society for Cardiovascular Magnetic Resonance endorsed by the American College of Chest Physicians and the Society of Critical Care Medicine. J Am Coll Cardiol 50(2):187–204
36. Stewart BF, Siscovick D, Lind BK et al (1997) Clinical factors associated with calcific aortic valve disease. Cardiovascular Health Study. J Am Coll Cardiol 29(3):630–634
37. Lindroos M, Kupari M, Heikkila J et al (1993) Prevalence of aortic valve abnormalities in the elderly: an echocardiographic study of a random population sample. J Am Coll Cardiol 21(5):1220–1225
38. Vahanian A, Iung B (2012) The new ESC/EACTS guidelines on the management of valvular heart disease. Arch Cardiovasc Dis 105(10):465–467
39. Calleja AM, Dommaraju S, Gaddam R et al (2010) Cardiac risk in patients aged >75 years with asymptomatic, severe aortic stenosis undergoing noncardiac surgery. Am J Cardiol 105(8):1159–1163
40. Vincentelli A, Susen S, Le Tourneau T et al (2003) Acquired von Willebrand syndrome in aortic stenosis. N Engl J Med 349(4):343–349
41. Royse CF, Canty DJ, Faris J et al (2012) Core review: physician-performed ultrasound: the time has come for routine use in acute care medicine. Anesth Analg 115(5):1007–1028
42. van Royen N, Jaffe CC, Krumholz HM et al (1996) Comparison and reproducibility of visual echocardiographic and quantitative radionuclide left ventricular ejection fractions. Am J Cardiol 77(10):843–850
43. Australian and New Zealand College of Anaesthetists Professional Standards PS46 - Guidelines on Training and Practice of Perioperative Cardiac Ultrasound in Adults. http://www.anzca.edu.au/resources/professional-documents/documents/professional-standards/professional-standards-46.html. Last accessed 10/7/2013
44. International expert statement on training standards for critical care ultrasonography (2011) Intensive Care Med 37(7):1077–1083
45. Godley RW, Green D, Dillon JC et al (1981) Reliability of two-dimensional echocardiography in assessing the severity of valvular aortic stenosis. Chest 79(6):657–662
46. Chang S, Clements S, Chang J (1977) Aortic stenosis: echocardiographic cusp separation and surgical description of aortic valve in 22 patients. Am J Cardiol 39(4):499–504
47. Lichtenstein DA, Meziere GA (2008) Relevance of lung ultrasound in the diagnosis of acute respiratory failure: the BLUE protocol. Chest 134(1):117–125
48. Kirkpatrick AW, Sirois M, Laupland KB et al (2004) Hand-held thoracic sonography for detecting post-traumatic pneumothoraces: the Extended Focused Assessment with Sonography for Trauma (EFAST). J Trauma 57(2):288–295
49. Vieillard-Baron A, Slama M, Cholley B et al (2008) Echocardiography in the intensive care unit: from evolution to revolution? Intensive Care Med 34(2):243–249

Chapter 5
Defining Postoperative Quality of Recovery

Andrea Bowyer and Colin F. Royse

Abstract Central to the practice of perioperative medicine and patient centred care are the abstract ideals of quality and recovery, neither of which have a universally accepted definition or assessment tool. Recovery definitions have progressed from a historical emphasis on short-term restitution of purely physiological parameters to a multidimensional concept, which includes physiological, nociceptive, emotive, functional and cognitive domains. With an ageing, ambulant perioperative population and an increasing understanding of the interplay between physiological stress and neurophysiology, there is increased importance of cognitive assessment. Current recovery assessment tools vary in their validation and differ in their assessment of recovery as a dichotomous vs. continuous variable, and at the individual vs. group level. Ideally, quality of recovery would be defined as a return of a patient's function to their own baseline or better and assessed in multiple domains and over multiple time points. The ideal recovery assessment tool would allow real-time assessment of recovery with facility to identify in which sub-domain recovery has failed, in both individual patients and at a population level, thus allowing both clinical time-critical interventions and research applications.

Keywords PostopQRS • Postoperative quality of recovery scores • Real-time assessment • Multi-dimensional assessment • Recovery scores • Post-operative cognitive dysfunction

---

A. Bowyer, MBBS, FANZCA, PG-CU
Department of Anaesthesia and Pain Management, Royal Melbourne Hospital, Melbourne, VIC, Australia
e-mail: Andrea.Bowyer@mh.org.au

C.F. Royse, MBBS, MD, FANZCA (✉)
Ultrasound Education Group, Department of Surgery, The University of Melbourne, Melbourne, VIC, Australia

Department of Anaesthesia and Pain Management, The Royal Melbourne Hospital, Melbourne, VIC, Australia
e-mail: Colin.Royse@heartweb.com

K. Stuart-Smith (ed.), *Perioperative Medicine – Current Controversies*,
DOI 10.1007/978-3-319-28821-5_5

5.1 Introduction

Quality and recovery are two abstract ideals that are integral to the practice of perioperative medicine. With the developing emphasis on patient centred care in both clinical and research settings, there is an increasing need to both define and measure these ideals in real time. Traditionally, assessment of recovery was uni-dimensional and short-term, focusing primarily on restitution of physiological parameters in the immediate postoperative period. The breadth of recovery assessed was originally derived from expert opinion or patient focus groups, with only some tools subsequently being systematically validated [1]. The maturation of perioperative medicine into an independent speciality mandated the development of multidimensional tools that addressed recovery in terms of functional activity [2–9], psychological recovery [2, 3, 5, 8, 9], satisfaction [2, 3, 6, 10, 11] and the presence of adverse symptomatology [2–4, 7]. With an increasingly elderly and ageing surgical population, there has been renewed interest in perioperative neurocognitive pathology and cognitive recovery. Cognitive recovery has thus developed from simple assessments of orientation and comprehension [3, 5, 6], to a broader assessment derived from neurocognitive testing [2, 12–14]. Validation of recovery tools was often incomplete and limited to the patient population and clinical context from which it was developed, with only some subsequently being systematically validated [1]. Thus, there remains a need for uniformity with respect to defining recovery, the timing of its assessment and recovery tool validation. The ideal recovery tool would assess multidimensional recovery at the individual or patient level, both overall and at the subdomain level. It would be validated for repeated measures (assessing recovery at sequential time points) in both the clinical and research setting and provide real-time recovery data. This latter aspect of a tool has the potential to impact clinically on individual patient outcome via timely early intervention as well as be applied to the research setting.

5.2 Historical Definitions of Recovery

The abstract concepts of quality and recovery are quantities that we are increasingly attempting to measure. A recent review of 11 unique recovery assessment tools highlighted their differences in definition of recovery, its breadth of assessment, timing and patient cohort [15]. The definition of recovery has evolved since the initial attempt at its quantification, which focused purely on restitution of physiological parameters in the immediate postoperative period [16]. Historical indicators of poor recovery have been the presence of postoperative pain and nausea [2–4, 7], emotional and psychological distress [2, 3, 5, 8, 9] or dissatisfaction [2, 3, 6, 10, 11]. Other fundamental clinical parameters assessed included cardio-respiratory status, basic functional assessment and nociceptive (pain/nausea) [17–19]. These, along with postoperative complications, have been traditionally used as indicators of recovery as they impact on readiness for discharge and hence resource utilisation. The progression to defining recovery as a multidimensional construct has seen the development of multidimensional tools that addressed functional activity [2–9], psychological recovery [2, 3, 5, 8, 9], satisfaction [2, 3, 6, 10, 11] and the presence of adverse symptomatology [2–4, 7]. With an ageing perioperative population there has been renewed interest in the interaction between surgical stress and neurocognitive decline. This has driven the development of cognitive assessment from simple assessments of orientation and comprehension [3, 5, 6], to more formal neuropsychologically based assessment [2, 12–14]. More recent clinical recovery assessment tools, such as the Postoperative Quality of Recovery Scale (PostopQRS) [2] assess patient recovery in relation to individual baseline function, in real time and over multiple time points.

5.3 Recovery Versus Process and Outcome Measures

At an institutional and legislative level, recovery is often defined by clinical performance indicators that affect either resource utilisation, funding or indicate potential substandard care [15]. Using clinical indicators, variations in patient recovery (however this is defined) can be due to patient case-mix, measurement, chance or true differences in quality of clinical care [20]. However, these performance indicators are inherently different from patient focused recovery as they focus mainly on complications and potential clinical errors rather than patient function [21]. Their utility is in identifying sentinel events or changes in performance indicators that may reflect deficits in patient care, rather than recovery itself. Furthermore, ultimate patient recovery may not reflect quality of patient care, as the latter is only one of many perioperative influences on the former.

A systematic review [21] identified 108 contemporary quality and safety (Q&S) anaesthetic related clinical performance indicators, 53 of which assessed postoperative outcomes. Only 40 % demonstrated more than purely face validity and, of these, 42 % used process, 57 % outcome and 1 % structural measures. A process indicator is one that focuses on the implementation of a component of the clinical care pathway (i.e. administration of DVT prophylaxis, instituting early postoperative ambulation). Process measures are thus direct measures of *quality of care*, but are only measures of *recovery* if these clinical components have been a priori identified as significantly impacting on ultimate patient recovery. They are also easy to interpret and sensitive to changes in quality of care [20], thus being ideal in a research setting assessing specific aspect of care. They do not, however, confirm causality between quality of care and recovery, nor do they assess the impact of any other direct or indirect variables on ultimate patient outcome. In comparison, outcome measures are indirect measures of *quality of care* but are direct measure of *patient recovery*. Recovery itself is inherently an 'outcome' variable and thus is the result of known and unknown variables. Outcome variables correlate more closely with quality of care when that quality has substantial influence on patient outcome, particularly when technical expertise is involved. They also account for both known and unknown variables influencing patient recovery. It is important to acknowledge that outcome variables such as presence/absence of nausea or pain are also important in their own right irrespective of causality. They have direct implications to both the patient and the institution, impacting on patient experience and discharge readiness respectively. They are, however, often underpowered to detect rare clinical events [20] and lack sensitivity and specificity [22].

Q&S clinical indicators address primarily patient safety (83 %) and effectiveness (68 %). Current National Safety and Quality Health Service Standards [23] require reporting on 13 anaesthetic related indicators, including 5 pertaining to the recovery room (respiratory distress, PONV, hypothermia, severe pain and stay >2 h) and 3 to the immediate postoperative period (unplanned ICU admission within 24 h, severe pain and review of epidural analgesia). As stated previously, these indicators are often erroneously substituted for measures of both quality of care and recovery. For example, PONV and pain, often identified as surrogates for anaesthetic recovery and patient satisfaction, are themselves interrelated and immeasurably intertwined with surgical and patient factors [24]. Unidimensional outcome measures do not account for the contextual and multifactorial components of quality of recovery (i.e. transient PONV and pain may be crucial to a patient's concept of recovery in the palliative setting but less so in the curative setting). Furthermore, quality and safety outcomes, when used in isolation, are rarely shown to improve patient outcome [25, 26]. Thus, it is erroneous to use purely outcome measures to assess quality of patient recovery without also incorporating measures of patient, anaesthetic and surgical complexity.

5.4 Influences on Recovery

Recovery as a general construct is influenced by identifiable factors, with surgical, anaesthetic and patient factors implicated in poor recovery. Laparoscopic technique has been associated with improved patient function at day 7 when compared to open surgery [8, 27]. The hypermetabolic perioperative state with its associated insulin resistance and glucose derangement has also been implicated in delayed recovery [28–30], as has general anaesthesia in early delayed recovery in the elderly [31, 32]. Furthermore, early poor recovery is predictive of later dysfunction, with a correlation between both delayed patient discharge and reduced global recovery at 30 days [5], and total in-hospital stay and failure to recovery at 60 days. Recovery also varies between patient subpopulations, with physical recovery being delayed in those >75 years old compared to those <65 years old [5].

Cognitive recovery is both a subset, and an integral component, of global recovery. It also has identifiable risk factors and is predictive of long-term morbidity and mortality. Univariate associations of early incomplete cognitive recovery are increasing age, duration of anaesthesia, reoperation, postoperative infections and respiratory complications [33]. On multivariate logistical regression, independent risk factors for long-term cognitive decline are advanced age, lower education level, a history of cerebral injury without residual deficit, and POCD at hospital discharge [14, 33, 34]. At discharge, cognitive dysfunction affects all age-groups (36.6 % young, 30.4 % middle-aged and 41.4 % elderly, $p > 0.05$) and is more prevalent than in corresponding age and ability matched non-surgical controls [34]. Persistent cognitive dysfunction, however, is more prevalent in the elderly population (12.7 %) compared to both age-matched non-surgical controls ($p < 0.001$) and younger postoperative patients (5.7 % young and 5.6 % middle aged, $p = 0.001$). Early and late POCD are both predictive of short and long term mortality. The presence of POCD at discharge is associated with increased 3 month mortality (6.5 % POCD vs. 3.4 % no POCD, $p = 0.02$) [34] whilst POCD at 3 months is associated with increased mortality at 12 months. Furthermore, persistent POCD (i.e. POCD at both discharge and 3 months) is associated with significantly higher 1 year mortality (10.6 %) compared to patients with either no POCD or transient POCD at a single postoperative time point such as at discharge or 3 months (mortality 2.1 % & 2.86–3.9 % respectively) [34]. Whether there is a causal or confounding relationship between POCD and mortality is unknown, but the association is consistent with the observed increase in mortality in non-surgical patients displaying cognitive decline. Depression and deep-hypnotic time are also known predictors of long-term mortality but do not always correlate with long-term POCD [35]. Thus, whilst early POCD affects patients of all ages, it is more persistent in the elderly and confers a higher short and long-term mortality risk.

5.5 Recovery Assessment Tools

A recent review [15] of current recovery assessment tools identified differences in their target patient populations, timing, validation and breadth of assessment. They also differed in their definition of recovery and its assessment as a continuous vs. dichotomous outcome at a group vs. individual level.

The review highlighted the difference in patient populations used to derive recovery tools, with exclusively inpatients [2, 5, 7, 16], ambulatory patients [4, 6, 8, 36], a combination of ambulatory and inpatients [3, 37] or non-surgical patients [8] variably utilised. Timing of recovery assessment also differed between tools, with short- [3, 4, 6, 8, 9, 16, 36, 37] or intermediate-term [2, 5, 7] single [3, 6, 8, 9, 16, 36, 37] or multiple [2, 4, 5] time points commonly used. Comparison of recovery also differed in its assessment at the groups [2–9, 16, 36, 37] or at an individual level [2, 4] using either composite [3–9, 16, 36, 37] or dichotomous [2] variables.

As recovery tools differed in their time periods of assessment, they therefore differed in their focus of recovery. Modern ambulatory anaesthesia recovery assessment tools [3, 4, 6, 8, 36] assess recovery over short term intervals (24 h [3, 6] and 7 days [4, 8, 9]) and focus on physiological and early functional recovery important for discharge (physiological safety, pain, nausea, return of gastrointestinal function). Tools that assess intermediate recovery (postoperative day 28 [7], day 60 [5]), focuses on nociceptive, emotive, functional and cognitive recovery. Assessment of late recovery (6 weeks [38] and 3 months) [2] seeks to quantify persisting pain, nausea, poor functional recovery and cognitive decline. The time interval for measuring recovery is important as early recovery is influenced by physical symptoms (pain, nausea, anxiety), perioperative medications (anaesthesia and analgesia) and coexisting delirium [31]. In addition, the utility of a recovery assessment tool is limited to the post-operative time-period and patient population for which it has been previously validated.

Recovery lacks a universal definition and, as such, recovery assessment tools have been designed and validated using a variety of approaches. Most simply, pre-existing non-operative patient assessment tools were applied to the postoperative period [8] or were extended and validated to the postoperative period [3, 5, 6, 37]. More comprehensive was the development of new [2, 4, 7, 9, 36] recovery assessment tools, with their definitions and scope of recovery measured originally being determined by patient [4, 6, 7] and/or clinical focus groups [4, 6, 7, 9]. A more structured approach was the Delphi method [39], a structured group communication whereby expert opinion was collected, collated and refined using sequential rounds of questionnaires and feedback.

Historically, validation of recovery tools according to existing criteria [1, 40] has been variable. Typically, initial feasibility was confirmed in either a pilot study, followed by assessment of face validity and feasibility in a large multi-group study [2]. Alternatively, all validation was performed in a single definitive study [3–7, 9]. Although the lack of a gold standard for quality of recovery hindered assessment of criterion validity, content and construct validity were uniformly reported. Content validity, the tool's ability to sample the domain of interest (i.e. recovery), was determined in the original development process. Construct validity, the tool's subdomains being consistent with anticipated recovery, was confirmed using either human volunteer [2, 41] or test-retest [3] studies. Surprisingly, discriminant validity, the important measure of the tool's ability to differentiate between groups of patients, was often performed using the original study population [4–7, 9]. Rarely was this important attribute confirmed in a meta-analysis [42] or subsequent trials. Discriminant validation using multiple trials, different surgical populations and time frames is a very important aspect that is seldom performed. Additional validation criteria assessed included internal consistency, reproducibility, responsiveness, interpretability, and floor and ceiling effects.

5.6 Post-operative Cognitive Assessment

It is predicted that by 2051 almost half of all anaesthetics administered will be to patients >65 years old [43]. This, combined with renewed interest in the effects of anaesthetic and surgical practices on cognitive performance, has highlighted the issue of postoperative cognitive recovery. ***Cognitive recovery*** has been traditionally poorly defined and often used interchangeably with other cognitive conditions such as postoperative delirium and postoperative cognitive dysfunction (POCD). It is, however, a discrete entity. Cognitive recovery and global recovery are also interrelated, with impaired recovery in non-cognitive domains being associated with an increased risk of postoperative cognitive dysfunction [14, 44]. Assessment of cognition using a comprehensive yet brief tool is integral to perioperative management as it has both clinical and prognostic implications.

5.7 Perioperative Neuropathophysiology

Postoperative cognitive changes have been long reported in initially cardiac, and then non-cardiac, surgical populations. This observed perioperative neurophysiological decline is hypothesised to be the combination of a reduction in cognitive reserve with an aggravation of underlying neurodegenerative disease. The concept of an individual's cognitive reserve originated from the observed protective effect of education and intelligence against postoperative cognitive dysfunction [33, 34]. Cognitive reserve can be both passive (an increase in synaptic number and neuronal mass) and active (activation of redundant neuronal networks) [45], and results in maintenance of cognitive function in the event of global or focal subclinical insults. Ageing, even in the absence of disease, is associated with a degree of neurodegeneration, both at the cellular (reduced neuronal cell number and synapse [46] and subcellular (reduced synaptic density and cerebral microvasculature, accumulation of oxidative stress [47] levels. Furthermore, increased levels of IL-6 and CRP in both elderly human brain [48] and in perioperative animal CSF [49] suggest a pro-inflammatory cause for reduced cognitive reserve and/or overt cognitive decline in both the ageing process and the perioperative period. Thus, even in the absence of a underlying neurodegenerative disease, an advanced age and a pro-inflammatory state, both of which are present in the ageing perioperative population, are risk factors for declining cognitive reserve.

In addition its effects on cognitive reserve, there is concern that the perioperative period may potentiate any underlying neurodegenerative disease processes. Both anaesthetic and surgical factors have been implicated. The inhalation agents sevoflurane and isoflurane are both associated with altered amyloid precursor protein (APP) synthesis [50, 51] and reduced memory function in elderly, compared to younger, rodents [50]. This infers a possible potentiation by inhalation anaesthetic agents of the Alzheimer's disease process. Clinically, however, inhalation anaesthetic agents are not associated with an increase in cognitive dysfunction, with the incidence of POCD in patients undergoing regional vs. general anaesthesia [52] being comparable. This suggests that surgical inflammation, present with both regional and non-regional anaesthesia, may be an important confounder contributing to cognitive decline.

Surgery, and in particular its pro-inflammatory state, has also been implicated in accelerating underlying neurodegenerative disease processes. Surgical stress induces an acute systemic inflammation, the latter being independently associated with a twofold increase in incidence of cognitive decline at 1 year in non-surgical patients [53]. The proposed mechanisms for systemic inflammation causing neurodegeneration are twofold. Firstly, neuro-inflammation mirrors systemic inflammation, with increased plasma IL-1beta and IL-6 correlating with increased levels in the hippocampus and associated microgliosis [49]. Secondly, microglial priming, a process whereby diseased microglia exhibit an exaggerated neuroinflammatory response to a systemic inflammatory stimulus [54, 55], contributes to tau protein accumulation. Thus, surgical stress induces a neuroinflammtory state and subsequent neuronal injury, a process that is amplified in the presence of chronic inflammatory insults and neurodegeneration.

5.8 Delirium

Delirium is a well-defined medical condition and, when present in the perioperative period, has prognostic implications. Delirium is defined by the Diagnostic and Statistical Manual IV (DSM-IV) as an alteration of clarity of awareness of environment, fluctuation of orientation and an inability to focus, sustain or shift attention [56–58]. It has demonstrable pathology of altered cortical function and diffuse EEG slowing, with associated increases in central cytokine production. Its perioperative incidence ranges from 5–15 % to 62 % in high risk patients and is more prevalent in patients with advanced age (>70 yo), urea:creatinine ratio >18, multiple co-morbidities and surgical complexity [59]. Pre-existing cognitive and functional decline, even when subtle, may predict delirium at discharge but not necessarily at 3 months [60, 61]. Postoperative delirium (POD) is that which occurs acutely in a previously well patient after a normal emergence (i.e. is not emergence delirium), typically on postoperative day 1–3. It is independently associated with an increase hospital length of stay, risk of POCD (at discharge but not 3 months), mortality (in-hospital and beyond 12 months [59], institutional placement and progression of pre-existing neurocognitive disease [59, 62]. However, when compared to delirium occurring in the non-operative setting, it is associated with an increased incidence of complete recovery, surmised to be due to transient perioperative factors.

5.9 Postoperative Cognitive Dysfunction

In comparison, postoperative cognitive dysfunction (POCD) is defined as a neurocognitive disorder that presented in the first few postoperative weeks as a subtle alteration in patient memory and/or cognitive function [56–58]. It is a neuropsychological construct that arose as an attempt to quantify the observed subtle decline in cognition post cardiac, and subsequently non-cardiac, surgery. It is defined as a statistically significant deterioration in performance on objective neurocognitive testing that is inferred to have clinical significance. There is discrepancy of self-reporting functional decline between patients exhibiting POCD in executive, but not memory, domains [63]. This, combined with the poor correlation between objective neurological impairment and self reported deficits [33, 64] leads to an under-reporting of POCD, and emphasises the need for objective, rather than subjective, measures when testing cognition. Accurate assessment is vital due to the correlation of POCD with short and long term mortality [34, 63]. Its assessment, however, is complicated by the heterogeneity of tests used, the inherent inability to assess their sensitivity, the dependence of test accuracy on number of tests and the impact of learning effect on neurocognitive performance.

Early POCD is predictive of late (3 month) POCD. Conversely, however, prediction of late POCD is limited by the subset of patients displaying 'new' POCD at 3 months i.e. patients who did not display cognitive decline at discharge. This suggests that late POCD is heterogeneous in origin and comprises patients that have persistent postoperative cognitive decline ('true' POCD), a progression of an underlying neuro-degenerative disease or a new cerebral insult. Furthermore, early POCD may also be heterogeneous in aetiology and confounded by factors limited to the short-term perioperative period. POCD at discharge is associated with ASA score, surgical complexity, hospital LOS, operative duration and complexity and an episode of delirium. Whilst delirium must be excluded before POCD may be assessed, screening tests for such are not always powered to detect subclinical delirium and may result in patients being diagnosed with early POCD when an underlying delirium may be present. This hypothesis is consistent with the current understanding of the effects of neuro-inflammatory mediators, perioperative pharmacology and sleep deprivation on cognitive performance [34]. Thus, as cognitive decline is associated with increased short and long-term mortality in both surgical and non-surgical populations, the ability to predict long-term cognitive dysfunction would have direct clinical utility.

Arising as a neurophysiological statistical construct, there is as yet no consensus as to the definition of POCD. A systematic review [57] reported POCD as being variably defined as a difference in group change scores, a change of greater than one (or two) standard deviation from preoperative score in 2 or more tests or the same change in one or more tests [65, 66], a 20 % change in at least 20 % of tests or identification of clinically significant differences between groups [67, 68]. In a research setting, comparing neurocognitive performance between two groups involves comparing each group's mean change score, i.e. the average difference in individual postoperative and preoperative performances. This assesses whether the difference in cognitive performance between the two groups is statistically significant but does not discern clinical significance or quantify this in relation to the expected variation in cognitive performance in the non-surgical population. In addition, it does not directly indicate recovery at the individual patient level or within specific domains.

The initial definition of POCD was a change of greater than one standard deviation (1SD) in an individual's postoperative performance. This was a measure of an individual's postoperative function compared to their own preoperative reference, with the magnitude of this change standardised in relation to the variation in the study population's preoperative value. Assessment of group performance was thus the mean individual standardised change score, i.e. the average individual change in function, divided by a measure of the group variation. A limitation of this approach is that, as intra-group variation is unique to each study population, the absolute difference represented by 1 SD differs between studies. This difference in magnitude may have clinical implications.

5.10 Sensitivity and Specificity of POCD Testing

POCD is defined as change in cognitive performance (in neurocognitive testing) beyond a predetermined threshold deemed clinically significant. Assessment tools thus have inherent difficulties with regard to evaluating their sensitivity and specificity, with the lack of a 'gold standard' rendering it impossible to assess true test sensitivity. Furthermore, the measured incidence of POCD is dependent upon the magnitude of cognitive change defined as significant and the number of tests comprising assessment. Neuropsychological tests were originally derived to detect gross changes in cognitive function in overall populations and thus often lack the sensitivity to detect the more subtle changes of POCD at the individual level. However, the incidence of POCD to varies from 13.3 to 49.4 % in accordance with increasing the number of tests [69]. Increasing the number of neuropsychological tests within a test battery increases the sensitivity of the overall test for POCD (i.e. ability to detect cognitive dysfunction) but concurrently increases the random (Type 1) error and decreases overall test specificity.

Assessment of PCOD is prone to type 1 error arising from both random (inter-patient variation) and systematic (learning effect) sources. The increase in random error associated with increasing the number of tests used can be ameliorated by increasing the threshold deemed significant, such as increasing the threshold from 1SD to 2SD or a mandating a deficit to be in more than one test. Systematic error associated with repeated testing of patients over time (learning effect) can be minimised by increasing the time interval between testing (not relevant when timing of tests are determined by clinical significance), performing the tests in random order or using parallel forms. In addition, the use of a control group that has been exposed to repeated testing is essential to distinguish POCD from Type 1 error as it quantifies learning effects, inter-patient variation (variation between patients at each testing time point) and intra-patient variation (variation in cognitive performance present patients that have not been exposed to the potential cognitive insult of surgery) [70].

5.11 Cognitive Recovery

Postoperative cognitive recovery differs from both delirium and POCD in that it is defined as a return to baseline (or better), with each patient providing their own preoperative benchmark. Thus, it differs from POCD in that it represents a failure to return to preoperative cognitive function as opposed to a new cognitive deficit [2]. Important for future research is the finding that only 86 % patients attain cognitive recovery at postoperative day 3 [41]. Thus, previously diagnosed 'early POCD' may actually represent delayed cognitive recovery or, as recovery delay is predictive of future development of POCD, this early cognitive and global recovery assessment may itself identify patients at risk. Irrespective of the underlying cause, identification of failure of early cognitive recovery allows the identification of those patients at future risk or who may benefit from early intervention.

5.12 Timing of Cognitive Assessment

Cognitive function has been historically assessed at clinically relevant time points, namely that corresponding to readiness for discharge (early assessment <7 days postoperatively) and at an assumed recovery plateau (late assessment usually at 3 months). The type, prevalence and severity of cognitive dysfunction, as well as patient population affected, differ between the two assessment periods. Price et al assessed cognitive recovery in young, middle-aged and elderly patients at discharge and at 3 months [63]. At discharge, the majority of patients (44.3 %) showed no decline (>1SD) in cognitive function compared to baseline. However, when present, deficits were equal in terms of domain affected and severity (23.5 % executive, 18.3 % memory, 14.9 % combined executive and memory dysfunction). The presence and type of cognitive deficit at discharge was predictive of presence and type of cognitive dysfunction at 3 months and, most importantly, any associated functional impairment. Of those patients with executive cognitive dysfunction at discharge, 60 % displayed no deficit on cognitive testing at 3 months, whilst in the remaining 40 % the persistent cognitive deficit was 3 times more likely to be in executive function (17 % executive vs. 8 % memory vs. 1.3 % combined dysfunction). Importantly, this persistent deficit was associated with reduced ability in commuting, shopping, meal preparation, housework and management of medicine and finances. In comparison, patients with memory cognitive dysfunction at discharge were less likely to have persistent deficit at 3 months (68.9 % of patients with a memory deficit at discharge displayed no cognitive dysfunction at 3 months) whilst of those that did show a deficit, it was predominately persistently in the memory domain. (19.7 %, 4.9 % and 1.6 % deficit in memory, executive and combined function respectively). Overall, the dominant cognitive defect at 3 months was in memory function, due primarily to either persistence of early postoperative memory dysfunction or development of memory dysfunction in patients in whom early executive dysfunction had resolved. Furthermore, persistent deficits in patients with only a single domain affected (executive or memory) were deemed mild in severity and not associated with functional decline, compared to those with a combined (executive and memory) impairment. Patient insight was also reduced in those with executive, but not memory, impairment. These findings have direct clinical implications as, although less common than memory deficits, persistent deficits in executive and combined cognitive domains are more severe and are associated with reduced patient insight and functional impairment.

5.13 Recovery Assessment Methodological Issues

5.13.1 Recovery as a Dichotomous Versus Continuous Variable

A fundamental issue in comparing recovery is the variation in scoring. Recovery assessment can be at the group or individual level, with the comparator being another group or the individual themselves, and can be measured as a continuous or dichotomous outcome. For recovery as a continuous variable, each patient's composite recovery score results from summing the values from analogue scores or from multiple Likert scales to produce a single number [2–9, 16, 36, 37], with or without a cut-off 'threshold' value defining acceptable recovery [16]. Assessing (or restitution) of a group recovery as a continuous variable can then be expressed as a 'mean change score'; the group's mean difference in scores postoperatively compared to preoperatively, which describe differences in *overall* recovery at the group level. Subsequent comparison of mean change scores between groups can determine any statistical significance. For an individual patient, recovery is assessed as an individual's change score compared to a reference population (the mean change score of the group to which they belong or a historical control) and a predefined acceptable threshold in variation (Fig. 5.1). Recovery as a continuous variable thus preserves the power of parametric analysis, but does not allow assessment of recovery at the subdomain level. It also assigns equal clinical weight to each domain assessed (e.g. persistent pain equal to cognitive decline) and, when comparing group recovery, is susceptible to undue influence by those patients with extreme values- i.e. those that fail demonstrably in all domains or those whose performance is enhanced by practice [12]. Thus, although convenient, composite scores reduce a data rich assessment tool into a data poor output and are more suited to comparison of recovery between groups, rather than individual patients.

Defining recovery as a dichotomous variable enables analysis at the group and individual level, with some advantages. In dichotomous recovery, a patient is recovered if their postoperative score equates to or exceeds a threshold value (usually their pre-surgical value), either overall or in each domain. Recovery can be compared between groups (by comparison of the prevalence, and hence odds ratio, of recovery) or at the individual level. Importantly, it can assess recovery *within* differing domains. Unlike composite scores, which collapse all domains into one overall score, dichotomous scores have the ability to 'drill down' and identify in which (an how many) domains a patient failed to recover. It preserves the ability to assess recovery at four levels; overall recovery, within each domain to see where recovery has failed, within each subdomain to identify the particular problem (e.g. is the failure of functional

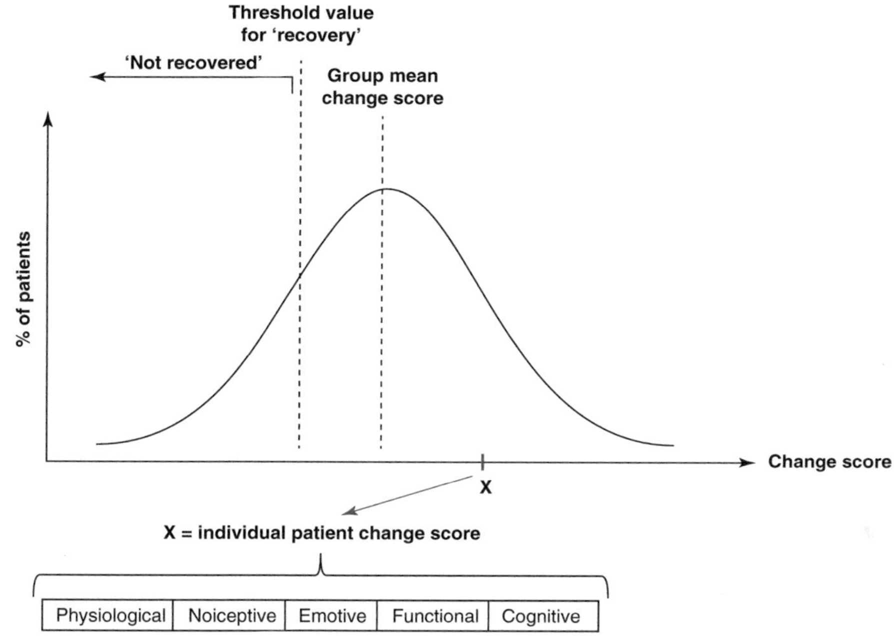

Fig. 5.1 Recovery as a continuous variable. X=individual patient change score, which is a measure of individual patient recovery as a continuous variable derived from summation of multiple domains

recovery due to eating, dressing, standing or walking), and the severity of the problem indicated by the scores, or by the number of tests within each subdomain that fail to recover (Fig. 5.2), all of which are clinically relevant. This has implications in both the clinical and research setting, as it enables targeted clinical intervention or assessment of impact of intervention respectively.

Dichotomisation of a continuous variable must always be substantiated and none more so than with recovery. Recovery may be seen, as with many physiological variables, to be a continuum (i.e. 50 % recovered, 75 % recovered etc.). Thus, to maintain statistical rigor and minimise bias, it is argued that its data collection and statistical assessment should follow suit. In research, recovery data is often collected as a continuous variable which preserves data integrity and maximises internal validity and power with the lowest sample size [71]. Comparison of recovery between groups is by comparing the difference in mean change score between intervention and control groups in order to audit or research an intervention. However, the utility of recovery as a continuous variable for clinical decision-making is limited. Whilst recovery may be a continuum, the outcome of a clinical decision is binary – the patient has or has not recovered enough to finish the crossword or safely return to work (patient focus), or to be discharged (institution focus). A difference in mean change scores (derived from analysing continuous variables) does not directly indicate a tangible clinical effect, whilst recovery as a dichotomised

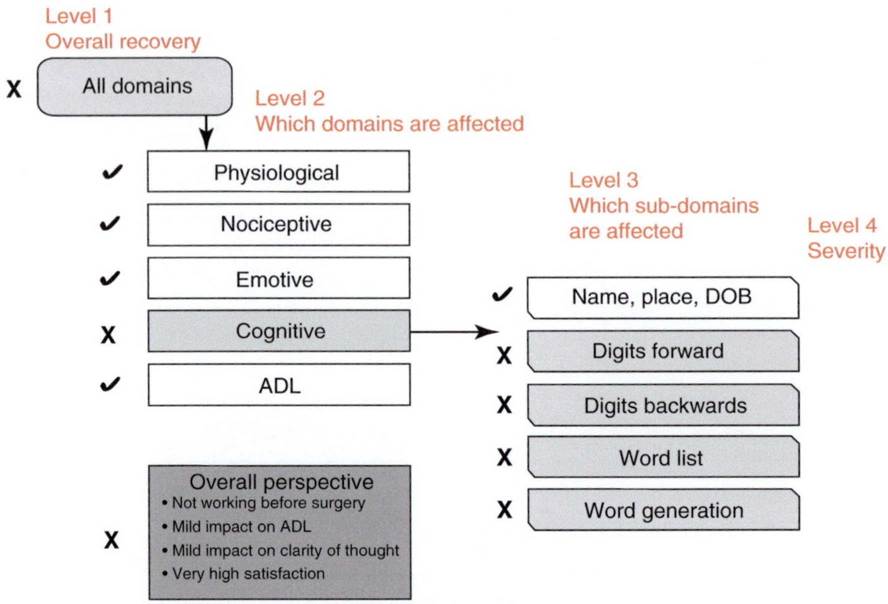

Fig. 5.2 Recovery as a dichotomous score. This approach to domain analysis allows multiple levels of data interrogation to pinpoint the problem areas contributing to failed recovery, as well as allowing an indicator of severity of the problem. √ is recovered and X is not-recovered

outcome has direct clinical utility. When comparing recovery between two groups as a continuous variable, a difference in mean change scores may indicate a statistical difference in recovery between the two groups, but does not indicate whether this has clinical implications. By comparison, when comparing recovery between two groups as a dichotomous outcome, a difference in incidence or odds ratio of recovery has direct clinical consequences in terms of resource utilisation. In addition, the issue regarding the increased sample size resulting from dichotomisation is irrelevant in the clinical setting where the assessment is on an individual patient. This reinforces a limitation of translating research tools, with their definition of recovery as a continuous variable, into the clinical setting.

It is argued that by dichotomising a continuous variable, data richness is lost i.e. dichotomisation does not indicate the magnitude by which a patient may fail to recover but rather just that the postoperative threshold value for recovery has not been achieved. However, this limitation can be mitigated by the collection of raw data in its continuous form (where applicable) and subsequent secondary dichotomization for analysis. This enables analysis of a clinically relevant outcome (i.e. has this patient recovered or not), with secondary analysis of the raw data to determine the magnitude (i.e. by how much has the patient failed to recover?).

Dichotomisation of recovery requires a postoperative performance threshold to be defined, under which a patient is deemed to have failed to recover. The external validity of the recovery tool is thus limited by the duration in which this threshold is valid. However, by defining recovery as 'a return to a patient's own baseline or better', this allows the recovery threshold definition to be independent upon future changes in nomenclature. This is because the patient's postoperative recovery is derived using their own preoperative performance as a control *for that specific* postoperative event.

5.13.2 Objective vs. Subjective Testing

Although recovery has inherent implications to individual patients, it is essential that it be measured objectively to avoid undue influence of reduced postoperative patient insight [3, 38, 72, 73], impaired cognition or patient satisfaction. There is poor correlation between objective neurological impairment and self reported deficits [33, 64] and functional ability [63]. Although important in its own right, satisfaction is limited as a recovery assessment due to its lack of uniform definition, ease of measurement and rapid early recovery [74], with satisfaction rates of above 95 % [2] postoperatively reducing its discriminant utility. Satisfaction is also correlated more with patient expectation, patient experience and quality of therapeutic relationship [74] than with professional competence and anaesthetic outcome [74].

5.13.3 General Methodological Issues

The ideal recovery assessment tool is objective, provides both quantitative and qualitative data and is independent of bias. The introduction of baseline data and calculation of individual patient change scores addresses the initial differences between each group, thus reducing Type 1 error. A composite score allows quantification of the magnitude of recovery but collapses the multiple aspects of recovery into a single numerical value. By comparison, a dichotomised score allows assessment of recovery at the group, individual and sub domain level.

Both scoring types are susceptible to ceiling and floor effects. Ceiling bias results from an assessment question being answered uniformly independent upon actual recovery. Patient satisfaction and orientation rapidly uniformly recover [2, 38, 75] and thus are poor discriminators of recovery beyond the immediate postoperative recovery room. Ceiling bias may also result from inappropriate timing of testing. Floor bias occurred when patients respond poorly in baseline data such that it is mathematically impossible for them to achieve a 'not recovered' or 'back to baseline' score postoperatively. This occurs if the test is inappropriately difficult or administered prematurely in the recovery period.

Assessment of recovery over time requires tools to be resistant to learning effects, which can be addressed by subtracting the mean change score of a healthy population from that of the study population [64] or by introducing a tolerance factor [41].

There are specific methodological issues pertaining to assessment of cognitive recovery [57, 64, 70]. Measured cognitive performance is dependent upon the ability of the patient and the ease of the test. A patient with high cognitive ability may score above the population average on an easy test irrespective of whether there is any decline in cognitive function. Similarly, a patient with limited cognitive ability may fail to show an adequate performance (relative to a reference population) on a difficult test even when no cognitive decline is present. In addition, a tool that assesses cognitive recovery is biased when applied to a study population that has lower cognitive baseline scores than in the validation studies, as these patients are more likely to show increased recovery compared to the validation population. This exemplifies the importance of comparing a patient's postoperative function with their individual preoperative baseline. Cognitive function scores were derived using age matched, healthy study populations without dementia. However, such study populations often exhibit a higher IQ, less cognitive deficits and a different learning effect and recovery time course than the general postoperative population [31, 64], thus reducing the external validity of the score [31].

5.14 The Future of Recovery and its Assessment: Patient Centred Care (PCC) and Real Time Recovery (RTR)

Patient Centred Care (PCC) has emphasised the importance of quantifying and optimising postoperative recovery in real time. PCC is defined as care that 'is responsive to the preferences, needs and values of patients and consumers [76]. It is a concept that is foremost in healthcare worldwide, is advocated by the WHO and is associated with reduced mortality, length of stay, readmission and health-care cost. It is also associated with an improvement in functional status, patient satisfaction and real-time outcomes. PCC highlights the need for real time identification and management of patient's with poor recovery and the corresponding need for routine qualitative and quantitative data collection and feedback.

Real-time recovery (RTR) is a concept ubiquitous in any arena that involves systems (i.e. business, computing) and is directly relevant to assessing a patient's course of recovery. RTR is the ability of a system to recover from any error (human induced or infrastructure) in a time frame that has minimal impact. It thus requires first identification of a failure (i.e. poor recovery) and then secondarily implementation of an intervention known to counter this failure (i.e. implementation of an early intervention known to improve recovery). This highlights the difference between assessment of recovery in the research vs. clinical setting. Research recovery tools compare patient selection, anaesthetic and surgical technique and their impact on postoperative factors (i.e. pain management, function and discharge destination). It is *retrospective* and thus induces a time delay in the identification of poor recovery (and associated factors), which is prohibitive of early intervention for the original study patients. However, a recovery assessment tool that identifies RTR can serve both research and clinical functions. A RTR assessment tool analyses data as it is entered at clinically significant timepoints and, through processes such as automated alerts, alerts the clinician *at the time* to patients that have sub-optimal recovery. Therapeutic interventions known to improve recovery can thus be implemented in a timely manner. Resources can thus be directed to those patients that would benefit most in a time frame that may have the most clinical impact, and in addition, resource use may be decreased in patients with good early recovery and facilitate fast-track care. RTR recovery tools also have research utility in that the databank can retrospectively assess recovery at the group level with respect to pre, intra and postoperative variables.

Integral to RTR assessment is the assumption that a patient's recovery can be modified by early intervention. This requires not only identification of the presence of sub-optimal recovery but in which domain. Composite recovery scores are sensitive to the former, but not the latter. Multidimensional dichotomised assessment tools, however, allow a drill down of recovery assessment. They first identify poor recovery (a failure of performance in relation to a preoperative baseline) and then allow assessment of *the domains involved* and *in which individual patients*. This allows individual targeted intervention relevant to the clinical issue in an individual patient. For example, failure to recover due to persistent nausea requires different clinical interventions compared to cognitive decline. Postoperative recovery sub domains are interrelated (i.e. persistent pain may result in concomitant persistent nausea due to increased opiate use) but, by drilling down of recovery at the sub domain level, identification of the primary cause can target clinical intervention (i.e. use of non-opiate analgesia to reduce the persistent pain and subsequent nausea).

5.15 Conclusion

Defining the abstract ideals of quality and recovery is of paramount importance in perioperative medicine. Postoperative quality of recovery has developed from a unidimensional construct focused on restitution of physiological parameters, to one that is multidimensional and assesses recovery at both the individual patient and group level. Recovery can be assessed at clinically relevant postoperative time intervals, with early, intermediate and late recovery having both prognostic implications. The increasing awareness of the effects of surgical stress on the neurophysiology of an increasingly elderly ambulant surgical population highlights the need to include robust assessment of cognitive recovery. Recovery is ideally assessed as a dichotomous outcome overall and at a subdomain level, with real-time assessment at the individual and group level having both clinical and research utility.

References

1. Terwee CB, Bot SD, de Boer MR, van der Windt DA, Knol DL, Dekker J et al (2007) Quality criteria were proposed for measurement properties of health status questionnaires. J Clin Epidemiol 60(1):34–42
2. Royse CF, Newman S, Chung F, Stygall J, McKay RE, Boldt J et al (2010) Development and feasibility of a scale to assess postoperative recovery: the post-operative quality recovery scale. Anesthesiology 113(4):892–905
3. Myles PS, Weitkamp B, Jones K, Melick J, Hensen S (2000) Validity and reliability of a post-operative quality of recovery score: the QoR-40. Br J Anaesth 84(1):11–15
4. Wong J, Tong D, De Silva Y, Abrishami A, Chung F (2009) Development of the functional recovery index for ambulatory surgery and anesthesia. Anesthesiology 110(3):596–602
5. Paddison JS, Sammour T, Kahokehr A, Zargar-Shoshtari K, Hill AG (2011) Development and validation of the Surgical Recovery Scale (SRS). J Surg Res 167(2):e85–e91
6. Hogue SL, Reese PR, Colopy M, Fleisher LA, Tuman KJ, Twersky RS et al (2000) Assessing a tool to measure patient functional ability after outpatient surgery. Anesth Analg 91(1):97–106
7. Talamini MA, Stanfield CL, Chang DC, Wu AW (2004) The surgical recovery index. Surg Endosc 18(4):596–600
8. Swan BA, Maislin G, Traber KB (1998) Symptom distress and functional status changes during the first seven days after ambulatory surgery. Anesth Analg 86(4):739–745
9. Oakes CL, Ellington KJ, Oakes KJ, Olson RL, Neill KM, Vacchiano CA (2002) Assessment of postanesthesia short-term quality of life: a pilot study. AANA J 70(4):267–273
10. Caljouw MA, van Beuzekom M, Boer F (2008) Patient's satisfaction with perioperative care: development, validation, and application of a questionnaire. Br J Anaesth 100(5):637–644
11. Capuzzo M, Gilli G, Paparella L, Gritti G, Gambi D, Bianconi M et al (2007) Factors predictive of patient satisfaction with anesthesia. Anesth Analg 105(2):435–442
12. Murkin JM, Stump DA, Blumenthal JA, McKhann G (1997) Defining dysfunction: group means versus incidence analysis--a statement of consensus. Ann Thorac Surg 64(3):904–905
13. Murkin JM, Newman SP, Stump DA, Blumenthal JA (1995) Statement of consensus on assessment of neurobehavioral outcomes after cardiac surgery. Ann Thorac Surg 59(5):1289–1295
14. Newman MF, Kirchner JL, Phillips-Bute B, Gaver V, Grocott H, Jones RH et al (2001) Longitudinal assessment of neurocognitive function after coronary-artery bypass surgery. N Engl J Med 344(6):395–402
15. Bowyer A, Jakobsson J, Ljungqvist O, Royse C (2014) A review of the scope and measurement of postoperative quality of recovery. Anaesthesia 69(11):1266–1278
16. Aldrete JA, Kroulik D (1970) A postanesthetic recovery score. Anesth Analg 49(6):924–934
17. Wu CL, Rowlingson AJ, Partin AW, Kalish MA, Courpas GE, Walsh PC et al (2005) Correlation of postoperative pain to quality of recovery in the immediate postoperative period. Reg Anesth Pain Med 30(6):516–522
18. White PF, Sacan O, Tufanogullari B, Eng M, Nuangchamnong N, Ogunnaike B (2007) Effect of short-term postoperative celecoxib administration on patient outcome after outpatient laparoscopic surgery. Can J Anaesth 54(5):342–348
19. Sun T, Sacan O, White PF, Coleman J, Rohrich RJ, Kenkel JM (2008) Perioperative versus postoperative celecoxib on patient outcomes after major plastic surgery procedures. Anesth Analg 106(3):950–958
20. Mant J (2001) Process versus outcome indicators in the assessment of quality of health care. Int J Qual Health Care 13(6):475–480
21. Haller G, Stoelwinder J, Myles PS, McNeil J (2009) Quality and safety indicators in anesthesia: a systematic review. Anesthesiology 110(5):1158–1175
22. Veloski J, Boex JR, Grasberger MJ, Evans A, Wolfson DB (2006) Systematic review of the literature on assessment, feedback and physicians' clinical performance: BEME Guide No. 7. Med Teach 28(2):117–128

23. Australian Council on Healthcare Standards (ACHS) (2012) Australasian Clinical Indicator Report 2004–2011, 13th ed. ACHS, Sydney
24. Macario A, Weinger M, Carney S, Kim A (1999) Which clinical anesthesia outcomes are important to avoid? The perspective of patients. Anesth Analg 89(3):652–658
25. Bahtsevani C, Uden G, Willman A (2004) Outcomes of evidence-based clinical practice guidelines: a systematic review. Int J Technol Assess Health Care 20(4):427–433
26. Grimshaw JM, Russell IT (1993) Effect of clinical guidelines on medical practice: a systematic review of rigorous evaluations. Lancet 342(8883):1317–1322
27. Abraham NS, Byrne CM, Young JM, Solomon MJ (2007) Meta-analysis of non-randomized comparative studies of the short-term outcomes of laparoscopic resection for colorectal cancer. ANZ J Surg 77(7):508–516
28. Nygren J, Soop M, Thorell A, Efendic S, Nair KS, Ljungqvist O (1998) Preoperative oral carbohydrate administration reduces postoperative insulin resistance. Clin Nutr 17(2):65–71
29. Soop M, Nygren J, Myrenfors P, Thorell A, Ljungqvist O (2001) Preoperative oral carbohydrate treatment attenuates immediate postoperative insulin resistance. Am J Physiol Endocrinol Metab 280(4):E576–E583
30. Gustafsson UO, Ljungqvist O (2011) Perioperative nutritional management in digestive tract surgery. Curr Opin Clin Nutr Metab Care 14(5):504–509
31. Krenk L, Rasmussen LS (2011) Postoperative delirium and postoperative cognitive dysfunction in the elderly – what are the differences? Minerva Anestesiol 77(7):742–749
32. Bryson GL, Wyand A (2006) Evidence-based clinical update: general anesthesia and the risk of delirium and postoperative cognitive dysfunction. Can J Anaesth 53(7):669–677
33. Moller JT, Cluitmans P, Rasmussen LS, Houx P, Rasmussen H, Canet J et al (1998) Long-term postoperative cognitive dysfunction in the elderly ISPOCD1 study. ISPOCD investigators. International Study of Post-Operative Cognitive Dysfunction. Lancet 351(9106):857–861
34. Monk TG, Weldon BC, Garvan CW, Dede DE, van der Aa MT, Heilman KM et al (2008) Predictors of cognitive dysfunction after major noncardiac surgery. Anesthesiology 108(1):18–30
35. Monk TG, Saini V, Weldon BC, Sigl JC (2005) Anesthetic management and one-year mortality after noncardiac surgery. Anesth Analg 100(1):4–10
36. Kleinbeck SV (2000) Self-reported at-home postoperative recovery. Res Nurs Health 23(6):461–472
37. Stark PA, Myles PS, Burke JA (2013) Development and psychometric evaluation of a postoperative quality of recovery score: the QoR-15. Anesthesiology 118(6):1332–1340
38. Myles PS, Reeves MD, Anderson H, Weeks AM (2000) Measurement of quality of recovery in 5672 patients after anaesthesia and surgery. Anaesth Intensive Care 28(3):276–280
39. Rowe G, Wright G (1999) The Delphi technique as a forecasting tool: Issues and analysis. Int J Forecast 15(4):353–375
40. Aaronson N, Alonso J, Burnam A, Lohr KN, Patrick DL, Perrin E et al (2002) Assessing health status and quality-of-life instruments: attributes and review criteria. Qual Life Res 11(3):193–205
41. Royse CF, Newman S, Williams Z, Wilkinson DJ (2013) A human volunteer study to identify variability in performance in the cognitive domain of the postoperative quality of recovery scale. Anesthesiology 119(3):576–581
42. Gornall BF, Myles PS, Smith CL, Burke JA, Leslie K, Pereira MJ et al (2013) Measurement of quality of recovery using the QoR-40: a quantitative systematic review. Br J Anaesth 111(2):161–169
43. Australian Institute of Health and Welfare (ed) (2010) Australia's Health 2010. The Institute, Canberra
44. Stygall J, Newman SP, Fitzgerald G, Steed L, Mulligan K, Arrowsmith JE et al (2003) Cognitive change 5 years after coronary artery bypass surgery. Health Psychol 22(6):579–586
45. Stern Y (2002) What is cognitive reserve? Theory and research application of the reserve concept. J Int Neuropsychol Soc: JINS 8(3):448–460

46. Pakkenberg B, Gundersen HJ (1997) Neocortical neuron number in humans: effect of sex and age. J Comp Neurol 384(2):312–320
47. Riddle DR, Sonntag WE, Lichtenwalner RJ (2003) Microvascular plasticity in aging. Ageing Res Rev 2(2):149–168
48. Yaffe K, Lindquist K, Penninx BW, Simonsick EM, Pahor M, Kritchevsky S et al (2003) Inflammatory markers and cognition in well-functioning African-American and white elders. Neurology 61(1):76–80
49. Cibelli M, Fidalgo AR, Terrando N, Ma D, Monaco C, Feldmann M et al (2010) Role of interleukin-1beta in postoperative cognitive dysfunction. Ann Neurol 68(3):360–368
50. Xie Z, Dong Y, Maeda U, Alfille P, Culley DJ, Crosby G et al (2006) The common inhalation anesthetic isoflurane induces apoptosis and increases amyloid beta protein levels. Anesthesiology 104(5):988–994
51. Dong Y, Zhang G, Zhang B, Moir RD, Xia W, Marcantonio ER et al (2009) The common inhalational anesthetic sevoflurane induces apoptosis and increases beta-amyloid protein levels. Arch Neurol 66(5):620–631
52. Avidan MS, Evers AS (2011) Review of clinical evidence for persistent cognitive decline or incident dementia attributable to surgery or general anesthesia. J Alzheimer's Dis: JAD 24(2):201–216
53. Holmes C, Cunningham C, Zotova E, Woolford J, Dean C, Kerr S et al (2009) Systemic inflammation and disease progression in Alzheimer disease. Neurology 73(10):768–774
54. Cunningham C, Wilcockson DC, Campion S, Lunnon K, Perry VH (2005) Central and systemic endotoxin challenges exacerbate the local inflammatory response and increase neuronal death during chronic neurodegeneration. J Neurosci: Off J Soc Neurosc 25(40):9275–9284
55. Perry VH, Cunningham C, Holmes C (2007) Systemic infections and inflammation affect chronic neurodegeneration. Nat Rev Immunol 7(2):161–167
56. Association AP (2000) Diagnostic and statistical manual of mental disorders, 4th ed, txt rev. American Psychiatric Association, Washington DC
57. Newman S, Stygall J, Hirani S, Shaefi S, Maze M (2007) Postoperative cognitive dysfunction after noncardiac surgery: a systematic review. Anesthesiology 106(3):572–590
58. Deiner S, Silverstein JH (2009) Postoperative delirium and cognitive dysfunction. Br J Anaesth 103(Suppl 1):i41–i46
59. Rudolph JL, Marcantonio ER (2011) Review articles: postoperative delirium: acute change with long-term implications. Anesth Analg 112(5):1202–1211
60. Jankowski CJ, Trenerry MR, Cook DJ, Buenvenida SL, Stevens SR, Schroeder DR et al (2011) Cognitive and functional predictors and sequelae of postoperative delirium in elderly patients undergoing elective joint arthroplasty. Anesth Analg 112(5):1186–1193
61. Leung JM, Tsai TL, Sands LP (2011) Brief report: preoperative frailty in older surgical patients is associated with early postoperative delirium. Anesth Analg 112(5):1199–1201
62. Demeure MJ, Fain MJ (2006) The elderly surgical patient and postoperative delirium. J Am Coll Surg 203(5):752–757
63. Price CC, Garvan CW, Monk TG (2008) Type and severity of cognitive decline in older adults after noncardiac surgery. Anesthesiology 108(1):8–17
64. Funder KS, Steinmetz J, Rasmussen LS (2010) Methodological issues of postoperative cognitive dysfunction research. Semin Cardiothorac Vasc Anesth 14(2):119–122
65. Shaw PJ, Bates D, Cartlidge NE, French JM, Heaviside D, Julian DG et al (1987) Neurologic and neuropsychological morbidity following major surgery: comparison of coronary artery bypass and peripheral vascular surgery. Stroke 18(4):700–707
66. Treasure T, Smith PL, Newman S, Schneidau A, Joseph P, Ell P et al (1989) Impairment of cerebral function following cardiac and other major surgery. Eur J Cardiothorac Surg 3(3):216–221
67. Williams-Russo P, Sharrock NE, Mattis S, Szatrowski TP, Charlson ME (1995) Cognitive effects after epidural vs general anesthesia in older adults. A randomized trial. JAMA 274(1):44–50

68. Campbell DN, Lim M, Muir MK, O'Sullivan G, Falcon M, Fison P et al (1993) A prospective randomised study of local versus general anaesthesia for cataract surgery. Anaesthesia 48(5):422–428
69. Lewis MS, Maruff P, Silbert BS, Evered LA, Scott DA (2006) Detection of postoperative cognitive decline after coronary artery bypass graft surgery is affected by the number of neuropsychological tests in the assessment battery. Ann Thorac Surg 81(6):2097–2104
70. Rasmussen LS, Larsen K, Houx P, Skovgaard LT, Hanning CD, Moller JT et al (2001) The assessment of postoperative cognitive function. Acta Anaesthesiol Scand 45(3):275–289
71. Streiner DL (2002) Breaking up is hard to do: the heartbreak of dichotomizing continuous data. Can J Psychiat Revue canadienne de psychiatrie 47(3):262–266
72. Myles PS, Williams DL, Hendrata M, Anderson H, Weeks AM (2000) Patient satisfaction after anaesthesia and surgery: results of a prospective survey of 10,811 patients. Br J Anaesth 84(1):6–10
73. Newman S, Klinger L, Venn G, Smith P, Harrison M, Treasure T (1989) Subjective reports of cognition in relation to assessed cognitive performance following coronary artery bypass surgery. J Psychosom Res 33(2):227–233
74. Heidegger T, Saal D, Nubling M (2013) Patient satisfaction with anaesthesia – Part 1: satisfaction as part of outcome – and what satisfies patients. Anaesthesia 68(11):1165–1172
75. Royse CF, Chung F, Newman S, Stygall J, Wilkinson DJ (2013) Predictors of patient satisfaction with anaesthesia and surgery care: a cohort study using the Postoperative Quality of Recovery Scale. Eur J Anaesthesiol 30(3):106–110
76. Australian Commission on Safety and Quality in Health Care (ACSQHC) (ed) (2011) National Safety and Quality Health Service Standards. Australian Commission on Safety and Quality in Health Care, Sydney

Chapter 6
Enhanced Recovery for Colorectal Surgery

William John Fawcett

Abstract The adoption of evidenced-based, multimodal pathways for the perioperative management of patients undergoing colorectal surgery has led to dramatic changes for all those involved in perioperative care. There is improved cooperation between primary and secondary care and thus the preoperative period ensures that patient's physiological status is optimized and that the most appropriate operation and postoperative care is planned. Intraoperatively, the emphasis is twofold: minimizing the surgical stress response (with subsequent organ dysfunction) and reducing complications (both medical and surgical). There are many aspects to this area, such as small incision/laparoscopic surgery, individualized IV fluid management, multimodal opioid sparing analgesia together with minimizing tubes and drains and avoidance of hypothermia. In the postoperative period early enteral feeding and mobilization are key to continuing to reducing stress and avoiding complications. The results can be impressive, with not only significant reductions in length of stay and complications but also patients able to more quickly resume their usual lifestyle or can undertake other therapy (e.g. chemotherapy or radiotherapy) earlier. The process requires the development of, and adherence to, a number of pathways agreed by the healthcare team with regular audit of the results. Yet at the core of this process are the patients and their families and/or carers, as they move from a purely passive role to being actively involved in every aspect of the care, knowing well in advance what is planned at every stage.

Keywords Anaesthesia, General • Colorectal Surgery • Preoperative care • Analgesia, Postoperative • Fluid therapy • Complications, postoperative • Stress, physiological

---

W.J. Fawcett, MBBS, FRCA, FFPMRCA
Department of Anaesthesia, Royal Surrey County Hospital NHS Foundation Trust, Egerton Road, Guildford GU2 7XX, UK

Postgraduate Medical School, University of Surrey, Guildford GU2 7XH, UK
e-mail: wfawcett@nhs.net

K. Stuart-Smith (ed.), *Perioperative Medicine – Current Controversies*,
DOI 10.1007/978-3-319-28821-5_6

The care of patients undergoing major colorectal surgery has been transformed in the last two decades. Large incisions, preoperative fasting and bowel preparation, the use of drains and tubes, several days of postoperative 'nil by mouth' whilst the patient received several litres of iv fluids, all within a background of opioid analgesia, had been the cornerstone of colorectal surgery for many years, and was viewed as a necessity for a successful outcome.

Within a few centres in Europe, popularised by Henrik Kehlet, a Surgeon from Denmark, the situation began to change as patients could be discharged home after just a few days in hospital. This began under a number of names – Fast Track Surgery, Accelerated Recovery and only later was the term Enhanced Recovery used, and has now become global practice. The potential for this concept was appreciated by all: patients had a reduced length of stay in hospital and resumed their normal activities more quickly, healthcare staff could increase activity, hospital managers were able to increase throughput or reduce the number of inpatients beds, and the purchasers of healthcare envisaged large savings as a result [1].

What were the key areas in transforming surgical, anaesthetic and nursing practice that had been largely unchanged since 1970s into what is now taken for granted as a standard of care? Three key areas are relevant in understanding the success of Enhanced Recovery After Surgery (ERAS). At the very heart of ERAS is the modification of the physiological stress response to surgery, the swift resumption of the physiological status quo with minimum organ dysfunction. In addition, the rate of complications – many of which are catastrophic in terms of morbidity and mortality – were also reduced. In order to do this, almost every step of the pathway had to examined and challenged. Many areas were steeped in tradition and were part of surgical dogma, and were regarded as set in stone. Alongside these two key areas is a third concept: the aggregation of marginal gains.

The stress response to surgery began to be characterized over 85 years ago by Cuthbertson [2]. Since that time further work on this fundamental response to surgery and injury as well as attempts to modify this response and the effects on outcome have extensively studied. Broadly the stress response consists of both a neuroendocrine activation of both the pituitary and the sympathetic nervous system (with consequent metabolic sequelae such as catabolism and hyperglycaemia) and an inflammatory response (which may evolve into a systemic inflammatory response syndrome – SIRS) [3]. There may also be immunosuppression and other non specific but important symptoms such as malaise and tiredness. Within the context of a successful ERAS programme, catabolism (particularly loss of muscle mass), SIRS response, malaise and fatigue are key areas to target. Whilst modifying neuroendocrine activation with regional anaesthesia is well described, there are many other areas that will limit the magnitude of the stress response [4], such as minimally invasive surgery without tubes/drains, various drugs, fluid management and oral nutrition. These areas are key and will be explored in more detail later.

The avoidance of complications – both medical and surgical – is a key area in colorectal surgery, but it is only in the last 10 years that the magnitude of complications has become apparent. Khuri has demonstrated that complications not only affect short-term morbidity and mortality, but that they markedly affect long

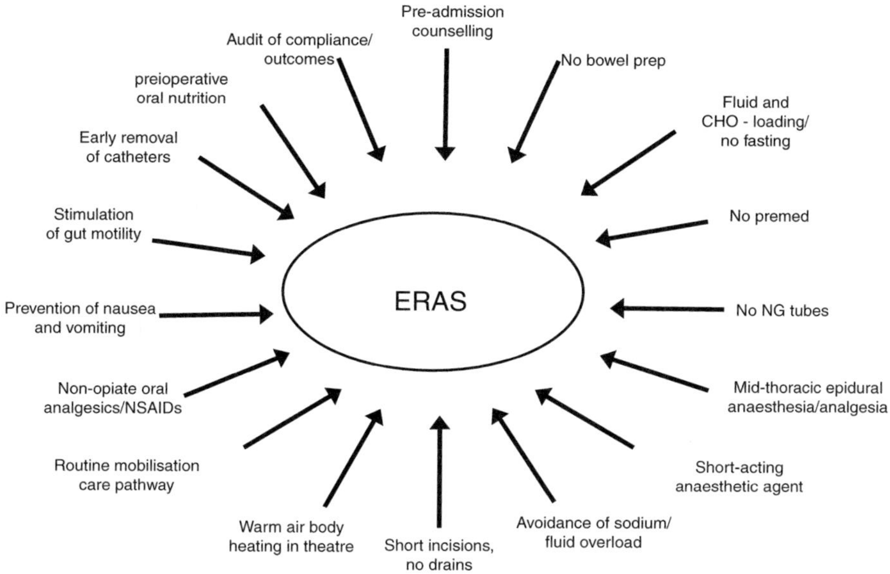

Fig. 6.1 Original ERAS Elements described for colorectal surgery. From Fearon KCH et al. *Clinical Nutrition* 2005

term outcomes for many years too [5]. Intuitively, one might guess that any complications merely impacted on that hospital stay alone and, if successfully treated, these patients would then re-join the cohort of patients who had not had complications. His paper demonstrated that this is not the case and that complications within 30 days of surgery are more important than both preoperative risk and intraoperative factors in determining survival, with complications in this timeframe reducing survival by 69 % – from 18.4 years to 5.6 years.

The final area is the concept of the "aggregation of marginal gains", a concept credited to a chess player, William Steinitz at the end of the nineteenth century, and originally known as The Steinitz Accumulation Theory. By painstakingly improving small areas of practice, the overall impact or aggregation of these gains can produce large changes in practice. These concepts are familiar to many sportsmen but now this concept has been translated into improved outcomes in perioperative care too [6].

Enhanced recovery is a procedure specific, protocolised, multistep pathway (Fig. 6.1). Some workers have been tempted to merely adopt the ones that suit them or their institution. Does this matter? Increasing adherence to ERAS pathways improves outcomes [7]. Conversely, not all the steps in the pathway are equally weighted, which would clearly be a huge mathematical coincidence, so are there some more important than others? It would appear that of all the many steps described oral carbohydrate preload and avoidance of fluid overload are the two independent predictors of improved outcome and patient wellbeing (see below).

Whilst some the aspects can be challenged or, updated, figure one which is now 10 years old, was remarkably prescient and still largely relevant as the basis of

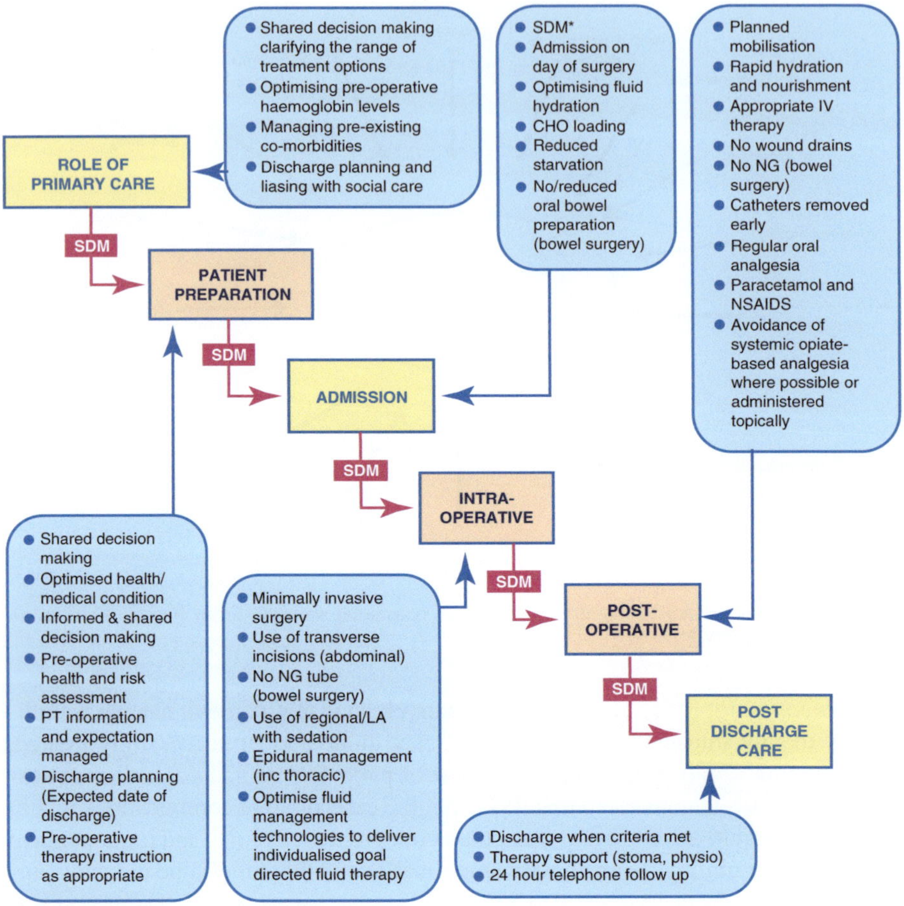

Fig. 6.2 Updated ER pathway for Colorectal Surgery. *SDM* shared decision making

modern ER in colorectal surgery. It has recently been updated by the UK's Department of Health (Fig. 6.2)

From a practical perspective, the setting up of an ERAS programme in a unit not currently practising it is not always straightforward. A number of hurdles have to be overcome, with buy in from managers, the appointment of a local champion (often an experienced nurse) who will then coordinate regular multidisciplinary meetings with surgeons, anaesthetists, physiotherapists, and others e.g. dieticians and occupational health. In addition, full participation and understanding from the primary care providers is fundamental too. It is very important that at these regular meetings problems are addressed and progress reported on for the next meeting. The cornerstone of this process is the collection of good quality audit data, permitting analysis of success and failures for clinicians as well as also allowing the managers to see the effects on bed usage and cost (Table 6.1).

Table 6.1 Stress response to surgery

Neuroendocrine activation	Metabolic consequences	Inflammatory reponses	Other responses
Sympathetic nervous system ↑Catecholamines	Salt and water retention Potassium loss	Cytokine release e.g. Interleukins (especially IL-1, IL-6) Tumour Necrosis Factor alpha Interferons VEGF	Malaise
Pituitary activation ↑GH, ADH, ACTH, Prolactin, Beta-endorphin	Fat metabolism Lipolysis	SIRS response	Fever
Other endocrine glands (e.g. adrenal, pancreatic) ↑Renin, Aldosterone, Cortisol	Carbohydrate metabolism Insulin resistance, hyperglycaemia, glycogenolysis, gluconeogenesis	Immunosuppression	
	Protein metabolism Muscle breakdown, nitrogen loss		

GH growth hormone, *ADH* antidiuretic hormone, *ACTH* corticotrophin, *VEGF* vascular endothelial growth factor, *SIRS* systemic inflammatory response syndrome

6.1 Preoperative Preparation

The preoperative preparation is a fundamental part of the ERAS pathway. Various issues need to be addressed.

6.1.1 Assessment of the Functional Physiological Status and Reserve

The assessment of the patient's ability to withstand and respond to surgery is a key determinant in risk stratification for predicting outcome, providing informed consent planning appropriate surgery and the level of anticipated postoperative care. Major elective colorectal surgery carries significant risk (particularly in the elderly or those with comorbidities) and imposes significant perioperative demands on the patients, in particular causing a catabolic response and an increase in oxygen consumption. Patients who are unable to meet this increase in oxygen consumption with a concomitant rise in oxygen delivery are at risk of oxygen debt, anaerobic metabolism and the build up of lactic acid. Moreover, in patients undergoing colorectal surgery the situation is compounded by the fact that haemodynamic upset (e.g. from hypovolaemia) will divert splanchnic blood other organs, which may put anastomotic healing at particular risk.

For many years, intensivists have focussed on the improved outcomes of patients who are able to increase oxygen delivery (DO_2) in the intraoperative and postoperative period to supranormal levels e.g. a DO_2 of >600 $ml \cdot min^{-1} \cdot m^{-2}$ [8]. In addition a number of risk scores have also been described to aid in the preoperative assessment. Goldman's original risk score [9] has been superseded by a number others, the most well known is Lee's Revised Cardiac Risk Index [10]. There others that are widely used including Physiological and Operative Severity Score for the enUmeration of Mortality and Morbidity (POSSUM) scores and data from large surgical databases such as the American College of Surgeons National Surgical Quality Improvement Program (NSQIP).

More recently, there has been a paradigm shift to dynamic assessment of patients. This may includes relatively simple assessment such as estimating metabolic equivalents (MET) which is the resting oxygen uptake (approx. 3 ml/kg/min), with, for example walk up a flight of steps or a brisk walk on level ground representing 4 METs and strenuous sports representing >10 METs. The gold standard is increasingly being viewed as cardiopulmonary exercise testing (CPET), in which a number of measurements are made including maximal oxygen consumption (VO_2 peak) or more usefully the estimation of anaerobic threshold (AT), which is effort independent. Although the evidence is still growing a commonly used threshold is Older's study in elderly patients undergoing abdominal surgery, with a mortality in a patients with an AT of >11 ml/kg/min was 1 %, increasing to 18 % with an AT of 8–11 ml/kg and over 50 % with an AT of <8 ml/kg/min [11]. More recently, studies for colorectal surgery patients have broadly agreed with this, such as West's group who demonstrated the risk threshold to be an AT of <10.1 ml/kg/min and VO_2 peak < 16.7 ml/kg/min [12].

What is the value of this assessment of physiological status? There is compelling evidence that patients with poor reserve have a poorer outcome, both in terms or mortality but also morbidity. The latter, in the form of complications has the potential to cause chronic ill health, and given the volume of colorectal surgery undertaken worldwide annually clearly represents not only huge personal burdens, but also widespread social and economic burdens too. Above all, good preoperative assessment permits risk stratification, guiding consent, appropriate surgery and the postoperative care. For example some patients may not be able to withstand the cardiovascular and respiratory sequelae of prolonged pneumoperitoneum for a laparoscopic resection and may be best served by an open procedure. Other patients may be deemed at such high risk that stenting may be the safest alternative, or require a planned admission to intensive are for targeted organ support.

6.1.2 Optimization of Patients' Health

There is little controversy for patients with an ERAS programme should have their health optimized preoperatively. Much of this involves coordination with primary care physicians and includes optimization of haemoglobin levels (e.g. with iron therapy), and of comorbidities such as diabetes, angina hypertension, atrial fibrillation asthma etc. For some patients, preoperative assessment may unmask other serious comorbidities and these may need surgical intervention prior to surgery, such as coronary stenting.

An exciting area of perioperative medicine emerging is the concept of prehabilitation. Given that there seems to be an improved outcome in with increasing physiological reserve, it is logical to take steps to improve this reserve preoperatively. Indeed, the parallel between some of the physiological changes of major surgery and a major sporting endeavour has been made [13] and there exists a wealth of data in exercise physiology as to how training may improve exercise performance as measured by an increase in VO_2 max or lactate threshold [14]. Therefore if exercise can improve the results of CPET for athletes, it is very tempting to suppose that this might have the same effect in patients and thus be translated into an improved outcome. This is an emerging field with no unified agreement as to an exercise regime. However there seems to be a dose–response effect with increased benefits occurring following increased exercise. Prehabilitation is a wider, multimodal discipline, encompassing not only exercise regimes, but nutritional supplementation (see below) and psychological preparation too. Although in some circumstances effective prehabilitation may be impractical, either because of the relative urgency of surgery or because of other comorbidities, this area is key within ERAS programme, as patients with poor preoperative functional capacity and grip strength have a slower recovery, with increased complications, sleep disturbance and fatigue persisting weeks or even months. Moreover the least fit patients appear to have the most to gain [15].

6.1.3 Other Areas of Preparation for Surgery

Nutritional supplementation and advice and has been mentioned above, and is one of the most exciting areas in the preoperative care and indeed we have coined the phrase "nutritional perioperative medicine" [13]. The use of nutritional support and supplements for those with a high degree of nutritional risk may need to escalated so to formal enteral (and in some cases parenteral) for 7–10 days prior to surgery, even if surgery may have to be delayed. More recently the effects of immunonutrients and perhaps even other nutrients that improve oxygen efficiency (such as beetroot and other nitrates) are awaited with interest [13, 15].

In addition, simple advice (but often hard for the patients to implement) such as weight loss and cessation of smoking and management of excessive alcohol intake must nevertheless be enthusiastically given to patients, extolling their virtues. In particular smoking cessation has a greater effect the longer the patient is able to stop, and there is a reduction in postoperative complications after just 3 weeks of abstinence. Furthermore, multidisciplinary support, counselling and nicotine replacement patches may aid in this process.

The patient, their family and their carers need to be given advice and prepared for postoperative management and what is expected of the patient on a daily basis in terms of nutrition and activity, as well as and psychological support where necessary. There are several ways of imparting this information such as literature, or DVD or a website. In addition, the *routine* use of bowel preparation has largely been abandoned (although there is a degree of personal choice from surgeons and some advocate for certain circumstances, such when intraoperative colonoscopy is contemplated) as it has often resulted in patients arriving for surgery in a state of dehydration and electrolyte imbalance. Finally the admission of patients the day before surgery (or even earlier) has largely been abandoned as patients should continue to have a normal level of activity until just prior to surgery.

6.2 Oral Carbohydrate Loading

A key area within ERAS pathways is the prevention of prolonged fasting, with dehydration (compounded if a bowel preparation is used). If the analogy of major surgery and sport is accepted, then clearly one would not expect an athlete to perform at their best dehydrated and with little or no attention paid to nutrition (Table 6.2).

Anaesthetists have traditionally required fasting of 6 h for solids and 2 h for clear liquids to minimize the risks of pulmonary aspiration of gastric contents, but in practice this has often resulted in patients being fasted for much longer. However whilst dehydration may be prevented or rapidly reversed intraoperatively, major interest has focussed on the immediate nutritional optimization of patients, in a similar fashion to athletes.

For nearly 50 years, carbohydrate loading in athletes has improved performance during endurance exercise (75–85 % VO_2 max). This is attributed to an increase in muscle glycogen stores. For elective surgical patients, carbohydrate drinks not only prevents dehydration, but also reduces insulin resistance, catabolism and preserves muscle function [13].

The mixture consists of maltodextrins, which are complex carbohydrates and empty readily and predictably from the stomach, in contrast to glucose or milk. Generally two doses are given, 100 g of carbohydrate in a volume of 800 ml approximately 12 h before surgery and half this (50 g carbohydrates in 400 ml) 2–3 h before surgery.

Oral carbohydrate preload has been a key elements of the ERAS protocol for colorectal surgery for over 10 years, and is the first and probably the most simple

Table 6.2 Benefits of oral carbohydrate preload

Improves patient comfort and wellbeing
Patient in metabolically conditioned or fed state prior to surgery
Attenuates insulin resistance
Attenuates protein loss and catabolism, preserves muscle strength
Reduces complications
No effect on gastric emptying/reflux

element to reduce surgical stress and metabolically condition a patient to aid early return to oral diet, mobility and recovery. In a recent review of all ERAS elements, preload was one of two ERAS elements that independently had a significant effect on reducing complications and improving wellbeing [7].

Safety is clearly paramount and the first concern some will have is that the risk of pulmonary aspiration may be more likely. Whilst not appearing to increase this risk, given the relative rarity of this condition (1 in 7000 anaesthetics and a mortality of 1;1000000) a very large study would be needed to provide further information. Indirect information from other studies which have estimated gastric emptying from ultrasound and/or co-administering with paracetemol (them measuring the change in serum paracetemol concentrations) also suggests that stomach emptying should have occurred within 2 h of the second preload. Another area of concern is its use in diabetics, who are at risk of both hyperglycaemia and (in the presence of autonomic neuropathy) gastric stasis. There is a little evidence to support its use in type 2 diabetics, when a carbohydrate preload with paracetemol (the latter to determine gastric emptying) did not delay gastric emptying nor risk hyperglycaemia or pulmonary aspiration. The current recommendations are to give diabetic patients a preload along with their usual diabetic medication [16]. Further, larger studies, and its use on Type I diabetics, are required. Finally, in the emergency patient (see below) where there is like to be poor gastric motility, oral preload should be avoided.

6.3 Intraoperative Anaesthesia

The anaesthetist has a pivotal role in the patient's ERAS pathways. There are many aspects to this (see Figs. 6.1 and 6.2), but the two major areas are fluid management and analgesia, which have a large evidence base and are dealt with in detail below. Indeed, so important are these areas that we have introduced the trimodal approach – fluids, analgesia and "all the rest" [17]. The last term refers to many of the fundamental elements which are often protocolised and in which there either exists little controversy in their use, or there is little compelling evidence for an alternative.

Most ERAS anaesthetists would support the idea of no sedative premedication and the use of short acting anaesthetics. The ideal agents for induction and maintenance of anaesthesia may not be proven, but most anaesthetists would avoid nitrous oxide as it may cause bowel distension and increase the risk of postoperative nausea and vomiting (PONV). Short acting inhaled agents such as desflurane or sevoflurane are commonly used although a total intravenous regime with propofol is preferable where there is high risk of PONV. An area that will undoubtedly receive more attention in the future is cerebral monitoring e.g. bispectral index (BIS) and anaesthetic depth, particularly in the elderly, as excessive anaesthetic depth is associated with delirium and postoperative cerebral dysfunction [18]. Most recently and rather more speculatively, an editorial examines some of the evidence for the potential role in which volatile anaesthetic agents may play in cancer call growth [19].

The use of neuromuscular blocking drugs is currently of interest as there is some data that shows that deep neuromuscular blockade (DNB) as defined as a Post tetanic twitch count of 1–2, may aid laparoscopic access (e.g. for rectal surgery) and allow a marked reduction in insufflation pressures [20], perhaps also reducing postoperative pain and mitigating some of the cardiorespiratory effects of laparoscopy. DNB is most easily achieved by the use of rocuronuium which is easily and reliably reversed by suggammadex. Whatever type of neuromuscular blocking drug is used, confirmation of restoration of neuromuscular function at the end of surgery is paramount. In spite of postoperative residual curarization (PORC) having been described over 35 years ago [21], and is commonplace even with modern neuromuscular blocking drugs [22]. PORC is associated not only with patient distress, but delayed post anaesthetic care unit discharge, and early and late postoperative pulmonary complications [23].

Tracheal intubation is almost universal for these patients and care is required to prevent of microaspiration by careful tracheal toilet at the end of the procedure. A newer area is that of lung protection strategies – well recognised by our intensivist colleagues – and the recent the focus of studies for (mainly) open abdominal surgery in which low tidal volumes (7–8 ml/kg) and PEEP (6–10 cmH_2O) reduced hospital stay and postoperative pulmonary complications compared to higher tidal volumes (9–12 ml/kg) and no PEEP [24, 25], although most recently, the PROVHILO trial, found no benefit in PEEP and recruitment manoeuvres, which merely resulted in more hypotension requiring vasoactive support [26].

6.4 Fluid Management

There is probably no area in perioperative medicine that generates more controversy than fluid management, particularly in patients undergoing major colorectal surgery, where the potential for fluid losses and shifts is enormous. Both the protocol for administration (fixed, liberal, zero-balance, restrictive, individualized, goal directed etc.) and the type of fluid – crystalloid and colloid (and the various different types) - and have all been the subject of numerous studies and reviews.

6.4.1 *Volume of Fluid Administered*

The fundamentals fluid management are best appreciated by understanding two fundamental concepts. Firstly the importance of adequate cellular oxygen delivery and secondly the consequences of poor fluid management.

Adequate tissue oxygenation is a basic physiological principle and over 40 years ago the link between higher levels of oxygen delivery (DO_2) and consumption (VO_2) with increased survival in septic patients [27]. Many, but not all, later trials have supported confirmed these findings in both high risk and septic patients [28, 29] and the concept of ensuring adequate or indeed "supranormal" DO_2 to patients undergoing major surgery to reduce complications and death is now fundamental modern perioperative practice. DO_2 is usually indexed (i.e. expressed per m^2) and expressed as DO2I using the well known equation:

$$DO2I = \frac{SV \times HR \times SaO2 \times [Hb] \times 1.34}{BSA}$$

Where SV = stroke volume, HR = heart rate, SaO2 = oxygen saturation (expressed as fraction), [Hb] = haemoglobin concentration and BSA = body surface area.

Commonly used targets are cardiac index (CI) of >4.5 $l\ min^{-1}$ and a DO2I of >600 $ml{\cdot}min^{-1}{\cdot}m^{-2}$ [30]. This forms the basis of modern fluid management as there is a move away from pressure guided fluid therapy (such as blood pressure or central venous pressure – CVP) to flow guided fluid therapy. Given that HR, SaO2 and [Hb] are often not markedly deranged, the major focus is directed at stroke volume optimisation (SVO) according to Frank-Starling curve.

Secondly, an appreciation is required of the consequences of poor fluid management, either with inadequate fluid or excess fluid. The former will ultimately result in a reduced SV, CO and DO_2, and morbidity and mortality will be increased, secondary to inadequate tissue perfusion and organ dysfunction. The latter – excess fluid administration – will result in salt and water overload and thus oedema, which will not only effect major organs such as the lung, but critically in colorectal surgery, gut tissues and in particular any anastomosis as well as potentially increasing myocardial workload. Both these extremes of fluid management will therefore ultimately impair tissue oxygenation, leading to anaerobic metabolism, an oxygen debt, lactic acidosis and ultimately organ dysfunction and anastomotic failure. In particular excess fluids increases complications: for every litre of excess increased complications by 32 % [7] and prolonged hospital stay by 1 day [31].

Thus if we accept that excess or inadequate fluid for the patient is harmful and leads to increased morbidity and mortality, the key question is how to find the "sweet spot" of just the right amount of fluid, i.e. achieve SVO. It is just over a 100 years since Starling was credited with the concept that the more a heart muscle is stretched the greater the power of contractility (although the concept was first described 50 years prior to that) [32]. The process of SVO involves measurement and then the administration of an IV fluid bolus of 200–250 ml. The SV is measured again 10–15 min later. If there has been >10 % rise in SV, further boluses are given until SV no longer rises (i.e. optimisation has occurred). Further boluses of fluid are potentially detrimental as SV can reduce with excess of fluid. SVO is sometimes used synonymously with (Individualized) Goal Directed Fluid Therapy (GDFT), but the latter is not restricted to fluids but other methods too (e.g. inotropes) to achieve various targets or goals.

Historically, formulaic methods based on weight, standard 25 years ago (e.g. 5–10 ml/kg/h intraoperatively followed by 3 l in the first 24 h) are generally seen as inferior as they take no account of the patients starting fluid status or cardiac function. Moreover the widely held belief of third space loss and the concomitant requirement to replace these losses is probably flawed [33]. A significant improvement occurred with the adoption of routine CVP monitoring and the use of intermittent estimation of lactate, base excess and central venous oxygen saturation (as a surrogate marker for mixed venous oxygen saturation and hence oxygen extraction) to guide fluid therapy provide a more individualized approach. However, CVP measurement assumes pressure (right atrial pressure) and volume (left ventricular end diastolic volume) measurements are directly correlated which is clearly not always the case e.g. with a poorly compliant myocardium. Arterial and central venous biomarkers are of use but may show a delay of several hours. Thus flow-derived variables such as SVO, whilst not having universal acceptance, are nevertheless regarded as probably the superior method of fluid management.

Nevertheless, how can we make sense of the large amount of studies, some conflicting, in this all-important area? Firstly an appreciation is required in the changes in perioperative care that have occurred which rapidly outdates studies even only a few years old. The ERAS philosophy represents one of the most major changes in modern perioperative care and many of the physiological stresses such as prolonged starvation, the use of bowel preparation have produced a huge change improvement in the fluid status of patients arriving for surgery. Secondly the advent of laparoscopic surgery (vide infra) has dramatically reduced the physiological upset and fluid shifts associated with open surgery. Thirdly, there are various different available methods of measurement of cardiac output (with the majority of good quality data using the oesophageal Doppler monitor (ODM) though some more recent use other method such as arterial waveform analysis e.g. LiDCO rapid). This area has been recently reviewed [34]. Finally, many studies are on various abdominal surgeries and not specific to colorectal.

The key, current area seem are exemplified in the OPTIMISE trial, looking at 734 high-risk patients aged 50 years or older undergoing a variety of major gastrointestinal surgeries [35]. SVO and inotropic support was compared to a control group (CVP guided therapy) during surgery and for 6 h postoperatively. Although complications and 30-day mortality was different to the control group, when included in an updated meta-analysis their intervention group was associated with a reduction in complication rates. Thus we can postulate that SVO probably does no harm, probably reduces complications and probably has no effect on mortality. Two other major reviews would broadly support these findings. The former, most definitive review, studied the effects of increasing perioperative blood flow using fluids (with or without inotropes/ vasoactive drugs) and found that mortality was not reduced but three complications – renal failure, respiratory failure, and wound infections – were reduced, with 13 out of every hundred patients treated expected to avoid having complications. The length of stay was also reduced by just over 1 day [36]. The latter systematic review, whilst specifically looing at colorectal surgery had only 691 patients and found no difference in outcome (morbidity and mortality), particularly if GDFT was conducted within modern 'optimized perioperative care' [37].

It would seem that the benefits of modern, flow-targeted therapies, using fluids and where necessary inotropes had benefits that have been rather easier to demonstrate in older studies, prior to the adoption of ERAS pathways, and any benefits, are now much harder to prove. Indeed, a recent editorial – Stroke volume Optimization is the fairy tale over? – makes this very point [38]. SVO probably does no harm, and in aerobically fit patients many be unnecessary, but some patients will benefit more than others, especially the very frail and those in whom there is significant blood loss (7 ml/kg), or other large fluid shifts. Overall with the reduction of variability of patient care that comes with protocolised ERAS pathways, preoperative optimization and the reduction of physiological upset have all probably served to negate the earlier benefits of SVO.

This author's approach to fluid management takes a pragmatic approach of the available evidence. The key areas are [39]

- Patients should arrive euvolaemic for surgery, aided by avoiding prolonged fasting, mechanical bowel preparation and ingestion of carbohydrate drinks
- Intraoperative fluid therapy to maintain volume status, usually guided by SVO particularly in high risk patients and/or high risk surgery, with the use of vasopressors (and not fluids) to treat hypotension from anaesthetic vasodilation (e.g. regional blockade) to maintain arterial pressure as they preserve gastrointestinal blood flow) [40, 41]. For low risk patients undergoing low risk surgery a zero-balance approach can be adopted (i.e. merely replacing maintenance fluids of 1–3 mL/kg/h)
- Maintenance postoperative IV maintenance fluids, avoiding both fluid and saline excess, (e.g. <2.5 L with <100 mmol sodium per 24 h) aiming to transfer to oral fluids as soon as possible. Moreover, "permissive oliguria" [39] should be tolerated, and not chased with IV fluids, unless other signs of hypovolaemia co-exist.

The approach to laparoscopic surgery is dealt with later.

6.4.2 *Type of Fluid*

The type of fluid administered is also important and has generated a large number of trials, reviews and debates spanning four decades. A recent area of interest as centred on the endothelial gylcocalyx, where excessive volumes of fluid may disrupt this barrier promoting vascular leakage in to the interstitial space [39].

There are a variety of crystalloids available, but there is little doubt that physiologically balanced crystalloids (e.g. Hartmann's solution and latterly Plasma-Lyte) are superior to 0.9 % saline, which has consistently been reported to cause hyperchloraemic acidosis, itself predisposing to reduced kidney perfusion, reduced gastric blood flow and motility and reduced myocardial contractility [42]. If 0.9 % saline has a place, it is for the treatment of hypocholoraemic alkalosis, such as prolonged vomiting or large volumes of nasogastric aspirate.

Colloids have been preferred by some as the intravascular half-life is improved, providing that the endothelial glycocalyx is intact. Theoretically, a ratio of crystalloid: colloid might be expected to be about 3:1 to meet the same targets, given the early distribution of the two fluids, but a recent meta analysis demonstrated that in practice although there was much variation amongst the studies, the overall ratio was about 1.5:1 [43]. Of the major synthetic colloids – gelatins and starches – the latter have recently fallen into disuse because of the risk of death and renal failure in critically ill patients, although whether or not this can be extrapolated to elective patients is debatable. Many still use gelatin derivatives, although potential disadvantages include cost, allergy and coagulopathy. Albumin, whole blood and blood products are used to replace pre-existing and/or losses deficiencies.

Thus for critically ill patients who may be septic and/or have renal injury, in whom the endothelial glycocalyx may not be intact, many would avoid colloids to exacerbate any renal dysfunction. For other patients, many still use colloids for SVO to limit excess crystalloid (and saline) load, which is a major predictor for adverse outcomes for ERAS patients [7].

6.5 Analgesia

Analgesic requirements for patients undergoing colorectal surgery will clearly depend on the type of surgery – laparoscopic or open. Although they differ, there are nevertheless some principles common to providing analgesia for both types of surgery. Good analgesia is of course a fundamental humanitarian objective for all doctors, but, correctly undertaken, there are other potential benefits too. These include earlier mobilization, reduced organ dysfunction, reduced stress response, an earlier return to a diet and ultimately an earlier discharge from hospital, all of which are key ERAS principles. Moreover, if the analgesic modality reduces early complications, this will also impact on outcome both in the short term and long term (see above). On the horizon, is the prospect of how analgesic technique may impact on survival following major cancer surgery – see below. Thus modern pain medicine has the potential not only to improve patient's well being in the short term but may also have an effect on long-term survival as well.

The core philosophy is the provision of effective, opioid sparing analgesia. Whilst no drug is free from side effects, opioids, whilst effective analgesics, are widely regarded as undesirable with ERAS pathways, being particularity associated with PONV and slower return of gastrointestinal function, as well as other undesirable effect such as sedation, dysphoria and cough suppression. The provision of opioid sparing analgesia is best undertaken with multimodal analgesia, a concept that is over 20 years old and popularised by Kehlet [44]. In essence it is achieved by combining separate analgesics that act by different mechanisms, resulting in additive or synergistic analgesia with lowered adverse effects compared to sole administration of analgesics individually. Thus a combination of non-opioid analgesic drugs can still result in effective analgesia and a reduction in opioid consumption.

There are a large and growing number of analgesics described for use multimodal use. They can be conveniently classified into local anaesthetic (LA) techniques and systemic analgesics. The former involves blocking the pain pathway between the operation site and onward transmission within the spinal cord, and includes central neuraxial block, (spinal and epidural), nerve plexus blockade, TAP blocks and lastly local anaesthetic into the wound itself. The latter classification includes some very familiar drugs such non steroidal anti inflammatory drugs (NSAIDs), paracetemol, tramadol and perhaps some less familiar agents such as iv lidocaine, steroids, ketamine and magnesium.

6.5.1 Central Neuraxial Block for Open Surgery

Open colorectal surgery usually necessitates a large incision and requires intensive pain relief, usually with either regional blockade (usually an epidural) or parenteral opioids. For many years a thoracic epidural anaesthesia (TEA) was viewed as the gold standard for these patients rather than opioids as it provided a number of well-documented advantages including not only excellent pain relief, but in addition a swifter return of GI function, both as a result of a reduced in opioid requirements but also due to the sympathectomy and a relative increase in parasympathetic action on the gut. Other advantages from other forms of abdominal and pelvic surgery included a reduced VTE rate, reduced blood loss, and a reduction in some aspects of the stress response (pituitary, adrenocortical and sympathetic responses).

However in recent years enthusiasm for epidurals has declined. There are several reasons for this including a significant failure rate with quoted ranges between 13 and 47 %, with a large study describing an incidence of 32 % for thoracic epidurals and 27 % for lumbar epidural [45]. Although in many cases re-siting or adding adjuvants (such as epidural diamorphine) may rectify the situation, a failed epidural not only leaves the patient in pain, but it may also restrict other alternatives (such as systemic opioids) as these drugs cannot be co-administered if the patient is receiving epidural opioids. In addition, although epidurals are widely perceived as safe, the permanent neurological damage can occur, either due to vertebral canal haematoma, abscess and direct trauma. The NAP3 project highlighted the risks associated with postoperative epidurals having the highest incidence of permanent neurological harm, estimated between 1:5 700 – 1:12 200 [46]. Finally patients with epidurals often receive a greater volume of fluids to combat hypotension, which predisposes to fluid overload and can increase hospital stay [31]. These patients should receive vasoactive drugs rather than fluids to combat the sympathectomy [40, 41], but if so the patient will require nursing in an ICU/HDU environment as these agents cannot safely be administered on a ward and without intrarterial monitoring. Whilst epidurals will nevertheless continue to have a place for open surgery, other techniques have been used successfully used such as bilateral Transversus Abdominis Plane (TAP) catheters [47], rectus sheath catheters, and wound catheters which infuse local anaesthetic into the wound edges.

6.5.2 Central Neuraxial Block for Laparoscopic Surgery

Whilst the optimum analgesic technique for open colorectal is well studied and has a good evidence base to support the use of epidurals (even if latterly other modalities nerve blocks are gaining popularity), there is a paucity of data for laparoscopic surgery.

Generally these patients' analgesic requirements are much more modest in both intensity and in duration and although epidurals have been used, they are at best probably not necessary and at worst detrimental. A logical alternative to more the modest analgesic requirements is spinal anaesthesia with local anaesthetic and opioid. This has been used with a degree of success by a number of centres, and was used for patients in the first 23-h stay colorectal resections paper [48]. It offers good intraoperative analgesia extending several hours postoperatively with postoperative opioid sparing, following which a regime of regular simple oral analgesics suffices. Compared to epidurals, spinals permit greater mobilization and with no on-going sympathectomy, results in patients receiving less iv fluids, with their attendant complications [31], although others have found its benefits purely limited to opioid sparing [49]. Whilst spinals are approximately 2½ safer than epidurals [46], they are nevertheless not risk free for neurological damage. In addition, the onset of the sympathectomy is much more rapid than an epidural and caution is required, particularly in the elderly. Also the spinal must have "fixed" before adopting any head down position, so it is advisable to leave 10–15 min before adopting this position. Other forms of one-shot local anaesthetics have also been used, such as TAP blocks and wound infiltration, also with success. As with epidurals for open surgery, after initial success with spinals for laparoscopic surgery, many clinicians are opting for a simpler alternative, such as TAP blocks. Spinals are of little use for very prolonged surgery or if conversion to open surgery occurs.

6.5.3 Peripherally Placed Blocks

6.5.3.1 TAP Blocks

Transversus abdominus plane (TAP) blocks have been used with good effect in both laparoscopic and open surgery for nearly 10 years. A large volume (20 ml) of local anaesthetic is instilled between internal oblique and transversus abdominus muscles, in the lumbar triangle of Petit, whose borders are the external oblique muscle anteriorly, the latissimus dorsi muscle posteriorly and the iliac crest inferiorly. Originally a blind technique using anatomical landmarks was used with a "double pop" as a blunted needle passes through first the external and then the internal oblique muscles (posterior approach). The procedure is probably more safely undertaken using ultrasound and can be undertaken midaxillary line (lateral approach), with latterly a laparoscopic technique having been described. A TAP block provides analgesia from approximately T10-L1 dermatomes covering the incision for the specimen and port sites for laparoscopic bowel surgery, but they may also be used for open surgery too. Bilateral catheters may be used to extend the block duration. A number of studies have shown reductions in morphine use, time to tolerating diet, PONV and length of hospital stay have been described [50, 51]. The posterior approach appears to give more reliable effects [52]. Moreover, preoperative blocks provide better overall outcomes, with a dose response effect of larger dose of LA producing a greater effect on pain scores and reduced opioid consumption [53]. They are generally safe although liver trauma has been described [54].

6.5.3.2 Rectus Sheath Catheters

Rectus sheath catheters (RSC) have been used to provide effective analgesia for major open surgery. They can be inserted by a loss of resistance or ultrasound, but a logical way is for the surgeon to insert them under direct vision too. The catheter is left in situ and LA can be administered either by bolus dosing or via infusion. It has been used successfully particularly in urological surgery, and early work suggests they may provide comparable analgesia to TEA, without the associated problems of hypotension and limited mobility. The results of an ongoing study are awaited comparing TEA versus RSC (TERSC study) for open midline incisions in major abdominal surgery within an enhanced recovery programme [55].

6.5.3.3 Wound Catheters or Surgical Site Analgesia (SSA)

For many years, surgeons have infiltrated long acting local anaesthetic into the wound edges at the end of surgery and it is therefore logical to extend this by the placement of a multi-holed catheter to provide continuous infusion into the postoperative period.

They have been used in many types of surgery, including major orthopaedic surgery, abdominal and pelvic surgery. For open abdominal surgery, these catheters need to be the correct length of the incision and placed in the pre-peritoneal space by the surgeon and so are not suitable for mass closure techniques. Recently there has been renewed interest with good postoperative analgesia described for colorectal surgery [56], and a meta-analysis also demonstrated that they provided broadly similar results to epidural analgesia, with less urinary retention [57]. Two recent studies for open colorectal surgery have provided conflicting results. In one study SSA produced good pain control, faster recovery of postoperative ileus and bowel function with a lower incidence of PONV, and improved night sleep [58]. Another study however, performed within an ERAS protocol, did not replicate these results, with the epidural group having better pain control on the first day, better sleep quality, reduced time to both normal gut function and to hospital discharge [59]. SSAs are a promising, especially as they are relatively simple to manage postoperatively. There were concerns about an increased in wound infection, bit this appears to be unsubstantiated.

6.5.3.4 Intraperitoneal Local Anaesthetic

This technique, which has been used intermittently for over 60 years, has been studied more enthusiastically with the marked increase in laparoscopic operations such as gastric surgery, cholecystectomy, colonic resection and gynecological procedures. It is a promising technique, using for example 20 ml of 0.5 % bupivacaine, resulting in modest reductions in pain scores, opioid consumption and even a reduced stress response activation described. Part of the mechanism may be due to the systemic effect of the local anaesthetic.

6.5.4 *Systemic Analgesics*

6.5.4.1 Opioid Analgesia

A key area of ERAS analgesic pathways is a reduction in systemic opioid consumption as described above, yet morphine is nevertheless used both as rescue analgesia, and as a comparison to other modalities of analgesia. We demonstrated that in patients randomized to receive either morphine patient controlled analgesia (PCA), spinal anaesthesia or thoracic epidural analgesia, the length of stay was identical in the morphine PCA group and the spinal groups, with the epidural group required to stay an extra day in hospital [31]. A later study, whilst demonstrating better analgesia with spinal LA and opioid compared to PCA morphine showed that although the spinal group had better analgesia and reduced morphine consumption, there were no other differences in time taken to the return of bowel function, dietary intake and length of hospital stay [49]. This paradox is not easy to explain, but the above studies and others suggest that early morphine (i.e. in the first 24 h) probably has little effect on outcome, whereas prolonged use is very likely to cause problems such as PONV and constipation etc. Tramadol, which works via opioid receptors and other mechanisms (inhibition of reuptake of serotonin and norepinephrine) is frequently used without apparent problems.

6.5.4.2 Paracetemol

Paracetemol is familiar to all anaesthetists. Given regularly (either orally or iv as appropriate) it has a well-documented opioid sparing effect and is commonly used for at least 48 h following both open and laparoscopic surgery. It is a safe drug provided the maximum dose so not exceeded (commonly by prescription errors or giving more than 4 g/day in those under 50 kg). There is evidence to support an oral loading dose in adults with faster onset of paracetemol levels and thus pain relief with no increase in toxicity.

6.5.4.3 Anti-inflammatory Drugs

These commonly used drugs are also familiar to all anaesthetists, and working via cyclooxygenase (COX) inhibition. They are usually subdivided into whether they are non-selective cyclooxygenase (COX) inhibitors, inhibiting both the constitutive COX isoform (COX-1) and inducible isoform (COX-2), and the COX-2 inhibitors, which primarily inhibit the COX-2 isoform.

The non-selective COX drugs have a number of side effects such as gastrointestinal bleeding, allergy and bronchospasm and renal dysfunction, and whilst COX-2 inhibitors are safer (particularly for gastrointestinal bleeding) both groups of drugs are associated with thrombotic events in those with cerebrovascular and cardiac disease. However, of particular relevance for anaesthetists caring for colorectal patients, is the concern surrounding anti-inflammatory drugs and their effect on anastomotic breakdown. Both non-selective COX inhibitors, (especially diclofenac – [60]) and COX-2 inhibitors have been implicated in a higher risk of anastomotic leakage after colonic surgery. Currently they are still widely used, but growing evidence implicates the non-selective COX inhibitors more than the COX-2 inhibitors [61]. In addition, patients undergoing emergency surgery are more at risk than those undergoing elective colorectal resections [62]. Thus although these drugs are still widely used, their potential detrimental effects on healing, both in GI surgery and in other surgery (orthopaedics), has led to increasing reluctance by some to continue to prescribe them.

6.5.4.4 Intravenous Lidocaine

IV lidocaine infusions were described over 50 years ago as a systemic analgesic, and has recently been reintroduced. Following a bolus dose (e.g. 1.5–2 mg/kg) prior to the surgical incision an infusion an infusion (1.5–2 mg/kg/h) is administered for up to 24 h postoperatively. Principally used for open surgery, a number of beneficial effects have been reported: reduced opioid requirements (by up to two thirds), reduced duration of PONV and ileus and also hospital stay [63, 64]. In addition there are systemic effects reported including a reduction in stress response markers such as total leukocyte count, C-reactive protein and interleukin-6 (IL-6) [65], and excitingly, possibly an anticancer effect too [66]. There is good evidence to support its use, certainly as a second line drug [67] and it is perhaps surprising it is not more commonly used, although vigilance us required given that some studies have observed toxic levels and others did not look for signs of toxicity [63].

6.5.4.5 Gabapentinoids

Both gabapentin and pregabalin, which are commonly used in treatment of chronic pain conditions, have been successfully used as part of a multimodal analgesia postoperatively. Moreover, it is tempting to speculate that their use may be associated with a reduction in the frequency of occurrence of chronic pain syndromes. There is little information specifically for colorectal surgery patients, but studies in other areas demonstrate they do confer a degree of opioid sparing, and a reduction in PONV. For pregabalin, there is a dose related reduction in morphine consumption, but it may cause sedation, dizziness and visual disturbances. The dose and duration of these drugs is debated, but typically a dose of 150–300 mg/day of pregabalin preoperatively is used.

6.5.4.6 N-methyl-d-aspartate (NMDA) Receptor Antagonists

The central glutamatergic system, and in the particular its NMDA receptor, when stimulated plays a pivotal role in pain, including acute tolerance and opioid induced hyperalgesia, as well as in neuroplasticity and the progression to chronic pain syndromes. There are several antagonists at these excitatory receptors, including ketamine and magnesium, and both have been used to provide analgesia.

Ketamine, is well known to anaesthetists as an analgesic, but has been used for postoperative pain relief either mixed with morphine PCA (where it appears to be more efficacious for chest rather than abdominal surgery) [68], and also when given intraoperatively, (both as a bolus and infusion). For the latter, the intraoperative dose and duration is undecided, but it appears that when administered at low doses (2 mcg/kg/min after a 0.5 mg/kg bolus), morphine consumption was halved and adverse effects such as sedation, delusions and nightmares, were not seen [69]. Part of this effect may be related to its effects on reducing the inflammatory response, as measured by IL-6 [70]. There is little data to date to support its use in colorectal surgery.

Magnesium has also a documented place as an analgesic, but its other widespread effects including potentiation of neuromuscular blockade, hypotension and effects on cardiac conduction limit its usefulness [71].

6.5.4.7 Other Drugs

Alpha-2 agonists (clonidine and dexmedetomidine) are theoretically attractive for pain management, as they reduce sympathetic outflow and norepinephrine release. Clonidine has been administered via various routes: topically, systemically orally, intravenously, as well as into the epidural space and intrathecally. There is very little data relevant to colorectal patients, and although generally effective at reducing opioid consumption, pain and nausea its side effect profile particularly hypotension and bradycardia [72] will probably limit its use. The Perioperative Ischaemic Evaluation-2 (POISE-2) Trial confirmed these effects [73] and if clonidine has a place it has yet to be elucidated.

Glucocorticoids have a number of actions, some of which may be useful within ERAS programmes and others that may be harmful, with the exact dosing regimen for when benefit outweighs harm still undetermined. They reduce inflammation and pain (partly by reducing production of inflammatory mediators from arachidonic acid) and also via other methods such as anti-nociception at spinal cord level. They also have a well-documented effect on reducing PONV. However concern surrounds their use, as prolonged, high doses may cause wound infections delay healing, and hyperglycaemia [74]. However, a single dose of dexamethasone produces slight hyperglycaemia, reduces opioid consumption (but in a non-dose dependent manner) without apparent adverse effects [75]. There is little to specifically recommend steroids to colorectal surgery, but there is nevertheless interest from orthopaedics that high dose steroids may cause less fatigue from reduced inflammatory responses [76].

The concept of peripheral opioid antagonists, such as alvimopan, is an exciting idea as it permits the benefits of central opioid analgesia, with the peripheral side effects (e.g. ileus) to be reversed. Data is still limited, but there are encouraging results from both open and laparoscopic surgery [77, 78].

There are a number of other techniques on the horizon in this constantly changing field, including long acting local anaesthetic infiltration, lidocaine patches, beta-blockers, capsaicin, and other opioids such as Tapentadol. In addition, non-pharmacological methods are of interest to some, with sporadic evidence for acupuncture, music therapy, hypnosis and transcutaneous electrical nerve stimulation (TENS) [74].

6.5.5 *Analgesic Summary*

How are we to make sense of the plethora of analgesic techniques on offer? It will depend to some extent on local expertise and experience. Multimodal analgesia remains the mainstay of any analgesic approach. For open surgery, epidurals are favoured still by many, but more peripheral neural blockade is gaining in evidence and popularity, such as TAP and RSC and SSC catheters. For laparoscopic surgery, with more modest analgesic requirements, epidurals are generally regarded as unnecessary, as maybe spinal anaesthesia too, as again more peripheral techniques – e.g. TAP blocks – are gaining popularity.

With regard to systemic analgesics, paracetemol has an assured place with caution for anti-inflammatory drugs. Of the second line drugs IV lidocaine has a good evidence base, with other analgesics yet to find their confirmed place in colorectal surgery, either alone or in combination.

6.6 Laparoscopic or Open Surgery?

It may seem self evident that laparoscopic surgery is prerequisite for any ERAS programme in colorectal surgery, and indeed the widespread adoption of laparoscopic surgery is arguably the biggest single advance within the ERAS pathway. Yet it must not be forgotten that fast track surgery predated the widespread adoption of laparoscopic techniques and it is still possible to achieve impressive results with open surgery within an ERAS pathway. Moreover, laparoscopic surgery itself poses a considerable intraoperative physiological demand on patients, and has led to the phrase "minimal access surgery, maximum cardiopulmonary stress" to remind clinicians that small incision surgery may not always be appropriate for patients [79]. A pneumoperitoneum (especially at pressures of >20 mmHg) and head down, causes marked cardiovascular changes, including reduced venous return, increased afterload and a reduction in cardiac output, with reduced perfusion to the intra-abdominal organs (gut, liver and kidney), and eventually a lactic acidosis. The lungs have reduced compliance and functional residual capacity, ventilation-perfusion mismatch and atelectasis (albeit relatively short-lived). In addition to these cardiopulmonary changes, there may also be reduced cerebral drainage and cerebral oedema. Some patients with very poor physiological reserve may be unable to

Table 6.3 Physiological changes during laparoscopy

Cardiovascular	Respiratory	Other organs
Increased SVR	Reduced pulmonary compliance	Reduced visceral perfusion (gut/liver/kidney)
Reduced venous return	Reduced FRC	Reduced cerebral venous drainage (may lead to cerebral oedema)
Reduction in CO	Pulmonary atelectasis	Metabolic (lactic) acidosis if prolonged surgery
BP changes variable (very often reduces with reverse Trendelenberg)	VQ mismatch	
	CO_2 absorption with respiratory acidosis	

SVR systemic vascular resistance, *CO* cardiac output, *FRC* functional residual capacity, *VQ* ventilation: perfusion

tolerate these changes and occasionally patients may need to be refused laparoscopic surgery in favour of an open procedure (Table 6.3).

However a major advantage of laparoscopic surgery is that a key area of the stress response – the inflammatory changes – is dramatically reduced. These inflammatory changes are broadly proportional to the tissue injury, and driven by cytokines (e.g. interleukin 6 and tumour necrosis factor-alpha) cause a SIRS response, which can have a number of undesirable effects including increased oxygen consumption with activation of sympathetic nervous system, as well as other changes such as malaise and fatigue. There are other advantages to laparoscopic surgery too: bowel handling, fluid shifts, ileus are reduced and postoperative pulmonary function is improved.

The two major areas for the perioperative physician to address for laparoscopic patients are analgesia and fluid management, as it is not to extrapolate results form open surgery to laparoscopic. The more modest analgesic requirements for these patients have been dealt with above, and the evidence base is growing. However, fluid management in laparoscopic surgery is still a matter of debate with a poor evidence base. Generally the cardiovascular effects of the laparoscopy itself are far greater than the relatively small changes associated with SVO fluid management. Whereas a DO_2 of >600 $ml{\cdot}min^{-1}{\cdot}m^{-2}$ is commonly used goal for open surgery, we have shown for laparoscopic surgery an intraoperative DO_2 of 400 $ml{\cdot}min^{-1}{\cdot}m^{-2}$ as a threshold to minimize postoperative anastomotic breakdown, with an incidence of over 22 % below this DO_2 and less than 2 % above this level [80]. A pragmatic approach is to use the lowest intra-abdominal pressures possible (perhaps assisted by DNB – see above), use vasopressors to assist in maintaining laparoscopic-induced changes in blood pressure rather than large volumes of crystalloids and most importantly in the interval between the end of the pneumoperitoneum and the end of surgery to ensure the patient leaves theatre euvolaemic.

The LAFA trial provided us with further comparisons between open and laparoscopic surgery. This trial randomized patients to receive LAparoscopy and/or FAst track (i.e. ERAS) multimodal management versus standard care for colectomy. There were four groups of patients:

- open colectomy with standard care
- open colectomy with fast track care
- laparoscopic colectomy with standard care
- laparoscopic surgery with fast track care

Perhaps unsurprisingly, as we go down this group of patients, the length of stay in hospital became progressively shorter, with regression analysis showing that laparoscopy was the only independent predictive factor to reduce hospital stay and morbidity [81]. A further paper analysed the impact of LAFA intervention on inflammatory markers and immune competence (the latter helping to protect against potential consequences of tumour cell seeding and hence metastases) on these 4 groups of patients and showed that both inflammatory markers (as measured by IL-6 and CRP) and immune competence (as measured by HLA-DR expression on monocytes) were least disrupted in the laparoscopic within an ERAS programme [82] However of equal importance is that whilst laparoscopic and ERAS programme is probably the gold standard, open surgery within an ERAS programme showed marked improvements compared to open surgery and standard care. The practical significance of the LAFA trail is thus: both laparoscopic surgery and ERAS programmes confer the most benefit as measured by reducing length of stay, inflammatory markers and immune function, but there is still substantial benefits for patients who receive open and surgery within an ERAS programme. Moreover laparoscopic surgery alone (i.e. without ERAS integration) is suboptimal.

6.7 Other ERAS Elements

There are many areas of the ERAS pathway that may seem self evident, and are usually incorporated into care bundles. These include:

6.7.1 PONV

Prophylaxis using a multimodal approach is needed. The guidelines for PONV have recently been updated [83]. Many components of the ERAS pathway themselves contribute to reducing PONV, including euvolaemia, carbohydrate loading and opioid sparing where possible. In addition, the use of total intravenous anaesthesia (TIVA) and avoidance of nitrous oxide is also beneficial (Table 6.4).

Table 6.4 Analgesic methods and PONV medication

Peroperative medication	Post operative regular medication	Post operative rescue medication
Local anaesthetic: Open surgery: epidural/ TAP block Laparoscopic surgery: spinal/TAP block/wound infiltration	Local anaesthetic Open surgery: epidural infusion/TAP infusion/RSC/ SSA for 48 h	Tramadol 50–100 mg oral or intravenously 6 hourly (if not receiving it already) Open surgery: consider iv Lidocaine (With appropriate monitoring) Consider ketamine
Remifentanil infusion	Paracetemol 1 g oral or intravenously 6 hourly (Open surg. 96 h, Lap surgery 48 h)	Morphine Up to 10 mg intramuscularly or intravenously
Paracetemol 1 g iv 15 min prior to end of surgery	Anti-inflammatories e.g. Diclofenac with CAUTION 50 mg oral 8 hourly once eating OR Tramadol 50–100 mg oral 6 hourly if diclofenac contraindicated	Cyclizine 25 mg intramuscularly or intravenously 6 hourly
Ondansetron 4 mg prior to end of surgery	Omeprazole 20 mg oral daily (with diclofenac)	Ondansetron 4 mg intravenously 6 hourly

TAP transversus abdominis plane block, *RSC* rectus sheath catheter, *SSA* surgical site analgesia (wound catheter)

6.7.2 Avoidance of Hypothermia

In spite of many papers highlighting this issue over the last 30 years, inadvertent hypothermia is still common. The maintenance of normothermia is crucial as hypothermia adversely affects nearly every physiological system. Patients with a core temperature of <35°C have poor drug metabolism, shivering (with concomitant increased oxygen consumption), reduced immune overall function with increased wound infection rates, increased bleeding, reduced renal blood flow and increased adverse cardiac events (persisting for up to 6 months postoperatively) and an increased length of stay in hospital Some have suggested preoperative warming strategies to help prevent hypothermia. Normothermia (>36°C) is a thus a fundamental requirement. A recent review demonstrated hypothermic patients undergoing elective operations had a four-fold increase in mortality and a doubling of complication rate (particularly stroke and sepsis) [84].

In addition, a protocol to reduce wound infections showed that with normothermia, appropriate prophylactic antibiotics (see below), good glycaemic control and subcutaneous drains in patients with a body mass index (BMI) of >25, surgical site infections (SSI) were reduced by nearly 40 % [85] Whilst we may debate the usefulness of drains, the other factors are well established.

6.7.3 Antibiotic Prophylaxis and Thromboprophylaxis

These are fundamental and are usually included in care pathways for the prevention of SSI, systemic infections and pulmonary thromboembolism respectively. Antibiotics should be repeated as required for prolonged surgery and/or major blood loss.

6.7.4 Drains/Tubes

The routine use of nasogastric tubes (NGT) for elective abdominal surgery has now been replaced for use in selective patients only, as various studies over 20 years following laparotomy have shown an increase in pulmonary complications and a delay in the return of bowel function. However a more recent meta-analyses for colorectal surgery patients showed the benefits of not placing an NGT was limited to reduced respiratory infection and pharyngolaryngitis [86], and may even increase the rate of SSRI, although the reason for this are not clear [87]. It is wise to avoid bag and mask ventilation prior to intubation to avoid having to decompress the stomach afterwards.

Urinary catheters are inserted for maybe a day but then can be removed even if a TEA is in situ. It is not necessary to leave a patient catheterized for the duration of a TEA as it results in an increased of urinary infections and an increased length of stay [88]. Moreover routine surgical drainage should not routinely inserted as they have confer no advantage and may impair mobilization.

6.7.5 Serum Glucose Control

Hyperglycaemia from insulin resistance is a well-recognised feature of patients undergoing major surgery and is part of the classic hormonal and metabolic response. However, given that a cornerstone of ERAS pathways is to reduce the stress response, it is reasonable to assume that this should become less common. Intensivists have achieved improved outcomes for their patients by tight blood glucose control but the relevance of this to for elective colorectal patients is uncertain, and of course the risks of poorly administered insulin are high. However diabetic patients are known to have a higher incidence of complications, and patients not known to be diabetics with high glycosylated haemoglobin (HbA1c of more than 6 %) had a higher CRP and postoperative complications [89].

6.7.6 Mobilization

Early mobilization is a fundamental issue within ERAS, and indeed one should always ask "why is my patient still in bed" or indeed "why are they still in hospital" [76]. There are many reasons for poor mobilization, include various symptoms such as pain weakness, dizziness (including hypotension) nausea and vomiting, and just feeling "unwell" which may be multifactorial but at least a part of it will be secondary to inflammatory mediators. In addition there may be physical reasons, such attachment to drips and tubes etc. Finally, there may be psychological reasons from the patient, their relatives or indeed healthcare staff that the patient needs to be bedbound, rather than dressed and mobile. Prolonged immobility may predispose to pulmonary thromboembolism, poor respiratory activity encouraging atelectasis, sputum retention and chest infection, and may ultimately have an adverse effect on work capacity and muscle strength, all a potential harbinger for complications and a prolonged hospital stay.

6.7.7 Early Feeding and Gut Motility

For many years nil by mouth operatively until the onset of bowel sounds or flatus was standard surgical care, with concerns of vomiting, abdominal distension and anastomotic dehiscence. This is not the case with early enteral feed generally well tolerated and reducing length of stay and infections [90].

The concept of perioperative nutritional medicine was coined to highlight preoperative supplementation where necessary, carbohydrate preload and postoperative feeding. Whilst normal food is currently regarded as optimal there is much interest on nutritional supplements and also immunonutrients (such as n-3 polyunsaturated fatty acids, glutamine, arginine or nucleotides) to improve wound healing and reduce inflammatory markers [13].

Other are of interest is whether or not stimulating gut motility by other means is useful. Alvimpopan is mentioned above; in addition, chewing gum shortens the duration of postoperative ileus

6.7.8 Audit

Good quality data collection and its analysis, together multidisciplinary meetings involving relevant healthcare workers (surgeons, anaesthetists, nursing, physiotherapists, pharmacists) are essential to address outcomes (such as LOS, complications, mortality and quality of life) with benchmarking through audit and re-audit. Continuous vigilance is required to ensure all aspects of the pathway are being followed (e.g. following changeover of staff) as well as monitoring the introduction of new evidence-based changes.

6.8 Emergency Colorectal Surgery

Thirty per cent of patients present for their colorectal surgery as emergencies [91]. Yet, on the face of it emergency surgery may have little to do with the concept of ERAS, but much of the philosophy of ERAS is applicable to emergency surgery and perhaps acute medicine too. Therefore many clinicians now implement evidenced-based care bundles both from ERAS and surviving sepsis combined with timely senior clinician involvement and intensive care input postoperatively.

It has long been recognised that the results from emergency abdominal surgery are both poor and highly variable between hospitals. There is an average mortality of nearly 15 % rising to nearly 25 % in the over eighty patient over eighty group [92]. Moreover younger patients (less than eighty) are ten times more likely to be discharged home to live independently compared to their more elderly counterparts. These poor outcomes, together with marked variations between hospitals has led to the introduction of quality improvement initiatives, which have improved results [93]. There is currently the EPOCH (Enhanced Perioperative Care for high risk patients) study is underway. This is a randomised stepped-wedge cluster trial, investigating whether fewer patients die within 90 days of surgery in hospitals where the quality improvement intervention has taken place. This landmark study is thus applying some ERAS principles to these patients with poor outcomes.

Of the many steps for patients undergoing elective surgery within an ERAS pathway, a major area of change is preoperatively, where any optimization of health status has to take place in hours rather than days or weeks. Nevertheless, preoperative oxygen therapy, fluid management (sometimes with inotropes), IV antibiotics may rapidly improve patients' status. Early access to imaging is essential. Many patients are elderly, have significant chronic comorbidities and may be receiving drugs that may adversely affect surgery (particularly anticoagulants) or not absorbing drugs that are essential to their condition (e.g. for rate control in atrial fibrillation, heart failure or angina). The acute situation of fluid imbalance and sepsis exacerbates the situation.

Many of the intraoperative steps can be applied such as SVO, although open surgery is much more common. Rapid sequence induction and nasogastric drainage are almost universal. Caution is required as arm-brain circulation may be very slow and rapid administration of an induction agent may lead to excessive drug doses and hypotension. The patient is ventilated usually with air and oxygen and a short acting volatile agent such as sevoflurane or desflurane. Full monitoring is required in all but the smallest cases (arterial and central venous access) with cardiac output monitoring and SVO. Once fluid optimised, if cardiac output is low inotropes such as adrenaline may be commenced; if mean arterial pressure is low in the face of normal or high cardiac output (such as in sepsis) then vasoconstrictors such as noradrenaline can be used. Analgesia may be difficult and epidurals may be contraindicated due to sepsis, hypotension, or coagulopathy. NSAIDS are often contraindicated especially in the elderly or those with renal or cardiovascular dysfunction. Monitoring of neuromuscular blockade is essential, as is the insertion of a urinary catheter.

Postoperatively intensive care is ideal, with some patients able to have their tracheas extubated, but those with marked metabolic derangement such as a base deficit of >10 mmol/l or serum lactate >5 mmol/l or with severe pre-existing disease or the requiring a high inspired oxygen with hypothermia suggest that the patient may not cope without ventilatory support.

However, once stable and extubated, a planned care pathway planned ensures that their analgesic (with or without epidural) nutritional (oral or parenteral) and mobilization demands can be addressed.

A key decision is not only that the patients may require surgery, but also whether or not surgery is appropriate because they are so frail or whether a lesser procedure may be appropriate (e.g. defunctioning or stenting the patient) prior to a definitive procedure at a later date.

6.9 The Future

ERAS is not merely limited to the short-term hospital discharge. The benefits of a reduction in early complications rates have also been addressed and the long-term benefits that these confer. However, interest has also been in other benefits [1]. For example ERAS patients may get better sooner and there is evidence that life expectancy may be improved as they are better for adjuvant therapy (chemotherapy or radiotherapy). Another topical area is the effect of anaesthetic technique on cancer outcome. This is still largely unproven [94] but emerging evidence gives food for thought, as both morphine and volatile agents may later prove to be desirable agents to minimize [19] and although unproven for abdominal [95] and especially laparoscopic colorectal surgery [96] epidurals may nevertheless prove to have a beneficial effects, perhaps by opioid sparing, perhaps by immune mechanisms. The effect of other drugs awaits clarification e.g. NSAIDS and dexamethasone and IV lidocaine. Finally the LAFA trial and its enhanced effects on immunocompetence, also may act as a stimulus to investigate further any benefits on the seeding of tumour cells for patients on ERAS pathways.

6.10 Summary

ERAS represents a sea-change for those practising perioperative medicine. The rigid adoption of protocolised pathways, whilst reducing clinical freedom, serve to maintain the most evidenced based approach to colorectal surgery. With these pathways accounted for, the two other major areas of the trimodal approach are fluid management and analgesia. However, as our speciality moves into pre and postoperative care, we can shape patients progress throughout the entire journey as well as learn lessons from poor practice. Emergency surgery is a huge challenge, but given adequate resources, major improvements should be possible. As our clinical activities and influence moves from the operating theatre to clinics and wards, we hope to see good practice not only rewarded with benefits not limited to the short term, but to the long term as well.

References

1. Fawcett WJ, Mythen MG, Scott MJP (2012) Enhanced Recovery – more than just reducing length of stay? Br J Anaesth 109:671–674
2. Cuthbertson DP (1932) Observations on the disturbance of metabolism produced by injury to the limbs. Quart J Med 25:233–246
3. Desborough JP (2000) The stress response to trauma and surgery. Br J Anaesth 85:1–9–117
4. Kehlet H, Mythen M (2011) Why is the surgical high-risk patient still at risk? Br J Anaesth 106:289–291
5. Khuri SF, Henderson WG, DePalma RG et al (2005) Determinants of long-term survival after major surgery and the adverse effects of postoperative complications. Ann Surg 242:326–343
6. Durrand JW, Batterham AM, Danjoux GR (2014) Pre-habilitation (i): aggregation of marginal gains. Anaesthesia 69:403–406
7. Gustafsson UO, Hausel J, Thorell A et al (2011) Adherence to the enhanced recovery after surgery protocol and outcomes after colorectal cancer surgery. Arch Surg 146:571–577
8. Shoemaker WC, Appel PL, Kram HB (1990) Measurement of tissue perfusion by oxygen transport patterns in experimental shock and in high-risk surgical patients. Intensive Care Med 16(Suppl 2):S135–S144
9. Goldman L, Caldera DL, Nussbaum SR et al (1977) Multifactorial index of cardiac risk in noncardiac surgical procedures. N Engl J Med 297:845–850
10. Lee TH, Marcantonio ER, Mangione CM et al (1999) Derivation and prospective validation of a simple index for prediction of cardiac risk of major noncardiac surgery. Circulation 100:1043–1049
11. Older P, Smith R, Courtney P et al (1993) Preoperative evaluation of cardiac failure and ischemia in elderly patients by cardiopulmonary exercise testing. Chest 104:701–704
12. West MA, Lythgoe D, Barben CP et al (2014) Cardiopulmonary exercise variables are associated with postoperative morbidity after major colonic surgery: a prospective blinded observational study. Br J Anaesth 112:665–671
13. Scott MJ, Fawcett WJ (2014) Oral carbohydrate preload drink for major surgery – the first steps from famine to feast. Anaesthesia 69:1308–1313
14. Hurley BF, Hagberg JM, Allen WK et al (1984) Effect of training on blood lactate levels during submaximal exercise. J Appl Physiol 56:1260–1264
15. Carli F, Scheede-Bergdahl C (2015) Prehabilitation to enhance perioperative care. Anesthesiol Clin 33:17–33
16. Gustafsson UO, Nygren J, Thorell A et al (2008) Pre-operative carbohydrate loading may be used in type 2 diabetes patients. Acta Anaesth Scand 52:946–951
17. Levy BF, Scott MJP, Day A, Fawcett WJ, Rockall TA (2011) Optimizing patient outcomes in laparoscopic surgery. Colorectal Dis 13:8–11
18. Chan MT, Cheng BC, Lee TM (2013) BIS-guided anesthesia decreases postoperative delirium and cognitive decline. J Neurosurg Anesthesiol 25:33–42

19. Durieux ME (2015) Time to dial down the vaporizer? Br J Anaesth 114:715–716
20. Martini CH, Boon M, Bevers RF et al (2014) Evaluation of surgical conditions during laparoscopic surgery in patients with moderate vs. deep neuromuscular block. Br J Anaesth 112:498–505
21. Viby-Mogensen J, Jørgensen BC, Ording H (1979) Residual curarization in the recovery room. Anesthesiology 50:539–541
22. Fawcett WJ, Dash A, Francis GA et al (1995) Recovery from neuromuscular blockade: residual curarisation following atracurium or vecuronium by bolus dosing or infusions. Acta Anaesth Scand 39:288–293
23. Murphy GS, Brull SJ (2010) Residual neuromuscular block: lessons unlearned Part I: definitions incidence and adverse physiologic effects of residual neuromuscular block. Anesth Analg 111:120–128
24. Severgnini P, Selmo G, Lanza C et al (2013) Protective mechanical ventilation during general anesthesia for open abdominal surgery improves postoperative pulmonary function. Anesthesiology 118:307–1321
25. Futier E, Constantin JM, Paugam-Burtz C et al (2013) A trial of intraoperative low-tidal-volume ventilation in abdominal surgery. N Engl J Med 369:428–437
26. Hemmes SN, de Abreu MG, Pelosi P et al (2014) High versus low positive end-expiratory pressure during general anaesthesia for open abdominal surgery (PROVHILO trial): a multicentre randomised controlled trial. Lancet 384:495–503
27. Shoemaker WC, Montgomery ES, Kaplan E et al (1973) Physiologic patterns in surviving and nonsurviving shock patients. Arch Surg 106:630–636
28. Shoemaker WC, Appel PL, Kram HB (1993) Hemodynamic and oxygen transport responses in survivors and nonsurvivors of high-risk surgery. Crit Care Med 21:977–990
29. Boyd O, Grounds RM, Bennett ED (1993) A randomized clinical trial of the effect of deliberate perioperative increase of oxygen delivery on mortality in high-risk surgical patients. JAMA 270:2699–2707
30. Tote SP, Grounds RM (2006) Performing perioperative optimization of the high-risk surgical patient. Br J Anaesth 97:4–11
31. Levy BF, Scott MJ, Fawcett WJ et al (2011) Randomized clinical trial of epidural spinal or patient-controlled analgesia for patients undergoing laparoscopic colorectal surgery. Br J Surg 98:1068–1078
32. Katz AM (2002) Ernest Henry Starling his predecessors and the "Law of the Heart". Circulation 106:2986–2992
33. Minto G, Mythen MG (2015) Perioperative fluid management: science art or random chaos. Br J Anaesth 114:717–721
34. Thiele RH, Bartels K, Gan TJ (2015) Inter-device differences in monitoring for goal-directed fluid therapy. Can J Anaesth 62:169–181
35. Pearse RM, Harrison DA, MacDonald N et al (2014) Effect of a perioperative cardiac output – guided hemodynamic therapy algorithm on outcomes following major gastrointestinal surgery: a randomized clinical trial and systematic review. JAMA 311:2181–2190
36. Grocott MPW, Dushianthan A, Hamilton MA et al (2013) Perioperative increase in global blood flow to explicit defined goals and outcomes after surgery: a Cochrane Systematic Review. Br J Anaesth 111:535–548
37. Srinivasa S, Lemanu DP, Singh PP et al (2013) Systematic review and meta-analysis of oesophageal Doppler-guided fluid management in colorectal surgery. Br J Surg 100:1701–1708
38. Minto G, Struthers R (2014) Stroke volume optimisation: is the fairy tale over? Anaesthesia 69:291–296
39. Miller TE, Roche AM, Mythen M (2015) Fluid management and goal-directed therapy as an adjunct to Enhanced Recovery After Surgery (ERAS). Can J Anaesth 62:158–168
40. Gould TH, Grace K, Thome G et al (2002) Effect of thoracic epidural anaesthesia on colonic blood flow. Br J Anaesth 89:446–451
41. Hiltebrand LB, Koepfli E, Kimberger O et al (2011) Hypotension during fluid-restricted abdominal surgery: effects of norepinephrine treatment on regional and microcirculatory blood flow in the intestinal tract. Anesthesiology 114:557–564

42. Raghunathan K, Singh M, Lobo DN (2015) Fluid management in abdominal surgery: what, when, and when not to administer. Anesthesiol Clin 33:51–64
43. Cortés DO, Barros TG, Njimi H et al (2015) Crystalloids versus colloids: exploring differences in fluid requirements by systematic review and meta-regression. Anesth Analg 120:389–402
44. Kehlet H, Dahl JB (1993) The value of " multimodal" or "balanced analgesia" in postoperative pain treatment. Anesth Analg 77:1048–1056
45. Hermanides J, Hollmann MW, Stevens MF et al (2012) Failed epidural: causes and management. Br J Anaesth 109:144–154
46. National Audit of Major Complications of Central Neuraxial Block in the United Kingdom. http://www.rcoa.ac.uk/system/files/CSQ-NAP3-Full_1.pdf (2009) Accessed 30 May 2015
47. Rao KV, Van Wijk RM, Moran JI et al (2013) Epidural versus continuous transversus abdominis plane catheter technique for postoperative analgesia after abdominal surgery. Anaesth Inten Care 41:476–481
48. Levy BF, Scott MJ, Fawcett WJ et al (2009) 23-hour-stay laparoscopic colectomy. Dis Colon Rectum 52:1239–1243
49. Wongyingsinn M, Baldini G, Stein B et al (2012) Spinal analgesia for laparoscopic colonic resection using an enhanced recovery after surgery programme: better analgesia but no benefits on postoperative recovery: a randomized controlled trial. Br J Anaesth 108:850–856
50. Zafar N, Davies R, Greenslade GL et al (2010) The evolution of analgesia in an 'accelerated' recovery programme for resectional laparoscopic colorectal surgery with anastomosis. Colorectal Dis 12:119–124
51. Johns N, O'Neill S, Ventham NT et al (2012) Clinical effectiveness of transversus abdominis plane (TAP) block in abdominal surgery: a systematic review and meta-analysis. Colorectal Dis 14:e635–e642
52. Abdallah FW, Laffey JG, Halpern SH et al (2013) Duration of analgesic effectiveness after the posterior and lateral transversus abdominis plane block techniques for transverse lower abdominal incisions: a meta-analysis. Br J Anaesth 111:721–735
53. De Oliveira Jr GS, Castro-Alves LJ, Nader A et al (2014) Transversus abdominis plane block to ameliorate postoperative pain outcomes after laparoscopic surgery: a meta-analysis of randomized controlled trials. Anesth Analg 118:454–463
54. Lancaster P, Chadwick M (2010) Liver trauma secondary to ultrasound-guided transversus abdominis plane block. Br J Anaesth 104(4):509–510
55. Wilkinson KM, Krige A, Brearley SG et al (2014) Thoracic Epidural analgesia versus Rectus Sheath Catheters for open midline incisions in major abdominal surgery within an enhanced recovery programme (TERSC): study protocol for a randomised controlled trial. Trials 15:400
56. Ozer A, Yılmazlar A, Oztürk E et al (2014) Preperitoneal catheter analgesia is an effective method for pain management after colorectal surgery: the results of 100 consecutive patients. Local Reg Anesth 7:53
57. Ventham NT, Hughes M, O'Neill S et al (2013) Systematic review and meta-analysis of continuous local anaesthetic wound infiltration versus epidural analgesia for postoperative pain following abdominal surgery. Br J Surg 100:1280–1289
58. Bertoglio S, Fabiani F, De Negri P et al (2012) The postoperative analgesic efficacy of preperitoneal continuous wound infusion compared to epidural continuous infusion with local anesthetics after colorectal cancer surgery: a randomized controlled multicenter study. Anesth Analg 115:1442–1450
59. Jouve P, Bazin JE, Petit A et al (2013) Epidural versus continuous preperitoneal analgesia during fast-track open colorectal surgery. Anesthesiology 118:622–630
60. Klein M, Gögenur I, Rosenberg J (2012) Postoperative use of non-steroidal anti-inflammatory drugs in patients with anastomotic leakage requiring reoperation after colorectal resection: cohort study based on prospective data. BMJ 345:e6166
61. Gorissen KJ, Benning D, Berghmans T et al (2012) Risk of anastomotic leakage with non-steroidal anti-inflammatory drugs in colorectal surgery. Br J Surg 99:721–727
62. Hakkarainen TW, Steele SR, Bastaworous A et al (2015) Nonsteroidal anti-inflammatory drugs and the risk for anastomotic failure: a report from Washington State's Surgical Care and Outcomes Assessment Program (SCOAP). JAMA Surg 150:223–228

63. Vigneault L, Turgeon AF, Côté D (2011) Perioperative intravenous lidocaine infusion for postoperative pain control: a meta-analysis of randomized controlled trials. Can J Anesth 58:22–37
64. Sun Y, Li T, Wang N et al (2012) Perioperative systemic lidocaine for postoperative analgesia and recovery after abdominal surgery: a meta-analysis of randomized controlled trials. Dis Colon Rectum 55:1183–1194
65. Sridhar P, Sistla SC, Ali SM et al (2014) Effect of intravenous lignocaine on perioperative stress response and post-surgical ileus in elective open abdominal surgeries: a double-blind randomized controlled trial. Aust NZ J Surg. doi:10.1111/ans.12783 (Epub ahead of print)
66. Lirk P, Berger R, Hollmann MW et al (2012) Lidocaine time-and dose-dependently demethylates deoxyribonucleic acid in breast cancer cell lines in vitro. Br J Anaesth 109:200–207
67. Joshi GP, Bonnet F, Kehlet H (2013) Evidence-based postoperative pain management after laparoscopic colorectal surgery. Colorectal Dis 15:146–155
68. Carstensen M, Moller AM (2010) Adding ketamine to morphine for intravenous patient-controlled analgesia for acute postoperative pain: a qualitative review of randomized trials. Br J Anaesth 104:401–406
69. Zakine J, Samarcq D, Lorne E et al (2008) Postoperative ketamine administration decreases morphine consumption in major abdominal surgery: a prospective randomized double-blind controlled study. Anesth Analg 106:1856–1861
70. Dale O, Somogyi AA, Li Y et al (2012) Does intraoperative ketamine attenuate inflammatory reactivity following surgery? A systematic review and meta-analysis. Anesth Analg 115:934–943
71. Fawcett WJ, Haxby EJ, Male DA (1999) Magnesium: physiology and pharmacology. Br J Anaesth 83:302–320
72. Blaudszun G, Lysakowski C, Elia N et al (2012) Effect of perioperative systemic α2 agonists on postoperative morphine consumption and pain intensity: systematic review and meta-analysis of randomized controlled trials. Anesthesiology 116:1312–1322
73. Devereaux PJ, Sessler DI, Leslie K et al (2014) Clonidine in patients undergoing noncardiac surgery. N Engl J Med 370:1504–1513
74. Tan M, Law LSC, Gan TJ (2015) Optimizing pain management to facilitate enhanced recovery after surgery pathways. Can J Anaesth 62:203–218
75. Waldron NH, Jones CA, Gan TJ et al (2013) Impact of perioperative dexamethasone on postoperative analgesia and side-effects: systematic review and meta-analysis. Br J Anaesth 110:191–200
76. Kehlet H (2015) Enhanced Recovery After Surgery (ERAS): good for now but what about the future? Can J Anaesth 62:99–104
77. Vaughan-Shaw PG, Fecher IC, Harris S et al (2012) A meta-analysis of the effectiveness of the opioid receptor antagonist alvimopan in reducing hospital length of stay and time to GI recovery in patients enrolled in a standardized accelerated recovery program after abdominal surgery. Dis Colon Rectum 55:611–620
78. Kelley SR, Wolff BG, Lovely JK et al (2013) Fast-track pathway for minimally invasive colorectal surgery with and without Alvimopan (Entereg): which is more cost-effective? Am Surg 79:630–633
79. Minto G, Scott MJ, Miller TE (2015) Monitoring needs and goal-directed fluid therapy within an enhanced recovery program. Anesthesiol Clin 33:35–49
80. Levy BF, Fawcett WJ, Scott MJP et al (2012) Intra-operative oxygen delivery in infusion volume-optimized patients undergoing laparoscopic colorectal surgery within an enhanced recovery programme: the effect of different analgesic modalities. Colorectal Dis 14:887–892
81. Vlug MS, Wind J, Hollmann MW et al (2011) Laparoscopy in combination with fast track multimodal management is the best perioperative strategy in patients undergoing colonic surgery: a randomized clinical trial (LAFA-study). Ann Surg 254:868–875
82. Veenhof AAFA, Vlug MS, Van Der Pas MHGM et al (2012) Surgical stress response and postoperative immune function after laparoscopy or open surgery with fast track or standard perioperative care: a randomized trial. Ann Surg 255:216–221

83. Gan TJ, Diemunsch P, Habib AS et al (2014) Consensus guidelines for the management of postoperative nausea and vomiting. Anesth Analg 118:85–113
84. Billeter AT, Hohmann SF, Druen D et al (2014) Unintentional perioperative hypothermia is associated with severe complications and high mortality in elective operations. Surgery 156:1245–1252
85. Hedrick TL, Heckman JA, Smith RL et al (2007) Efficacy of protocol implementation on incidence of wound infection in colorectal operations. J Am Coll Surg 205:432–438
86. Rao W, Zhang X, Zhang J et al (2011) The role of nasogastric tube in decompression after elective colon and rectum surgery: a meta-analysis. Int J Colorectal Dis 26:423–429
87. Verma R, Nelson RL. Prophylactic nasogastric decompression after abdominal surgery. The Cochrane Library. 2007:CD004929
88. Zaouter C, Kaneva P, Carli F (2009) Less urinary tract infection by earlier removal of bladder catheter in surgical patients receiving thoracic epidural analgesia. Reg Anesth Pain M 34:542–548
89. Gustafsson UO, Thorell A, Soop M, Ljungqvist O et al (2009) Haemoglobin A1c as a predictor of postoperative hyperglycaemia and complications after major colorectal surgery. Br J Surg 96:1358–1364
90. Gustafsson UO, Scott MJ, Schwenk W et al (2013) Guidelines for perioperative care in elective colonic surgery: Enhanced Recovery After Surgery (ERAS®) Society recommendations. World J Surg 37:259–284
91. Scott NA, Jeacock J, Kingston RD (1995) Risk factors in patients presenting as an emergency with colorectal cancer. Br J Surg 82:321–323
92. Saunders DI, Murray D, Pichel AC et al (2012) Variations in mortality after emergency laparotomy: the first report of the UK Emergency Laparotomy Network. Br J Anaesth 109:368–375
93. Huddart S, Peden CJ, Swart M et al (2015) Use of a pathway quality improvement care bundle to reduce mortality after emergency laparotomy. Br J Surg 102:57–66
94. Buggy DJ, Borgeat A, Cata J et al (2015) Consensus statement from the BJA Workshop on Cancer and Anaesthesia. Br J Anaesth 114:2–3
95. Myles PS, Peyton P, Silbert B et al (2011) Perioperative epidural analgesia for major abdominal surgery for cancer and recurrence-free survival: randomised trial. BMJ 342:d1491
96. Day A, Smith R, Jourdan I et al (2012) Retrospective analysis of the effect of postoperative analgesia on survival in patients after laparoscopic resection of colorectal cancer. Br J Anaesth 109:185–190

Chapter 7
Perioperative Beta-Blockade, the Pros and Cons. The Story of Beta-Blockade and Cardiac Protection

Helen Higham and Pierre Foëx

Abstract Once considered to confer cardiac protection, perioperative beta-blockade is now regarded as potentially harmful as reduction of cardiac complications is associated with increased all-cause mortality and increased risk of strokes. Over the past decades relatively small studies were not powered enough to detect relatively rare complications, thus all-cause mortality and increased risk of stroke never reached statistical significance, and were discounted. It needed a large randomised controlled trial (RCT) POISE with more than 8,000 patients for these clinically important complications to reach statistical significance. POISE did not address the issue of acute goal-directed administration of beta-blockers when clinically indicated for the management of cardiovascular instability, such as tachycardia, hypertension, myocardial ischaemia, thus practice needs not change. However, initiating beta-blockade for cardiac protection needs to consider the balance of risk and benefits for each individual patient. With considerable caution in respect of beta-blockers, other pharmacological agents need to be considered. While showing promise in previous relatively small studies, clonidine and aspirin have been found in POISE 2, a RCT with over 10,000 patients, to offer no cardiac protection. Clonidine caused hypotension and aspirin increased bleeding. There is limited data and evidence for cardiac protection in respect of calcium channel blockers, ACE inhibitors, angiotensin receptor antagonists and nitroglycerin. However, observational studies and limited RCTs, mostly in cardiac surgery, suggest that statins offer perioperative protection and should be initiated in patients who need them for medical reasons.

Keywords Beta-blockers • Perioperative cardiac protection • Controversy • Guidelines • Clonidine • Aspirin • Statins • Calcium channel blockers • Nitroglycerin • ACE inhibitors • Angiotensin receptor antagonists • Major cardiovascular events • Strokes • All-cause mortality

H. Higham (✉) • P. Foëx
Nuffield Division of Anaesthetics, University of Oxford, Oxford, UK
e-mail: Helen.higham@ndcn.ox.ac.uk; Pierre.foex@nda.ox.ac.uk

K. Stuart-Smith (ed.), *Perioperative Medicine – Current Controversies*,
DOI 10.1007/978-3-319-28821-5_7

7.1 Introduction

For the best part of two decades perioperative beta-blockade was regarded as the best approach to cardiac protection in patients at risk for, or with, coronary heart disease. Large benefits were claimed in some studies [1, 2], and firm recommendations made in guidelines [3] to the extent that perioperative beta-blockade was regarded, in the USA, as an index of quality of care for surgical patients with coronary artery disease.

However, doubts as to the efficacy of perioperative beta-blockade were expressed because some studies did not confirm benefits, and a systematic review failed to demonstrate significant cardiac protection by beta-blockers [4]. This led to the POISE (PeriOperative Ischemic Study Evaluation) trial, a randomised controlled trial (RCT) of beta-blockade versus placebo, far larger than any previous RCT in perioperative cardiac protection [5]. The trial confirmed that beta-blockade protected the heart, yet showed that all-cause mortality, and the rate of stroke, were significantly higher in beta-blocked patients [6].

7.2 The Story of Perioperative Beta-Blockade

7.2.1 The 1960s

In the early 1960s, Sir James Black developed molecules to block the beta-adrenergic receptors. This class of drug was hailed as the most important advance in clinical pharmacology and earned him a share, with Gertrude B Elion and George H Hitchings, of the 1988 Nobel Prize for Medicine. Over and above their role in angina, beta-blockers started to be used in many clinical conditions including arterial hypertension. This meant that by the late 1960s many patients presenting for surgery were receiving beta-blockers. The majority of patients were suffering from coronary heart disease and arterial hypertension. In the late 1960s beta-blockers were considered extremely dangerous in patients undergoing surgery. It was recommended to stop their administration two weeks before surgery whenever possible. This was based on anecdotal evidence of cardiovascular collapse, and a small case series of serious complications [7].

7.2.2 The 1970s

The risks of anaesthesia in patients on beta-blockers remained a major concern in the early 1970s. In the absence of clear data on the interactions between beta-blockade and anaesthesia a detailed haemodynamic study was warranted. An observational study, carried out in 1972, included two arms: (1) the intravenous administration of a beta-blocker (practolol) after induction of anaesthesia, and (2) an oral course of practolol for two days before surgery. All the patients suffered from arterial hypertension [8]. The results of this small study (with a total of 23 patients) were encouraging: beta-blockade had no detrimental effect on blood pressure and cardiac output; there was no evidence of cardiovascular collapse. Moreover, the hypertensive and tachycardic responses to laryngoscopy and intubation were blunted; there were also fewer arrhythmias and less evidence of myocardial ischaemia in the beta-blocked patients [8]. The study was purely observational, yet it was very influential in that the recommendation to stop beta-blockers two weeks before surgery was rapidly abandoned in favour of maintenance of treatment.

7.2.3 The 1980s

In 1988, a small RCT with 128 patients showed that administration of a single dose of a beta-blocker, given to mildly hypertensive patients with their premedication, substantially reduced the risk of perioperative myocardial ischaemia [9]. However, some patients developed hypotension needing treatment. Thus single-dose beta-blockade given with the premedication could not be recommended without a warning about the risk of hypotension.

7.2.4 The 1990s

While the benefits of perioperative beta-blockade were recognised, it is only in 1996 that a randomised controlled trial with 200 patients addressed the question of *cardiovascular outcome* (overall survival and cardiovascular morbidity) after the perioperative administration of a beta-blocker [1]. The patients were at risk for coronary disease and received atenolol for seven days starting the day of surgery. At one and two years, patients who had received the beta-blocker atenolol showed a significantly reduced mortality. This study led to the publication of a guideline strongly recommending perioperative beta-blockade, stating that atenolol should be given to all patients with coronary disease, or risk factors for coronary disease, presenting for non-cardiac surgery [3].

Further support for the benefits of perioperative beta-blockade came from another RCT [2]. Patients with coronary heart disease, confirmed by the demonstration of reversible myocardial ischaemia on dobutamine echocardiography, were randomly allocated to receive the beta-blocker bisoprolol, or a placebo, for at least one week (in practice one month) before surgery; they were maintained on this medication for 30 days postoperatively. This RCT included 112 patients, 59 in the bisoprolol and 53 in the placebo arm of the study, but the study was not blinded. The trial was stopped early because the benefits of beta-blockade were so large that it was considered unethical to continue enrolling patients. This decision was not surprising as, in the bisoprolol treated patients, there was a 100 % reduction of the risk of acute myocardial infarction, and a 90 % reduction of the risk of cardiac death. Retrospectively, the magnitude of the benefits of beta-blockade was considered improbably large [10].

7.2.5 2000–2010

By contrast with the above study, other studies were less supportive of beta-blockade or showed no benefits [11–14]. Some meta-analyses were supporting the benefits of beta-blockade [15, 16], while the largest meta-analysis did not show statistically significant cardiac protection [4]. Thus guidelines became more cautious in their recommendations of perioperative beta-blockade [17].

Doubts about the cardiac protective role of beta-blockers led to POISE [6]. With over 8,400 patients, this RCT had four times as many patients as all the RCTs carried out previously. While the cardiac benefits of slow-release metoprolol given for 30 days starting the day of surgery were unquestionable, the increased all-cause mortality and rate of stroke were also unquestionable. Unsurprisingly, the study showed significantly more episodes of bradycardia and hypotension in the beta-blocked patients than in the placebo group. Hypotension was a correlate of both all-cause mortality and debilitating strokes. Congestive heart failure was not observed in POISE or in other studies [18]. Bradycardia and hypotension, but not heart failure, have been observed whether beta-blockade started before or during surgery [18].

Almost immediately after publication of the results of POISE an updated guideline on the management of patients with heart disease presenting for non-cardiac surgery was published by the American College of Cardiology Foundation/American Heart Association (ACCF/AHA); it restricted the recommendations for initiation of perioperative beta-blockade mostly to high risk patients undergoing high risk surgery [19]. The European Society of Cardiology (ESC) also issued new guidelines [20]. The guideline panel, under the leadership of Dr Poldermans, was much more liberal in its recommendations for perioperative beta-blockade than the ACCF/AHA panel. There were class I recommendations for beta-blockers to be given to patients who have known ischaemic heart disease (IHD) or myocardial ischaemia according to preoperative stress testing (I B), and to patients scheduled for high-risk surgery (I B). Moreover, it stated that beta-blockers should be considered for patients scheduled for intermediate-risk surgery (IIa B), and even those presenting for low-risk surgery (IIb B).

Both sets of guidelines gave a Class I recommendation for maintaining long-term beta-blocker therapy during the perioperative period as beta-blocker withdrawal is known to increase the risk of postoperative adverse cardiac events [21–23]. The differences between the two sets of guidelines are seen in Table 7.1.

POISE was the subject of intense debate and controversy, generating many editorials and comments [24–27]. The main criticisms of POISE were that: (1) the dose of slow-release metoprolol was too high (though no patients in the study

Table 7.1 Recommendations on perioperative beta-blockade according to the 2009 guidelines

2009 ACCF/AHA guideline recommendations	2009 ESC/ESA guideline recommendations
Class I Beta blockers should be continued in patients undergoing surgery who are receiving beta-blockers for treatment of conditions with ACCF/AHA Class I guideline indications for the drugs **C**	**Class I** Beta-blockers are recommended in patients who have known IHD or myocardial ischaemia according to pre-operative stress testing **B** Beta-blockers are recommended in patients scheduled for high-risk surgery **B** Continuation of beta-blockers is recommended in patients previously treated with beta-blockers because of ischaemic heart disease, arrhythmias, or hypertension **C**
Class IIa Beta blockers titrated to heart rate and blood pressure are probably recommended for patients undergoing vascular surgery who are at high cardiac risk owing to coronary artery disease or the finding of cardiac ischemia on preoperative testing **B** Beta blockers titrated to heart rate and blood pressure are reasonable for patients in whom preoperative assessment for vascular surgery identifies high cardiac risk, as defined by the presence of more than 1 clinical risk factor **C** Beta blockers titrated to heart rate and blood pressure are reasonable for patients in whom preoperative assessment identifies coronary artery disease or high cardiac risk, as defined by the presence of more than 1 clinical risk factor, who are undergoing intermediate-risk surgery **B**	**Class IIa** Beta-blockers should be considered for patients scheduled for intermediate-risk surgery **B** Continuation in patients previously treated with beta-blockers because of chronic heart failure with systolic dysfunction should be considered **C**
Class IIb The usefulness of beta blockers is uncertain for patients who are undergoing either intermediate risk procedures or vascular surgery in whom preoperative assessment identifies a single clinical risk factor in the absence of coronary artery disease **C** The usefulness of beta blockers is uncertain in patients undergoing vascular surgery with no clinical risk factors* who are not currently taking beta blockers **B**	**Class IIb** Beta-blockers may be considered in patients scheduled for low-risk surgery with risk factor(s) **B**
Class III Beta blockers should not be given to patients undergoing surgery who have absolute contraindications to beta-blockade **B** Routine administration of high-dose beta-blockers in the absence of dose titration is not useful and may be harmful to patients not currently taking beta blockers who are undergoing noncardiac surgery **B**	**Class III** Perioperative high-dose b-blockers without titration are not recommended **A** Beta-blockers are not recommended in patients scheduled for low-risk surgery without risk factors **B**

Recommendations as classes I, IIa, IIb, III depending upon their strength, and A, B, and C according to the level of evidence

received the maximum dose that was permissible according to the protocol) [25]; (2) the treatment with the beta-blocker should have been started at least 1 week, preferably 1 month before surgery (this is endorsed by the 2009 guidelines [19, 20]); (3) and its effect should have been titrated more carefully in terms of heart rate and blood pressure to prevent excessive bradycardia and hypotension.

A potentially important factor of increased all-cause mortality and debilitating strokes in the metoprolol arm of POISE may have been acute, severe, surgical anaemia. Indeed, in POISE, bleeding was an independent factor of adverse outcome. In a single-center, propensity-matched cohort study, Beattie and colleagues [28] found that adverse cardiac events occurred in 6.5 % of the beta-blocked patients, as opposed to only 3 % in patients who had not received a beta-blocker. The increased rate of adverse outcome in the beta-blocked patients occurred in the face of a greater than 35 % reduction of the baseline haemoglobin concentration. Similarly, in another single-centre study, Le Manach and colleagues found preoperative beta-blocker therapy to be associated with a reduction in postoperative cardiac events, as could be expected. However, in patients who experienced severe bleeding, there was a higher all-cause mortality and an increased frequency of multiorgan dysfunction [29]. An adverse effect of anaemia in the face of beta-blockade may be expected as oxygen delivery to the tissues relies on an increase in cardiac output that is impaired by beta-blockade. Experimentally, beta-blockade reduces cerebral oxygenation and increases the level of the hypoxia inducible factor-1α (HIF-1α) protein [30].

7.2.6 2010 and Beyond

Dr Poldermans was particularly critical of POISE [24, 25]. The debate could have continued for a long time but in 2011 Erasmus University Medical Center dismissed Dr Poldermans because of scientific misconduct [31], as several of the studies from his group were considered potentially unsound [32]. Yet, to this day, no paper has been formally retracted. Dr Poldermans rejected the case made against him [33]. More recent meta-analyses have excluded the papers from Dr Poldermans' group as they were considered insecure [18, 34, 35]. The meta-analysis by Bouri and colleagues highlighted the increased risk of death where beta-blockade is initiated preoperatively and asked for guidelines to be urgently rewritten [34]. Shortly after the on-line publication of this meta-analysis a joint statement was published by ACC/AHA and ESC [35]. It emphasised that beta-blockers should only be given perioperatively after a careful evaluation of risks and benefits in individual patients.

Several recent meta-analyses have tried to identify groups of patients likely to benefit from perioperative initiation of beta-blockade, drawn from the available RCTs, often including the "insecure trials". For example the benefits of beta-blockade were seen only in studies with high risk of bias [36]. Benefits were seen in studies with up-titration of beta-blockade [36], as the best control of heart rate provided the most benefits [37], but with the caveat of risk of side effects such as bradycardia and congestive heart failure [37]. However, another meta-analysis did not find an association between heart rate control and outcome [38]. Separating high risk patients from moderate- and low-risk patients shows that all-cause mortality is significantly increased in 10,367 low and medium risk patients, while there is a trend towards lower mortality in 1,096 high risk patients [18].

Despite all the controversy surrounding the introduction of a beta-blocker at the time of surgery, there was (and still is) unanimity on the need to continue the chronic administration of beta-blockers perioperatively based on observational studies, some very large. There is also evidence that stopping chronic beta-blocker therapy is harmful, causing a three-fold increase in the risk of adverse cardiovascular events [21–23].

7.2.7 Timing of Beta-Blockade

How long before surgery should beta-blockade be started is still unclear. Various durations have been used, or proposed, ranging from the day of surgery as in POISE [6] and most previous RCTs, to 30 days or more in some studies [2, 39]. While previous guidelines recommended initiating beta-blockade, where indicated, 7 to 30 days before surgery [19, 20], in the absence of clear evidence to support a particular duration, the latest guidelines state that treatment should be started at least one day before surgery [40], or, when possible at least 1 week and up to 30 days before surgery, starting with a low dose [41]. This reflects a study using population-based administrative databases by Wijeysundera and colleagues [42]. They analysed data on a cohort of 48,103 patients. The majority (92 %) were on long-term beta-blockade (more than 31 days), 5.5 % had taken a beta-blocker for 8 to 30 days, and 2.3 % for 1 to 7 days. Compared with those treated for more than 31 days, patients in the 1 to 7 days group had a significantly increased 30-day mortality (OR 1.49; CI 1.03–2.6; $p=003$). However there was no association with increased non-fatal myocardial infarction, stroke, and one-year mortality. This suggests that starting beta-blockade only one week before surgery may not be sufficient to reduce the risk of adverse events. Indeed, based on a meta-analysis of 8 RCTs, Dai and colleagues found that starting beta-blockade more than 7 days before surgery improved outcomes [43], but their analysis was not concerned with starting treatment more than 30 days before surgery.

7.2.8 Type of Surgery and Level of Cardiovascular Risk

Because of the risks uncovered by POISE, several authors have tried to consider the type of surgery and the level of cardiovascular risk that would identify patients likely to benefit from perioperative beta-blockade. In a retrospective cohort evaluation of over 136,000 patients, just over 55,000 were exposed to beta-blockers at the time of surgery, mostly patients undergoing vascular surgery. Exposure to beta-blockade was much higher in patients with 4 or more risk factors according to the revised cardiac risk index (RCRI [44]) [45]. In the propensity-matched cohort, exposure to beta-blockade was associated with a significantly reduced 30-day risk of death. The relative risk was much lower in the higher RCRI classes than in those with no or only one risk factor who showed no significant benefit. This confirmed the findings of a previous study by Lindenauer and colleagues [46]. In over 780,000 patients, 122,000 of whom received a beta-blocker, beta-blockade was protective in those with RCRI 2 and above, while it could cause harm in those with RCRI 0 or 1. Importantly and unexpectedly, the study by London and colleagues showed no benefits of beta-blockade in patients undergoing vascular surgery, yet, intuitively, such patients may have been expected to benefit most from beta-blockade [45]. This is important data as previous guidelines recommended initiating beta-blockade in patients undergoing vascular surgery [19]. Finally, a large Danish Nationwide cohort study with over 28,000 patients showed that perioperative beta-blockade benefits only patients with heart failure or recent myocardial infarction [47]. Thus analyses of observational studies to identify patients likely to benefit from beta-blockade rather than be put at risk are likely to continue.

7.2.9 Type of Beta-Blocker

The question of the best beta-blocker for perioperative protection is debated. In the study by London and colleagues atenolol was associated with better outcomes than metoprolol (the drug used in POISE) [45]. This confirms data from Wallace and colleagues [48] who showed greater protection by atenolol than metoprolol in a cohort study with nearly 39,000 Veterans Association Hospitals patients. Of the beta-blockers used in RCTs, only metoprolol is associated with increased all-cause mortality in 10,027 patients [18]. Atenolol appears to reduce mortality and myocardial infarction when compared to metoprolol [43]. This is not surprising since POISE used metoprolol, found an increased mortality, and the study represents the vast majority of the patients in any meta-analysis. Another single-centre cohort study with over 44,000 [49] showed that the risk of perioperative stroke was lower in patients receiving bisoprolol than atenolol or metoprolol. However, in most studies patients were receiving long-term treatment with beta-blockers, thus the conclusions may not apply to the initiation of beta-blockade prior to surgery.

7.2.10 Latest Meta-analyses

The recent meta-analysis by Guay and Ochroch [50] was based on 27 RCTs including those by Mangano et al. [1], Poldermans et al. [2], and Dunkelgrun et al. [39] even though these studies have been criticised for their methodology [1], or concerns about scientific integrity [2, 39]. The authors concluded that addition of a beta-blocker did not increase the risk of 30-day mortality, yet the data showed 167 deaths in the beta-blocked patients and 146 in the placebo group (158 deaths and, respectively, 126 deaths if the criticised studies are excluded). These differences are not statistically significant but demonstrate an adverse trend. The authors found a significantly reduced mortality at one year in those given a beta-blocker (14 deaths vs 28 in the placebo group) but most of deaths (7 vs 16) were in the study by Mangano et al. [1]. The latter was criticised because it allowed patients on long-term beta-blocker therapy to be randomised to placebo, thus generating a group of patients with beta-blocker withdrawal, a known risk factor. Moreover all-cause mortality during the administration of atenolol (or placebo) was not included in the final analysis (4 in the atenolol and 2 in the placebo groups). Thus the data was intrinsically biased in favour of beta-blockade. If this study is excluded then the difference between 7 and 12 deaths in the remaining 3 studies is not significant. Unsurprisingly the meta-analysis confirmed the increased risk of stroke and the reduction of the risk of non-fatal myocardial infarction. Based on three studies, the latest published in 1983, and a total of 97 patients, the authors found no effect of stopping beta-blockers preoperatively but add that the small numbers "preclude drawing any conclusions" [50]. In view of advances in the management of patients with coronary artery disease over the last 25 years a meta-analysis based on three studies [51–53] published more than 30 years ago is unlikely to be relevant to current practice.

The latest meta-analysis by Neng and colleagues [43] shows that in eight RCTs including 11,180 patients, perioperative atenolol was associated with lower mortality. Similarly starting treatment more than one week before surgery was beneficial. Unfortunately this meta-analysis includes one of the insecure studies from Dr Poldermans' group [39].

As the body of evidence from observational studies is quite large, it is worth mentioning the systematic review and meta-analysis of cohort studies by Wan and colleagues [54]. Eight cohort studies were included with a total of 470,000 participants of whom 180,000 were in the beta-blocker group. Perioperative beta-blockers were not associated with significantly reduced risks of mortality (RR 0.88; CI 0.75–1.04), postoperative myocardial infarction (RR 1.30; CI 0.76–2.33), and postoperative stroke (RR 1.17; CI 0.53–2.57). With such large number of patients the absence of cardiac protection, and the known risks seen in RCTs, indicates that a thorough risk/benefit analysis is needed before initiating beta-blockade preoperatively.

As the controversy surrounding POISE was settling new guidelines were prepared and they deserve to be discussed in detail.

7.3 The 2014 Guidelines

In August 2014, the American College of Cardiology, the American Heart Association, and the American Society of Anesthesiologists (ACC/AHA) published their new guidelines on the management of patients with heart disease undergoing

Table 7.2 Recommendations on perioperative beta-blockade according to the 2014 guidelines

2014 ACC/AHA guideline recommendations	2014 ESC/ESA guideline recommendations
Class I Beta-blockers should be continued in patients undergoing surgery who have been on beta-blockers chronically **B**	**Class I** Perioperative continuation of beta-blockers is recommended in patients currently receiving this medication **B**
Class IIa It is reasonable for the management of beta-blockers after surgery to be guided by clinical circumstances, independent of when the agent was started **B**	
Class IIb In patients with intermediate- or high-risk myocardial ischemia noted in preoperative risk stratification tests, it may be reasonable to begin perioperative beta-blockers **C** In patients with 3 or more RCRI risk factors (e.g., diabetes mellitus, heart failure, coronary artery disease, renal insufficiency, cerebrovascular accident), it may be reasonable to begin beta-blockers before surgery **B** In patients with a compelling long-term indication for beta-blocker therapy but no other RCRI risk factors, initiating beta blockers in the perioperative setting as an approach to reduce perioperative risk is of uncertain benefit **B** In patients in whom beta-blocker therapy is initiated, it may be reasonable to begin perioperative beta-blockers long enough in advance to assess safety and tolerability, preferably more than 1 day before surgery **B**	**Class IIb** Pre-operative initiation of beta-blockers may be considered in patients who have known ischaemic heart disease or myocardial ischaemia **B** Pre-operative initiation of beta-blockers may be considered in patients scheduled for high-risk surgery and who have 2 clinical risk factors or ASA status 3 **B** When oral beta-blockade is initiated in patients who undergo non-cardiac surgery, the use of atenolol or bisoprolol as a first choice may be considered **B**
Class III Beta-blocker therapy should not be started on the day of surgery **B**	**Class III** Initiation of peri-operative high-dose beta-blockers without titration is not recommended **B** Pre-operative initiation of beta-blockers is not recommended in patients scheduled for low-risk surgery **B**

Recommendations as classes I, IIa, IIb, III depending upon their strength, and A, B, and C according to the level of evidence

non-cardiac surgery [40]. At the same time the European Society of Cardiology and the European Society of Anesthesiology (ESC/ESA) published theirs [41]. Both comment extensively on perioperative beta-blockade, with some differences in their approach (Table 7.2).

Both sets of guidelines recommend maintaining long-term beta-blocker therapy during the perioperative period (Class I recommendation, Evidence level B). The ACC/AHA guideline adds that the management of beta-blockade after surgery should be guided by clinical circumstances independent of when the agent was started (IIa B). The ESC/ESA guideline recommends initiating beta-blockers in patients scheduled for high-risk surgery who have at least two risk factors or are ASA 3, and in those with known ischaemic heart disease or myocardial ischaemia (Class IIb B). The ACC/AHA considers it reasonable to begin beta-blockers before surgery in patient with 3 or more RCRI risk factors (IIb B), and in patients with high-risk myocardial ischaemia in preoperative tests (IIb C). Thus there are no longer Class I indications for starting beta-blocker therapy before non-cardiac surgery.

If there are compelling long-term indications for beta-blocker therapy, it may seem legitimate to institute treatment before elective surgery. However, the ACC/AHA guideline considers that, in the absence of other RCRI risk factors, initiating beta-blockers to reduce the perioperative risk is of uncertain benefit (IIb B). Moreover, at variance with previous guidelines, vascular surgery is not mentioned as a particular reason for initiating beta-blockade.

As in 2009, guidelines recommend commencing beta-blockers, where indicated, far enough in advance of surgery to assess safety and tolerability and preferably more than one day before surgery (ACC/AHA 2014, IIb B). Without including this as a formal recommendation, the ESC/ESA guideline states that, ideally, treatment, where indicated, should be started between 30 and 2 days before surgery; in addition the target heart rate should be 60–70 bpm and systolic arterial blood pressure more than 100 mmHg. By contrast the ACC/AHA guideline does not indicate targets for titration because of insufficient evidence.

The ESC/ESA guideline recommends, when oral beta-blockade is initiated, using atenolol or bisoprolol as the first choice (IIb B). By contrast, the ACC/AHA guideline makes no specific recommendation.

Both sets of guidelines include class III recommendations, highlighting the risk of harm: beta-blockers should not be started the day of surgery (ACC/AHA, III B). High-dose without titration is not recommended (ESC/ESA III B), and initiation of beta-blockers before low risk surgery is not recommended (ESC/ESA III B).

What the new guidelines highlight again is that continuing long-term beta-blockers is important as there is evidence, in the perioperative and in the non-operative setting, that abrupt withdrawal carries significant risks [21, 22].

By comparison with the 2009 recommendations, it is in the ESC/ESA guideline that the changes relevant to beta-blockers are the most striking. This is not surprising as in 2009 the ESC/ESA guideline there was much emphasis on the results of the DECREASE studies [2, 39] from Dr Poldermans' group, leading the panel to promote perioperative beta-blockade despite the risks highlighted by POISE.

In the light of recent evidence, the new guidelines recommend that only patients at substantial cardiac risk undergoing intermediate or high-risk surgery should be considered for initiation of perioperative beta-blockade. If indicated it makes sense to start treatment well ahead or surgery to assess its safety and adjust the dose as a function of the patient's response. While the ESC/ESA guideline indicates targets for heart rate and blood pressure, these are not based on strong evidence. Indeed it has been shown that a low target for heart rate is associated with cardiac protection but carries the risk of excessive bradycardia, congestive heart failure [37] and hypotension. A target blood pressure higher than 100 mmHg before the next dose of beta-blocker is recommended, yet it is not evidence-based. It was the limit in POISE and it can be argued that the limit was too low. In high cardiac risk patients 100 mmHg systolic blood pressure, as one size fits-all, is difficult to justify. Even today many patients present with arterial hypertension that is not fully controlled to less that 140 mmHg systolic and less that 80 mmHg diastolic blood pressure. The latest JNC (JNC8)[55] proposes 150 mmHg as the goal for patients above the age of 65 in daily clinical practice. In such patients is it legitimate to give another dose of a beta-blocker when their systolic blood pressure is only 100 mmHg? One could easily argue that it is not legitimate and that the target blood pressure should take into consideration the patient's everyday blood pressure. This is why the ACC/AHA guideline does not recommend particular targets for heart rate and blood pressure, but draws attention to the need to prevent hypotension.

As the indications for using beta-blockers to protect the heart against perioperative cardiovascular complications of non-cardiac surgery are now very limited, it may be appropriate to consider briefly what is known in cardiac surgery, based on a recent Cochrane Collaboration review of 53 randomised controlled trials [18]. The meta-analysis showed that beta-blockade had no influence on all-cause mortality, acute myocardial infarction, and cerebrovascular accidents, based on studies including between 1700 and 3983 patients. For other adverse events such as myocardial ischaemia, bradycardia, hypotension and congestive heart failure beta-blockade did not influence these outcomes, but the number of patients included in the available studies was low (between 166 and 660). Therefore firm conclusions cannot be drawn. On the positive side, beta-blockade reduced the risk of ventricular and supraventricular arrhythmias and reduced the length of hospital stay in studies with between 2292 and 6420 patients. Thus, the role of beta-blockade in cardiac surgery is relatively limited but at least there is no evidence of increased all-cause mortality.

7.3.1 Influence of the Passing of Time

It is not surprising that the protective effects of beta-blockers have changed over the past four decades. Over that period treatment of arterial hypertension and coronary artery disease, including myocardial infarction, has changed beyond recognition. Thus the context of recent studies is very different from that of earlier studies. Two studies that showed cardiac protection [1, 2] even led to the view that preoperative investigations may not be essential, as using a clinical risk index (such as the RCRI) to identify the very high risk patients, and initiating beta-blockade preoperatively may reduce the risk of adverse cardiac events substantially [56]. However, these very influential studies were published 20 [1] and respectively 16 [2] years ago and have been justifiably criticised. Moreover recent studies have been much larger than older studies. Thus, it is not surprising that a class of drug that appeared highly protective has now been found to put patients at risk. Thus, today the risks and benefits need to be evaluated for each patient, with early start where indicated and clear protocols for administration of successive doses.

7.3.2 Goal Directed Beta-Blockade

All RCTs, including POISE, and observational studies were concerned with patients receiving beta-blockers because they were at risk for, or suffered from, coronary artery disease. These studies were not concerned with the administration of a beta-blocker *where it is indicated* by haemodynamic instability, or ischaemia, attributable to sympathetic overactivity. In such circumstances, the indications for beta-blockers remain therefore unchanged and their use is legitimate as beta-blockers are titrated to effect by the anaesthetist in the theatres or recovery wards, hence under close medical supervision [57].

As beta-blockade has only limited indications in non-cardiac surgery, what are the other pharmacological options?

7.4 Alpha$_2$-Adrenoceptor Agonists

Alpha$_2$-adrenoceptor agonists reduce post-ganglionic noradrenaline output, thereby reducing the catecholamine output surge during the perioperative period. Thus they could be expected to protect the heart. Over the years several studies have shown benefits of this class of drugs. Studies used clonidine, dexmetedomidine, or mivazerol. The largest trial was the European mivazerol trial. It randomized 1,897 patients with ischaemic heart disease who underwent intermediate- or high-risk surgery. It is only in the subpopulation of 904 patients undergoing vascular surgery that mivazerol significantly reduced the risk of postoperative death or myocardial infarction [58]. As protection was seen only in patients undergoing vascular surgery, mivazerol was not commercialized.

A meta-analysis comprising 23 trials enrolling a total of 3,395 patients who were randomised to receive clonidine, dexmedetomidine, mivazerol or a placebo showed overall both mortality [RR 0.64 (CI 0.42-0.99)] and myocardial ischaemia [RR 0.76 (CI 0.63–0.91)] to be significantly reduced. Mortality and myocardial infarction were reduced in vascular surgical patients, and mortality in cardiac surgery patients [59]. In 2004 a relatively small RCT with190 patients showed clonidine to reduce perioperative myocardial ischaemia, and to reduce long-term mortality over the two years following surgery [60]. In view of the relatively limited evidence, based mostly on small trials, with the exception of the mivazerol trial, a large randomized controlled trial was needed.

POISE-2 randomized 10,010 patients undergoing non-cardiac surgery to low-dose clonidine or placebo, or aspirin or placebo, in a 2x2 factorial design. In the clonidine arm of the study, patients received 0.2 mg of oral clonidine or placebo and also a transdermal clonidine patch which releases 0.2 mg per day and has physiological effects within 24 h, or a patch placebo, on the day of surgery. The patch remained in place until 72 h after surgery. Clonidine did not reduce the risk of death or non-fatal myocardial infarction. However, low dose clonidine significantly increased the risk of important hypotension and non-fatal cardiac arrest [61].

Not surprisingly, both the 2014 ACC/AHA [40] and ESC/ESA [41] guidelines conclude that alpha$_2$-agonists for the prevention of cardiac perioperative cardiac events are not recommended (Class III B).

7.5 Aspirin

The benefits of aspirin in the prevention of myocardial infarction and stroke in daily life are well known. A meta-analysis of data from large randomized controlled trials in primary and in secondary prevention, involving more than 110,000 patients confirmed the efficacy of aspirin [62]. In primary prevention trials the use of aspirin was shown to reduce the risk of serious cardiovascular events by 12 %, and myocardial infarction by nearly 20 %. In secondary prevention trials the benefits were 18 % reductions in serious vascular events, strokes and coronary events. As could be expected aspirin increased the risk of gastro-intestinal bleeding, thus reducing its benefits in primary prevention.

Among patients undergoing non-cardiac surgery, there is uncertainty regarding the risks and benefits of aspirin [63, 64]. A meta-analysis, including 41 studies with 49,590 patients, which compared peri-procedural withdrawal vs. bleeding risks of aspirin, concluded that the risk of bleeding with aspirin therapy was increased by 50 %, yet without greater severity of bleeding complications [64].

Relatively limited evidence for the management of aspirin therapy in non-cardiac surgical patients was the rationale for the aspirin arm of POISE 2 [65]. Patients started to take a dose of aspirin of 200 mg just before surgery and continued with a daily dose of 100 mg for 30 days. If they were already on aspirin they continued the trial regime for 7 days after which they resumed their regular regimen. The primary outcome (composite of death or non-fatal myocardial infarction 30 days after randomization) and secondary outcome (myocardial infarction) were the same in both arms of the trial and, unsurprisingly, there was more bleeding in the patients who had received aspirin. The recent guidelines do not recommend initiating aspirin administration with the aim of reducing the risk of cardiac events. Moreover they even question the value of maintaining aspirin therapy unless there is a clear indication such as carotid endarterectomy. Thus the ACC/AHA guideline considers that initiation or continuation of aspirin is not beneficial in patients undergoing elective non-cardiac, non-carotid surgery, and those who have not received coronary stents, unless the risk of ischaemic events outweighs the risk of surgical bleeding (III C) [40]. The ESC/ESA guideline is a little more supportive of aspirin and concludes that continuation of aspirin, in patients previously thus treated, may be considered in the perioperative period, and should be based on an individual decision that depends on the perioperative bleeding risk, weighed against the risk of thrombotic complications (IIb B), while recommending to consider the discontinuation of aspirin therapy, in patients previously treated with it, where haemostasis is anticipated to be difficult to control during surgery (IIa B) [41]. With the evidence currently available aspirin should not be initiated solely for cardiac protection and its perioperative maintenance should depend upon the balance of risk of thrombotic events (carotid endarterectomy, presence of coronary stents) and excess bleeding.

7.6 Statins

Lipid lowering with statin agents is highly effective for primary and secondary prevention of cardiac events even in patients considered at relatively low risk [66, 67]. The effectiveness of this class of agents in reducing cardiovascular events in high-risk patients suggests that they may improve perioperative cardiovascular outcomes. There are only three randomised controlled trials of the initiation of statins prior to non-cardiac surgery, two are from the group of Dr Poldermans and may be regarded as insecure. Two meta-analyses have reported the effects of the introduction of statins in cardiac and non-cardiac surgery, and in the pre-procedural period for coronary interventions [68, 69]. Taken all together, these RCTs show that initiation of statins before procedures, or surgery, reduces cardiovascular risks. However, for non-cardiac surgery, most of the evidence comes from observational studies suggesting that patients on long-term statin treatment suffer fewer perioperative cardiovascular complications than patients not on statins [70–74]. The mechanisms of statin therapy prescribed perioperatively to lower cardiac events are unclear and may relate to their pleiotropic, as well as cholesterol-lowering effects. However, the authors of a Cochrane Systematic Review on statins in patients undergoing vascular surgery found insufficient evidence to conclude that statins' use resulted in either a reduction or an increase in any of the outcomes examined [75].

The ACC/AHA guideline recommends that statins should be continued in patients currently taking statins and scheduled for noncardiac surgery (I B) and considers that perioperative initiation of statin use is reasonable in patients undergoing vascular surgery (IIa B), and may be considered in patients with clinical indications who are undergoing elevated-risk procedures (IIb C). The guideline adds that in patients meeting indications for statin therapy, starting this medication perioperatively may also be an opportunity to impact long-term health [40].

The ESC/ESA guideline makes a potentially important point. It recommends peri-operative continuation of statins and favours using statins with a long half-life, or extended-release formulation (I C). As there are no parenteral statins it is logical to prefer an extended-release formulation in patients who may not be able to take oral medication. Both the ESC/ESA guideline and the ACC/AHA guidelines recommend that preoperative initiation of statin therapy should be considered in patients undergoing vascular surgery, ideally at least 2 weeks before surgery (IIa B).

There is no evidence to decide how long before surgery statins should be started if there is an indication for their initiation. It is probably sensible to start 2–4 weeks before surgery. If initiated before surgery they should be continued for long-term cardiovascular protection.

7.7 Calcium Channel Blockers

Calcium overload plays an important role in myocardial cell damage. It could be anticipated that blockers of myocardial calcium channels (verapamil, diltiazem) may offer cardiac protection. Calcium channel blockers have been shown, in a meta-analysis, to reduce myocardial ischaemia, supraventricular tachycardia, and the combined death and myocardial infarction rates, effects mostly attributable to diltiazem [76]. The vascular selective dihydropyridines are not expected to offer cardiac protection. On the contrary, a single-centre study in patients undergoing surgery of the abdominal aorta showed use of dihydropirydine to increase perioperative mortality [77].

The ACC/AHA guideline does not make recommendations in respect of calcium channel blockers in the perioperative period. The ESC/ESA guideline considers that the use of short-acting dihydropyridines, in particular, nifedipine capsules should be avoided. It adds that heart rate-reducing calcium channel blockers are not indicated in patients with heart failure and systolic dysfunction, as they have a substantial negative inotropic effect and can precipitate or worsen heart failure. However, continuation or introduction of heart rate-reducing calcium channel blockers may be considered in patients who do not tolerate beta-blockers.

7.8 ACE Inhibitors and Angiotensin Receptor Antagonists

Observational studies have shown that patients on ACE inhibitors or angiotensin receptor blockers (ARBs) have more frequent transient intraoperative hypotension, but no difference in important outcomes such as death, myocardial infarction, stroke or renal failure [78, 79]. In the absence of protection, or hazard, the ACC/AHA guideline recommends to continue treatment with ACE inhibitors and ARBs perioperatively (IIa B). If they are withheld before surgery it is recommended to restart as soon as clinically feasible postoperatively (IIa C).

The ESC/ESA guideline recognises that perioperative use of ACE inhibitors or ARBs carries the risk of severe hypotension, especially following induction of anaesthesia. This adverse effect is less frequent when ACE inhibitors are discontinued the day before surgery, hence ACE inhibitors withdrawal should be considered 24 h before surgery when they are prescribed for hypertension, and resumed after surgery as soon as blood volume and arterial pressure are stable (IIa C). With ARBs hypotension occurs at least as much as with ACE inhibitors, and the response to vasopressors may be impaired.

In patients with stable left ventricular systolic dysfunction it appears reasonable to continue treatment with ACE inhibitors under close monitoring (IIa C). By contrast if LV dysfunction is discovered during the pre-operative evaluation in untreated patients in a stable condition, surgery should, if possible, be postponed, to allow for diagnosis of the underlying cause and the introduction of ACE inhibitors.

7.9 Nitroglycerin

For many years, nitroglycerin has been used successfully to reverse myocardial ischaemia. However, the effect of perioperative intravenous nitroglycerin on perioperative ischaemia is debated. Nitroglycerin does not decrease perioperative myocardial infarction or cardiac death. The ESC/ESA does not make specific recommendations but stresses that its use may, through decreased preload, lead to tachycardia and hypotension, both of which are undesirable. The ACC/AHA guideline states that prophylactic intravenous nitroglycerin is not effective in reducing myocardial ischaemia in patients undergoing surgery and is of no benefit in non-cardiac surgery (III B). However, it states that this view is based on the absence of significant recent studies (within the last 10 years), while prior RCTs were small and yielded conflicting results. Thus there is no evidence for its use.

7.10 Conclusions

Several classes of drugs, beta-blockers, alpha$_2$-adrenoceptor agonists, aspirin, calcium channel blockers, and statins appeared to confer perioperative cardiac protection over the past four decades. Unfortunately RCTs were often small and, as a result, complications of treatment did not reach statistical significance, even though there may have been a clear trend toward them. With considerably larger studies, like POISE and POISE 2, the complications of treatment with beta-blockers, aspirin, and clonidine, were clinically and statistically significant. Thus the current guidelines can only recommend considering beta-blockers very cautiously in high-risk patients and to consider statins in patients who would normally receive this type of medication over the long-term. There may well be a shift of emphasis toward more systematic monitoring for perioperative myocardial ischaemia and myocardial injury in non-cardiac surgery (MINS)[80] using fourth generation troponin assays [81], with aggressive management of patients with clear demonstration of myocardial injury. However, there are not yet agreed protocols for the management of MINS.

References

1. Mangano DT, Layug EL, Wallace A, Tateo I (1996) Effect of atenolol on mortality and cardiovascular morbidity after noncardiac surgery. Multicenter Study of Perioperative Ischemia Research Group. N Engl J Med 335(23):1713–1720
2. Poldermans D, Boersma E, Bax JJ, Thomson IR, van-de-Ven LL, Blankensteijn JD et al (1999) The effect of bisoprolol on perioperative mortality and myocardial infarction in high-risk patients undergoing vascular surgery. Dutch Echocardiographic Cardiac Risk Evaluation Applying Stress Echocardiography Study Group. N Engl J Med 341(24):1789–1794
3. Palda VA, Detsky AS (1997) Perioperative assessment and management of risk from coronary artery disease. Ann Intern Med 127(4):313–328

4. Devereaux PJ, Beattie WS, Choi PT, Badner NH, Guyatt GH, Villar JC et al (2005) How strong is the evidence for the use of perioperative beta blockers in non-cardiac surgery? Systematic review and meta-analysis of randomised controlled trials. BMJ 331(7512):313–321
5. Devereaux PJ, Yang H, Guyatt GH, Leslie K, Villar JC, Monteri VM et al (2006) Rationale, design, and organization of the PeriOperative ISchemic Evaluation (POISE) trial: a randomized controlled trial of metoprolol versus placebo in patients undergoing noncardiac surgery. Am Heart J 152(2):223–230
6. Devereaux PJ, Yang H, Yusuf S, Guyatt G, Leslie K, Villar JC et al (2008) Effects of extended-release metoprolol succinate in patients undergoing non-cardiac surgery (POISE trial): a randomised controlled trial. Lancet 371(9627):1839–1847
7. Viljoen JF, Estafanous FG, Kellner GA (1972) Propranolol and cardiac surgery. J Thorac Cardiovasc Surg 64(5):826–830
8. Prys Roberts C, Foex P, Biro GP, Roberts JG (1973) Studies of anaesthesia in relation to hypertension. V. Adrenergic beta-receptor blockade. Br J Anaesth 45(7):671–681
9. Stone JG, Foex P, Sear JW, Johnson LL, Khambatta HJ, Triner L (1988) Myocardial ischemia in untreated hypertensive patients: effect of a single small oral dose of a beta-adrenergic blocking agent. Anesthesiology 68(4):495–500
10. Cole GD, Francis DP (2014) Perioperative beta blockade: guidelines do not reflect the problems with the evidence from the DECREASE trials. BMJ 349:g5210
11. Juul AB, Wetterslev J, Gluud C, Kofoed-Enevoldsen A, Jensen G, Callesen T et al (2006) Effect of perioperative beta blockade in patients with diabetes undergoing major non-cardiac surgery: randomised placebo controlled, blinded multicentre trial. BMJ 332(7556):1482–1485
12. Brady AR, Gibbs JS, Greenhalgh RM, Powell JT, Sydes MR (2005) Perioperative beta-blockade (POBBLE) for patients undergoing infrarenal vascular surgery: results of a randomized double-blind controlled trial. J Vasc Surg 41(4):602–609
13. Yang H, Raymer K, Butler R, Parlow J, Roberts R (2006) The effects of perioperative beta-blockade: results of the Metoprolol after Vascular Surgery (MaVS) study, a randomized controlled trial. Am Heart J 152(5):983–990
14. Zaugg M, Bestmann L, Wacker J, Lucchinetti E, Boltres A, Schulz C et al (2007) Adrenergic receptor genotype but not perioperative bisoprolol therapy may determine cardiovascular outcome in at-risk patients undergoing surgery with spinal block: the Swiss Beta Blocker in Spinal Anesthesia (BBSA) study: a double-blinded, placebo-controlled, multicenter trial with 1-year follow-up. Anesthesiology 107(1):33–44
15. Auerbach AD, Goldman L (2002) Beta-blockers and reduction of cardiac events in noncardiac surgery: scientific review. JAMA 287(11):1435–1444
16. Stevens RD, Burri H, Tramer MR (2003) Pharmacologic myocardial protection in patients undergoing noncardiac surgery: a quantitative systematic review. Anesth Analg 97(3):623–633
17. Fleisher LA, Beckman JA, Brown KA, Calkins H, Chaikof E, Fleischmann KE et al (2006) ACC/AHA 2006 guideline update on perioperative cardiovascular evaluation for noncardiac surgery: focused update on perioperative beta-blocker therapy: a report of the American College of Cardiology/American Heart Association Task Force on Practice Guidelines (Writing Committee to Update the 2002 Guidelines on Perioperative Cardiovascular Evaluation for Noncardiac Surgery): developed in collaboration with the American Society of Echocardiography, American Society of Nuclear Cardiology, Heart Rhythm Society, Society of Cardiovascular Anesthesiologists, Society for Cardiovascular Angiography and Interventions, and Society for Vascular Medicine and Biology. Circulation 113(22):2662–2674
18. Blessberger H, Kammler J, Domanovits H, Schlager O, Wildner B, Azar D et al (2014) Perioperative beta-blockers for preventing surgery-related mortality and morbidity. Cochrane Database Syst Rev 9:CD004476.
19. Fleisher LA, Beckman JA, Brown KA, Calkins H, Chaikof EL, Fleischmann KE et al (2009) 2009 ACCF/AHA focused update on perioperative beta blockade incorporated into the ACC/AHA 2007 guidelines on perioperative cardiovascular evaluation and care for noncardiac surgery: a report of the American college of cardiology foundation/American heart association task force on practice guidelines. Circulation 120(21):e169–e276

20. Poldermans D, Bax JJ, Boersma E, De Hert S, Eeckhout E, Fowkes G et al (2009) Guidelines for pre-operative cardiac risk assessment and perioperative cardiac management in non-cardiac surgery: the Task Force for Preoperative Cardiac Risk Assessment and Perioperative Cardiac Management in Non-cardiac Surgery of the European Society of Cardiology (ESC) and endorsed by the European Society of Anaesthesiology (ESA). Eur Heart J 30(22): 2769–2812
21. Shammash JB, Trost JC, Gold JM, Berlin JA, Golden MA, Kimmel SE (2001) Perioperative beta-blocker withdrawal and mortality in vascular surgical patients. Am Heart J 141(1):148–153
22. Hoeks SE, Scholte Op Reimer WJ, van Urk H, Jorning PJ, Boersma E, Simoons ML et al (2007) Increase of 1-year mortality after perioperative beta-blocker withdrawal in endovascular and vascular surgery patients. Eur J Vasc Endovasc Surg 33(1):13–19
23. Wallace AW, Au S, Cason BA (2010) Association of the pattern of use of perioperative beta-blockade and postoperative mortality. Anesthesiology 113(4):794–805
24. Fleisher LA, Poldermans D (2008) Perioperative beta blockade: where do we go from here? Lancet 371(9627):1813–1814
25. Poldermans D, Devereaux PJ (2009) The experts debate: perioperative beta-blockade for noncardiac surgery – proven safe or not? Cleve Clin J Med 76(Suppl 4):S84–S92 [Research Support, Non-U.S. Gov't]
26. Sear JW, Giles JW, Howard-Alpe G, Foex P (2008) Perioperative beta-blockade, 2008: what does POISE tell us, and was our earlier caution justified? Br J Anaesth 101(2):135–138
27. Foex P, Sear JW (2014) II. Beta-blockers and cardiac protection: 5 yr on from POISE. Br J Anaesth 112(2):206–210
28. Beattie WS, Wijeysundera DN, Karkouti K, McCluskey S, Tait G, Mitsakakis N et al (2010) Acute surgical anemia influences the cardioprotective effects of beta-blockade: a single-center, propensity-matched cohort study. Anesthesiology 112(1):25–33
29. Le Manach Y, Collins GS, Ibanez C, Goarin JP, Coriat P, Gaudric J et al (2012) Impact of perioperative bleeding on the protective effect of beta-blockers during infrarenal aortic reconstruction. Anesthesiology 117(6):1203–1211
30. Ragoonanan TE, Beattie WS, Mazer CD, Tsui AK, Leong-Poi H, Wilson DF et al (2009) Metoprolol reduces cerebral tissue oxygen tension after acute hemodilution in rats. Anesthesiology 111(5):988–1000
31. Center EM (2011) Erasmus MC dismisses professor. http://erasmusmc.nl/corp_home/corp_news-center/2011/2011-11/ontslag.hoogleraar/?lang=en
32. Center EM (2012) Report of the 2012 follow-up investigation of possible breaches of academic integrity 2012. http://www.erasmusmc.nl/5663/135857/3675250/3706798/integrity_report_2012-10.pdf?lang=en&lang=en
33. Poldermans D (2013) Scientific fraud or a rush to judgment? Am J Med 126(4):e5–e6 [Comment Letter]
34. Bouri S, Shun-Shin MJ, Cole GD, Mayet J, Francis DP (2014) Meta-analysis of secure randomised controlled trials of beta-blockade to prevent perioperative death in non-cardiac surgery. Heart 100(6):456–464 [Meta-Analysis Research Support, Non-U.S. Gov't Review]
35. Wijeysundera DN, Duncan D, Nkonde-Price C, Virani SS, Washam JB, Fleischer L; ACC/AHA Task force members (2014) Perioperative beta blockade in noncardiac surgery: a systematic review for the 2014 ACC/AHA Guideline on perioperative cardiovascular evaluation and management of patients undergoing noncardiac surgery: a report of the American College of Cardiology/American Heart Association Task Force on Practice Guidelines. Circulation 130(24):2246–2264
36. Bangalore S, Wetterslev J, Pranesh S, Sawhney S, Gluud C, Messerli FH (2008) Perioperative beta blockers in patients having non-cardiac surgery: a meta-analysis. Lancet 372(9654):1962–1976
37. Beattie WS, Wijeysundera DN, Karkouti K, McCluskey S, Tait G (2008) Does tight heart rate control improve beta-blocker efficacy? An updated analysis of the noncardiac surgical randomized trials. Anesth Analg 106(4):1039–1048

38. Biccard BM, Sear JW, Foex P (2008) Meta-analysis of the effect of heart rate achieved by perioperative beta-adrenergic blockade on cardiovascular outcomes. Br J Anaesth 100(1):23–28
39. Dunkelgrun M, Boersma E, Schouten O, Koopman-van Gemert AW, van Poorten F, Bax JJ et al (2009) Bisoprolol and fluvastatin for the reduction of perioperative cardiac mortality and myocardial infarction in intermediate-risk patients undergoing noncardiovascular surgery: a randomized controlled trial (DECREASE-IV). Ann Surg 249(6):921–926 [Comparative Study Meta-Analysis Randomized Controlled Trial Research Support, Non-U.S. Gov't]
40. Fleisher LA, Fleischmann KE, Auerbach AD, Barnason SA, Beckman JA, Bozkurt B et al (2014) 2014 ACC/AHA guideline on perioperative cardiovascular evaluation and management of patients undergoing noncardiac surgery: a report of the American College of Cardiology/American Heart Association Task Force on Practice Guidelines. J Am Coll Cardiol 64(22):e77–e137
41. Kristensen SD, Knuuti J, Saraste A, Anker S, Botker HE, Hert SD et al (2014) 2014 ESC/ESA guidelines on non-cardiac surgery: cardiovascular assessment and management: the Joint Task Force on non-cardiac surgery: cardiovascular assessment and management of the European Society of Cardiology (ESC) and the European Society of Anaesthesiology (ESA). Eur Heart J 35(35):2383–2431
42. Wijeysundera DN, Beattie WS, Wijeysundera HC, Yun L, Austin PC, Ko DT (2014) Duration of preoperative beta-blockade and outcomes after major elective noncardiac surgery. Can J Cardiol 30(2):217–223
43. Dai N, Xu D, Zhang J, Wei Y, Li W, Fan B et al (2014) Different beta-blockers and initiation time in patients undergoing noncardiac surgery: a meta-analysis. Am J Med Sci 347(3):235–244
44. Lee TH, Marcantonio ER, Mangione CM, Thomas EJ, Polanczyk CA, Cook EF et al (1999) Derivation and prospective validation of a simple index for prediction of cardiac risk of major noncardiac surgery. Circulation 100(10):1043–1049
45. London MJ, Hur K, Schwartz GG, Henderson WG (2013) Association of perioperative beta-blockade with mortality and cardiovascular morbidity following major noncardiac surgery. JAMA 309(16):1704–1713 [Research Support, Non-U.S. Gov't]
46. Lindenauer PK, Pekow P, Wang K, Mamidi DK, Gutierrez B, Benjamin EM (2005) Perioperative beta-blocker therapy and mortality after major noncardiac surgery. N Engl J Med 353(4):349–361
47. Andersson C, Merie C, Jorgensen M, Gislason GH, Torp-Pedersen C, Overgaard C et al (2014) Association of beta-blocker therapy with risks of adverse cardiovascular events and deaths in patients with ischemic heart disease undergoing noncardiac surgery: a Danish nationwide cohort study. JAMA Intern Med 174(3):336–344 [Research Support, Non-U.S. Gov't]
48. Wallace AW, Au S, Cason BA (2011) Perioperative beta-blockade: atenolol is associated with reduced mortality when compared to metoprolol. Anesthesiology 114(4):824–836 [Comparative Study Research Support, Non-U.S. Gov't Research Support, U.S. Gov't, Non-P.H.S.]
49. Ashes C, Judelman S, Wijeysundera DN, Tait G, Mazer CD, Hare GM et al (2013) Selective beta1-antagonism with bisoprolol is associated with fewer postoperative strokes than atenolol or metoprolol: a single-center cohort study of 44,092 consecutive patients. Anesthesiology 119(4):777–787
50. Guay J, Ochroch EA (2013) Beta-blocking agents for surgery: influence on mortality and major outcomes. A meta-analysis. J Cardiothorac Vasc Anesth 27(5):834–844
51. Kirsh MM, Behrendt DM, Jackson AP, Dhadphale P, Alseri S, Brymer J et al (1978) Myocardial revascularization in patients receiving long-term propranolol therapy. Ann Thorac Surg 25(2):117–121
52. Wechsler AS (1980) Assessment of prospectively randomized patients receiving propranolol therapy before coronary bypass operation. Ann Thorac Surg 30(2):128–136
53. Ponten J, Haggendal J, Milocco I, Waldenstrom A (1983) Long-term metoprolol therapy and neuroleptanesthesia in coronary artery surgery: withdrawal versus maintenance of beta 1-adrenoreceptor blockade. Anesth Analg 62(4):380–390

54. Wan YD, Zhang SG, Sun TW, Kan QC, Wang LX (2014) The effects of perioperative beta-blockers on mortality in patients undergoing non-cardiac surgery in real world: a meta-analysis of cohort studies. Int J Cardiol 176(3):605–610
55. James PA, Oparil S, Carter BL, Cushman WC, Dennison-Himmelfarb C, Handler J et al (2014) 2014 evidence-based guideline for the management of high blood pressure in adults: report from the panel members appointed to the Eighth Joint National Committee (JNC 8). JAMA 311(5):507–520 [Research Support, N.I.H., Extramural]
56. Kelion AD, Banning AP (1999) Is simple clinical assessment adequate for cardiac risk stratification before elective non-cardiac surgery? Lancet 354(9193):1837–1838
57. Stone JG, Khambatta HJ, Sear JW, Foex P (2009) Beta-blockers: must we throw the baby out with the bath water? Anesth Analg 108(6):1987–1990
58. Oliver MF, Goldman L, Julian DG, Holme I (1999) Effect of mivazerol on perioperative cardiac complications during non-cardiac surgery in patients with coronary heart disease: the European Mivazerol Trial (EMIT). Anesthesiology 91(4):951–961
59. Wijeysundera DN, Naik JS, Beattie WS (2003) Alpha-2 adrenergic agonists to prevent perioperative cardiovascular complications: a meta-analysis. Am J Med 114(9):742–752
60. Wallace AW, Galindez D, Salahieh A, Layug EL, Lazo EA, Haratonik KA et al (2004) Effect of clonidine on cardiovascular morbidity and mortality after noncardiac surgery. Anesthesiology 101(2):284–293
61. Devereaux PJ, Sessler DI, Leslie K, Kurz A, Mrkobrada M, Alonso-Coello P et al (2014) Clonidine in patients undergoing noncardiac surgery. N Engl J Med 370(16):1504–1513 [Research Support, Non-U.S. Gov't]
62. Baigent C, Blackwell L, Collins R, Emberson J, Godwin J, Peto R et al (2009) Aspirin in the primary and secondary prevention of vascular disease: collaborative meta-analysis of individual participant data from randomised trials. Lancet 373(9678):1849–1860
63. Neilipovitz DT, Bryson GL, Nichol G (2001) The effect of perioperative aspirin therapy in peripheral vascular surgery: a decision analysis. Anesth Analg 93(3):573–580
64. Burger W, Chemnitius JM, Kneissl GD, Rucker G (2005) Low-dose aspirin for secondary cardiovascular prevention – cardiovascular risks after its perioperative withdrawal versus bleeding risks with its continuation – review and meta-analysis. J Intern Med 257(5):399–414
65. Devereaux PJ, Mrkobrada M, Sessler DI, Leslie K, Alonso-Coello P, Kurz A et al (2014) Aspirin in patients undergoing noncardiac surgery. N Engl J Med 370(16):1494–1503
66. Mihaylova B, Emberson J, Blackwell L, Keech A, Simes J, Barnes EH et al (2012) The effects of lowering LDL cholesterol with statin therapy in people at low risk of vascular disease: meta-analysis of individual data from 27 randomised trials. Lancet 380(9841):581–590
67. Ridker PM, Wilson PW (2013) A trial-based approach to statin guidelines. JAMA 310(11):1123–1124
68. Winchester DE, Wen X, Xie L, Bavry AA (2010) Evidence of pre-procedural statin therapy a meta-analysis of randomized trials. J Am Coll Cardiol 56(14):1099–1109
69. Chopra V, Wesorick DH, Sussman JB, Greene T, Rogers M, Froehlich JB et al (2012) Effect of perioperative statins on death, myocardial infarction, atrial fibrillation, and length of stay: a systematic review and meta-analysis. Arch Surg 147(2):181–189 [Meta-Analysis Research Support, Non-U.S. Gov't Review]
70. Lindenauer PK, Pekow P, Wang K, Gutierrez B, Benjamin EM (2004) Lipid-lowering therapy and in-hospital mortality following major noncardiac surgery. JAMA 291(17):2092–2099
71. Kennedy J, Quan H, Buchan AM, Ghali WA, Feasby TE (2005) Statins are associated with better outcomes after carotid endarterectomy in symptomatic patients. Stroke 36(10):2072–2076
72. Hindler K, Shaw AD, Samuels J, Fulton S, Collard CD, Riedel B (2006) Improved postoperative outcomes associated with preoperative statin therapy. Anesthesiology 105(6):1260–1272; quiz 89–90
73. Desai H, Aronow WS, Ahn C, Gandhi K, Amin H, Lai HM et al (2010) Incidence of perioperative myocardial infarction and of 2-year mortality in 577 elderly patients undergoing noncardiac vascular surgery treated with and without statins. Arch Gerontol Geriatr 51(2):149–151

74. Raju MG, Pachika A, Punnam SR, Gardiner JC, Shishehbor MH, Kapadia SR et al (2013) Statin therapy in the reduction of cardiovascular events in patients undergoing intermediate-risk noncardiac, nonvascular surgery. Clin Cardiol 36(8):456–461
75. Sanders RD, Nicholson A, Lewis SR, Smith AF, Alderson P (2013) Perioperative statin therapy for improving outcomes during and after noncardiac vascular surgery. Cochrane Database Syst Rev 7:CD009971.
76. Wijeysundera DN, Beattie WS (2003) Calcium channel blockers for reducing cardiac morbidity after noncardiac surgery: a meta-analysis. Anesth Analg 97(3):634–641
77. Kertai MD, Westerhout CM, Varga KS, Acsady G, Gal J (2008) Dihydropiridine calcium channel blockers and perioperative mortality in aortic aneurysm surgery. Br J Anaesth 101(4):458–465
78. Rosenman DJ, McDonald FS, Ebbert JO, Erwin PJ, LaBella M, Montori VM (2008) Clinical consequences of withholding versus administering renin-angiotensin-aldosterone system antagonists in the preoperative period. J Hosp Med 3(4):319–325
79. Turan A, You J, Shiba A, Kurz A, Saager L, Sessler DI (2012) Angiotensin converting enzyme inhibitors are not associated with respiratory complications or mortality after noncardiac surgery. Anesth Analg 114(3):552–560
80. Botto F, Alonso-Coello P, Chan MT, Villar JC, Xavier D, Srinathan S et al (2014) Myocardial injury after noncardiac surgery: a large, international, prospective cohort study establishing diagnostic criteria, characteristics, predictors, and 30-day outcomes. Anesthesiology 120(3):564–578
81. Devereaux PJ, Chan MT, Alonso-Coello P, Walsh M, Berwanger O, Villar JC et al (2012) Association between postoperative troponin levels and 30-day mortality among patients undergoing noncardiac surgery. JAMA 307(21):2295–2304 [Research Support, Non-U.S. Gov't]

Chapter 8
Perioperative Management of the Diabetic Patient

Berta Soldevila, Anna M. Lucas, Roxana Zavala, and Didac Mauricio

Abstract Patients with diabetes mellitus are at increased risk of intra and post-operative morbidity and mortality. Poor glycaemic control in these patients is associated with worse surgical outcomes. Additionally, diabetes mellitus is often complicated by the presence of late chronic complications that may further increase the risk associated to hyperglycaemia. Therefore, diabetic patients require a careful preoperative assessment. However, there is not enough evidence concerning the optimal perioperative glycaemic target and treatment strategy in diabetic patients. Current guidelines provide clinical recommendations on the management of glycaemia based on the limited available evidence. The treatment schedule may vary according to the patient's individual characteristics and the type and duration of the surgical procedure. The aim of this chapter is give an overview of the available, clinically relevant, evidence on the management of patients with diabetes mellitus undergoing any surgical procedure.

Keywords Diabetes mellitus • Surgical stress • Perioperative diabetes management • Surgical complications • Blood glucose target • Hypoglycemia

8.1 Introduction

Hyperglycaemia in hospitalized patients may result from stress or decompensated type 1, type 2 (previously known or undiagnosed) or other forms of diabetes mellitus (DM), and/or it may be iatrogenic due to the withholding of antihyperglycemic medications or the administration of hyperglycemic agents. This chapter focuses on the management of patients with DM during the perioperative period. Issues related to the treatment of hyperglycemia in nondiabetic patients are addressed in another chapter of this book. The specific management of critically ill diabetic patients will not be discussed herein.

B. Soldevila • A.M. Lucas • R. Zavala • D. Mauricio (✉)
Department of Endocrinology & Nutrition, University Hospital & Health Sciences Research Institute Germans Trias i Pujol,
Autonomous University of Barcelona, Carretera Canyet S/N, Badalona 08916, Spain
e-mail: bsoldevila.germanstrias@gencat.cat; alucas.germanstrias@gencat.cat; roxneth@gmail.com; didacmauricio@gmail.com

K. Stuart-Smith (ed.), *Perioperative Medicine – Current Controversies*,
DOI 10.1007/978-3-319-28821-5_8

DM is present in a proportion as high as one third to half of the patients in community hospitals, and this percentage will grow in the near future in industrialized countries due to the increasing prevalence of DM, which is mostly related to aging. Therefore, it is expected that the number of people with DM undergoing surgical procedures will also rise [1].

DM and perioperative hyperglycemia are associated with an increased in-hospital morbidity and mortality [2–5]. Thus, diabetic patients have a higher incidence of wound infection, postoperative renal failure, cardiovascular complications, longer hospital stay, higher mortality and greater health care resource utilization than non-diabetic subjects [4, 6]. This higher morbidity and mortality is due, at least in part, to the comorbidities that are frequently present in diabetic patients, including coronary heart disease and other cardiovascular complications, hypertension, and renal insufficiency.

Additionally, hyperglycemia is associated with an increased mortality rate and has been identified as an independent risk factor for perioperative surgical complications in patients with previously untreated diabetes, with a nearly two-fold higher risk of infection, as well as the burden of other operative complications [7]. Moreover, the risk of perioperative complications depends on the condition of the patient prior to surgery, the prevalence of comorbidities and the magnitude and duration of the surgical procedure.

This piece of work is not intended to serve as a guideline for practising clinicians in the area; alternatively, we will try to provide an overview on the matter and to comment on the most relevant issues related to the perioperative management of patients with previous diabetes mellitus.

8.2 Impact of Surgical Procedures in Diabetic Patients

Diabetes is associated with increased atherosclerosis, endothelial dysfunction and the activation of platelets as well as several proinflammatory cytokines. Surgical stress is associated with hemodynamic stress and vasospasm; it also increases the prothrombotic state and inhibits fibrinolysis. All of these changes could lead to the instability of pre-existing atherosclerotic plaques, promote thrombus formation and vessel occlusion and could even result in acute atherosclerotic vascular events, especially myocardial infarction. For these reasons, DM is an important risk factor for perioperative cardiac complications. Additionally, the long-term hyperglycemia in diabetic patients compromises one or more potential target organs, thus necessitating the identification and treatment of comorbidities to prevent other potential non-cardiac complications during the perioperative period [8, 9].

Multiple factors that are linked to the perioperative period may result in worsening of hyperglycemia in patients with diabetes, including the preoperative metabolic state of the patient, neuroendocrine stress response, and acute perioperative insulin resistance, as well as the intraoperative management. Patients with DM, metabolic syndrome, pre-existing insulin resistance or those with underlying β-cell dysfunction are more likely to develop perioperative hyperglycemia [10, 11]. Surgery induces a stress response with multiple hormonal effects; glucagon, epinephrine and cortisol (counter-regulatory hormones) are the primary secreted hormones. This increase in counter-regulatory hormones leads to a catabolic state that contributes to the hyperglycemia observed during the perioperative period. In addition, it could potentially lead to metabolic decompensation and diabetic ketoacidosis in patients with absolute insulin deficiency or to a hyperosmolar nonketotic state in those who are susceptible [12].

However, although these changes all contribute to acute hyperglycemia, changes in glucose homeostasis persist long after counter-regulatory hormones have returned to normal concentrations. The development of this prolonged period of stress hyperglycemia is related predominately to insulin resistance, which is defined as a decrease in the biological response to insulin relative to normal, thus resulting in hyperglycemia. The elevated counter-regulatory hormones play a role in the relative insulin resistance associated with surgery [11].

8.2.1 Potential Consequences of Surgical Stress in Diabetic Patients

The major risk factors for diabetic patients undergoing surgery are those associated with late chronic complications of diabetes. Chronic hyperglycemia causes significant microvascular and macrovascular disease, as has been demonstrated in multiples studies in which a link has been established between hyperglycemia and poor outcomes in diabetic patients [13]. These diabetic late complications are characteristically classified into microvascular (including diabetic retinopathy, neuropathy -peripheral and autonomic- and nephropathy) and macrovascular complications (coronary heart disease, cerebrovascular disease and peripheral vascular disease). For these reasons, the presence of such complications places the patient with diabetes at an increased risk of cardiovascular morbidity and mortality during the perioperative period [8].

The specific management of each of these complications is beyond the scope of this chapter. The reader is referred to those parts of this book that are dedicated to the specific management of complications relevant to perioperative management, such as cardiovascular complications and chronic renal disease. We will only comment on the main complications that may occur during the intra and postoperative period in patients with diabetes. The issue of hypoglycaemia, as an adverse event resulting from the pharmacological glucose management, will be addressed in a specific section.

8.2.1.1 Cardiovascular Events

Diabetic patients have an increased risk of various heart complications, including hypertension, coronary artery disease, systolic and diastolic cardiac dysfunction, and congestive heart failure. Furthermore, heart disease is the cause of death in 80 % of diabetic patients [14]. In fact, the recent updated guidelines on perioperative cardiac assessment of patients undergoing non-cardiac surgery place diabetic patients, especially those receiving insulin, at a minimum of an intermediate risk [9].

Another concern is the specific cardiomyopathy that occurs in diabetic patients, which contributes to the increased cardiovascular risk. Diabetic cardiomyopathy progresses from impaired ventricular relaxation to diastolic dysfunction, with high left ventricular filling pressures and finally to evident heart failure [15].

In a recent systematic review of 12 trials investigating the perioperative management of diabetes, cardiovascular events were reported in 7 trials. These included myocardial infarction, cerebrovascular accidents, stroke, and arrhythmias such as atrial fibrillation or heart blockade. Actually, cardiovascular events were not prevented by intensive perioperative blood glucose control when compared with conventional blood glucose control [16].

8.2.1.2 Perioperative Infections

Diabetic patients show an increased susceptibility to perioperative infections. In fact, they present more than twice the risk of surgical site infections after surgery compared with non-diabetic subjects [17]. Hyperglycaemia per se impairs the different mechanisms involved in wound healing and host defence against infection [18]. Hyperglycaemia plays an important role in altering leukocyte function, including that of polymorphonuclear neutrophil activity, degranulation, and phagocytosis; the expression of cellular adhesion molecules; and the promotion of neutrophil adherence and sequestration in peripheral tissues [19, 20]. These changes lead to the impaired ability of the host to control pathogens and a higher perioperative rate of infection. The reversal of hyperglycemia with insulin partially or completely prevents or reverses these hyperglycemic effects [21]. Interestingly, a recent work has shown that a combination of intravenous (IV) and subcutaneous (SC) insulin treatment provides identical beneficial effects compared with IV insulin therapy alone on reducing infection rates during the postoperative period [22].

8.2.1.3 Renal Failure

A relevant proportion of patients with diabetes mellitus present diabetic nephropathy. This chronic complication of DM is characterized by the development of albuminuria and progressive impairment of renal function. In general, patients carrying this complication are at increased risk of morbidity and mortality, which is even more pronounced during the perioperative period. Therefore, special attention should be paid to the existence of this condition in any diabetic patient. Albuminuria should be determined in these patients to further assess their risk of acute renal failure [23].

Finally, it should be noted that diabetic autonomic neuropathy is a common complication of diabetes that is frequently undiagnosed. This complication can affect the gastrointestinal, genitourinary, and cardiovascular systems. The major clinical manifestations of diabetic autonomic neuropathy include resting tachycardia, exercise intolerance, orthostatic hypotension, constipation, gastroparesis, pseudomotor dysfunction, impaired neurovascular function, and hypoglycaemic autonomic failure [24]. In addition to cardiac complications, autonomic neuropathy may increase the risk of perioperative complications by decreasing oesophageal motility and causing gastroparesis that may lead to vomiting and aspiration of gastric content [25]. Autonomic neuropathy may also increase the risk of urinary tract infection in the presence of neurogenic bladder. Therefore, the presence of these clinical manifestations should alert the managing physician about the existence of this complication.

8.2.2 *Anaesthesia in Diabetic Patients*

The drugs that are used during surgery could also influence the severity of hyperglycemia in the diabetic patient. Anaesthetic and sedative agents may affect glucose homeostasis by modulating the sympathetic tone [10]. Indeed, some anaesthetic agents may decrease the secretion of catabolic hormones or alter insulin secretion in patients with type 2 DM with residual insulin secretion.

Regional anaesthesia such as an epidural block could decrease the exacerbated activation of the sympathetic nervous system and reduce the impairment of glucose homeostasis during the perioperative period [26]. Although there is insufficient evidence in the literature, some authors suggest that regional is superior to general anaesthesia in terms of mortality or major complications in surgical patients with DM.

8.2.2.1 General Anaesthesia

General anaesthesia could mask the usual signs and symptoms of hypoglycaemia, which is one of the main concerns of anaesthetists in the perioperative period and may explain some of the “permissive hyperglycemia” that occurs in the care of these patients [27]. However, the availability of accurate and easy-to-use glucose meters facilitates the monitoring of blood glucose in the operating room and the prevention of hypoglycaemia [28].

The choice of anaesthetic agent can affect glucose homeostasis. High doses of benzodiazepines and gamma-aminobutyric acid agonists decrease the secretion of adrenocorticotropic and cortisol hormones and may decrease the hyperglycemic response to surgery. Etomidate inhibits adrenal steroid synthesis [29], and midazolam infusion decreases cortisol and insulin secretion and increases GH secretion. Clonidine reduces sympathetic tone and norepinephrine release from nerve terminals. The effect of clonidine on the pituitary–adrenocortical axis is a controversial topic, although a decrease in the release of adrenocorticotropic hormone and cortisol has been reported. The use of clonidine during surgery has been proposed as a way to improve perioperative haemodynamics and decrease anaesthetic requirements [30]. High doses of opiates appear to decrease the hyperglycemic response to surgery by decreasing catabolic hormones [31]. Therefore, the use of these drugs could be beneficial in diabetic patients. The effect of propofol on insulin secretion is not known. However, the use of propofol as an induction agent has no adverse effects in diabetic patients, and no changes in the pharmacokinetics or pharmacodynamics of this drug have been observed in an experimental study conducted in diabetic rats [32].

In conclusion, it is currently uncertain which types of general anaesthetic agents could facilitate glucose control and hemodynamic stability during the perioperative period in patients with DM.

8.2.2.2 Regional Anaesthesia

Regional anaesthesia, including spinal, epidural, and other regional blocks, may modulate catabolic hormones and insulin secretion. High spinal anaesthesia (dermatome level T2–T6) could induce a reduction in the acute insulin response to glucose, whereas low spinal anaesthesia (dermatome level T9–T12) results in no reduction of this response [33]. Activation of the sympathetic nervous system and hypothalamic pituitary axis, which are induced by surgical stress under general anaesthesia, could be prevented by regional and epidural anaesthesia. Some authors have demonstrated that epidural anaesthesia and analgesia can decrease the degree of insulin resistance in the immediate postoperative period in patients with insulin resistance before surgery compared with general anaesthesia [34].

High doses of local anaesthetics have been reported to induce nerve injury and to reduce nerve blood flow, which likely results from an activity of ischemic origin [35]. Fibres in diabetic nerves may be more susceptible to ischemic injury because they are already stressed by chronic ischemic hypoxia. Caution must therefore be exerted in diabetic patients regarding the total dosage of local anaesthetics used to avoid peripheral nerve injury.

8.3 General Concepts on the Perioperative Management of Diabetic Patients

The perioperative management of patients with DM is defined herein as all of the procedures during the period of a surgical intervention in a patient, including the preoperative evaluation, admission (preoperative phase), anaesthesia and surgery (intraoperative phase), and recovery (immediate postoperative phase).

The current review focuses on the perioperative management of diabetic patients based on the available systematic reviews, randomized controlled trials and meta-analyses, when available, and the following recently-issued practice guidelines are employed for the discussion of the recommendations of two scientific bodies: the National Health Service (NHS) diabetes guideline for the perioperative management of adult patients with diabetes in the United Kingdom [36], and the Society for ambulatory anaesthesia (SAMBA) consensus statement in perioperative blood glucose management in diabetic patients undergoing ambulatory surgery [37].

The scope of this chapter does not include a review of the currently available antidiabetic medications. It is well-known that patients with type 1 diabetes are most commonly treated with multiple insulin injections that include basal (long-acting insulins) – bolus (rapid-acting insulins) regimens; alternatively, these patients are treated with a continuous subcutaneous insulin infusion that is also based on basal insulin administration and meal-coupled boluses. However, in type 2 diabetic patients, current treatment algorithms include the use of different types of oral hypoglycaemic agents, non-insulin injectable medications and insulin. There is a lack of evidence regarding the use of the newest medications during the perioperative period.

Type 1 diabetic patients are insulin-deficient and prone to developing ketosis and acidosis. Therefore, the treating clinician should also remember the concept that type 1 diabetic subjects require a continuous exogenous supply of insulin. A common mistake is to manage these patients like type 2 diabetic patients who are not prone to ketosis and, for example, omit short-acting insulin administration if the glucose level is in the normal range. In contrast, patients with type 2 diabetes are susceptible to the development of a hyperosmolar hyperglycemic state (also known as a nonketotic hyperosmolar state), which may lead to severe volume depletion and neurologic complications, although they may also develop ketoacidosis in response to extreme stress conditions. Another important issue that should be considered in patients with diabetes is the duration of the fasting period. We must also bear in mind that the perioperative management needed for a programmed surgical procedure versus an urgent one is not equivalent.

The main objectives of the perioperative management of diabetic patients undergoing surgery are to prevent hypo and hyperglycemia and their consequences. However, it is not possible to provide strong recommendations due to the lack of high quality studies. There are many review articles based on clinical experience and few on prospective trials; these often suggest different management strategies or procedures. Therefore, recent guideline recommendations for the perioperative management of diabetic patients [36, 37] are based on general glucose control in DM, data obtained from surgical populations, and reviews and expert opinions, without a strong level of evidence.

8.4 Preoperative Evaluation of Patients with Diabetes Mellitus

In the preoperative evaluation of diabetic patients, it is mandatory to assess the type of diabetes and the presence of chronic diabetic complications, especially cardiac and renal disease. It is also essential to identify the pharmacological treatment, including the type and dose of antidiabetic medications [38]. The preoperative evaluation should ideally include the evaluation of previous glycaemic control, as determined by the concentration of glycated haemoglobin (HbA_{1c}), and the occurrence and frequency of hypoglycaemia. In addition, information on hospital admissions due to diabetic decompensations should be obtained [37, 39]. Other relevant parameters that should be measured include body mass index, blood pressure and the estimated glomerular filtration rate [36, 38].

The early preoperative assessment should be performed to determine a perioperative DM management strategy and to optimize the control of glycaemia and diabetic comorbidities. High-risk patients, defined as those with poor glycaemic control or with chronic complications of DM, should be identified early to manage their risk [36, 38]. It is also important to identify high-risk patients who could potentially require admission to critical care units [38].

The guidelines recommend that patients with DM should be identified and prioritized in the operation list such that they are the first intervention in the morning, if possible, to minimize their treatment modifications [37, 38]. Therefore, evening surgeries are not recommended for these patients due to the associated longer fasting periods required and need for a variable rate of intravenous insulin infusion (VRIII). There are no studies demonstrating the safety of elective evening surgeries in patients with DM. Currently, routine overnight admission for the preoperative management of DM is not recommended [38].

Concerning the preoperative determination of HbA_{1c}, there is evidence that good preoperative glycaemic control, as assessed by HbA_{1c}, is associated with a lower incidence of complications, decreased mortality and a short duration of hospital stay [5]. Indeed, an elevated preoperative HbA_{1c} is associated with adverse outcomes following different types of surgery, irrespective of a previous diagnosis of DM [40, 41]. However, there is no consensus regarding the indications of preoperative HbA_{1c} determination in patients who will undergo surgery. Some authors suggest determining HbA_{1c} preoperatively in all patients scheduled to undergo a major surgery to diagnose diabetes in those with unknown DM and prior to elective surgery in all patients with diabetes, to better evaluate glucose control [42]. Others recommend the routine measurement of HbA_{1c} at the preoperative assessment of non-diabetic patients with risk factors for DM [38] and in diabetic patients for whom an HbA_{1c} value has not been determined in the previous 3 months. Nevertheless, a recent systematic review concluded that preoperative blood glucose and HbA_{1c} testing are not required in non-diabetic patients undergoing elective non-cardiac surgery unless there are clinical symptoms indicative of diabetes. HbA_{1c} should only be screened in patients who will undergo vascular and orthopaedic surgery, which carry an elevated risk and thus may justify preoperative screening for blood glucose or HbA_{1c} [43].

There are no prospective randomized investigations in relation to blood glucose control during the preoperative period in patients with diabetes. In diabetic patients, the major studies concluded that admission or preoperative glucose >200 mg/dl is correlated with an increased mortality, an increase of hospitalization days (conventional and intensive care unit), an increased infection rate and increased incidence of stroke and myocardial infarction. Also, a concentration of haemoglobin HbA_{1c} >7 % was associated with increased risk of infection and morbidity after different types of surgeries [41].

The HbA_{1c} values may also influence the timing of surgery. An elective major surgical procedure in patients with high HbA_{1c} values could be delayed until improved glycaemic control is achieved to decrease the risk of perioperative morbidity and mortality [42]. However, there is insufficient evidence to recommend an upper limit of HbA_{1c} prior to elective surgery [36, 38], either for preoperative fasting blood glucose or HbA_{1c}, above which elective ambulatory surgery should be postponed [37]. Therefore, clinical judgment is essential, including the possibility of seeking the opinion of a diabetologist/endocrinologist.

The risks associated with poor glycaemic control should be balanced against the necessity for surgery. Frequently, preoperative hyperglycemia is a consequence of the inappropriate discontinuation of preoperative antidiabetic treatment and/or a preoperative stress response. In general, surgery should be postponed in patients with significant complications of hyperglycemia, such as dehydration, ketoacidosis or hyperosmolar syndrome [10, 37]. However, surgery should be acceptable in patients with preoperative hyperglycemia but with adequate long-term glycaemic control [37]. An upper HbA_{1c} limit between 8 and 9 % is acceptable according to some authors, depending on the individual circumstances [38]. The NHS of the United Kingdom recommends that patients consult a diabetes specialist if they present an HbA_{1c} level above 8.5 % or they present with hypoglycaemia unawareness [38]. However, this recommendation does not seem to justify the delay of surgery until HbA_{1c} improvement as lowering of HbA1c could take weeks/months [36]. A high HbA_{1c} indicates poor glycaemic control during the previous months, and it would be useful to intensify the diabetic therapy in such cases to improve glycaemic control prior to surgery.

8.5 Perioperative Blood Glucose Targets

Blood glucose targets during the immediate preoperative period are controversial due to the absence of sufficient clinical studies in patients undergoing general surgery. The NHS guidelines recommend the implementation of the World Health Organization (WHO) surgical safety target: blood glucose should be maintained between the range from 108 to 180 mg/dl (6–10 mmol/l), if this can be achieved safely, with a range from 72 to 216 mg/dl (4–12 mmol/l) considered acceptable [36, 38]. This recommendation is concordant with the position statement on glycaemic control for in-patients of the American Diabetes Association (ADA) / American Association of Clinical Endocrinologists (AACE) [44], and the more recent clinical practice guidelines on the management of hyperglycemia in hospitalized patients in non-critical care settings of the Endocrine Society [45]. However, some authors do not agree with this target and argue that an upper limit of 12 mmol/l is similar to the concentration related to endothelial and immunological changes that induce inflammation and increase the risk of infection in vitro studies, and that the lower limit of 4 mmol/l is close to values that provoke a hypoglycaemic symptomatology in some patients [42].

In the case of blood glucose levels greater than 216 mg/dl (12 mmol/l), the NHS guideline recommendation is to assess the capillary or urinary ketones. If the capillary ketone levels are greater than 3 mmol/l or the urinary ketones are equal or greater than +++, it is recommended that surgery be cancelled, guidelines followed and a diabetes specialist be contacted. In the case of blood glucose above 216 mg/dl (12 mmol/l) with capillary ketones less than 3 mmol/l or urinary ketones less than +++, it is recommended that subcutaneous rapid-acting insulin analogues be administered and blood glucose levels be rechecked after 1 h to ensure their decline. If surgery cannot be delayed, a variable rate intravenous insulin infusion (VRIII) should be initiated [36], as explained in the following section.

In relation to the intraoperative blood glucose level, the NHS guidelines also recommend implementation of the WHO safety target blood glucose concentration ranging from 108 to 180 mg/dl (6–10 mmol/l), with an acceptable concentration ranging from 72 to 216 mg/dl (4–12 mmol/l) [36, 38]. There is insufficient evidence in the literature to determine the optimal intraoperative blood glucose level for ambulatory surgical diabetic patients. The recommendations of the SAMBA guidelines [37] are based on the ADA/AACE practice guidelines for inpatient glycaemic control [44]. For diabetic patients with good metabolic control, the intraoperative glucose levels should be maintained below 180 mg/dl (10 mmol/l); for patients with poor metabolic control, the blood glucose levels should be carefully managed according to their preoperative blood glucose values [37]. Chronically elevated blood glucose levels should not be acutely decreased or normalized due to the potential for hypoglycaemia [46], and indeed, significant fluctuations in blood glucose levels can increase the perioperative morbidity and mortality [10].

Due to the uncertain optimal perioperative glycaemic targets in patients with DM, the recommendations of the available guidelines are not supported by high quality evidence. Some recent systematic reviews and meta-analyses have attempted to identify the benefits of intensive glucose control in patients with diabetes undergoing surgery. A published meta-analysis concluded that moderate glucose control, defined as a glycaemic target between 150 and 200 mg/dl (8.3-11-1 mmol/l), during or immediately after surgery is associated with a decrease in mortality and stroke risk in patients with DM compared with a liberal glycaemic control defined as a glycaemic target >200 mg/dl (>11.1 mmol/l). The results of this meta-analysis also demonstrated no differences in relation to outcomes between moderate and strict blood glucose control, defined as glycaemic targets between 90 and 150 mg/dl (5.6–8.3 mmol/l) [47]. However, this meta-analysis include only five randomized controlled trials. In a recent Cochrane systematic review, Buchleitner et al. concluded that there are no differences between intensive, targeting near normal blood glucose, and conventional glucose control in relation to postsurgical outcomes, except for an increase in hypoglycaemic events that occurs in patients treated with intensive glucose control [16]. In conclusion, there is no clear evidence that intensive glycaemic control in patients with DM leads to reduced intra and postoperative complications in these patients.

8.6 Perioperative Glycaemic Management

In this section, we discuss those issues related to management of blood glucose in patients with DM according to patient and surgery characteristics.

8.6.1 Perioperative Management of Patients on Oral Antidiabetic and Non-insulin Injectable Therapy

There is insufficient evidence regarding the optimal perioperative management of oral antidiabetic treatment. Hypoglycaemia rarely occurs with these medications, except with sulfonylureas and meglitinides [46]. There is limited evidence concerning the perioperative management of metformin and the association of metformin with an increased risk of perioperative lactic acidosis [48, 49]. Therefore, there are controversial recommendations among different guidelines regarding the perioperative use of metformin. Some of them recommend the discontinuation of metformin 24–48 h before surgery [50], while others suggest continuing metformin in patients undergoing surgeries associated with a short fasting period [36, 38]. In relation to the new incretin-based treatments, dipeptidyl peptidase-IV inhibitors and glucagon-like peptide-1 receptor agonists, no studies have investigated their effects on perioperative management in diabetic patients. The general recommendations for each non-insulin diabetic drug management strategy during the perioperative period of surgeries with a short fasting period (no more than one missed meal) are summarized in Table 8.1. There is no experience on the perioperative management of patients treated with the most recently introduced sodium-glucose transporter 2 inhibitors; this group of oral agents may have an impact on the extracellular volume, and we would recommend at least stopping this treatment the day before surgery.

Some authors suggest that it is not necessary to discontinue oral antidiabetic and non-insulin injectable medications before the day of the surgical procedure, but these medications should not be taken on the day of surgery until normal food intake is restored [37, 39]. Furthermore, a recent study suggests that a preoperative adjustment of home diabetic medications and systematic treatment of perioperative hyperglycemia are safe and could be more effective than the usual care for patients with diabetes undergoing ambulatory surgery [51]. However, the recent Endocrine Society clinical practice guideline recommends the discontinuation of oral and non-insulin injectable antidiabetic agents for the management of hyperglycemia in most non-critical hospitalized diabetic patients, including those undergoing surgery, and the management of hyperglycemia with insulin during admission, including during the perioperative period [45].

Table 8.1 Perioperative management of patients treated with oral antidiabetic and non-insulin injectable therapies for surgeries with a short fasting period

Type of oral antidiabetic and non-insulin injectable therapy	Day before surgery	Day of surgery	
		Patient for AM surgery	Patient for PM surgery
Acarbose	No dose change	Omit morning dose	Give morning dose if eating
Meglitinide			
Metformin	No dose change unless contraindications		
Sulphonylurea	No dose change	Omit morning dose	Omit morning and evening dose
Pioglitazone		No dose change	
DPP-IV inhibitor		DPP-IV inhibitor and GLP-1 recpetor agonist should be omitted on day of surgery	
GLP-1 receptor agonist		Omit on day of surgery	
SGLT-2 inhibitor	Omit on day before surgery		

Adapted from Ref. [36–39]
There is limited evidence regarding the optimal perioperative management of oral antidiabetic and non-insulin injectable treatment. There is no consensus among different guidelines. The recent Endocrine Society guideline (Ref. [45]) recommends the discontinuation of oral antidiabetic and non-insulin injectable agents for the management of hyperglycemia in diabetic patients undergoing surgery
Surgeries with a short fasting period are defined as surgeries that required no more than one missed meal
DPP-IV dipeptidyl peptidase-4, *GLP-1* glucagon-like peptide-, *SGLT-2* sodium glucose transporter type 2, *AM* ante meridiem, *PM* post meridiem

8.6.2 Perioperative Management of Patients Treated with Insulin

It is important to note that the current management of insulin treatment in hospitalized patients is no longer based on the concept of correcting the blood glucose of the patient. This out-dated concept includes the use of the well-known, and still used in many settings, sliding scale subcutaneous regular insulin (SSI) regimen. Current practices should include the anticipation of blood glucose excursions of the patient blood glucose profile. In the perioperative setting, this anticipatory action implies the use of the variable rate intravenous insulin infusion (VRIII) in patients who are anticipated to have a long fasting period (more than one missed meal) and in patients with poorly controlled diabetes [36]. The VRIII includes the infusion of intravenous rapid-acting insulin at a variable rate according to regular capillary blood glucose measurements with the aim of controlling serum glucose levels within a specified range, as well as its tendency to increase or decrease. Regular insulin is the most commonly used in VRIII. However, other rapid-acting insulins as aspart, glulisine or lispro insulin, can also be used as alternative and have licence for intravenous use, although these are more expensive alternatives [52]. The VRIII is usually accompanied by an infusion of fluid that contains glucose to prevent insulin-induced hypoglycaemia [52].

There is also insufficient evidence regarding the perioperative management of insulin-treated diabetic patients undergoing surgery. Consequently, the recommendations are based on safety, the avoidance of hypoglycaemia, and the maintenance of adequate glucose control [37]. Recent guidelines recommend maintenance of the usual patient insulin regimen, regardless of a reduction or not of the insulin dose; however, they do not recommend the use of an SSI four times daily as a preoperative treatment [36, 38]. General recommendations for each type of insulin treatment management during the perioperative period for surgeries with a short fasting period (defined as no more than one missed meal) are summarized in Table 8.2.

There are insufficient data from fair quality studies concerning the use of insulin treatment during the perioperative period in insulin-naïve patients with DM. In

Table 8.2 Perioperative management of patients treated with insulin for surgeries with a short fasting period

<table>
<tr><th rowspan="2">Type of insulin therapy</th><th rowspan="2">Day before surgery</th><th colspan="2">Day of surgery</th></tr>
<tr><th>Patient for AM surgery</th><th>Patient for PM surgery</th></tr>
<tr><td>Continuous subcutaneous insulin infusion</td><td colspan="3">No dose change
or
Reduce 20–30 % of basal rate if history of hypoglycaemia</td></tr>
<tr><td>Long-acting insulin analogues</td><td>No dose change
or
Reduce 20–30 % of evening dose if history of nocturnal or morning hypoglycaemia.</td><td colspan="2">No dose change
or
Reduce 20–30 % of morning dose if history of hypoglycaemia</td></tr>
<tr><td>Intermediate-acting insulins</td><td>No change of morning dose
or
Reduce 20–30 % of evening dose</td><td colspan="2">Reduce 25–50 % of morning dose
No change of meal evening dose</td></tr>
<tr><td>Prefixed combination of insulins</td><td>No dose change</td><td>Reduce 25–50 % of morning dose of intermediate-acting insulin component.
Omit short-acting insulin component in the morning
Omit lunchtime dose
No change of meal evening dose</td><td rowspan="2">No change of morning dose
Omit lunchtime dose
Leave the evening meal dose unchanged</td></tr>
<tr><td>Short and rapid-acting insulins</td><td>No dose change</td><td>Omit morning and lunchtime dose</td></tr>
</table>

Adapted from Ref. [36–39]
There is insufficient evidence regarding the perioperative management of insulin-treated diabetic patients undergoing surgery
There is no consensus among different guidelines
Surgeries with a short fasting period are defined as surgeries that required no more than one missed meal

ambulatory surgeries, the SAMBA guidelines suggest that perioperative insulin treatment should be considered in this type of patient in the case of significantly increased glucose levels [37].

8.6.2.1 Long-Acting Insulin Analogues

Long-acting insulin analogues, such as glargine, degludec or detemir insulin, are generally used to maintain blood glucose control between meals. Patients may not experience an increased risk of hypoglycaemia with these treatments even if meals are missed, as observed in preoperative and postoperative fasting.

It remains debated whether doses of long-acting insulin analogues in those patients treated with this type of insulin should be reduced. If the dose of insulin prior to surgery is regarded as adequate, the NHS and SAMBA guidelines recommend the administration of the usual morning or evening dose of long-acting insulin analogue both the previous day and the day of surgery unless there is a history of hypoglycaemia or limited food intake on the day before surgery [36–38]. Other authors recommend reducing the dose of long-acting insulin analogues by 20–30 % in the evening dose the day before surgery and in the morning dose on the day of surgery [39].

A recent randomized trial evaluated three different strategies of long-acting insulin analogue dosing on the evening before surgery (80 % of the usual dose, the dose recommended by a physician, or a dosing table depending on the fasting glucose range and the presence or absence of rapid/short-acting insulin in the daily regimen) and showed no differences among the three groups in the achievement of the fasting blood glucose target [53]. Although this study has several limitations, it is an important step in the management of patients with diabetes undergoing elective surgery. Thus, clinical protocols might be adapted according to local clinical practice.

8.6.2.2 Other Types of Insulin

Combined treatment using insulin and oral antidiabetic drugs, treatments using intermediate-acting or premixed insulins may cause hypoglycaemia while fasting. In relation to intermediate-acting insulins such as Neutral Protamine Hagedorn (NPH) or Neutral Protamine Lispro (NPL) insulin, it is not necessary to change the dose if the medication is taken in the daytime; however, some authors recommend a 25 % reduction of this dose if it is taken in the evening, especially if the patient has a history of hypoglycaemia. On the day of surgery, it is recommended that the morning dose be reduced by 25–50 % [37]. Premixed insulins are a type of fixed combination of rapid-acting and intermediate-acting insulin. It is not necessary to change the dose of premixed insulins the day before surgery. However, premixed insulins should be changed to an intermediate-acting and a rapid-acting insulin on the day of the surgery to prevent hypoglycaemia, with each type of insulin dose administered independently [39] and a reduction of the corresponding proportional morning dose of intermediate-acting insulin by 25–50 % [37–39]. In currently used insulin treatment schedules, short-acting insulin is used to manage meal-induced glycaemic excursions. In relation to these rapid-acting insulins, such as aspart, glulisine, lispro and regular insulin, it is recommended that the dose remain unchanged the day before surgery [37–39]. In addition, it is intuitive to avoid administering the dose while the patient is fasting during the day of surgery due to the postprandial glucose control and potential to increase the risk of hypoglycaemia during fasting.

8.6.3 Intraoperative Treatment to Maintain Glycaemic Control

Classically, glycaemic control has been achieved with the administration of intravenous infusions of insulin and glucose and later of glucose, insulin and potassium (GIK). More recently, insulin has been administered independently, using a VRIII, while glucose and potassium are administered via a volumetric pump to allow better blood glucose control and dose adjustment flexibility. However, the use of VRIII has potential risks, such as hyper and hypoglycaemia due to inappropriate insulin infusion rates or inadequate monitoring, rebound hyperglycemia and potential ketoacidosis if intravenous access is lost or VRIII is terminated inappropriately (especially in patients with type 1 diabetes), as well as infections related to the intravenous line [36, 38, 52]. To prevent these possible secondary effects, it is crucial to maintain strict glucose monitoring and to increase or decrease the insulin infusion rate, and even to stop it and reassess the glucose values in cases presenting a risk of severe hypoglycaemia. It is important to take into account that the interruption of insulin infusion in patients with type 1 compared with type 2 diabetes is associated with a greater potential risk of ketoacidosis due to the lack of a pancreatic insulin reserve. Because of these issues, many units currently manage the perioperative glucose control of diabetic patients by modifying their usual treatment, although successful modification of the usual treatment for DM can only be utilized if the fasting period is short, surgery can be carried out early, previous metabolic control of DM is acceptable (HbA_{1c} <8.5 % at admission) and the patient will be capable of self-managing his DM during the immediate post-operative period. If these criteria are not present then VRIII should be initiated [36, 38].

8.6.3.1 Surgeries with a Short Fasting Period

There is insufficient evidence to recommend a safe and effective treatment strategy to achieve target blood glucose levels in diabetic patients undergoing surgery with a short fasting period, which is defined as no more than one missed meal. This is the case in, for example, ambulatory surgical procedures [37].

The NHS guidelines recommend that patients undergoing surgeries with a short fasting period should be managed by a modification of their usual antidiabetic medication, including insulin, avoiding a VRIII regimen whenever possible [36, 38]. There are insufficient data to recommend the use of regular or rapid-acting insulin analogues in diabetic ambulatory surgical patients. However, the SAMBA guidelines suggest the use of subcutaneous rapid-acting insulin analogues rather than regular insulin [37] to correct hyperglycemia due to their pharmacokinetic properties and shorter duration of observation requirements in the postoperative period.

Although the intravenous infusion of rapid-acting insulin has been used to maintain optimal blood glucose levels in patients undergoing major surgeries, recent guidelines recommend the subcutaneous route for noncritical patients [44, 45]. The SSI regimens are commonly used in clinical practice, but they have been questioned recently [44, 45]. For all of these reasons, it is recommended that subcutaneous rather than intravenous insulin administration be applied in diabetic patients undergoing ambulatory surgery to achieve intraoperative glucose control [37]. In cases in which the blood glucose levels cannot be maintained adequately below 216 mg/dl (12 mmol/l), it is recommended to consider changing to a VRIII [38].

8.6.3.2 Surgeries with a Long Fasting Period

Surgeries that require hospital admission are typically associated with a long fasting period, which is defined as more than one missed meal, and VRIII treatment. An adequate glucose supply should be provided to prevent the induction of a catabolic state, starvation ketosis, and insulin-induced hypoglycaemia. VRIII is often poorly used in the perioperative setting and requires explicit guidelines such as the NHS guidelines [36, 38]. The initial insulin infusion should be determined by capillary blood glucose (CBG). There are multiple and a different recommendations of VRIII initiation and one example is described in Table 8.3. It is recommended that the CBG be measured hourly. If the CBG remains above 216 mg/dl (12 mmol/l) for three consecutive readings and is not dropping by 54 mg/dl/h (3 mmol/l/h) or more, the rate of insulin infusion should be increased. If the CBG is less than 72 mg/dl (4 mmol/l), then the rate of insulin infusion should be reduced [38]. The judgement of the clinician is also important in cases in which hypoglycaemia is considered to increase the risk of adverse events for the patient, and the alternative should include the decision to stop the insulin infusion for a short period (for instance, 30 min) with a subsequent re-assessment of CBG to determine the new infusion rate. The VRIII must be administered continuously with a concomitant glucose and electrolyte infusion substrate, using a pump adjacent to the substrate infusion. Insulin must be infused at a variable rate to maintain the blood glucose between 108 and 180 mg/dl (6–10 mmol/l). If the infusion is stopped, no exogenous insulin will be present in the circulation after 3–5 min.

Table 8.3 Suggested intraoperative variable rate intravenous insulin infusion recommendation for a diabetic patient

How to set up an intravenous insulin infusion:
1 mL of rapid-acting insulin (100 IU) + 99 mL of 0.9 % saline solution
Concentration of 1 IU rapid-acting insulin/mL of solution
Types of rapid-acting insulins:
Regular insulin is the most commonly used in VRIII but other types, as aspart, glulisine and lispro insulin, can also be used
Glucose target:
108–180 mg/dl (6–10 mmol/l), acceptable 72–216 mg/dl (4–12 mmol/l)

Different scales for variable rate intravenous insulin infusion			
Glucose mg/dl (mmol/l)	Insulins rates (ml/h)		
	Reduced rate	Standard rate	Increased rate
<72 mg/dl (<4 mmol/l)	0	0	0
72–144 mg/dl (4–8 mmol/l)	0.5	1	2
145–216 mg/dl (8.05–12 mmol/l)	1	2	4
217–288 mg/dl (12.05–16 mmol/l)	2	4	6
289–360 mg/dl (16.05–20 mmol/l)	3	5	7
361–432 mg/dl (20.05–24 mmol/l)	4	6	8
>432 mg/dl (>24 mmol/l)	6	8	10

Adapted from Ref. [52]

Three different scales of variable rate intravenous insulin infusion (VRIII) are recommended in relation to different degrees of insulin sensibility of patients: (a) a reduced rate for insulin sensitive patients, for example those treated previously with diet, oral antidiabetic therapy or insulin treatment with <24 daily IU; (b) a standard rate for most of diabetic patients; (c) an increased rate for insulin resistant patients, for example for patients treated with insulin therapy >100 daily IU. Capillary blood glucose (CBG) should be checked hourly. In case of CBG <72 mg/dl (4 mmol/l), hypoglycaemia should be treated and VRIII should be stopped for a short period (for instance, 30 min) with a subsequent re-assessment of CBG to determine the new infusion rate. In case of CGB >216 mg/dl (12 mmol/l) for three consecutive readings and is not dropping by 54 mg/dl/h (3 mmol/l/h) or more, the VRIII should be increase to next scale. In case that the patient is on the increased rate, a diabetes specialist team should be advised

8.6.3.3 Fluid Management

The aims of fluid management in the surgery of a patient with DM are to provide glucose as a metabolic substrate to maintain blood glucose levels, to optimize the intravascular volume status and to maintain serum electrolytes [38]. Adequate preoperative hydration and intraoperative fluid administration are necessary. The daily requirements of major ions and fluid in a healthy adult are 60–100 mmol of sodium, 40–80 mmol of potassium and 1.5–2.5 l of water. These requirements may change in disease states, and careful daily monitoring is needed. There is limited evidence on which to base recommendations for optimal fluid and insulin management in patients with DM undergoing surgery. The guidelines recommend different fluid solutions depending on the need for VRIII. The recommended first choice substrate solution for patients with DM who require VRIII is 0.45 % sodium chloride with 5 % glucose and 0.15–0.3 % potassium chloride (KCl). In the rare case that the patient develops hyponatremia, a 0.9 % saline infusion containing 5 % glucose and 0.15–0.30 KCl is recommended [36, 38]. The recommended substrate solution for diabetic patients who do not require VRIII is Hartmann's solution, which is preferred over 0.9 % saline sodium chloride to reduce the risk of hyperchloremic acidosis. It has been suggested that Hartmann's solution may lead to hyperglycemia in patients with type 2 diabetes. However, 1 l of Hartmann's solution increases the plasma glucose no more than 1 mmol/l [54]; thus, it is in fact not contraindicated in diabetic patients.

Glucose and insulin infusions can provide good glycaemic control but may lead to hyponatremia. Surgical patients with DM are not only at risk of complications associated with fluid and electrolyte management, but are also at an elevated risk of hyponatremia due to the use of hypotonic glucose solutions.

8.6.4 Perioperative Blood Glucose Monitoring

Adequate monitoring of the CBG concentration is necessary to maintain patient safety, correct hyperglycemia, detect hypoglycaemia and modify the insulin treatment if necessary [38]. Although the perioperative CBG ranges are controversial, as discussed above, the WHO recommends a surgical safety target of blood glucose between 108 and 180 mg/dl (6–10 mmol/l), with a range that is considered acceptable from 72 to 216 mg/dl (4–12 mmol/l). In ambulatory surgeries, it has been suggested that blood glucose levels should be evaluated prior to surgery when the patient arrives at the clinic and before his or her discharge home [37]. Although there are no recommendations for specific procedures, the guidelines recommend assessing CBG prior to the induction of anaesthesia and regularly monitoring it during the procedure and in the immediate postoperative period, at least hourly when the blood glucose is stable and in the target range, depending on the duration of surgery and the type of insulin administered [38]. If the blood glucose level changes or the patient is receiving intraoperative insulin, then more frequent glucose measurements might be required.

8.6.5 Immediate Postoperative Management

The aims of postoperative care include to ensure glycaemic control and maintain the fluid and electrolyte balance, optimize pain control (adequate pain relief decreases catabolic hormone secretion), encourage an early return to normal eating and drinking by managing nausea and vomiting, facilitate return to the usual diabetes treatment regimen and avoid iatrogenic effects.

Glucose control during the postoperative period may be unpredictable and difficult, and there is insufficient evidence to support the guideline recommendations.

The Endocrine Society clinical practice guideline on the management of hyperglycemia in hospitalized patients does not recommend prolonged use of an SSI regimen for hyperglycemic control during the postoperative period in patients with DM [45] based on the RABBIT 2 Surgery Study results. This prospective randomized study in patients with type 2 diabetes undergoing surgery revealed an improvement in glycaemic control and a reduction in postoperative complications with a bolus-basal treatment compared with an SSI [55].

The timing of surgery and the expected time to initiate a normal diet after surgery should be considered to minimize the risk of hypoglycaemia [37]. Additional fluid therapy may be required according to the specific needs of the patient for a given surgical procedure, and the utilization of Hartmann's solution is considered acceptable [38].

There is insufficient evidence regarding the switch from VRIII to the usual treatment in surgical patients with DM. Few studies have investigated the procedure used for the transition from intravenous insulin to subcutaneous insulin or to oral antidiabetic treatment. The NHS guidelines recommend that VRIII and glucose and fluid infusions should be continued until the patient is ready to eat and drink without nausea or vomiting, and that treatment should be re-initiated [38]. When it is time for a meal, it is recommended that the usual non-insulin diabetic treatment be initiated. To avoid rebound hyperglycemia, VRIII should not be stopped until at least 30 min after the usual treatment is re-initiated and the patient has eaten [52]. Regarding the re-initiation of the non-insulin treatment, the dose of sulfonylurea must be reduced or delayed if food intake is reduced. Additionally, the NHS recommends the re-introduction of metformin only if the glomerular filtration rate is greater than 50 ml/min/1.73 m^2 [38].

In patients who are treated with previous insulin regimens, the transition from intravenous to subcutaneous insulin should occur with the next meal-related subcutaneous insulin. It is recommended that the same preoperative regimen be re-initiated, although the doses will likely need to be adjusted. Indeed, the treatment requirements may differ from the usual ones in the immediate postoperative period due to glycaemic instability [38, 52]. The starting dose of basal insulin should be 50–80 % of the previous intravenous insulin total daily dose if stable glycaemic control had been achieved with intravenous insulin. Alternatively, a calculation called the "Miami 4/12 rule" can be used, in which the basal insulin replacement dose (in units) is equal to the weight of the patients in kilograms divided by 4 and the prandial coverage is calculated by dividing the weight of the patient in kilograms by 12 [4].

In patients who are receiving basal-bolus insulin treatment, VRIII should not be continued until at least 30–60 min after the first short-acting subcutaneous bolus of insulin has been administered to avoid rebound hyperglycemia [52]. If a long-acting insulin analogue has been continued, the only action should be to restart the short-acting insulin at the next meal. The NHS guidelines also recommend the continuation of long-acting insulin analogues, such as glargine or detemir, together with VRIII. This tactic has the advantage that no time is lost in re-establishing basal insulin following the discontinuation of VRIII [37, 39, 52]. In patients treated with a prefixed combination of insulin, the insulin should be re-introduced at the next meal requiring scheduled insulin, and VRIII should be maintained for 30–60 min after administration of the subcutaneous insulin [38, 52].

In the case of postoperative hyperglycemia, which is defined as a blood glucose level greater than 216 mg/dl (12 mmol/l) with blood ketones below 3 mmol/l or urinary ketones less than +++, it is recommended that subcutaneous rapid-acting insulin analogues be administered and the blood glucose level re-assessed after 1 h to ensure that it is decreasing. If the blood glucose remains above 216 mg/dl (12 mmol/l), subcutaneous insulin administration should be repeated after 2 h and the blood glucose re-evaluated after 1 h. If the blood glucose levels do not fall, it has been suggested to consider re-starting VRIII [38].

As mentioned previously, HbA_{1c} can identify diabetic patients with a higher risk of elevated perioperative blood glucose concentrations, as well as those that require modification of their previous therapeutic regimen at the time of discharge. In 2009, Moitra et al. showed that HbA_{1c} and the duration of surgery correlated with poor glucose control in the immediate perioperative period [56]. In our own experience, almost half (47 %) of the previously known hospitalized patients with type 2 diabetes had prior suboptimal chronic glycaemic control at the time of hospital admission in conventional surgical beds, according to the HbA_{1c} values [57]. These patients with poor metabolic control have an increased risk for poor outcomes and elevated perioperative glucose levels. In general, there are few dietary restrictions in the hospital. The goal is adequate nutrition and wound healing. A diet that does not conform entirely to the recommended outpatient diet is better than not eating at all [58]. Most patients are in the fasting state (nothing by mouth/ nil per os (NPO)) at the start of the postoperative period, at least temporarily, and some patients may need nutritional support such as total parenteral nutrition (TPN) or enteral nutrition (EN). EN is usually administered continuously over 24 h at a fixed rate; however, if EN is likely to be needed after discharge, it may be given intermittently. This EN is often administered overnight (e.g., from 6:00 PM to 6:00 AM), and patients are encouraged to eat during the day. In patients for whom a switch from continuous to night time enteral feeding is recommended, the implementation of this change a couple of days before discharge will allow time to make adjustments in the insulin administration schedule [4, 58].

Instructions must be provided for the administration of subcutaneous supplemental doses of insulin. The blood glucose management treatment schedule should then follow the usual guidelines for inpatient glycaemic control [3].

8.7 Special Circumstances

8.7.1 Emergency Surgery

There are limited data concerning the perioperative care of patients with DM undergoing urgent and emergency surgery, which constitute an important portion of the surgical patients in many hospitals. Due to the absence of studies, the guidelines are based on expert opinions [36–38]. In emergency surgery, there is no opportunity for preoperative management, and the patients are at a particularly high risk in the post-operative period because of the catabolic stress and infection that may be facilitated by hyperglycemia and ketogenesis. It is crucial to maintain glycaemic control to optimize outcomes. The guidelines recommend close monitoring of blood glucose; if it rises above 180 mg/dl (10 mmol/l), then VRIII should be started [38]. The latter does not apply to patients with type 1 diabetes; the treating physician should always bear in mind that these patients are completely insulin-deficient and require continuous supplementation with insulin, in this case via the intravenous route. Therefore, most type 2 and all type 1 diabetic patients will require VRIII perioperatively and will need subcutaneous insulin to ensure good glycaemic control in the post-operative period [38].

8.7.2 Patients Treated with Continuous Subcutaneous Insulin Infusion (CSII)

In many countries, an increasing proportion of type 1 diabetic patients are treated with continuous subcutaneous insulin infusion (CSII), also known as insulin pump treatment. CSII is based on a basal rate of rapid-acting insulin that encompasses basal needs and should not cause hypoglycaemia. In addition, variable boluses of rapid-acting insulin are used to cover meal-induced postprandial glycaemic excursions.

There are very few data on the use of CSII in the management of diabetic patients undergoing surgery [59, 60]. In the absence of consensus guidelines regarding the perioperative use of CSII, many institutions and societies have developed their own approaches with conflicting recommendations [37, 38, 61]. Some of them recommend the termination of CSII during surgery, while others permit the continuation of CSII treatment during surgical procedures [61]. However, the clinical experience of the staff in this type of technology and multidisciplinary management to minimize the risk of medical errors at different stages of the surgical management is required.

Some authors recommend maintaining the basal rate the day before surgery, while others recommend reducing the rate by 10–20 % to prevent hypoglycaemia [37]. General recommendations for perioperative management of patients treated with CSII for surgeries with a short fasting period are summarized in Table 8.2. If the fasting period is short (only one meal missed), the basal rate can be maintained [37, 38] or decreased by 20–30 % [39] on the day of surgery, with regular CBG testing. The mealtime boluses should be stopped until the patient is able to eat and drink normally. If the blood glucose cannot be maintained in the target range in the intra-operative or immediate post-operative period, VRIII should be initiated [38]. However, if the fasting period is long (more than one meal missed), the insulin pump should be stopped and VRII initiated. If the CSII has been discontinued and replaced with a VRIII, the subcutaneous insulin infusion should be started once the patient starts eating and drinking normally and the VRII may be discontinued 30 min after re-initiating the subcutaneous route [38]. For some authors, CSII should only be continued during general anaesthesia if the duration of surgery is 1–2 h, the infusion site is secure, there are no potential interferences and the blood glucose is checked every hour to ensure correct pump function [61]. The correction of elevated blood glucose is best treated with subcutaneous rapid-acting insulin analogues and not with the pump [39].

In relation to the transfer from VRII to CSII, the CSII should be recommenced at the basal infusion rate that is usually used by the patient. VRIII should be continued until the next meal bolus has been administered. It is not recommended to re-initiate CSII at bedtime [38].

8.8 Hypoglycaemia

Hypoglycaemia is an undesirable complication of the treatment of DM, and it can result in neurological damage or even death. Low glucose levels induce a compensatory stress response and a typical set of symptoms. However, in the perioperative period and during critical illness, the manifestations of hypoglycaemia may be masked, the compensatory response may be blunted and the affected patients may be incapable of communicating the symptoms. The occurrence of unrecognized hypoglycaemia can have deleterious consequences and has been associated with increased mortality. It may be a iatrogenic complication of the aggressive and fast treatment of hyperglycemia, especially if stringent glucose control is pursued. It must be recalled that insulin, when improperly used, may be dangerous to patients and is considered to be one of the five most common drugs associated with medication errors of clinical significance [10]. The prevention, identification and management of hypoglycaemia in diabetic patients undergoing surgery are important issues because anaesthesia and sedation can mask the hypoglycaemic symptomatology. An alert value for hypoglycaemia is generally considered a CBG level below 70 mg/dl, although symptoms may occur at lower blood levels below 45–55 mg/dl [62].

In surgical diabetic patients with symptoms of hypoglycaemia, the guideline recommendations differ regarding the treatment. The SAMBA guidelines recommend the consumption of 10–25 g of glucose in a repeated manner until the blood glucose increases and the symptoms disappear. In symptomatic patients who are unable to take oral glucose and do not have immediate intravenous access, 1 mg of subcutaneous glucagon should be administered. If intravenous access is already established, the treatment should be a glucose dose of 20–50 ml of 50 % dextrose. However, the increase in blood glucose levels after oral or intravenous glucose administration is transient, and an adequate supply of glucose is warranted. Blood glucose should be measured 15 min after treatment [37]. However, the NHS guidelines suggest that if CBG is 72–108 mg/dl (4–6 mmol/l) and the patient has hypoglycaemic symptoms, then the administration of an intravenous bolus of 50–100 ml of 10 % dextrose and repeated CBG after 15 min must be considered. If the CBG is less than 72 mg/dl (4 mmol/l), then intravenous 80–100 ml of 20 % glucose and repeated CBG after 15 min is recommended. The frequency of blood glucose monitoring should be increased until normoglycaemia is achieved [38].

8.9 Late Postoperative Period and Discharge

As described previously, the perioperative period has different phases, including the recovery phase (immediate postoperative phase). However, there is a period that can be referred to as the late postoperative phase, during which the management of patients with DM is also important. This period encompasses the time from the immediate postoperative phase (48–72 h after surgery) until discharge. There are some important issues to consider during the late postoperative period, such as the treatment of the patient during the final hours/days of admission, the establishment of treatment after discharge, the recommendations regarding the health care level to which the patient is referred (primary care, specialized care or assistance in the hospital), and communication with the professional team that will follow up with the patient. In this period, it is also crucial to remember to review and/or to initiate, if necessary, the education of patients and their families or caregivers regarding DM, which is a fundamental component of the correct treatment of this condition (insulin administration, home glucose monitoring techniques, establishment of blood glucose goals, management of hypo and hyperglycemia, among others). This education must include all aspects related to DM, including nutritional therapy. The messages must be clear and understandable and be provided in both oral and written manner.

To adjust the treatment in the postoperative period and to indicate the post-discharge treatment, the presence of chronic diabetic complications and the previous treatment and glycaemic control should be determined. These data would allow the establishment of personalized glycaemic control goals for the future [46]. A recent HbA1c can determine the previous glycaemic control of the patient. Therefore, it is recommended that this information be obtained from the clinical records of the patient. If this information is not available, or the values were obtained more than 3 months prior, then HbA_{1c} should be determined on the first or second admission day in all patients with DM regardless of the cause of admission [7, 46].

It is recommended to administer the preadmission insulin regimen or oral and non-insulin injectable antidiabetic drugs during the late perioperative period when food intake is restored, the previous glycaemic control was correct and contraindications for preadmission diabetic therapy are not present. In diabetic patients who were previously treated with an insulin basal-bolus regimen, it is suggested that this treatment be restarted with dose adjustments, if necessary, according to the CBG. In diabetic patients who were previously treated with different treatment schedules before admission, the late postoperative treatment depends on the CBG during hospitalization and current guidelines on inpatient management should be followed. In patients who were not treated previously with insulin and require it, the insulin administration must be started at least 1 day before discharge to permit assessment of the efficacy and safety of the transition. These patients must receive specific basic diabetes education prior to discharge [7].

As general recommendations [7], all patients with good previous glycaemic control (HbA_{1c} at specific goal level for each patient) must be returned to the same pre-admission treatment and health care level. If the level of HbA_{1c} is above the acceptable limits, the patient was treated with diet or with diet and oral hypoglycaemic drugs and the need for insulin during admission was less than 20 IU/day, then these drugs should be initiated, augmented or complemented with another hypoglycaemic oral drug or an insulin bedtime regimen at a dose of 0.2 IU/kg (NPH, NPL, glargine or detemir). In the case of an admission insulin requirement of more than 20 UI/day, the addition of a bedtime insulin regimen at a higher dose of 0.5 IU/kg is recommended. In these cases, the follow up must be carried out at the same health care level as previously established. In the event of any contraindication for the use of oral hypoglycaemic drugs, a bedtime insulin regimen should be prescribed at a dose of 0.2 UI/kg, and the follow up must again be conducted at the previous health care level. The same procedure is recommended for patients who have been treated with two or more oral drugs without optimal previous glycaemic control. In patients treated with multiple insulin dose regimens who present insufficient previous glycaemic control, the dose of insulin should be increased according to the CBG and the insulin requirements during hospitalization. Improved consultation of the diabetes team should be established to determine the appropriate treatment and follow up of each patient who presents a very high HbA_{1c} level (>9–9.5 %) during admission [7, 46].

It is important to note that hospital admission for any cause can be an opportunity for the optimal modification of hypoglycaemic treatment and improvement of the long-term glycaemic control in patients with DM, particularly in the subgroup requiring treatment intensification with insulin therapy. Indeed, hospitalization may aid in overcoming barriers to the initiation of insulin [63, 64]. In our experience, a substantial number of patients with type 2 diabetes show significant improvements in glycaemic control (with a mean decrease of HbA_{1c} of 1 %) over short and medium-term durations after admission in a conventional surgical bed for any cause and evaluation, with the implementation of therapeutic program modifications in specific cases [57].

8.10 Concluding Remarks

Patients with diabetes mellitus are at increased risk of intra and postoperative complications. Apart from hyperglycemia, the presence of diabetic micro and macrovascular complications often places the diabetic patient at increased risk of perioperative morbidity and mortality. Additionally, the surgical procedure produces an increased demand of insulin that could potentially lead to hyperglycemic decompensation in these patients. Therefore, a thorough assessment of the patient's characteristics, including antidiabetic treatment, is required before surgery. The perioperative management, especially the hypoglycaemic treatment, should be tailored according to the patient's treatment schedule, and the characteristics of the surgical procedure (mainly type and duration). Although there is insufficient evidence concerning the optimal blood glucose targets, current guidelines recommend the maintenance of blood glucose concentrations between 6 and 10 mmol/l.

It is clear that much more research in this field is necessary. We have previously assessed the available evidence on the effect of intensive blood glucose control in patients with diabetes in a systematic review of randomized trials [16]. The limited number of trials and patients participating in these trials did not show an effect of optimal glycaemic control in terms of postoperative relevant outcomes, including mortality. Therefore, there are still important relevant clinical questions that should be addressed by future clinical studies. We are in great need of well-designed randomized clinical trials addressing the question of how tight should be the glycaemic control in patients with diabetes mellitus. Future trials should also address which are the best treatment perioperative management schedules according to the previous hypoglycaemic treatment and other important characteristics of the diabetic patient, especially the influence of previous co-morbid conditions on the surgical outcome. This should be defined in a thorough and uniform way.

References

1. World Health Organization [homepage on the Internet] (2014) Fact sheet 312: diabetes. WHO, Geneva. Available from: http://www.who.int/mediacentre/factsheets/fs312/en/
2. Sampson MJ, Dozio N, Ferguson B, Dhatariya K (2007) Total and excess bed occupancy by age, specialty and insulin use for nearly one million diabetes patients discharged from all English Acute Hospitals. Diabetes Res Clin Pract 77:92–98
3. Clement S, Braithwaite SS, Magee MF et al (2004) Management of diabetes and hyperglycemia in hospitals. Diabetes Care 27:553–591
4. Smiley DD, Umpierrez GE (2006) Perioperative glucose control in the diabetic or nondiabetic patient. South Med J 99:580–589
5. Frisch A, Chandra P, Smiley D et al (2010) Prevalence and clinical outcome of hyperglycemia in the perioperative period in noncardiac surgery. Diabetes Care 33:1783–1788
6. Robertshaw HJ, Hall GM (2006) Diabetes mellitus: anaesthetic management. Anaesthesia 61:1187–1190
7. Umpierrez GE, Isaac SD, Bazargan N, You X, Thaler LM, Kitabchi AE (2002) Hyperglycaemia: an independent marker of in-hospital mortality in patients with undiagnosed diabetes. J Clin Endocrinol Metab 87:978–982
8. Kadoi Y (2010) Anaesthetic considerations in diabetic patients. Part I: preoperative considerations of patients with diabetes mellitus. J Anesth 24:739–747
9. Poldermans D, Bax JJ, Boersma E et al, European Society of Cardiology (ESC) and European Society of Anaesthesiology (ESA) (2010) Guidelines for pre-operative cardiac risk assessment and perioperative cardiac management in non-cardiac surgery: the Task Force for Preoperative Cardiac Risk Assessment and Perioperative Cardiac Management in Non-cardiac Surgery of the European Society of Cardiology (ESC) and endorsed by the European Society of Anaesthesiology (ESA). Eur J Anaesthesiol 27(2):92–137
10. Akhtar S, Barash PG, Inzucchi SE (2010) Scientific principles and clinical implications of perioperative glucose regulation and control. Anesth Analg 110:478–497
11. Bagry HS, Raghavendran S, Carli F (2008) Metabolic syndrome and insulin resistance: perioperative considerations. Anesthesiology 108:506–523
12. McAnulty GR, Robertshaw HJ, Hall GM (2000) Anaesthetic management of patients with diabetes mellitus. Br J Anaesth 85:80–90
13. Dokun AO (2010) Lessons learned from glycaemia control studies. Curr Diab Rep 10:133–138
14. Gu W, Pagel PS, Warltier DC, Kersten JR (2003) Modifying cardiovascular risk in diabetes mellitus. Anesthesiology 98:774–779

15. Amour J, Kersten JR (2008) Diabetic cardiomyopathy and anaesthesia: bench to bedside. Anesthesiology 108:524–530
16. Buchleitner AM, Martínez-Alonso M, Hernández M, Solà I, Mauricio D (2012) Perioperative glycaemic control for diabetic patients undergoing surgery. Cochrane Database Syst Rev (9):CD007315
17. Latham R, Lancaster AD, Covington JF, Pirolo JS, Thomas CS Jr (2001) The association of diabetes and glucose control with surgical-site infections among cardiothoracic surgery patients. Infect Control Hosp Epidemiol 22:607–612
18. McMahon MM, Bistrian BR (1995) Host defenses and susceptibility to infection in patients with diabetes mellitus. Infect Dis Clin North Am 9:1–9
19. Stegenga ME, van der Crabben SN, Dessing MC et al (2008) Effect of acute hyperglycaemia and/or hyperinsulinaemia on proinflammatory gene expression, cytokine production and neutrophil function in humans. Diabet Med 25:157–164
20. Sampson MJ, Davies IR, Brown JC, Ivory K, Hughes DA (2002) Monocyte and neutrophil adhesion molecule expression during acute hyperglycemia and after antioxidant treatment in type 2 diabetes and control patients. Arterioscler Thromb Vasc Biol 22:1187–1193
21. Turina M, Miller FN, Tucker CF, Polk HC (2006) Short-term hyperglycemia in surgical patients and a study of related cellular mechanisms. Ann Surg 243:845–851
22. Schmeltz LR, DeSantis AJ, Thiyagarajan V et al (2007) Reduction of surgical mortality and morbidity in diabetic patients undergoing cardiac surgery with a combined intravenous and subcutaneous insulin glucose management strategy. Diabetes Care 30:823–828
23. Mangano CM, Diamondstone LS, Ramsay JG, Aggarwal A, Herskowitz A, Mangano DT (1998) Renal dysfunction after myocardial revascularization: risk factors, adverse outcomes, and hospital resource utilization. The Multicenter Study of Perioperative Ischemia Research Group. Ann Intern Med 128:194–203
24. Vinik AI, Maser RE, Mitchell BD, Freeman R (2003) Diabetic autonomic neuropathy. Diabetes Care 26:1553–1579
25. Kong MF, Horowitz M, Jones KL, Wishart JM, Harding PE (1999) Natural history of diabetic gastroparesis. Diabetes Care 22:503–507
26. Kadoi Y (2012) Blood glucose control in the perioperative period. Minerva Anestesiol 78:574–595
27. Blanshard H (2006) Endocrine and metabolic disease. In: Allman KG, Wilson IH (eds) Oxford handbook of anaesthesia. Oxford University Press, Oxford, pp 15–190
28. Sebranek JJ, Lugli AK, Coursin DB (2013) Glycaemic control in the perioperative period. Br J Anaesth 111(Suppl 1):18–34
29. Fragen RJ, Shanks CA, Molteni A, Avram MJ (1984) Effects of etomidate on hormonal responses to surgical stress. Anesthesiology 61:652–656
30. Belhoula M, Ciébiéra JP, De La Chapelle A, Boisseau N, Coeurveille D, Raucoules-Aimé M (2003) Clonidine premedication improves metabolic control in type 2 diabetic patients during ophthalmic surgery. Br J Anaesth 90:434–439
31. Hall GM, Lacoumenta S, Hart GR, Burrin JM (1990) Site of action of fentanyl in inhibiting the pituitary-adrenal response to surgery in man. Br J Anaesth 65:251–253
32. Leal N, Calvo R, Agrad FZ, Lukas JC, de la Fuente L, Suarez E (2005) Altered dose-to-effect of propofol due to pharmacokinetics in rats with experimental diabetes mellitus. J Pharm Pharmacol 57:317–325
33. Halter JB, Pflug AE (1980) Effect of sympathetic blockade by spinal anaesthesia on pancreatic islet function in man. Am J Physiol 239:E150–E155
34. Donatelli F, Vavassori A, Bonfanti S et al (2007) Epidural anaesthesia and analgesia decrease the postoperative incidence of insulin resistance in preoperative insulin-resistant subjects only. Anesth Analg 104:1587–1593
35. Kalichman MW, Calcutt NA (1992) Local anaesthetic-induced conduction block and nerve fiber injury in streptozotocin-diabetic rats. Anesthesiology 77:941–947

36. Dhatariya K, Levy N, Kilvert A et al, Joint British Diabetes Societies (2012) NHS diabetes guideline for the perioperative management of the adult patient with diabetes. Diabet Med 29:420–433
37. Joshi GP, Chung F, Vann MA et al (2010) Society for Ambulatory Anaesthesia consensus statement on perioperative blood glucose management in diabetic patients undergoing ambulatory surgery. Anesth Analg 111:1378–1387
38. Diabetes UK [homepage on the Internet] (2011) Management of adults with diabetes undergoing surgery and elective procedures: improving standards. JBDS-IP, London. Available from: http://www.diabetologists-abcd.org.uk/JBDS/JBDS_IP_Surgery_Adults_Full.pdf
39. Vann MA (2014) Management of diabetes medications for patients undergoing ambulatory surgery. Anesthesiol Clin 32:329–339
40. Tennyson C, Lee R, Attia R (2013) Is there a role for HbA_{1c} in predicting mortality and morbidity outcomes after coronary artery bypass graft surgery? Interact Cardiovasc Thorac Surg 17:1000–1008
41. Underwood P, Askari R, Hurwitz S, Chamarthi B, Garg R (2014) Preoperative A_{1C} and clinical outcomes in patients with diabetes undergoing major noncardiac surgical procedures. Diabetes Care 37(3):611–616
42. Aldam P, Levy N, Hall GM (2014) Perioperative management of diabetic patients: new controversies. Br J Anaesth 113:906–909
43. Bock M, Johansson T, Fritsch G et al (2015) The impact of preoperative testing for blood glucose concentration and haemoglobin A_{1c} on mortality, changes in management and complications in noncardiac elective surgery: a systematic review. Eur J Anaesthesiol 32: 152–159
44. Moghissi ES, Korytkowski MT, DiNardo M et al (2009) American Association of Clinical Endocrinologists (AACE) and American Diabetes Association (ADA) consensus statement on inpatient glycaemic control. Diabetes Care 32:1119–1131
45. Umpierrez GE, Hellman R, Korytkowski MT et al, Endocrine Society (2012) Management of hyperglycemia in hospitalized patients in non-critical care setting: an endocrine society clinical practice guideline. J Clin Endocrinol Metab 97:16–38
46. American Diabetes Association. Standards of medical care in diabetes-2015 (2015) Diabetes Care 38(Suppl 1):S1–S89
47. Sathya B, Davis R, Taveira T, Whitlatch H, Wu HC (2013) Intensity of peri-operative glycaemic control and postoperative outcomes in patients with diabetes: a meta-analysis. Diabetes Res Clin Pract 102:8–15
48. Hanninen M, McAlister FA, Bakal JA, van Diepen S, Ezekowitz JA (2013) Neither diabetes nor glucose-lowering drugs are associated with mortality after noncardiac surgery in patients with coronary artery disease or heart failure. Can J Cardiol 29:423–428
49. Salpeter SR, Greyber E, Pasternak GA, Salpeter EE (2010) Risk of fatal and nonfatal lactic acidosis with metformin use in type 2 diabetes mellitus. Cochrane Database Syst Rev (14):CD002967
50. British Medical Association and Royal Pharmaceutical Society of Great Britain (2010) British National Formulary, London. Available from: http://www.bnf.org
51. DiNardo M, Donihi AC, Forte P, Gieraltowski L, Korytkowski M (2011) Standardized glycaemic management and perioperative glycaemic outcomes in patients with diabetes mellitus who undergo same-day surgery. Endocr Pract 17(3):404–411
52. Diabetes UK [homepage on the Internet] (2014) The use of variable rate intravenous insulin infusion (VRIII) in medical inpatients. JBDS-IP, London. Available from: http://www.diabetologists-abcd.org.uk/JBDS/JBDS_IP_VRIII.pdf
53. Rosenblatt SI, Dukatz T, Jahn R et al (2012) Insulin glargine dosing before next-day surgery: comparing three strategies. J Clin Anesth 24:610–617
54. Simpson AK, Levy N, Hall GM (2008) Peri-operative i.v. fluids in diabetic patients-don't forget the salt. Anaesthesia 63:1043–1045

55. Umpierrez GE, Smiley D, Jacobs S et al (2011) Randomized study of basal-bolus insulin therapy in the inpatient management of patients with type 2 diabetes undergoing general surgery (RABBIT 2 surgery). Diabetes Care 34:256–261
56. Moitra VK, Greenberg J, Arunajadai S, Sweitzer B (2010) The relationship between glycosylated haemoglobin and perioperative glucose control in patients with diabetes. Can J Anaesth 57:322–329
57. Lucas AM, Guanyabens E, Zavala-Arauco R et al. Breaking therapeutic inertia in type 2 diabetes: active detection of in-patient cases allows improvement of metabolic control at midterm. Int J Endocrinol. 2015;2015:381415
58. Hoogwerf BJ (2006) Perioperative management of diabetes mellitus: how should we act on the limited evidence? Cleve Clin J Med 73(Suppl 1):S95–S99
59. Nassar AA, Boyle ME, Seifert KM et al (2012) Insulin pump therapy in patients with DM undergoing surgery. Endocr Pract 18:49–55
60. Ma D, Chen C, Lu Y et al (2013) Short-term effects of continuous subcutaneous insulin infusion therapy in perioperative patients with diabetes mellitus. Diabetes Technol Ther 15:1010–1018
61. Boyle ME, Seifert KM, Beer KA et al (2012) Guidelines for application of continuous subcutaneous insulin infusion (insulin pump) therapy in the perioperative period. J Diabetes Sci Technol 6:184–190
62. Seaquist ER, Anderson J, Childs B et al, American Diabetes Association, Endocrine Society (2013) Hypoglycaemia and diabetes: a report of a workgroup of the American Diabetes Association and the Endocrine Society. J Clin Endocrinol Metab 98:1845–1859
63. Griffifth ML, Boord JB, Eden SK, Matheny ME (2012) Clinical inertia of discharge planning among patients with poorly controlled diabetes mellitus. J Clin Endocrinol Metab 97:2019–2026
64. Wexler DJ, Beauharnais CC, Regan S, Nathan DM, Cagliero E, Larkin ME (2012) Impact of inpatient diabetes management, education, and improved discharge transition on glycaemic control 12 months after discharge. Diabetes Res Clin Pract 98:249–256

Chapter 9
Perioperative Management of Non-diabetic Patients with Hyperglycaemia (Stress-Induced Hyperglycaemia)

Gabriel Giménez-Pérez, Isabel Salinas, Manel Puig-Domingo, and Dídac Mauricio

Abstract Stress-induced hyperglycaemia (SIH) is defined as an acute and transient elevation in blood glucose during illness in patients who do not have diabetes and in whom hyperglycaemia will presumably subside after discharge. Surgical patients with SIH have increased risks of surgical site infections, cardiovascular morbidity and total mortality. However, the causal relationship between SIH and poor outcome is still a matter of debate. Although previous studies have advocated strict control of hyperglycaemia to diminish SIH-related morbidity, recent studies have supported the use of a less strict approach, with the targeting of glucose levels to above 140 mg/dl (7.8 mmol/l) to minimize the risk of hypoglycaemic events. Treatment of SIH is based on the use of intravenous insulin therapy with variable rate infusion, adjusted according to the results of blood glucose monitoring during the operative period, with a switch to subcutaneous insulin as soon as possible during the recovery phase. Patients with perioperative SIH should be tested for previously undiagnosed diabetes using the HbA1c test and treated accordingly. Patients without diabetes should be followed by their primary care team because they probably have an elevated risk of developing diabetes in the future.

Keywords Stress-induced hyperglycemia • Perioperative glycemic control • Non-diabetic subjects • Complications • Treatment of stress hyperglycemia

---

G. Giménez-Pérez
Unit of Endocrinology, Department of Medicine, Hospital General de Granollers, Avda. Francesc Ribas S/N, Granollers 08402, Spain
e-mail: ggimenez@fhag.es

I. Salinas • M. Puig-Domingo • D. Mauricio (✉)
Department of Endocrinology & Nutrition, University Hospital & Health Sciences Research Institute Germans Trias i Pujol, Autonomous University of Barcelona, Carretera Canyet S/N, Badalona 08916, Spain
e-mail: isalinas.germanstrias@gencat.cat; mpuigd@gmail.com; didacmauricio@gmail.com

K. Stuart-Smith (ed.), *Perioperative Medicine – Current Controversies*,
DOI 10.1007/978-3-319-28821-5_9

9.1 Diagnosis and Prevalence of Stress-Induced Hyperglycaemia in Surgical Patients

9.1.1 *Definition and Diagnosis*

Hyperglycaemia in hospitalized patients may reflect previously known diabetes or previously undiagnosed diabetes or simply the presence of an illness [1]. Although in each of these cases, hyperglycaemia may be related to illness-induced stress [2], for the purposes of this review, stress-induced hyperglycaemia (SIH) is defined as the acute and transient elevation of blood glucose during illness in patients who do not have diabetes and in whom hyperglycaemia will presumably subside after discharge.

The usual cut-off glucose level for the diagnosis of SIH is above 140 mg/dl (7.8 mmol/l), based on the fact that patients with levels that are persistently above this level might require treatment [1]. However, other authors have used different cut-off levels, including that used for the definition of diabetes, i.e., two times the fasting blood glucose level of ≥126 mg/dl (7 mmol/l) or random blood glucose level of ≥200 (11.1 mmol/l) [3] or values higher than those usually recommended, for example, a peak value of above 360 mg/dl (20 mmol/l) [4].

Disregarding patients with known diabetes who are identified through their medical records, it is useful to distinguish those with SIH from those with previously undiagnosed diabetes. The measurement of glycosylated haemoglobin (HbA_{1c}) is useful for this purpose. HbA_{1c} refers to the proportion of glycosylated haemoglobin and reflects the mean blood glucose concentration over the past 2–3 months [5]. Levels of above 6.5 % (48 mmol/mol) in undiagnosed patients suggest that diabetes preceded hospitalization [6]. Notably, the presence of haemoglobinopathies and any condition that changes red cell turnover, such as haemolytic anaemia and major blood loss, in addition to the receipt of blood transfusions, will lead to spurious HbA1c results [7].

9.1.2 *Prevalence*

Generally speaking, SIH is considered to be a common disorder affecting hospitalized patients. However, its prevalence varies depending on the subgroup of patients being evaluated, the diagnostic criteria used and the ability to exclude undiagnosed diabetes.

The true prevalence of SIH in surgical patients is not known. The reported prevalence of SIH in surgical patients by some observational and retrospective studies is indicated in Table 9.1, and they generally range from 20 to 40 % of patients. Notably, the highest prevalence has been observed in pancreatic surgery patients, a finding that might be easily explained by the direct damage of insulin secreting cells.

Table 9.1 Prevalence of stress induced hyperglycaemia (SIH) in patients without known diabetes in observational retrospective studies

	Patients	n	Diagnostic criteria	Prevalence (%)
Shi et al. [3]	Pancreatic surgery	65	Fasting blood glucose ≥126 mg/dl (7 mmol/l) Or Random blood glucose ≥200 mg/dl (11.1 mmol/l)	63
Chen et al. [8]	Hip fracture	1257	Fasting plasma glucose level ≥126 mg/dl (7 mmol/l)	48[a]
Richards et al. [9]	Orthopaedic injuries	790	Random glucose value ≥200 mg/dl (11.1 mmol/l)	37
Shi et al. [3]	Biliary tract or liver surgery (excluding cholecystectomy)	48	Fasting blood glucose ≥126 mg/dl (7 mmol/l) Or Random blood glucose ≥200 mg/dl (11.1 mmol/l)	32
Shi et al. [3]	Cholecystectomy	593	Fasting blood glucose ≥126 mg/dl (7 mmol/l) Or Random blood glucose ≥200 mg/dl (11.1 mmol/l)	21
Richards et al. [10]	Operative orthopaedic injuries	171	Fasting glucose level ≥126 mg/dl (7 mmol/l) Or Random blood glucose level ≥200 mg/dl (11.1 mmol/l)	20[a]
Whitcomb et al. [11]	Mixed intensive care units (medical & surgical)	2713	Random blood glucose level ≥200 mg/dl (11.1 mmol/l)	18.5
Umpierrez et al. [12]	Community hospital (medical & surgical wards)	2030	Fasting glucose level ≥126 mg/dl (7 mmol/l) Or Random blood glucose level ≥200 mg/dl (11.1 mmol/l)	12
Doenst et al. [4]	Cardiac surgery	4701	Intraoperative peak glucose >360 mg/dl (20 mmol/l)	7
Kerby et al. [13]	Trauma level I patients	6852	Serum glucose ≥200 mg/dl (11.1 mmol/l)	4[a]

[a]Studies which exclude undiagnosed diabetes

The main studies addressing the clinical significance of SIH have been performed on patients with acute coronary syndrome with or without coronary surgery and in those in critical care settings, and they have reported prevalences ranging from 40 to 80 % [14]. However, most of these studies have not excluded patients with known diabetes [15]. We were unable to find any studies reporting the prevalence of SIH in non-diabetic subjects undergoing cardiac surgery except for that of Doenst et al. [4], which reported a very low prevalence (7 %). However, in this study, the cut-off value used for defining SIH was far higher than the commonly used values, resulting also in a low prevalence of hyperglycaemia in patients with diabetes (15 %).

Chen et al. [8] reported a high prevalence (48 %) of SIH, indicating that age is an important factor to consider when evaluating whether a particular patient may develop this condition. The mean age of the patients in the study of Chen et al. [8] was 75 years, whereas the mean ages in the remainder of the studies listed in Table 9.1 were between 38 and 60 years. In particular, in the study of Kerby et al. [13] of patients at a level 1 trauma centre, the mean age of those without diabetes was 38 years, and this value likely explains, along with the exclusion of patients with previously undiagnosed diabetes, the low prevalence of SIH found in this population (4 %).

Other factors that affect the prevalence of SIH during the perioperative period include the use of hyperglycaemic medications such as corticosteroids, although this factor may have a lesser impact than previously envisaged [16], and the use of total parenteral nutrition with a high infusion rate of glucose (>4 mg/kg/min) [17].

9.2 Pathophysiology of Stress-Induced Hyperglycaemia

9.2.1 The Response to Stress and Stress-Induced Hyperglycaemia

The response to stress is mediated through activation of the hypothalamic paraventricular nucleus and locus ceruleus, which in turn activate the pituitary-adrenal axis and sympatho-adrenal system. The responses of these systems are graded with higher levels of stress (e.g., the APACHE score, Glasgow Coma Scale, type of surgery, etc.), leading to higher levels of cortisol and catecholamines [18, 19]. A high cortisol level increases hepatic glucose production, stimulates protein catabolism and increases circulating amino acid concentrations, providing precursors for gluconeogenesis. Epinephrine stimulates glucagon secretion and inhibits insulin release by pancreatic β-cells.

Activation of the pituitary-adrenal and sympatho-adrenal axes acts synergistically with proinflammatory cytokines (TNF-α, IL-1 and IL-6) to induce SIH via an increase in hepatic glucose production and the promotion of insulin resistance. Insulin resistance contributes to increased hepatic glucose production and decreased peripheral glucose uptake [2]. SIH seems to mainly depend on increased hepatic glucose production through the activation of hepatic gluconeogenesis. The TNF-α-mediated increase in glucagon synthesis is thought to play a main role in the increased hepatic gluconeogenesis.

9.2.2 Physiopathologic Consequences of Stress-Induced Hyperglycaemia

The increased plasma glucose level that occurs during stress alters the body's homeostasis through multiple partially known mechanisms. The interpretation of how these changes affect homeostasis, either in a positive or negative manner, is the key to considering SIH either as a physiologic and protective adaptation [19] or as an exaggerated and deleterious response that must be reversed [14].

Some authors consider SIH as a means to improve the glucose gradient between the plasma and the intracellular medium in non-insulin-dependent tissues, especially the brain and the reticuloendothelial system, in situations of maldistributed microvascular flow [19]. In fact, a shift in membrane glucose transporters expression occurs during stress that favours the transport of glucose down a concentration gradient to brain cells and macrophages via up-regulation of GLUT-1 [20]. Conversely, expression of GLUT-4, which favours glucose transport to peripheral tissues, is down-regulated during stress [21].

Furthermore, it has been postulated that acute hyperglycaemia may protect against cell death following ischaemia by promoting antiapoptotic pathways and favouring angiogenesis [19]. Hence, endogenous cardioprotective mechanisms, although they have only been tested in animal models, have been demonstrated to become transiently activated by short-term hyperglycaemia through enhancement of vascular endothelial growth factor and nitric oxide synthase (eNOS) expression, NO formation, activation of cell survival signals, and decreased oxidative stress [22].

Conversely, increases in counterregulatory hormones and insulin resistance can be deleterious by increasing lipolysis and free fatty acid concentrations, which may be toxic to the ischaemic myocardium, leading to cardiac arrhythmias, sympathetic overactivity, increased blood pressure, oxidative stress and endothelial dysfunction [23]. In addition hyperglycaemia can impair immune system function by reducing phagocytic activity of macrophages, impairing chemotaxis of polymorphonuclear neutrophils, and increasing expression of adhesion molecules and free radical production in immune cells, which will ultimately increase the risk of infection [24].

In conclusion, the question of whether SIH is beneficial or detrimental cannot be definitively determined because evidence exists in support of both arguments. From a practical standpoint, it would be wise to permit a certain degree of hyperglycaemia, avoiding excessively high glucose levels that can affect immune function and induce osmotic diuresis, volume depletion and electrolyte loss.

9.3 Association Between Stress-Induced Hyperglycaemia and Outcomes and Benefits of Treatment in Surgical Patients

9.3.1 Stress-Induced Hyperglycaemia and Clinical Outcomes

Most observational and retrospective studies of hyperglycaemia conducted on patients during the perioperative period have demonstrated a strong association between SIH and poor clinical outcomes, including infections, cardiovascular morbidity, mortality, and overall complications.

9.3.1.1 Infections

The association of SIH, either preoperatively or postoperatively, with surgical site infections has been demonstrated across different surgical populations, including breast [25], colorectal [26], bariatric [27], hepatopancreaticobiliary [28], vascular [29], transplant [30] and orthopaedic patients [9], with adjusted odds ratios ranging from 1.4 to 9.4 [31]. Intraoperative SIH (glucose >200 mg/dl (11.1 mmol/l)) has also been associated with surgical site infections in liver transplantation patients [30]. This rather strong association is accompanied by evidence of a biological gradient and consistency, in support of hyperglycaemia as a causal factor for surgical site infections [31]. Moreover, there is also a strong biological plausibility for this association because hyperglycaemia impairs phagocyte function through different mechanisms, such as impairment of leukocyte chemotaxis [32] and reductions in superoxide formation [33], respiratory burst activity [34] and release of lysosomal enzymes [35].

Although the association of hyperglycaemia with infections other than surgical site infections has been well demonstrated in surgical diabetic patients [36], studies relating SIH with distant infections are scarce. However, an association between SIH and pneumonia has been reported [13].

9.3.1.2 Cardiovascular Morbidity

In patients with hip fracture, the incidence of acute myocardial infarction is almost doubled in those with SIH (12.5 % vs. 6.4 %) in the absence of diabetes [8]. In patients with SIH, the peak incidence of acute myocardial infarction parallels the peak in the blood glucose level. Moreover, SIH is the sole significant independent risk factor for the development of acute myocardial infarction. Hyperglycaemia increases platelet activation, thereby enhancing platelet-mediated thrombogenesis [37]. Hence, it is plausible that SIH increases the risk of acute coronary disease in at-risk patients. It must be noted, however, that hip fracture is a specific situation in which the risk of AMI is increased [38]. An association of SIH with cardiovascular events has also been demonstrated in other high-risk situations, such as cardiac surgery [4].

9.3.1.3 Mortality

SIH is clearly associated with mortality in critically ill patients with unstable angina, acute myocardial infarction, congestive heart failure, arrhythmia, ischaemic and haemorrhagic stroke, gastrointestinal bleeding, acute renal failure, pneumonia, pulmonary embolism, and sepsis [39], independent of the illness severity score and length of hospital stay. The mortality rate for each predefined glucose category of above 110 mg/dl is higher in subjects without diabetes compared with those with diabetes, highlighting the importance of SIH as a risk factor for these diseases. However, in this same study, the association between SIH and mortality was not found to be significant or was highly attenuated for most surgical procedures performed on patients admitted to intensive care units.

SIH has also been associated with mortality in non-critically ill surgical patients at a community hospital [12]. Again, the mortality rate was higher for patients with SIH without known diabetes.

In trauma patients [13], SIH has been associated with mortality in those who are non-diabetic (39.7 ± 18.1 years), also excluding those with undiagnosed diabetes, whereas no association between hyperglycaemia and mortality has been found in diabetic patients (mean age 57.5 ± 15.7).

Conversely, in a retrospective study of 1574 patients undergoing coronary artery bypass surgery, hyperglycaemia was not found to be associated with mortality after adjusting for diabetes status and the calculated preoperative mortality or mediastinitis risk scores [40].

Although most studies showing an association between SIH and mortality have adjusted for disease severity, a causal association between SIH and mortality has not been completely established. It can be argued that since the association between the glucose level and mortality depends on the presence or absence of previously diagnosed diabetes, it is not glucose *per se* that causes the increase in mortality because if hyperglycaemia was the cause, the mortality rates would be rather similar in patients with and without diabetes, if not worse in diabetic patients, who are usually older and have a worse health status. On the other hand, some authors have argued that this stronger association might reflect a less intensive treatment of hyperglycaemia in patients without diabetes [12]. However, studies of critically ill and sepsis patients have shown that the relationship between SIH and mortality is lost once the lactate level is taken into account [41, 42]. In a previously described study of trauma patients [13], lactate levels were found to be significantly higher in non-diabetic patients with SIH.

9.3.1.4 Other Outcomes

The modulating effect of hyperglycaemia on functional recovery in patients with ischaemic cerebral injury has been extensively studied [43]. In surgical patients, an association between intraoperative hyperglycaemia and neuropsychological (for glucose concentrations $\geq$129 mg/dl) and gross neurological (for glucose concentrations ≥152 mg/dl) recovery has been demonstrated in those undergoing cerebral aneurysm surgery [44]. However, it is difficult to discern from the data of this study whether a high level of glucose is a causal factor or is merely a marker of more extensive neurologic injury occurring during the procedure.

9.3.2 Benefits of Treatment of Stress-Induced Hyperglycaemia

Association does not indicate causality; therefore, before recommending a specific glucose-lowering strategy to surgical patients, the beneficial effects of this strategy must be assessed in randomized controlled trials.

The traditional view on glycaemic control in critically ill patients was that acute hyperglycaemia represented a normal, and perhaps beneficial, adaptive response that promoted cellular glucose uptake [45]. In fact, this view is still defended by some experts [19] and supported by some recent publications [46]. With this approach, the tolerant management of SIH (glucose <200 mg/dl) is recommended to avoid the adverse effects of hyperglycaemia and increased osmolarity on granulocyte function [45].

The Leuven trial, conducted in 2001 [47] on critically ill patients in a single centre intensive care unit with predominantly surgical patients, showed benefits of intensive insulin therapy (aiming for a glucose level of between 80 and 110 mg/dl) vs. a conventional approach (aiming for a glucose level of below 215 mg/dl) on mortality, mainly due to improvements in the rates of multiorgan failure with a proven septic focus and morbidity (bloodstream infections, acute renal failure, critical illness polyneuropathy and prolonged mechanical ventilation). However, these positive results have not been reproduced in subsequent trials [14]. Randomized controlled trials conducted after the Leuven trial have mainly evaluated the benefits and risks of strict blood glucose control (between 80 and 110 mg/dl) vs. somewhat more permissive blood glucose control (less than 180 mg/dl (10 mmol/l)) in a mixed surgical and medical population of patients showing no benefits of the intensive strategy [48] or even an increase in 90-day mortality [49]. Interestingly, another randomized, although not blinded, study [50] comparing the intensive approach vs. an even more permissive approach (blood glucose level of between 180 mg/dl and 200 mg/dl (10–11.1 mmol/l)) [10] has also found no benefit of the intensive strategy on either 28-day mortality or morbidity. Furthermore, all of these studies have uniformly shown robust increases in hypoglycaemic events in the intensive arm [51], a fact that might be important in view of the association found between hypoglycaemic events and mortality in critically ill patients [52].

Notably, all of these studies were performed on a mixed population of patients with and without previously diagnosed diabetes, with the proportion of non-diabetic patients ranging from 70 to 87 % [14]. This approach to a complex problem might cause confusion because the treatment effects cannot be assumed to be the same in both groups of patients. Patients with already established diabetes may harbour clinical or subclinical vascular damage that can affect the outcome of acute disease and the response to insulin treatment. Additionally, a certain degree of hyperglycaemia may reflect a substantially different degree of stress, depending on the glycaemic status before the onset of the acute disease. In fact, it can be inferred that patients with diabetes have a higher level of glycaemia than non-diabetic patients experiencing the same level of stress, making it difficult to interpret studies reporting that SIH is particularly detrimental in patients without previously established diabetes [12, 39].

9.4 Treatment of Stress-Induced Hyperglycaemia

9.4.1 Preoperative Control

Worldwide epidemiological data indicate that 30–50 % of the population has impaired glycaemic control (diabetes or prediabetes), and up to 30 % of individuals are unaware of their status [53, 54]. Thus, the glucose levels of patients undergoing preoperative evaluation must be checked because unrecognized impaired glycaemia may lead to hyperglycaemia during the intraoperative or postoperative period. Patients especially predisposed to diabetes should be carefully evaluated prior to surgery.

Nevertheless, there is no consensus among scientific societies with regard to the criteria to use to screen for diabetes or prediabetes in asymptomatic patients. The American Diabetes Association recommends screening for diabetes or prediabetes in all adults aged 45 years and at even younger ages in overweight patients with other risk factors such as physical inactivity, a first-degree relative with diabetes, a high-risk ethnicity (African-American, Pacific Islander, or Latino), delivery of a baby weighing >9 lb for females, or the presence of polycystic ovarian syndrome, dyslipidaemia, a history of impaired glycaemia or glycosylated haemoglobin (HbA1c), the presence of hypertension with other clinical conditions associated with insulin resistance, such as obesity or acanthosis, or a history of cardiovascular disease [55]. The US Preventive Services Task Force only recommends screening for diabetes in adults with a blood pressure of >135/80 mmHg [56]. The Endocrine Society Guidelines for the management of hyperglycaemia in hospitalized patients in non-critical care settings recommends laboratory blood glucose testing in all patients upon admission and bedside glucose monitoring with a point-of-care glucose metre in all non-diabetic patients with a blood glucose level of >140 mg/dl (7.8 mmol/l) for at least 24–48 h. The same recommendation applies to patients receiving therapies that promote hyperglycaemia, such as corticosteroids, octreotide and enteral or parenteral nutrition [57]. However, current perioperative recommendations do not advocate testing for diabetes mellitus prior to surgery, except for cardiac surgery [58].

Patients diagnosed with diabetes during the preoperative period must be evaluated by their general practitioner or diabetologist before surgery and treated during the perioperative period, as discussed in a specific chapter of this book on the perioperative management of patients with diabetes.

9.4.2 Intraoperative Control

There is an absence of prospectively controlled studies specifically comparing patients with and without SHI; thus, no consensus has been reached on the intraoperative management of this condition in non-critical patients because no studies have distinguished these patients from previously diabetic patients. Regular monitoring of blood glucose and intravenous insulin therapy with variable rate infusion (VRIII) are the best methods for controlling hyperglycaemia in non-critically ill surgical patients. These methods result in added responsibilities and tasks for anaesthesiologists and nurses during surgical procedures. Guidelines for perioperative glycaemic control in all types of patients recommend defining and implementing a standard protocol for glucose control that must be implemented only after intensive training of all relevant staff [55, 59, 60].

It must be considered that multiple transient factors, such as pain, the type of surgery and the use of some drugs and anaesthetic agents, could modify glycaemia [61]. This issue is addressed in the chapter on the perioperative management of diabetic patients.

9.4.3 Glucose Monitoring

Monitoring glycaemia during anaesthesia is the only method available for diagnosing SIH during surgery. Ideally, blood glucose monitoring should be started while the patient is awaiting elective surgery and repeated intraoperatively, although there are no recommendations on the frequency of glycaemic monitoring on the operative day if insulin therapy is not prescribed. When VRIII is used hourly, blood glucose monitoring is mandatory.

Glycaemia may be monitored using a point-of-care glucose metre, which tests for capillary blood glycaemia using a blood gas/glucose analyser, or in the laboratory by performing measurements in whole arterial or venous blood, which is the recommended method by some scientific societies for critically patients and can be used if an arterial line is placed intraoperatively [62]. Levels measured in capillary samples using glucometers may differ by up to 20 % from those reported in laboratory results, and such a discrepancy is acceptable for hyperglycaemia, but not for hypoglycaemia. The accuracy of blood glucose monitoring is affected by various potential patient conditions, such as anaemia, oedema, skin hypoperfusion or concomitant use of drugs, such as dopamine, mannitol or noradrenaline. In these circumstances, the laboratory blood glucose level must be checked regularly.

9.4.4 Glucose Targets in Stress-Induced Hyperglycaemia

Present recommendations for SIH during the perioperative period reflect the results of recent trials, which have demonstrated no clear benefit or even harm resulting from the strict control of blood glucose levels. Therefore, all existing guidelines stress the need for avoiding hypoglycaemia and the goal of a blood glucose level of above 140 mg/dl (7.8 mmol/l). Some differences exist in the upper blood glucose target and therefore in the level above which insulin treatment is recommended. Whereas the American Diabetes Association [55] and Diabetes UK guidelines [59] recommend an upper blood glucose target of less than 180 mg/dl (10 mmol/l), the American College of Physicians [63] recommends a target of less than 200 mg/dl (11.1 mmol/l).

Recommendations of surgical guidelines also differ. For non-cardiac surgery, the guidelines of the European Cardiology and Anaesthesiology state that intraoperative prevention of hyperglycaemia with insulin may be considered only as a IIb class recommendation (conflicting evidence) [64]. On the other hand, for cardiac surgery, the 2009 American Society of Thoracic Surgeons guidelines recommends treatment of non-diabetic patients with a glucose level of >180 mg/dl (10.0 mmol/l) with intravenous insulin during cardiopulmonary bypass to maintain the level at <180 mg/dl (10 mmol/l). Patients with a persistently elevated serum glucose level after intervention should receive continuous insulin infusion to maintain the serum glucose level below this value during their ICU stay [65]. However, some authors consider that these recommendations are not based on a high level of evidence [66, 67].

Although some studies have indicated that intraoperative SIH may be detrimental [4], trials evaluating the effects of the strict control of blood glucose levels during the operative period do not support the use of stringent goals during this period [68]. Presently, the recommended blood glucose targets do not differ for different populations of surgical patients [67]. In addition, some evidence indicates that these targets should be reached without sharp fluctuations due to the association of glucose variability with poorer outcome [69].

9.4.5 Insulin Therapy

Intraoperatively, rapid-acting insulin must be infused intravenously at a variable rate (VRIII) according to regular blood glucose measurements to maintain serum glucose levels within a specified range. The insulin infusion rate should be reviewed in cases in which the target blood glucose level is not achieved. All types of rapid-acting insulin may be used for VRII. Further, it is usually accompanied by infusion of glucose and electrolytes to prevent insulin-induced hypoglycaemia and must be administered using independent volumetric infusion pumps.

Different insulin infusion rates are used depending on the insulin sensitivity of patients, as reflected by the response of the blood glucose level to the initial infusion

Table 9.2 Suggested intraoperative variable rate intravenous insulin infusion (VRIII) recommendation for a non-diabetic patient

When to start
Capillary blood glucose (CBG) >180 mg/dl (10.0 mmol/L)
How to set up an intravenous insulin infusion:
1 ml of rapid-acting insulin (100 IU) + 99 ml of 0.9 % saline solution
Concentration of 1 IU rapid-acting insulin/ ml of solution
Types of rapid-acting insulins:
Regular insulin is the most commonly used in VRIII but other types, as a part,
Glulisine and Lispro insulin, can also be used
Glucose monitoring
Capillary blood glucose should be checked hourly
Glucose target:
108–180 mg/dl (6–10 mmol/l), acceptable 72–216 mg/dl (4–12 mmol/l)

Different scales for variable rate intravenous insulin infusion

	Insulins rates (ml/h)		
Glucose mg/dl (mmol/l)	Reduced rate	Standard rate	Increased rate
<72 mg/dl (<4 mmol/l)	0	0	0
72–144 mg/dl (4–8 mmol/l)	0.5	1	2
145–216 mg/dl (8.05–12 mmol/l)	1	2	4
217–288 mg/dl (12.05–16 mmol/l)	2	4	6
289–360 mg/dl (16.05–20 mmol/l)	3	5	7
361–432 mg/dl (20.05–24 mmol/l)	4	6	8
>432 mg/dl (>24 mmol/l)	6	8	10

Adapted from Ref. [59]

Three different scales of variable rate intravenous insulin infusion (VRIII) are recommended in relation to different degrees of insulin sensibility of patients: a reduced rate for insulin sensitive patients, seems the most indicated to start in SIH. In case of CBG <72 mg/dl (4 mmol/l), hypoglycaemia should be treated and VRIII should be stopped for a short period (for instance, 30 min) with a subsequent re-assessment of CBG to determine the new infusion rate. In case of CGB >216 mg/dl (12 mmol/l) for three consecutive readings and is not dropping by 54 mg/dl/h (3 mmol/l/h) or more, the VRIII should be increase to next scale. In case that the patient is on the increased rate, a diabetes specialist team should be advised

rate, which is adjusted to maintain the level at between 108 and 180 mg/dl (6–10 mmol/l). Table 9.2 illustrates an example adapted from the VRIII Diabetes UK guidelines. Because SIH is a transient disorder in non-diabetic patients, it seems reasonable to start with a low infusion rate.

9.4.6 Hypoglycaemia

Hypoglycaemia due to the use of an inappropriate insulin infusion rate or inadequate monitoring is one of the most prevalent complications of VRIII and must be avoided and treated urgently because severe episodes can result in neurological damage or death. During the perioperative period, signs of hypoglycaemia cannot be recognized due to anaesthetic procedures. For its prevention, all patients receiving VRIII must be monitored at least hourly. Specific trials on hypoglycaemia management in perioperative hyperglycaemia treatment are lacking. The most recent guideline on VRIII therapy [59] states that intravenous insulin infusion should be stopped if the blood glucose level reaches 72 mg/dl (4 mmol/l), and 75–100 ml of 20 % glucose should then be administered over 15 min (i.e., 300–400 ml/h) by an infusion pump if it is promptly available. Blood glucose monitoring should be repeated in 15 min, and if the level is still less than 70 mg/dl, glucose infusion must be repeated. Insulin infusion should not generally be stopped for more than 20–30 min. If hypoglycaemia has been corrected, VRIII should be restarted at a reduced rate or at the same rate while increasing the glucose infusion rate (i.e., using 10 % glucose).

9.4.7 Postoperative Care

Recovery instructions must consider maintenance of good glycaemic control because SIH may be prolonged. It is necessary to continue monitoring capillary blood glucose every hour until VRIII is suspended and every 4–6 h thereafter during the first 24 h to ensure that no rebound hyperglycaemia occurs. If hyperglycaemia persists, general guidelines that are not specific for surgical patients recommend refraining from restarting VRIII unless the patient is unwell [59]. Instead, subcutaneous insulin must be administered before discontinuation of intravenous insulin to patients without a history of diabetes who have hyperglycaemia and require more than 2 IU/h of i.v. insulin [57], and a diabetes specialist should be consulted. Patients with SIH seem to require a lower subcutaneous insulin dose than diabetic patients, as has been described in cardiac surgery patients [70]. Some patients, such as those undergoing kidney transplantation, have a high risk of developing new-onset diabetes after transplantation, and it has been reported that tight glycaemic control during the immediate postoperative period (3 weeks) might reduce this risk [71]. The use of some concomitant medications, such as corticosteroids, and enteral or parenteral nutrition contribute to maintenance of hyperglycaemia, and subcutaneous insulin must be administered and its dosage must be adjusted closely in relation to any dose change or discontinuation of these drugs.

9.4.8 Discharge

Measuring the HbA1c level during the hospital stay in patients who develop perioperative hyperglycaemia and have not undergone a preoperative glycaemic test helps to exclude diabetes. For those with an HbA1c level of >6.5 % who are diagnosed as diabetic, a diabetes specialist team must be contacted to establish the appropriate treatment and education to be provided at discharge. Those with normal glycaemia and HbA1c level at discharge are to be advised to be followed by the primary care team [55].

9.5 Concluding Remarks

Perioperative SIH is associated with significant increases in morbidity and mortality. However, the causal link of SIH with poor outcomes is still a matter of debate because pathophysiologic evidence exists for SIH both as an adaptive and beneficial response and as a detrimental condition that should be intensively treated. For the time being and in view of recent studies, a conservative approach targeting glucose levels of above 140 mg/dl (7.8 mmol/l) is advised to minimize the risk of hypoglycaemic episodes while avoiding the detrimental effects of hyperosmolarity and very high glucose levels on immune function.

References

1. American Diabetes Association (2015) Diabetes care in the hospital, nursing home, and skilled nursing facility. Sec 13 in Standards of medical care in diabetes 2015. Diabetes Care 38(Suppl 1):S80–S85
2. Dungan KM, Braithwaite SS, Preiser JC (2009) Stress hyperglycaemia. Lancet 373:1798–1807
3. Shi Z, Tang S, Chen Y et al (2013) Prevalence of stress hyperglycaemia among hepatopancreatobiliary postoperative patients. Int J Clin Exp Med 6:799–803
4. Doenst T, Wijeysundera D, Karkouti K et al (2005) Hyperglycaemia during cardiopulmonary bypass is an independent risk factor for mortality in patients undergoing cardiac surgery. J Thorac Cardiovasc Surg 130:1144–1150
5. Reynolds TM, Smellie WS, Twomey PJ (2006) Glycated (HbA_{1c}) monitoring. BMJ 333:586–588
6. Saudek CD, Herman WH, Sacks DE et al (2008) A new look at screening and diagnosing diabetes mellitus. J Clin Endocrinol Metab 93:2447–2453
7. International Expert Committee International Expert Committee report on the role of the A1C assay in the diagnosis of diabetes (2009). Diabetes Care 32:1327–1334
8. Chen Y, Yang X, Meng K et al (2013) Stress-induced hyperglycaemia after hip fracture and the increased risk of acute myocardial infarction in nondiabetic patients. Diabetes Care 36:3328–3332
9. Richards JE, Kauffmann RM, Zuckerman SL, Obremskey WT, May AK (2012) Relationship of hyperglycaemia and surgical-site infection in orthopaedic surgery. J Bone Joint Surg Am 94:1181–1186
10. Richards JE, Hutchinson J, Mukherjee K et al (2014) Stress hyperglycaemia and surgical site infection in stable nondiabetic adults with orthopedic injuries. J Trauma Acute Care Surg 76:1070–1075
11. Whitcomb BW, Pradhan EK, Pittas AG et al (2005) Impact of admission hyperglycaemia on hospital mortality in various intensive care unit populations. Crit Care Med 33:2772–2777
12. Umpierrez GE, Isaacs SD, Bazargan N, You X, Thaler LM, Kitabchi AE (2002) Hyperglycaemia: an independent marker of in-hospital mortality in patients with undiagnosed diabetes. J Clin Endocrinol Metab 87:978–982
13. Kerby JD, Griffin RL, MacLennan P, Rue LW 3rd (2012) Stress-induced hyperglycaemia, not diabetic hyperglycaemia, is associated with higher mortality in trauma. Ann Surg 256:446–452
14. Farrokhi F, Smiley D, Umpierrez GE (2011) Glycemic control in non-diabetic critically ill patients. Best Pract Res Clin Endocrinol Metab 25:813–824
15. Schmeltz LR, DeSantis AJ, Thiyagarajan V et al (2007) Reduction of surgical mortality and morbidity in diabetic patients undergoing cardiac surgery with a combined intravenous and subcutaneous insulin glucose management strategy. Diabetes Care 30:823–828

16. Abdelmalak BB, Bonilla AM, Yang D et al (2013) The hyperglycaemic response to major noncardiac surgery and the added effect of steroid administration in patients with and without diabetes. Anesth Analg 116:1116–1122
17. Rosmarin DK, Wardlaw GM, Mirtallo J (1996) Hyperglycaemia associated with high, continuous infusion rates of total parenteral nutrition dextrose. Nutr Clin Pract 11:151–156
18. Marik PE (2009) Critical illness related corticosteroid insufficiency. Chest 135:181–193
19. Marik PE, Bellomo R (2013) Stress hyperglycemia: an essential survival response! Crit Care 17:305
20. Maratou E, Dimitriadis G, Kollias A et al (2007) Glucose transporter expression on the plasma membrane of resting and activated white blood cells. Eur J Clin Invest 37:282–290
21. Lang CH, Dobrescu C, Mészáros K (1990) Insulin-mediated glucose uptake by individual tissues during sepsis. Metabolism 39:1096–1107
22. Ma G, Al-Shabrawey M, Johnson JA et al (2006) Protection against myocardial ischemia/reperfusion injury by short-term diabetes: enhancement of VEGF formation, capillary density, and activation of cell survival signalling. Naunyn Schmiedebergs Arch Pharmacol 373:415–427
23. Oliver MF, Opie LH (1994) Effects of glucose and fatty acids on myocardial ischaemia and arrhythmias. Lancet 343:155–158
24. Weekers F, Giulietti AP, Michalaki M et al (2003) Metabolic, endocrine, and immune effects of stress hyperglycemia in a rabbit model of prolonged critical illness. Endocrinology 144:5329–5338
25. Vilar-Compte D, Alvarez de Iturbe I, Martín-Onraet A et al (2008) Hyperglycaemia as a risk factor for surgical site infections in patients undergoing mastectomy. Am J Infect Control 36:192–198
26. Serra-Aracil X, García-Domingo MI, Parés D et al (2011) Surgical site infection in elective operations for colorectal cancer after the application of preventive measures. Arch Surg 146:606–612
27. Kwon S, Thompson R, Dellinger P et al (2013) Importance of perioperative glycaemic control in general surgery: a report from the Surgical Care and Outcomes Assessment Program. Ann Surg 257:8–14
28. Ambiru S, Kato A, Kimura F et al (2008) Poor postoperative blood glucose control increases surgical site infections after surgery for hepato-biliary-pancreatic cancer: a prospective study in a high-volume institute in Japan. J Hosp Infect 68:230–233
29. Ramos M, Khalpey Z, Lipsitz S et al (2008) Relationship of perioperative hyperglycaemia and postoperative infections in patients who undergo general and vascular surgery. Ann Surg 248:585–591
30. Park C, Hsu C, Neelakanta G et al (2009) Severe intraoperative hyperglycaemia is independently associated with surgical site infection after liver transplantation. Transplantation 87:1031–1036
31. Kao LS, Phatak UR (2013) Glycemic control and prevention of surgical site infection. Surg Infect (Larchmt) 14:437–444
32. Mowat A, Baum J (1971) Chemotaxis of polymorphonuclear leukocytes from patients with diabetes mellitus. N Engl J Med 284:621–627
33. Sato N, Kashima K, Ohtani K et al (1997) Epalrestat, an aldose reductase inhibitor, improves an impaired generation of oxygen-derived free radicals by neutrophils from poorly controlled NIDDM patients. Diabetes Care 20:995–998
34. Nielson C, Hindson D (1989) Inhibition of polymorphonuclear leukocyte respiratory burst by elevated glucose concentration in vitro. Diabetes 38:1031–1035
35. McManus LM, Bloodworth RC, Prihoda TJ et al (2001) Agonist-dependent failure of neutrophil function in diabetes correlates with extent of hyperglycaemia. J Leukoc Biol 70:395–404
36. Tsang ST, Gaston P (2013) Adverse peri-operative outcomes following elective total hip replacement in diabetes mellitus: a systematic review and meta-analysis of cohort studies. Bone Joint J 95-B:1474–1479

37. Shechter M, Merz CN, Paul-Labrador MJ et al (2000) Blood glucose and platelet-dependent thrombosis in patients with coronary artery disease. J Am Coll Cardiol 35:300–307
38. Chiang CH, Liu CJ, Chen PJ et al (2013) Hip fracture and risk of acute myocardial infarction: a nationwide study. J Bone Miner Res 28:404–411
39. Falciglia M, Freyberg RW, Almenoff PL et al (2009) Hyperglycemia-related mortality in critically ill patients varies with admission diagnosis. Crit Care Med 37:3001–3009
40. Estrada CA, Young JA, Nifong LW et al (2003) Outcomes and perioperative hyperglycemia in patients with or without diabetes mellitus undergoing coronary artery bypass grafting. Ann Thorac Surg 75:1392–1399
41. Green JP, Berger T, Garg N et al (2012) Hyperlactatemia affects the association of hyperglycemia with mortality in nondiabetic adults with sepsis. Acad Emerg Med 19:1268–1275
42. Kaukonen KM, Bailey M, Egi M (2014) Stress hyperlactatemia modifies the relationship between stress hyperglycemia and outcome: a retrospective observational study. Crit Care Med 42:1379–1385
43. Capes SE, Hunt D, Malmberg K et al (2001) Stress hyperglycemia and prognosis of stroke in nondiabetic and diabetic patients: a systematic overview. Stroke 32:2426–2432
44. Pasternak JJ, McGregor DG, Schroeder DR et al (2008) Hyperglycemia in patients undergoing cerebral aneurysm surgery: its association with long-term gross neurologic and neuropsychological function. Mayo Clin Proc 83:406–417
45. Egi M, Bellomo R, Stachowski E et al (2008) Blood glucose concentration and outcome of critical illness: the impact of diabetes. Crit Care Med 36:2249–2255
46. Tiruvoipati R, Chiezey B, Lewis D (2012) Stress hyperglycemia may not be harmful in critically ill patients with sepsis. J Crit Care 27:153–158
47. Van den Berghe G, Wouters P, Weekers F et al (2001) Intensive insulin therapy in critically ill patients. N Engl J Med 345:1359–1367
48. Preiser JC, Devos P, Ruiz-Santana S et al (2009) A prospective randomised multi-centre controlled trial on tight glucose control by intensive insulin therapy in adult intensive care units: the Glucontrol study. Intensive Care Med 35:1738–1748
49. NICE-SUGAR Study Investigators, Finfer S, Chittock DR et al (2009) Intensive versus conventional glucose control in critically ill patients. N Engl J Med 360:1283–1297
50. De La Rosa GC, Donado JH, Restrepo AH et al (2008) Strict glycaemic control in patients hospitalised in a mixed medical and surgical intensive care unit: a randomised clinical trial. Crit Care 12(5):R120
51. Griesdale DE, de Souza RJ, van Dam RM et al (2009) Intensive insulin therapy and mortality among critically ill patients: a meta-analysis including NICE-SUGAR study data. CMAJ 180:821–827
52. Egi M, Bellomo R, Stachowski E et al (2010) Hypoglycemia and outcome in critically Ill patients. Mayo Clin Proc 85:217–224
53. Cowie CC, Rust KF, Ford ES et al (2009) Full accounting of diabetes and-pre-diabetes in the US population in 1988–1994 and 2005–2006. Diabetes Care 32:287–294
54. Soriguer F, Goday A, Bosch-Comas A et al (2012) Prevalence of diabetes mellitus and impaired glucose regulation in Spain: the Di@bet.es Study. Diabetologia 55:88–93
55. American Diabetes Association (2015) Classification and diagnosis. Diabetes Care 38 (Suppl 1):S8–S16
56. Calonge N, Petitti DB, DeWitt TG et al (2008) Screening for type 2 diabetes mellitus in adults: U.S. Preventive Services Task Force recommendation statement. Ann Intern Med 148:846–854
57. Umpierrez GE, Hellman R, Korytkowski MT et al (2012) Management of hyperglycaemia in hospitalized patients in non-critical care setting: an Endocrine Society Clinical Practice Guideline. J Clin Endocrinol Metab 97:16–38
58. Fleisher LA, Beckman JA, Brown KA et al (2007) ACC/AHA 2007 guidelines on perioperative cardiovascular evaluation and care for noncardiac surgery: executive summary. J Am Coll Cardiol 50:1707–1732
59. Diabetes UK [homepage on the Internet] (2014) The use of variable rate intravenous insulin infusion (VRIII) in medical inpatients. JBDS-IP, London. Available from: http://www.diabetologists-abcd.org.uk/JBDS/JBDS_IP_VRIII.pdf

60. Alexanian SM, Mc Donnell ME, Akhtar S (2011) Creating a perioperative glycaemic control program. Anesthesiol Res Pract 2011:465974
61. Takana K, Kawano Y, Tsuysumi YM et al (2011) Differential effects of propofol and isoflurane on glucose utilization and insulin secretion. Life Sci 88(1–2):96–103
62. Ichai C, Preiser JC (2010) International recommendations for glucose control in adult non-diabetic critically ill patients. Crit Care 14:R166
63. Qaseem A, Humphrey LL, Chou R et al (2011) Use of intensive insulin therapy for the management of glycemic control in hospitalized patients: a clinical practice guideline from the American College of Physicians. Ann Intern Med 154:260–267
64. Poldermans D, Bax JJ, Boersma E et al (2009) Guidelines for pre-operative cardiac risk assessment and perioperative cardiac management in non-cardiac surgery. Eur Heart J 20:2769–2812
65. Lazar HL, Mc Donnell M, Chipkin SR et al (2009) The Society of thoracic Surgeons practice guideline series: blood glucose management during cardiac surgery. Ann Thorac Surg 87:663–669
66. Akhtar S, Barash PG, Inzucchi SE (2010) Scientific principles and clinical implications of perioperative glucose regulation and control. Anesth Analg 110:487–497
67. Duncan AE (2012) Hyperglycemia and perioperative glucose management. Curr Pharm Des 18:6195–6203
68. Gandhi GY, Nuttall GA, Abel MD et al (2007) Intensive intraoperative insulin therapy versus conventional glucose management during cardiac surgery: a randomized trial. Ann Intern Med 146:233–243
69. Egi M, Bellomo R (2009) Reducing glycemic variability in intensive care unit patients: a new therapeutic target? J Diabetes Sci Technol 3:1302–1308
70. Dungan K, Hall C, Shuster D et al (2011) Differential response between diabetes and stress induced hyperglycemia to algorithmic use of detemir and flexible mealtime as part among stable post-cardiac surgery patients requiring intravenous insulin. Diabetes Obes Metab 13:1130–1135
71. Kuo HT, Sampaio MS, Vicenti F et al (2010) Associations of pretransplant diabetes mellitus, new-onset diabetes after transplant and acute rejection with transplant outcomes: an analysis of the Organ Procurement and transplant network/United network for organ sharing database. Am J Kidney Dis 56:1127–1139

Chapter 10
Postoperative Pulmonary Complications

Paul Forrest

Abstract Postoperative pulmonary complications (PPCs) are amongst the most important causes of morbidity and mortality in patients undergoing major non-cardiothoracic surgery. This chapter will examine how the risk of PPCs may be assessed and minimised, along with some of the key perioperative considerations in patients with pre-existing respiratory disease.

Keywords Postoperative pulmonary complications • CPAP • Obstructive sleep apnoea • Chronic obstructive pulmonary disease

10.1 What are PPCs and Why are They Important?

Unlike cardiac complications, there are few widely accepted, standard definitions of PPCs. They can include common conditions that have relatively minor clinical significance (such as transient hypoxia or mild bronchospasm) to major, life-threatening complications such as pneumonia.

Postoperative pneumonia has a similar mortality to cardiovascular complications from major non-cardiothoracic surgery [1] and significantly prolongs both the duration and expense of hospitalisation. In fact, major respiratory complications are more expensive to manage than any other complications (including infective, cardiac and thromboembolic) in this surgical cohort [2].

10.2 What are the Risk Factors for PPCs?

Major PPCs such as pneumonia and prolonged ventilation have been the focus of some very large observational studies. For example, in a study of 315,000 major vascular or general surgical procedures in US Veterans, postoperative pneumonia

P. Forrest, MB ChB, FANZCA
Department of Anaesthetics, Royal Prince Alfred Hospital,
University of Sydney Medical School, Sydney, NSW, Australia
e-mail: paul.forrest@sydney.edu.au

K. Stuart-Smith (ed.), *Perioperative Medicine – Current Controversies*,
DOI 10.1007/978-3-319-28821-5_10

Table 10.1 Risk factors for the development of postoperative pneumonia and their approximate odds-ratios

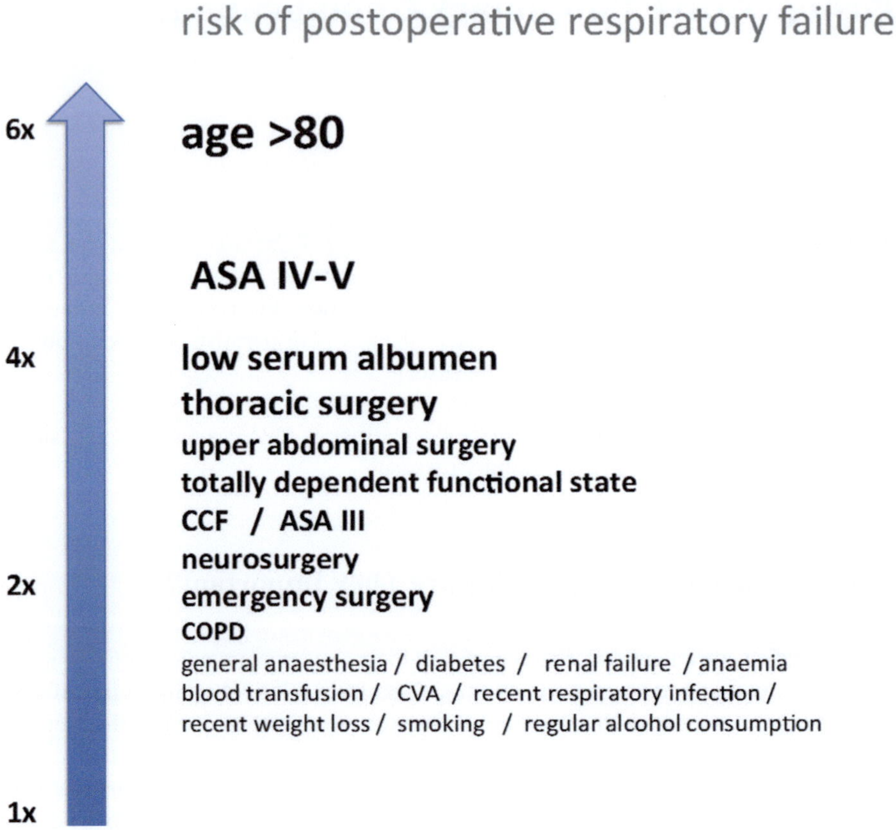

occurred in 1.5 % of patients, of whom 21 % died [3]. By far the most important risk factors were the patient's age, type of surgery and functional state.

In a subsequent study of 180,000 similar surgical cases that included a higher proportion of females, postoperative respiratory failure (defined as re-intubation or unplanned ventilation for more than 48 h) occurred in 3 % of patients, of whom more than 25 % died. By contrast, the mortality in patients who did not develop postoperative respiratory failure was only 1.4 % [4].

The major risk factors for postoperative respiratory failure that were identified in these two papers are represented in Table 10.1.

These authors also developed a multifactorial risk factor index for predicting postoperative respiratory failure. Postoperative respiratory failure occurred in <0.01 % of low -risk patients (who had a combined risk factor score <8), 0.8 % of medium -risk patients (score 8–12) and 6.8 % of high-risk patients. Very high-risk patients (score >25) had a 40 % incidence of postoperative respiratory failure.

More recently, an index for the risk of re-intubation was developed from a study of 34,000 patients. This comprised: ASA grade (III or IV: 3 points), emergency surgery (3 points), high-risk surgery (2 points), congestive cardiac failure (2 points) and COPD (1 point). The risk of re-intubation was significantly higher in patients with a score of 7–11 (5.9 %) than in patients with a score of 0 (0.1 %, $p<0.001$) [5].

These large trials provide insights into the multiplicity of factors that contribute to PPCs. In addition, respiratory risk indices are valuable research tools for investigating strategies to decrease their incidence. However, because of their low sensitivity, these indices have limited utility in the clinical management of individual patients. For example, it is unlikely that a patient would be denied essential surgery even if they had a high predicted risk of PPCs (i.e. more than 10 %). Similarly, a high predicted risk of re-intubation (i.e. about 6 %) is too insensitive to guide extubation and for the same reason, has limited utility in determining the allocation of intensive care resources. More research is needed to determine whether the sensitivity of risk indices for PPCs in high-risk patients can be improved by the addition of other variables, such as (for example) respiratory investigations or functional testing.

10.3 Patient-related Risk Factors for PPCs

While it should be no surprise that age and type of surgery are the most important risk factors for PPCs, the relative contribution of some other patient risk factors may be less widely appreciated.

10.3.1 Pulmonary Risk Factors

10.3.1.1 Chronic Obstructive Lung Disease

Chronic obstructive pulmonary disease (COPD) approximately doubles the risk of PPCs overall [4]. In one study, more than a third of patients with severe COPD ($FEV_1 \leq 1.2$ L and $FEV_1/FVC < 75$ %) who underwent major non-cardiothoracic surgery had major PPCs and nearly half of them died within 2 years of surgery [6]. However, well controlled mild to moderate asthma does not significantly increase the risk of PPCs [7].

10.3.1.2 Interstitial Lung Disease

The influence of other lung diseases on PPCs is much less well studied. A recent study showed that interstitial lung disease (which was mostly idiopathic pulmonary fibrosis) also approximately doubled the risk of PPCs. This risk was further increased in males (OR 1.8) and in patients with a low BMI (OR 2.5). However, in common with other studies, surgical factors (such as operative site and emergency procedures) had a much greater influence on the risk of PPCs [8].

10.3.1.3 Pulmonary Hypertension

Severe pulmonary hypertension has only relatively recently been recognised as a significant risk factor for postoperative respiratory failure (~25 %) and perioperative mortality. In a study of 145 patients undergoing major non-cardiac surgery with a mean pulmonary arterial systolic pressure (PAPs) of 68 mmHg, 16 patients (11 %) developed congestive heart failure and 5 patients (3.5 %) died from right ventricular (RV) failure [9]. In another study of 62 patients with more severe pulmonary hypertension (mean PAPs of 79 mmHg) undergoing similar surgery, mortality was 9.7 % vs. 0 % for case-matched controls [10]. Most of these deaths occurred due to acute RV failure.

While routine preoperative screening for pulmonary hypertension is not advocated, these studies highlight its clinical significance and the importance of preserving RV function in these patients perioperatively [11].

10.3.2 Non-pulmonary Risk Factors for PPCs

There are other important, but non-modifiable non-pulmonary patient risk factors that increase the risk of PPCs up to threefold. These include an ASA classification of >2, congestive cardiac failure and functional dependence [4].

10.3.2.1 Obesity

Contrary to popular perception, obesity is not strongly associated with an increased risk of PPCs. For example, in a recent review of 20 papers, obesity did not predict an increased risk of unplanned admission following day surgery [12].

Although morbid obesity does increase the risk of several postoperative complications (such as wound infections), it does not significantly increase the risk of postoperative respiratory failure [13, 14]. This is surprising, given the high incidence of obstructive sleep apnoea (OSA) in this population, which is more than 70 % in patients undergoing bariatric surgery [15].

However, super obese patients (BMI >50 kg/m^2) have an increased risk of obesity hypoventilation syndrome, heart failure and pulmonary hypertension. They also have an increased risk of perioperative complications from bariatric surgery compared with less obese patients, including airway difficulties, venous thromboembolism, prolonged hospitalisation and death [16].

Hence while day surgery appears to be well tolerated in patients with a BMI <40 kg/m^2, this may not be the case for super obese patients.

10.3.2.2 Obstructive Sleep Apnoea

Obstructive sleep apnoea is characterised by periods of partial or complete upper airway collapse and obstruction during normal sleep, which is aggravated by the respiratory depressant effects of general anaesthetic agents, sedatives and opiates [17]. Several studies have shown that OSA increases the risk of PPCs, especially hypoxaemia and airway difficulties [7, 18]. However, there remains considerable clinical uncertainty (and controversy) regarding the optimal perioperative management of OSA.

Recently published guidelines from the American Society of Anesthesiologists recommend that the decision to perform day surgery in patients with OSA should be based on its severity, the invasiveness of the surgery and the likely requirement for postoperative opioids [19]. However, these guidelines have not been validated clinically, are largely based on expert opinion and are poorly supported by the available literature.

For example, severity of OSA has not been proven to increase the risk of perioperative complications. In a prospective study of 797 patients undergoing bariatric surgery at the Mayo Clinic (all of whom had preoperative sleep studies), severity of OSA was not associated with an increased rate of any postoperative complications [20]. In an earlier study from the same institution, OSA was also not found to be a significant risk factor for unplanned admissions (or other adverse events) in patients undergoing outpatient non-ENT surgery [21]. However, OSA had been identified preoperatively in all patients in these two studies, so the perioperative risk of undiagnosed or untreated OSA was not examined. It is also possible that patients diagnosed with severe OSA may have been subject to selection bias and may have received different management (e.g. more intensive monitoring or lower opiate doses).

While the above studies examined the perioperative impact of OSA in previously diagnosed patients, Stierer et al. [22] assessed its prevalence and significance in previously undiagnosed patients. A screening survey for OSA (the Maislin Index Score) was administered to 3553 patients having day surgery at Johns Hopkins Hospital. One hundred and three patients (4.8 %) were identified as having a high propensity for OSA, which had not been previously diagnosed in three quarters of these cases. The survey results were not disclosed to their anesthesiologists. Patients with a high propensity for OSA did not have an increased rate of unplanned hospital admission, however they did have an increased incidence of airway difficulties and a greater requirement for postoperative oxygen. In addition, these patients were more likely to be on antihypertensive medications and to receive intraoperative pressors and beta-blockers, which is consistent with the increased incidence of hypertension and cardiac rhythm disturbances found in OSA patients [23].

Although there is good evidence that careful patient management in experienced centres mitigates the potential risks of OSA in patients undergoing outpatient surgery, these findings are not generalisable to less well resourced facilities, such as small private hospitals and stand -alone day-surgical units. This is an important distinction, because inadequate postoperative monitoring (and excessive opioid administration) have been implicated in the deaths of patients with OSA following relatively minor surgical procedures [24].

10.3.3 Why Assessing the Risk of PPC May (or May Not) Be Useful

10.3.3.1 Informed Patient Consent

The preoperative identification of patients who are at increased risk of PPCs facilitates informed consent for surgery, and gives patients an opportunity to discuss the limits of therapy that they are willing to receive. For example, some patients may not be willing to undergo surgery if it entailed a significant risk of prolonged post-operative ventilation. However, as discussed above, because of their poor sensitivity, the available risk indices have limited clinical utility in predicting this complication.

10.3.3.2 Targeting Interventions in High-Risk Patients

Another reason that may be advocated for assessing the risk of PPCs preoperatively is that preventative strategies can be implemented in high-risk patients. However, strategies that have been proven to be effective in decreasing PPCs (such as lung expansion techniques) are cheap, safe and easily implemented, hence they should be used in all patients undergoing major surgery, regardless of their level of baseline risk.

10.4 Preoperative Respiratory Investigations

Preoperative respiratory investigations have a very limited role in the assessment of patients undergoing major non- cardiothoracic surgery, which is primarily based on clinical history and examination. This is because routine preoperative screening investigations are poor predictors of PPCs.

For example, spirometry is no better than clinical assessment by history and examination in predicting the risk of PPCs and is therefore not recommended as a routine screening test [4]. In contrast to patients undergoing lung resections, there is also no lower limit of FEV_1 that reliably precludes non-thoracic surgery.

Similarly, in the absence of other risk factors, a routine preoperative chest X-ray is not indicated in patients less than 70 years of age, because it mostly detects pre-existing, chronic conditions. Routine preoperative CXR screening has also not been shown to significantly alter perioperative management or improve outcome [4, 25].

Cardiopulmonary exercise (CPEX) testing is a non-invasive method of assessing combined cardiovascular and pulmonary function. Unlike other screening tests, the anaerobic threshold that is derived from CPEX testing may provide a quantitative means of assessing "fitness for surgery," as it has been shown to independently predict morbidity, mortality and duration of hospitalisation after major abdominal surgery [26]. In a trial currently underway in London, perioperative care guided by CPEX testing is being compared with standard management in patients undergoing colorectal surgery.

10.4.1 Preoperative Strategies to Decrease PPCs

10.4.1.1 Smoking Cessation

Cigarette smokers have a higher incidence of mortality and major cardiac and respiratory complications than non-smokers [27, 28].

Smokers are commonly advised to stop smoking at least 4–8 weeks preoperatively, as this has been thought to decrease the incidence of PPCs. However, stopping smoking within this period of time has also been suggested to increase PPCs, possibly due to increased sputum production [29]. More recently, a meta-analysis failed to demonstrate any difference in postoperative complications between patients who ceased smoking within 8 weeks of surgery and those who did not [30]. By contrast, patients who stop smoking at least 1 year preoperatively have the same perioperative mortality as non-smokers and also have a lower risk of cardiac and respiratory complications than active smokers [21].

While ceasing smoking within 8 weeks of surgery may do little to mitigate the risk of PPCs, patients should still be encouraged to do so, as it decreases the incidence of wound infections [31] and may promote long term abstinence.

10.4.1.2 Preoperative Steroids for Reactive Airways Disease

While well-controlled reactive airways disease (asthma and COPD) does not significantly increase the risk of PPCs, this is not the case for poorly controlled disease. In a study of patients with reversible, but poorly controlled reactive airways disease (FEV1 <70 % of predicted and not on chronic bronchodilator therapy) who receive salbutamol preoperatively, there was a high incidence of bronchospasm following intubation if it was given either preoperatively for 5 days (7/9 patients) or 10 min before induction (8/10 patients). Of note, intraoperative bronchospasm still occurred even though spirometry improved following salbutamol therapy in these patients. However, in a third group of patients who received oral methylprednisolone (40 mg) for 5 days preoperatively in addition to salbutamol, intraoperative bronchospasm was significantly decreased compared to those patients who received salbutamol alone (1 / 10 patients, $p<0.001$) [32].

While there have been concerns that a short course of high-dose steroids may increase the risk of other surgical complications such as wound infections, several studies have shown that this does not occur.

10.5 Intraoperative Strategies to Decrease PPCs

10.5.1 Protective Lung Ventilation

There is recent evidence that the long established anaesthetic practice of mechanical ventilation with high tidal volumes (TV) of 7–10 ml/kg and low, or zero positive end-expired pressure (PEEP) should be re-evaluated in favour of "protective" lung ventilation strategies that minimise airway pressures and prevents atelectasis.

10.5.1.1 Background

The adverse effects of general anaesthesia and mechanical ventilation on the lung are well recognised. In particular, atelectasis rapidly occurs in the dependent regions of the lung. This is largely caused by compression of the lungs due to changes in the shape of the chest and the position and function of the diaphragm.

High inspired oxygen concentrations also contributes to alveolar collapse through absorption atelectasis [33]. Atelectasis from any cause increases intrapulmonary shunt and promotes bacterial growth, which in turn contributes to the development of postoperative pneumonia [34].

Traditionally, the rationale for mechanically ventilating patients with high tidal volumes during anaesthesia was to overcome atelectasis. However, this may cause lung injury by over-distending alveoli (volutrauma) in non-dependent regions of the lung, which receive a greater proportion of the tidal volume than dependent regions during mechanical ventilation. Volutrauma damages endothelial and other cells in the lung, which causes both local injury and damage to other organs through the release of systemic inflammatory mediators [35].

The most effective strategies for reducing atelectasis are avoiding high concentrations of inspired oxygen, the use of PEEP and recruitment manoeuvres.

While 100 % inspired oxygen rapidly promotes atelectasis during anaesthesia, this can be mitigated by even a small reduction in inspired oxygen concentration. For example, pre-oxygenation with 80 % oxygen causes much less atelectasis than 100 % oxygen (0.8 % vs. 6.8 % respectively), although it also decreases the time to oxygen desaturation [36]. Positive end-expired pressure (PEEP) should be routinely applied when breathing 100 % oxygen [37], as it reliably prevents atelectasis occurring.

While PEEP helps prevents atelectasis during general anaesthesia, it does not reliably re-expand areas of lung that have already collapsed. Atelectasis also occurs rapidly if PEEP is interrupted even briefly. Atelectasis can be effectively treated and prevented with the application of PEEP after a lung recruitment manoeuvre [33].

A recruitment manoeuvre is the delivery of a brief period of high inspiratory airway pressure. While a recruitment manoeuvre can completely reverse atelectasis [38], sufficiently high inflating pressures need to be delivered. For example, 20 cm H_2O has almost no effect, but 40 cm H_2O sustained for 7–8 s is completely effective [39].

While prolonged recruitment manoeuvres can cause haemodynamic compromise, repeated brief recruitment manoeuvres have not been shown to cause either adverse cardiovascular effects or lung damage [40].

10.5.1.2 Protective Lung Ventilation: The Clinical Evidence

Protective lung ventilation has an established role in intensive care following the landmark ARDSNet trial. In this trial, patients with acute lung injury and acute respiratory distress syndrome had lower mortality when ventilated with low tidal volumes (TV) and plateau pressure (6 ml/kg and <30 cm H_2O) compared with high TV and plateau pressures (12 ml/kg and <50 cm H_2O) [41].

While the mode of ventilation is unlikely to significantly affect outcome in patients with normal lungs having low risk surgery, there is recent evidence that this is not the case in other patient populations. For example, in a recent trial of 400 moderate to high risk patients undergoing abdominal surgery, ventilation with low TV (6–8 ml/kg), moderate PEEP (6–8 cm H_2O) and recruitment manoeuvres (30 cm H_2O for 30 s, every 30 min) significantly reduced the incidence of postoperative respiratory failure compared with ventilation with high TV (10—12 ml/kg), no PEEP and no recruitment manoeuvres (5 % vs. 17 %, $p=0.001$) [42]. However, by contrast, in a large retrospective study, the use of low TV ventilation with minimal PEEP (2.2–5 cm H_2O) was associated with increased 30-day mortality and prolonged hospitalisation [43]. One possible explanation for these contradictory findings is that that low TV ventilation might only be "protective" when adequate levels of PEEP are used to prevent atelectasis from occurring.

Furthermore, while moderate levels of PEEP decrease PPCs, higher levels of PEEP do not. In a large, prospective trial of abdominal surgical patients who were all ventilated with the same TV (8 ml/kg), the use of high PEEP (12 cm H_2O) did not decrease the incidence of PPCs compared with low PEEP (0–2 cm H_2O) [44]. The high PEEP group also required more vasopressors for hypotension intra-operatively.

In summary, while low TV ventilation during general anaesthesia decreases PPCs when used with moderate PEEP and recruitment manoeuvres, this benefit is not found when it is used with no or low PEEP.

10.5.1.3 Oxygen: Friend or Foe?

While the old clinical injunction to "avoid hypoxia" needs no further elaboration, there is emerging evidence that too much oxygen may also be harmful. This is leading to a re-evaluation of oxygen therapy in resuscitation and may have implications for the perioperative management of patients with acute lung injury (ALI).

When mitochondrial oxygen levels are higher than are required to maintain aerobic metabolism, the excess oxygen forms reactive oxygen species (otherwise known as free radicals). Oxygen free radicals contain one or more unpaired electrons, which makes them much more reactive and damaging to cellular components [45].

The toxic cellular effects of oxygen have been shown to be clinically relevant in a variety of conditions. For example, non-hypoxic patients with ischaemic stroke and acute myocardial infarction oxygen were routinely administered oxygen in the belief that it minimised the risk of cellular injury and that it did not in itself cause any harm. However, routine oxygen therapy has subsequently been shown to increase the mortality in both of these conditions [46]. Similarly, neonatal resuscitation with 100 % oxygen has been shown to increase mortality compared with room air. It is now recommended that room air is used initially, and if oxygen is required, that it be titrated to the lowest level that that prevents hypoxia [47].

High inspired oxygen concentrations not only promote atelectasis (see preceding), but also have direct toxic effects on the lung. These are related to the inspired concentration and to the duration of exposure. While brief perioperative exposure to high oxygen concentrations does not appear to be harmful in patients with healthy lungs, it does exacerbate pre-existing acute lung injury [48]. Hence patients with acute lung injury who require surgery should be ventilated with the lowest inspired oxygen concentration that is tolerable and the inspired oxygen should be titrated to the lowest level that provides acceptable oxygenation (SaO_2 >90 %).

Finally, several studies have examined whether the use of high inspired oxygen concentrations perioperatively decreases the incidence of wound infections. While early trials were favourable, a meta-analysis of seven trials (involving 2728 patients) did not show an overall benefit [49], nor did a large randomised controlled trial [50].

10.5.1.4 Intraoperative Bronchospasm

Intraoperative bronchospasm can easily be detected in patients who are mechanically ventilated with modern anaesthetic machines. It is characterised by end –expiratory flow that does not return to baseline at end-expiration, increased inspiratory pressures (peak and plateau) and an upward-sloping end-tidal CO_2 waveform during expiration. These signs can also be used to assess the response to therapies such as bronchodilators and increasing the duration of expiration (by decreasing respiratory rate and the I:E ratio).

10.5.1.5 Anaesthesia and Surgery

There is consensus opinion that substituting neuraxial or regional anaesthesia for general anaesthesia decreases PPCs. However, there is little data to support this supposition. There is better evidence to support the use of short vs. long-acting muscle relaxants and for laparoscopic vs. open surgery [9].

For patients undergoing general anaesthesia, the use of a laryngeal mask airway decreases the risk of laryngeal spasm compared with endotracheal intubation, although it has probably minor effects on other PPCs [51]. The use of volatile anaesthetics is theoretically attractive because of their bronchodilating properties, although there is no evidence that they decreases PPCs compared with total intravenous anaesthesia.

The use of opiate-sparing techniques (such as multi-modal analgesia, local and regional anaesthesia) may limit PPCs due to respiratory depression. Several meta-analyses have shown that thoracic epidural analgesia reduces respiratory complications and duration of ventilation in high- risk patients undergoing major surgery [52, 53]. However, the magnitude of these benefits has been lower in more recent studies [54].

10.5.2 Postoperative Strategies to Decrease PPCs

Following abdominal surgery, the selective use of nasogastric tubes in patients with abdominal distension or symptoms of nausea or vomiting decreases PPCs. However, nasogastric tubes also increase the risk of gastric aspiration and increase PPCs when used routinely [55].

The most effective proven postoperative strategies to reduce PPCs are lung expansion techniques. These include incentive spirometry, breathing exercises and CPAP. These techniques probably have roughly similar efficacy in reducing the incidence of PPCs (by about one third). However, CPAP is particularly useful for patients who are unable to perform deep breathing exercises [56].

10.6 Summary

By far the most important risk factors for PPCs are patient age and the type, site and acuity of surgery. Pre-existing respiratory disease is a relatively minor risk factor for PPCs and is seldom a contraindication to essential surgery. However, elective surgery should be postponed in patients with poorly controlled reactive airways disease until they have completed a course of high-dose steroids.

There are only a few strategies that have been proven to reduce the incidence of pulmonary complications following major surgery. These include protective lung ventilation with moderate levels of PEEP and recruitment manoeuvres, postoperative lung expansion techniques and the selective use of nasogastric tubes. Other strategies that may be effective include avoidance of general anaesthesia and the use of opiate-sparing analgesic techniques (such as regional anaesthesia). These strategies should be used in all patients who are at increased risk of PPCs, even in the absence of respiratory disease.

References

1. Fleischmann K et al (2003) Association between cardiac and non-cardiac complications in patients undergoing non-cardiac surgery. Am J Med 115:515–520
2. Dimick J et al (2004) Hospital costs associated with surgical complications. J Am Coll Surg 199:531–537

3. Azorullah AM et al (2001) Development and validation of a multifactorial risk index for predicting postoperative pneumonia after major noncardiac surgery. Ann Intern Med 135:847
4. Johnson R, Arozullah A, Neumayer L et al (2007) Multivariate predictors of postoperative respiratory failure after general and vascular surgery: results from the patient safety in surgery study. J Am Coll Surg 204:1188–1198
5. Brueckmann B et al (2013) Development and validation of a score for prediction of postoperative respiratory complications. Anesthesiology 1118(6):1276
6. Wong D, Weber E, Schell M et al (1995) Factors associated with postoperative complications in patients with severe chronic obstructive pulmonary disease. Anesth Analg 80:276–284
7. Liao P, Yegneswara B, Vairavanathan S (2009) Postoperative complications in patients with obstructive sleep apnea: a retrospective matched cohort study. Can J Anaesth 56:819–828
8. Choi S, Lee J, Park Y et al (2014) Postoperative pulmonary complications after surgery in patients with interstitial lung disease. Respiration 87:287–293
9. Ramakrishna G, Sprung J, Barugar S, Chandrasekaran K, McGoon M (2005) Impact of pulmonary hypertension on the outcomes of noncardiac surgery. J Am Coll Cardiol 45(10): 1691–1699
10. Lai H, Lai H, Wang K, Lee W, Ting C, Liu T (2007) Severe pulmonary hypertension complicates outcome of non-cardiac surgery. Br J Anaesth 99(2):184–190
11. Forrest P (2009) Anaesthesia and right ventricular failure. Anaesth Intensive Care 37(3): 37–386
12. Joshi G, Ahmad S, Riad W et al (2013) Selection of obese patients undergoing ambulatory surgery: a systematic review of the literature. Anesth Analg 117:1082–1091
13. Thomas EJ, Goldman L, Mangione C et al (1997) Body mass index as a correlate of postoperative complications and resource utilization. Am J Med 102:277–283
14. Qaseem A et al (2006) Risk assessment and strategies to reduce perioperative pulmonary complications for patients undergoing noncardiothoracic surgery: a guideline from the American College of Physicians. Ann Intern Med 144:575–580
15. Frey W, Pilcher J (2003) Obstructive sleep-related breathing disorders in patients evaluated for bariatric surgery. Obes Surg 13:676–683
16. Abdulla H, Chung F (2014) Perioperative management for the obese outpatient. Curr Opin Anaesthesiol 27:576–582
17. Gold A, Marcus C, Dipalo F et al (2002) Upper airway collapsibility during sleep in upper airway resistance syndrome. Chest 121:1531–1540
18. Vasu T (2012) Obstructive sleep apnea syndrome and perioperative complications: a review of the literature. J Clin Sleep Med 8(2):199
19. American Society of Anesthesiology Task Force on Perioperative Management of Obstructive Sleep Apnea (2014) Practice guidelines for the perioperative management of patients with obstructive sleep apnea. Anesthesiology 120:1081–93
20. Weingarten T et al (2011) Obstructive sleep apnea and perioperative complications in bariatric patients. Br J Anaesth 106(1):131–139
21. Sabers C, Plevak DJ, Schroeder DR, Warner DO (2003) The diagnosis of obstructive sleep apnea as a risk factor for unanticipated admissions in outpatient surgery. Anesth Analg 96:1328–1335
22. Stierer T, Wright C, George A et al (2010) Risk assessment of obstructive sleep apnea in a population of patients undergoing ambulatory surgery. J Clin Sleep Med 6(5):467–472
23. Alonso-Fernandez A, Garcia-Rio F, Racionero M et al (2005) Cardiac rhythm disturbances and ST-segment depression in patients with obstructive sleep apnea-hypopnea and its mechanisms. Chest 127:15–22
24. South Australian Coronial Inquest: http//bit.ly/1xMstcl
25. Joo H et al (2005) The value of screening preoperative chest x-rays: a systematic review. Can J Anaesth 52(6):568–574
26. Snowden S, Prentis J, Anderson H et al (2010) Submaximal cardiopulmonary exercise testing predicts complications and hospital length of stay in patients undergoing major elective surgery. Ann Surg 251(3):535–541
27. Brooks-Brunn JA (1997) Predictors of postoperative pulmonary complications following abdominal surgery. Chest 111(3):564–571

28. Musallam K, Rosendaal F, Zaatari G et al (2013) Smoking and the risk of mortality and vascular and respiratory events in patients undergoing major surgery. JAMA Surg 148(8):755–762
29. Warner MA, Offord KP, Warner ME, Lennon RL, Conover MA, Jansson-Schumacher U (1989) Role of preoperative cessation of smoking and other factors in postoperative pulmonary complications: a blinded prospective study of coronary artery bypass patients. Mayo Clin Proc 64(6):609–616
30. Myers K, Hajek P, Hinds C, McRobbie H (2011) Stopping smoking shortly before surgery and postoperative complications. A systematic review and meta-analysis. Arch Intern Med 171(11):983–989
31. Sørensen L (2012) Wound healing and infection in surgery. The clinical impact of smoking and smoking cessation: a systematic review and meta-analysis. Arch Surg 147(4):373–383
32. Silvanus M (2004) Corticosteroids and inhaled salbutamol in patients with reversible airways obstruction markedly decrease the incidence of bronchospasm after tracheal intubation. Anesthesiology 100:1052–1057
33. Magnusson L, Spahn D (2003) New concepts of atelectasis during general anaesthesia. Br J Anaesth 91(1):61–72
34. van Kaam AHLC, Lutter R, Lachmann RA, Haitsma JJ, Herting E, Snoek M, De Jaegere A, Kok JH, Lachmann B (2005) Effect of ventilation strategy and surfactant on inflammation in experimental pneumonia. Eur Respir J 26:112–117
35. Slutsky AS, Ranieri VM (2013) Ventilator-induced lung injury. N Engl J Med 369:2126–2136
36. Edmark L, Enlund M, Kostova-Aherdan K, Hedenstierna G (2001) Atelectasis formation and apnoea tolerance after pre- oxygenation with 100 %, 80 %, or 60 % oxygen. Anesthesiology 95:A1330
37. Neumann P, Rothen HU, Berglund JE et al (1999) Positive end- expiratory pressure prevents atelectasis during general anaesthesia even in the presence of a high inspired oxygen concentration. Acta Anaesthesiol Scand 43:295–301
38. Rothen HU, Sporre B, Engberg G, Wegenius G, Hedenstierna G (1993) Re-expansion of atelectasis during general anaesthesia: a computed tomography study. Br J Anaesth 71:788–795
39. Rothen HU, Neumann P, Berglund JE et al (1999) Dynamics of re- expansion of atelectasis during general anaesthesia. Br J Anaesth 82:551–556
40. Magnusson L, Tenling A, Lemoine R et al (2000) The safety of one, or repeated, vital capacity maneuvers during general anesthesia. Anesth Analg 91:702–707
41. The Acute Respiratory Distress Syndrome Network (2000) Ventilation with lower tidal volumes as compared with tradition al tidal volumes for acute lung injury and the acute respiratory distress syndrome. N Engl J Med 342:1301–1308
42. Futier E et al (2013) A trial of intraoperative low tidal –volume ventilation in abdominal surgery. N Engl J Med 369:428–437
43. Levin M et al (2014) Low intraoperative tidal volume ventilation with minimal PEEP is associated with increased mortality. Br J Anaesth 113(1):97–108; published online March 12
44. The PROVE Network Investigators (2014) High versus low positive end-expiratory pressure during general anesthesia for open abdominal surgery. Lancet 384(9942):495–503; published online June 1
45. Martin D, Grocott M (2013) Oxygen therapy in anaesthesia. Br J Anaesth 111(6):867–871
46. Ronning O, Guldvog B (1999) Should stroke victims routinely receive supplemental oxygen? A quasi-randomized controlled trial. Stroke 30:2033–2037
47. Wyllie J, Perlman JM, Kattwinkel J et al (2010) Part 11: neonatal resuscitation: 2010 international consensus on cardiopulmonary resuscitation and emergency cardiovascular care science with treatment recommendations. Resuscitation 81(Suppl 1):260–287
48. Hackney JD, Evans MJ, Spier CE et al (1981) Effect of high concentrations of oxygen on reparative regeneration of damaged alveolar epithelium in mice. Exp Mol Pathol 34:338
49. Qadan M, Akca O, Mahid SS, Hornung CA, Polk HCJ (2009) Perioperative supplemental oxygen therapy and surgical site infection: a meta-analysis of randomized controlled trials. Arch Surg 144:359–366; discussion 366–367
50. Meyhoff CS, Wetterslev J, Jorgensen LN et al (2009) Effect of high perioperative oxygen fraction on surgical site infection and pulmonary complications after abdominal surgery: the PROXI randomized clinical trial. JAMA 302:1543–1550

51. Yu S (2010) Laryngeal mask airways have a lower incidence of airway complications compared with endotracheal intubation: a systematic review. J Oral Maxillofac Surg 68(10): 2359–2376
52. Liu SS (2007) Effect of postoperative analgesia on major postoperative complications: a systematic update of the evidence. Anesth Analg 104:689–702
53. Nishimori M (2012) Epidural pain relief versus systemic opioid –based pain relief for abdominal aortic surgery. Cochrane Database Syst Rev (3):CD005059
54. Popping D (2008) Protective effects of epidural analgesia on pulmonary complications after abdominal and thoracic surgery: a meta-analysis. Arch Surg 143(10):990–999
55. Nelson R et al (2007) Prophylactic nasogastric decompression after abdominal surgery. Cochrane Database Syst Rev 18(3):CD004929
56. Ferreyra G (2008) Continuous positive airway pressure for treatment of respiratory complications after abdominal surgery. Ann Surg 247:617–626

Chapter 11
Hemostatic Resuscitation for Perioperative Bleeding

Philip C. Spinella, Heather F. Pidcoke, and Andrew P. Cap

Abstract The perioperative resuscitation approach to hemorrhagic shock is changing due to recent military experience in managing traumatic injury. For the past few decades, resuscitation of perioperative hemorrhagic shock has most commonly consisted of crystalloids and colloids, followed by RBCs, after which, if indicated according to laboratory results, plasma and platelets. If used, empiric ratios of RBC to plasma units ranged between 10 and 4 units of RBCs to 1 unit of plasma, without platelets. This approach was the foundation of Advanced Trauma Life Support (ATLS).

Recent combat experience re-ignited interest in whole blood for the resuscitation of hemorrhagic shock. US Army data indicates that warm whole blood transfusion is associated with improved or comparable survival compared to blood components. This data compliments that of pediatric cardiac surgery trials which found that cold-stored whole blood reduced blood loss and was associated with improved platelet function compared to blood components in a 1:1:1 ratio using platelets were stored at 22 °C.

Whole blood availability is limited in civilian hospitals due to perceived logistical barriers, but recent developments suggest that these barriers are obsolete. Data indicates it is safer to provide group O whole blood than ABO-specific whole blood; whole blood can be leukocyte-reduced with platelet-sparring filters; and whole blood stored at 4 °C retains platelet function during 15 days of storage. This approach will improve the availability of whole blood and facilitate its use for peri-operative hemorrhagic shock. There are also logistic and economical advantages of whole blood based resuscitation compared to components for hemorrhagic shock.

Keywords Trauma • Coagulopathy • Shock • Hemorrhage • Transfusion

---

P.C. Spinella, MD, FCCM (✉)
Division of Critical Care, Department of Pediatrics, Washington University in St Louis, St. Louis, MO, USA

U.S. Army Institute of Surgical Research, FT Sam Houston, TX, USA
e-mail: Spinella_p@kids.wustl.edu

H.F. Pidcoke, MD, PhD • A.P. Cap, MD, PhD, FACP
U.S. Army Institute of Surgical Research, FT Sam Houston, TX, USA
e-mail: heatherpidcoke@yahoo.com; andrew.p.cap.mil@mail.mil

K. Stuart-Smith (ed.), *Perioperative Medicine – Current Controversies*,
DOI 10.1007/978-3-319-28821-5_11

11.1 Hemostatic Resuscitation for Perioperative Bleeding

Damage Control Resuscitation (DCR) and hemostatic resuscitation are concepts recently developed to optimize resuscitative and transfusion approaches to patients with hemorrhagic shock and immediately life-threatening injuries. DCR has many components all of which are aimed at preventing or treating shock and coagulopathy and thereby reducing morbidity and mortality from severe traumatic injuries causing massive hemorrhage [1]. Hemostatic resuscitation is the central tenant of DCR. This is a concept that developed with the recognition that a blood based transfusion strategy would be optimal for severe bleeding and that crystalloid or colloid-based resuscitation cause hemodilution and acidosis, which aggravate the underlying coagulation and metabolic disorders that evolve following injury and blood loss. The ultimate hemostatic resuscitation product is whole blood, but since currently it is not commonly available in the developed world many substitute the transfusion of high ratios of plasma and platelets to RBCs that range between 1:1:2 and 1:1:1 units, respectively. Goal-directed hemostatic resuscitation is being explored as a method to alter empiric ratios of blood products and provide specific therapies based upon the rapid results from point-of-care coagulation and shock monitoring.

Recent conflicts in Iraq and Afghanistan re-ignited interest in the concept of transfusing whole blood or "reconstituted whole blood" with components in a 1:1:2–1:1:1 unit ratio as a result of many factors. The large volume of patients presenting in hemorrhagic shock with high mortality forced physicians to think of alternative methods to improve outcomes. Prior to 2005, platelet units were not available in Iraq or Afghanistan and even after that they were in limited supply. There was increased awareness that shock can directly lead to coagulopathy and that a balanced resuscitation that corrected both conditions might reduce death from hemorrhage. Physicians with prior combat experience maintained institutional knowledge of prior successful deployments of whole blood in previous conflicts and the ease with which such a program could be implemented and maintained. As a result, the use of whole blood at large combat support hospitals increased in 2003–2004 from a "rescue" therapy in the initial phases to a more widely adopted strategy that was employed earlier in the resuscitation of patients with life-threatening hemorrhagic shock. It was during the period of 2004–2006 that the concepts of DCR were solidifying into one bundle of care, with hemostatic resuscitation as the centerpiece of clinical practice guidelines (CPGs) for the treatment of severe traumatic hemorrhagic shock. The DCR concept and specific use of whole blood were codified into on-line CPGs and routinely re-evaluated and updated by the US Department of Defense Joint Trauma System.

The use of whole blood has a long history in military medicine, which began almost 100 years ago in World War I. Briefly, in World War I British and American forces supplied the front line hospitals with low titer Group O whole blood as the universal donor product. In WWII the lessons from WWI were initially side-lined and US forces planned for a plasma-based resuscitation strategy with freeze-dried plasma and albumin. This approach, though practical, was found to be inadequate to the task of preventing or reversing shock physiology. The British Forces primarily transfused whole blood during WWII based on the concept that both shock and coagulopathy needed to be equally addressed for optimal outcomes. After the US leadership recognized the successes the British Forces were having with this approach they also shifted back to a whole blood based resuscitation strategy [2].

US Forces in the World War II established what were termed "Field Blood Banks," where fresh whole blood was collected from immediately available donors and either used on site immediately, or packaged and delivered as far forward as possible for resuscitation near point of wounding. Units actively engaged in combat used freeze-dried plasma and any available whole blood until casualties could be evacuated to surgical facilities. This approach was replicated successfully in Korea where Group O low titer (<1:200 by saline dilution) whole blood was used almost exclusively [3]. In Vietnam, universal donor whole blood was the primary resuscitation agent until apheresis derived fresh frozen AB plasma was introduced in 1968. Eventually type-specific whole blood became available, but universal donor whole blood was the only blood product used in forward, pre-hospital settings [2].

During and after the Vietnam War Era, crystalloids and colloids became the primary resuscitative solution for hemorrhagic shock, replacing blood products including whole blood. This was in part due to the risks of infectious disease transmission with blood products, but also due to research performed by Carrico and Shires that indicating that the interstitial compartment or "third space" needed to be resuscitated with 1–2 L crystalloids in order to perfuse the tissues, after which the use transfusion of whole blood would be indicated only if hemodynamic instability persisted [4]. Their philosophy for resuscitation in patients with traumatic bleeding was misapplied and led to the overuse of crystalloids, to the detriment of patients with severe bleeding who commonly received 5–10 L of crystalloids prior to any blood product administration. The result was dilutional coagulopathy and severe interstitial edema. In addition, as whole blood became unavailable the use of RBCs took the place of whole blood. The lack of recognition that RBC units were not as efficacious at treating both shock and coagulopathy further exacerbated dilutional coagulopathy. The misapplication of the Carrico and Shires data and the disappearance of whole blood from the clinical armamentarium ushered in an era of ARDS, abdominal compartment syndrome, multi-organ failure and anasarca in ICUs [5, 6]. These outcomes were actually predicted by Shoemaker in 1976, where he challenges the notion that the interstitial compartment required resuscitation and emphasized the need for whole blood for the treatment of significant bleeding when the hematocrit fell below 30 % [7]. Subsequent research has demonstrated that a crystalloid-based resuscitation strategy, leads to increased inflammation and vascular permeability compared to whole blood [8, 9].

The pendulum has now swung back to focusing on a blood-based strategy for patients in hemorrhagic shock. The concept of hemostatic resuscitation as a component of DCR continues to be controversial, with unprecedented attention given to establishing the optimal resuscitation strategy for patients with hemorrhagic shock. Investigations continue to seek novel methods to further improve outcomes by establishing goal-directed hemostatic resuscitation strategies with point-of-care monitoring of shock and functional hemostatic parameters. Perioperative bleeding occurs in many different settings or in patients with varying disorders, requiring caution when extrapolating concepts derived primarily from patients with traumatic injury. Although there are increasing reports of improved outcomes when hemostatic resuscitation principles are used in patients with ruptured aortic aneurysms and in post-partum hemorrhage for example, data outside the trauma literature is still scant [10–13]. Recognizing the lack of robust data available to support the development of optimal resuscitative strategies in a wide variety of perioperative patients with hemorrhagic shock, the following will review the evidence regarding the use of whole blood and blood components in these patient populations.

11.1.1 Whole Blood Definitions

Whole blood is collected into 450–500 ml bags and then transfused or stored instead of being separated into components. When stored cold at 4 °C it has a shelf life of 21 days if collected in CPD, CP2D or ACD-A anticoagulant and 35 days if collected in CPDA-1. When collected in this manner and standard transfusion-transmitted disease testing is performed it is an approved product. The FDA Circular of Information for Use of Human Blood and Blood Components states that whole blood is indicated for symptomatic anemia with large volume deficits [14]. The FDA also states that whole blood must be type-specific. To differentiate this product from whole blood used in military settings, it is called Cold Whole Blood (CWB). When cold whole blood is used within 48 h of collection it is Cold Fresh Whole Blood (CFWB).

In military or austere settings, where formal testing for TTD's is not possible, whole blood is collected from pre-tested donors (tested pre-deployment and every 90 days thereafter) and stored warm at 22 °C for up to 8 h and is then stored for a maximum of 24 h at 4 °C. It is typically transfused immediately. This product is termed Warm Fresh Whole Blood (WFWB). Since WFWB is transfused before results of TTD testing are available (testing of each unit is retrospective since samples are sent from expeditionary forces to US labs for testing); it is thus not approvable by FDA and presents a higher risk of disease transmission and is reserved for situations in which tested blood products are unavailable or ineffective. It is important to understand the differences between cold and warm storage of whole blood since there are reports of both in the literature.

The use of the term "fresh" when storage is less than 48 h is arbitrary and not based on data regarding the loss of RBC, plasma or platelet function. It is related to the timing of the removal of platelets from the circulation and the perceived lack of function after this amount of time. The focus on timing of platelet removal from the circulation is a result of decades-old emphasis on circulation time as a measure of function instead of assessing function as a measure of hemostatic capacity. Arguably hemostatic capacity is a more relevant measure of function in patients with severe bleeding who are at high risk of death within hours than circulation time.

Whole blood, stored cold or transfused as WFWB can be leukocyte-reduced with platelet-sparing filters [15]. Leukoreduction decreases the white blood cell count in the whole blood unit to less than 5×10^6.

11.1.2 Cold Whole Blood

The two largest RCTs that have exclusively compared whole blood to blood components were performed in pediatric cardiac surgery patients. These two trials with significantly different methods reported contradictory results. In a double-blinded study, Manno et al. evaluated postoperative blood loss in 161 children undergoing open-heart surgery with cardiopulmonary bypass whose immediate postoperative transfusion requirements were met with either whole blood or reconstituted whole blood (packed RBCs, FFP, and platelets) [16]. The transfusion of whole blood was associated with significantly less postoperative blood loss than the transfusion of packed RBCs, FFP, and platelets in children under 2 of age who underwent complex cardiac surgery (mean 24-h post-operative blood loss in milliliters per kilogram 52.3 ± 10.8 vs. 96.2 ± 10.7, $P = .001$). In that study, whole blood was associated with improved platelet function according to aggregometry results. Mou et al. compared the use of fresh whole blood with the use of a combination of packed RBCs and FFP (reconstituted blood) for priming of the cardiopulmonary bypass circuit in children less than 1 year of age undergoing cardiac surgery [17]. Children were randomly assigned to receive either fresh whole blood (96 subjects) or reconstituted blood (104 subjects) for bypass-circuit priming. This study did not continue randomization of the intervention in the ICU, where outcomes were measured. The use of fresh whole blood for cardiopulmonary bypass priming had no advantage over the use of a combination of packed RBCs and FFP during surgery for congenital heart disease. Moreover, circuit priming with fresh whole blood was associated with an increased length of stay in the ICU and increased perioperative fluid overload. Unfortunately, this secondary outcome was not adjusted for confounding variables.

The only available prospective evidence in trauma patients that utilized whole blood came from a single small RCT of 107 patients requiring massive transfusion [18]. One arm received leukoreduced whole blood (with platelets removed during filtration) supplemented with platelets, and the other was treated exclusively with components (RBCs, plasma, platelet units). While the primary outcome of 24-h transfusion volume was equivalent between groups, a secondary analysis excluding patients with severe TBI demonstrated reduced 24-h blood use for the modified whole blood group. It is difficult to know whether the addition of platelets to whole blood sufficiently changed the results to limit generalizability, thus the study was insufficient to definitively establish the efficacy of whole blood.

A small RCT published in 1990 that compared WFWB to CFWB enrolled 36 adult patients requiring cardiac surgery and cardiopulmonary bypass. The hemostatic effect of fresh non-leukoreduced WFWB was compared to CFWB stored at

4 °C for both 5 and 24 h, and investigators found that blood loss increased postoperatively in those transfused with whole blood stored at 4 °C, suggesting that WFWB may be superior to 4 °C-stored WB [19]. Since WFWB is not clinically available, except for in austere or military environments, these results may be relevant only for this population of patients.

In vitro data indicates that the hemostatic capacity of whole blood is higher than that of blood reconstituted from individual components in a 1:1:1 ratio [19]. In one analysis in which platelets were stored at 22 °C for up to 5 days, and RBCs were stored at 4 °C for up to 35 days, investigators compared reconstituted whole blood to WFWB and found that the hemostatic capacity of platelets was reduced as storage duration increased [20]. Conversely, as RBC and platelet storage age increased more thrombin-antithrombin complexes were detected, indicating increased thrombin generation, which was confirmed in a thrombin generation assay. Similarly, WFWB platelet aggregation response to collagen was consistently better than that of reconstituted whole blood. A similar *in vitro* study compared warm and cold whole blood with and without additional platelets to reconstituted whole blood with components in a 1:1:1 unit ratio. No differences in platelet function or viscoelastic monitoring results were noted between warm and cold whole blood and reconstituted whole blood when RBCs and platelets were at their maximum storage duration [21]. Differences in study methods reported for these two studies may account for the conflicting results.

In vitro studies also indicate that considerable CWB hemostatic capacity is maintained for at least 14 days. One analysis of non-leukoreduced whole blood stored at 4C reported normal platelet function as reflected by TEG-MA values for all 21 samples out to 14 days of storage, with the median TEG-MA values within normal range out to 31 days of storage [22]. In addition light transmission aggregometry analysis of platelet function exhibited no change from Day 1 to Day 21 with adenosine diphosphate and epinephrine stimulation, but did report a decline in response to collagen (Day 7) and ristocetin (Day 17). These results were reported without adjusting a dramatic reduction in platelet count over time, which dropped to a median value of less than 100×10^9/L by day 7 of storage.

In vitro measures of hemostasis studied in whole blood units stored at 4 °C versus 22 °C over 21 days reported superior PT/PTT, impedance platelet aggregation, and TEG results in cold-stored blood [23]. In this study non-leukocyte reduced whole blood was stored for up to 21 days. Whole blood at 4 °C was superior to those units stored at 22 °C based on PT/PTT, impedance platelet aggregation, and TEG parameters. Refrigeration of whole blood units also increased shear-induced platelet aggregation and ristocetin-induced platelet agglutination as well as the proportion of GPIb-expressing platelets. Furthermore, platelet function in stored whole blood measured by shear- and Von Willebrand factor-dependent methodology is retained at 4 °C for up to 7 days [24].

These findings are similar to those regarding platelet units stored at 4 °C. Refrigerated PLTs perform better than room temperature–stored PLTs in aggregation assays as has been reported by multiple investigative teams [25–29]. This includes an RCT demonstrating that platelets at 4 °C were superior in reducing bleeding time in both patients on aspirin and in thrombocytopenic patients when compared to platelets at 22 °C [30].

11.1.3 Warm Whole Blood

Retrospective data in adults, further support the use of whole blood compared to exclusive use of components, including an analysis of US casualties with 100 % follow up which reported an independent association between survival and the addition of whole blood as an adjunct to resuscitation [31]. Additionally, a study of US and non-US casualties reported no difference in outcomes for those transfused whole blood as part of their resuscitation compared to those who only received blood components; however, results should be viewed with caution as approximately 33 % of patients were lost to follow-up [32]. These two studies of combat casualties are unique in that the whole blood transfused in combat was given immediately after collection or was stored very briefly (<24 h). They also differ from civilian data in that whole blood accounted for only a portion of the total blood used for resuscitation during the initial 24 h of admission. In the study that found an association with survival, 30 % of the 24-h blood volume transfused was whole blood, whereas in the study that demonstrated equivalence, only 20 % of the 24-h blood volume transfused was whole blood.

In a retrospective civilian study, the use of whole blood as part of a resuscitation was associated with similar outcomes compared to the exclusive use of blood components. In patients with multiple etiologies of hemorrhagic shock requiring massive transfusion, there were no differences in blood utilization rates and clinical outcomes [33].

Two small RCTs have compared WFWB to platelet concentrates stored at 22°C. One published in 1989, of 24 patients requiring cardiopulmonary bypass compared WFWB to platelet concentrates stored at 22°C in patients undergoing cardiac surgery [34]. A scanning electron microscope was used to assess platelet aggregation. Their results indicated that one unit of WFWB had improved hemostatic effect compared to 8–10 whole blood derived platelet units. In another small randomized study, published in 1988, of 27 patients requiring cardiopulmonary bypass, WFWB was also found to be associated with improved platelet function compared to platelet units stored at 22 °C as assessed with aggregation studies [35].

11.1.4 Whole Blood Availability

Whole blood is widely available in much of the world, particularly in developing nations such as those of Sub-Saharan Africa. For example, in Kenya at the Kijabe Hospital, CWB is commonly provided, where it is stored at 4 °C for up to 35 days, is typically group O, and is not leukocyte reduced (Kibuchi J and Letchford S, 2015, oral communication, 15th August).

In the developed world, the clinical availability of whole blood since the 1970s to 1980s has been dramatically reduced. WFWB essentially disappeared from US hospitals during the early days of the HIV epidemic. CWB, though fully tested for TTDs also vanished from blood banks. This limitation of whole blood availability has been driven due to four main concerns about whole blood.

The first is that regulatory bodies have mandated that whole blood be ABO-specific due to risks of ABO incompatibility, where ABO incompatible RBCs will cause severe hemolytic reactions and ABO incompatible plasma might uncommonly cause mild to moderate hemolytic reactions. Providing ABO specific WB introduces a significant logistical concern since maintaining sufficient inventory to ensure a ready supply of every ABO type would result in significant cost and waste.

The second barrier has been the lack of a platelet-sparing leukoreduction filter. Since leukoreduction has become either mandatory in most countries or widely used in others, the fact that whole blood could not be leukocyte reduced while maintaining platelets reduced the supply of whole blood.

The third barrier is due to the concern that the platelets in CWB stored at 4 °C might not be functional or viable, calling into question whether WB is truly a platelet-containing product. Due to this concern when whole blood is used in most developed countries it is used within 48–72 h of storage, even though it is licensed for storage between 21 and 35 days. Even in the rare US hospitals where WB is used, most transfusions occur within 7 days of collection. This is in contrast to hospitals in the developing world that transfuse whole blood until the date of expiration if necessary.

A fourth belief that has limited whole blood availability is the notion that blood components are equally efficacious compared to whole blood for patients with shock and coagulopathy. This concept has not been studied well, except in one pediatric cardiac surgery trial, where the use of whole blood did reduce blood loss secondary to improved platelet function compared to blood components [16]. As mentioned above, all other trials either added platelets to whole blood when compared to blood components, or whole blood was only given during a portion of the study period, potentially limiting generalizability [17, 18].

These four issues have caused concern regarding the efficacy, safety and logistic feasibility in providing whole blood for patients with life-threatening bleeding and have led to its dramatic reduction in use in the developed world. As a result of these concerns or beliefs, whole blood is rarely available in the developed countries such as Australia, the UK or Canada (Devine D, Doughty H, Irving D, 2015, oral communication, 15th August). While the concepts above have been challenged in an effort to provide patients with a better, more efficacious resuscitation product, acceptance and implementation have been slow, thus whole blood availability remains limited. Some developed countries do still use whole blood for pediatric transfusion, including France (Sailliol A, 2015, oral communication, 15th August), and in 15 % of children's hospitals in the US [36]. In both the US and France despite licensing for up to 35 days CWB storage duration is limited to 2 to 7 days in clinical settings (Jobes D and Sailliol A, 2015, oral communication, 15th August).

11.1.4.1 Group-Specific Versus O-Negative Low Titer Cold Whole Blood

Transfusion of type O low titer CWB is an alternative solution to group-specific CWB, but despite being the standard of care for treating hemorrhagic shock up to and during the Vietnam War, it is currently not permitted by AABB standards, although it is permitted in military scenarios during combat operations in a few countries, such as the US, United Kingdom, France, Australia, and Norway [37]. The rationale for this prohibition is difficult to understand. Hundreds of thousands of units of type O whole blood, with low anti-A and anti-B titers to reduce the risk of plasma incompatible ABO reactions, were used during both World Wars, the Korean War, and the war in Vietnam [38–40]. The threshold for low titer set by the US military in the Korean War was <1:256 [41]. A report by Nessen et al. indicated that, at US military forward surgical bases, in a subset of patients, the use of type O whole blood was associated with improved outcomes when compared to RBCs and plasma alone [42]. In addition to decades of use with few reports of adverse complications, there are multiple reasons to suppose that type O low titer CWB is safer than type-specific CWB. While a small risk of severe hemolytic reaction to type O low titer CWB does exist, it is similar to that of transfusing type O RBCs, which are universally available and the standard of care for emergency release at major medical centers. It is the plasma contained in the blood that causes the greatest concern for ABO incompatibility, but the associated hemolytic reaction is typically mild to moderate and the incidence, only 1:120,000 transfusions of type-incompatible platelet units (which contain an equivalent amount of plasma as that found in a unit of WB) in the UK SHOT database, is low [37]. In contrast to this non-fatal risk, type-specific WB transfusion carries a higher risk (1:80,000) of often-fatal severe hemolytic reactions, mainly due to human error in matching donor and recipient appropriately. Therefore, high-quality hemovigilance data reveals that the risks of hemolytic transfusion reactions are likely to be lower, and of much less clinical severity with Type Specific CWB. The use of type O low titer CWB would further reduce the risk of mild to moderate transfusion reactions since the UK shot database data on ABO incompatible plasma risks incorporated platelet transfusions that were not tittered for ABO antibodies. It's also likely that the risks of human error and ABO mismatch would be higher in the clinical circumstances surrounding massive transfusion protocol activation in a severely bleeding patient. In summary, our current standard practice of requiring whole blood to be ABO specific puts patients at higher risk of severe transfusion reactions.

Given the prevalence of type O donors, type O low titer CWB is a viable alternative to type-specific WB, but to date it continues to be largely unavailable. Implementation of Type O LTWB would require evaluation of current methods for determining titers, since multiple assays are available, but evidence-based standards are lacking. In addition, a commonly accepted, data-driven threshold that defines "low titer" anti-A and anti-B IgM and IgG is needed to replace the arbitrary values used in previous military conflicts. In the report by Nessen et al. described above, the authors also found that use of untitered type O whole blood was not associated with increased adverse events, but, outside of emergent cases or very austere environments, few would advocate this strategy given the small sample size and the relatively low cost of establishing titers.

11.1.4.2 Platelet Sparing Leukoreduction Filters

The misperception that whole blood cannot be leukocyte-reduced without removing platelets along with the leukocytes, leads clinicians to the erroneous conclusion that WB either carries the risk of graft-versus-host disease or if filtered contains an unacceptably low number of platelets and thus is not clinically useful. The FDA recently approved a WB leukoreduction filter that is platelet-sparing, paving the way for leukoreduced WB as a platelet-containing product [15].

11.1.4.3 Hemostatic Function of Platelets in Whole Blood Stored at 4 °C

Another concern is that the platelets in whole blood stored at 4 °C are not functional or viable, once again calling into question whether WB is truly a platelet-containing product. This misconception stems from the assumption that platelets that have undergone shape-change in response to refrigeration and are spherical are also non-functional. The opposite is true: cold-stored platelets aggregate better than those stored at 22 °C, are able to form a thrombus, and are associated with stronger clot formation when compared to the current standard of RT-stored platelet concentrates [22–30]. Most trials comparing a cold platelet containing blood product to warm stored products have indicated improved platelet aggregation, clotting times, and less blood loss with cold storage [16, 30].

While the standard approach for whole blood has been to limit its storage duration to 48–72 h due to concerns regarding platelet function, a few centers in the US have increased the storage duration past this point. The Children's Hospital of Philadelphia transfuses leukocyte-reduced, ABO specific, whole blood stored for up to 7 days to children with significant bleeding secondary to craniofacial surgery. The University of Pittsburgh Trauma Program is using leukocyte reduced, low titer group O whole blood, stored at 4 °C for up to 10 days for severely bleeding patients upon admission. The Mayo Clinic Trauma Program is planning to use non-leukocyte reduced low titer group O whole blood, stored at 4 °C for up to 10 days for severely bleeding patients in the prehospital and hospital setting. Data from these trauma centers will supplement *in vitro* studies completed by the US Army and Norwegian Navy that demonstrated adequate preservation of hemostatic function over 2 weeks of storage [23, 43]. Additional safety data in massively transfused patients is needed to determine if the use of low titer group O whole blood affects outcomes. Lastly, studies are also needed to determine if leukoreduction is necessary to improve outcomes in this patient population. While it is standard of practice in most developed countries, the evidence that it is associated with improved outcomes is controversial.

WB is generally stored without agitation, but military experience suggests that at least daily manipulation or gentle agitation is required to reduce platelet clumping and adherence to bag surfaces. Investigations are ongoing to determine optimal WB storage and handling protocols (Cap A, 2015, oral communication, 15th August).

11.1.4.4 Efficacy and Safety of Whole Blood Compared to Components

The human, and *in vitro* data evaluating the efficacy of whole blood has been summarized in the sections on CWB and WWB data. The belief that there is no benefit to whole blood compared to components is not based on data, but rather rooted in decades-old theory that patients with severe bleeding would not benefit from early administration of platelets or platelet containing products such as whole blood. This, in combination with the idea that CWB does not contain functional platelets strengthened the misperception that there is no potential increased efficacy with CWB. Current *in vitro* data in combination with the scant *in vivo* data of CWB have led to a reconsideration of whether it would be more efficacious than the use of blood components for patients with life-threatening bleeding. Furthermore, the convenience of administering only one product, particularly in the pre-hospital setting, has made CWB very attractive to military and civilian trauma system directors. Safety concerns regarding the use of a cold platelet-containing product have centered on the idea that the platelets would be irreversibly activated and cause thrombosis. This concern has not been noted in the few trials performed nor has this been observed in the clinical use of CWB at children's hospitals.

11.1.5 Rationale for Whole Blood

There is biological, logistical, and economic rationale that supports the use of whole blood rather than components for patients with severe life-threatening hemorrhagic shock. Biologically, whole blood provides a balanced amount of RBCs, plasma, and PLTs, as well as an increased concentration of cellular components and improved function compared with stored components in a 1:1:1 unit ratio [31, 44]. Each of the separate component units, RBCs, plasma, and PLTs, contains a considerably increased amount of anticoagulants and additives that will contribute to a dilutional coagulopathy compared with 1 U of FWB [31]. These characteristics of whole blood are important since patients with severe life-threatening bleeding are at high risk of shock and coagulopathy. Shock and coagulopathy potentiate each other leading to organ failure and death if not reversed rapidly. Therefore a resuscitative product that provides a balanced approach at preventing or treating both shock and coagulopathy is essential. The quality of the RBCs, plasma and platelets provided for patients in hemorrhagic shock is also important. This highly vulnerable patient population requires a resuscitative approach that is highly efficacious at improving oxygen delivery to treat shock, and that is hemostatic to prevent or treat coagulopathy-related bleeding. The safety of the resuscitative solution is also important. The adverse effects of resuscitation-induced dysregulation of immune, coagulation, and endothelial function can obviate potential benefits.

Among the potential advantages of WB compared to the use of components are that if the WB is limited to 10–15 days of storage at 4 °C, there are less "storage lesion" effects on RBCs, and platelet hemostatic function may actually be improved compared to platelet units stored at 22 °C. It is apparent that RBC quality is not directly related to storage duration based on the results of the ARIPI, RECESS, and ABLE studies that did not reveal improved outcomes for critically ill premature neonates or adults who received fresh RBCs compared to older RBC units [45–47]. These trials did not include patients who required large amounts of blood products, and did not include a significant number of patients with traumatic injury, who may be most vulnerable to storage lesion effects [48]. In addition, due to donor variability in RBC quality over storage time, RBC age may not be the best surrogate for RBC quality. Other RBC quality measures such as cell deformability may be more accurate measures of quality. In addition, the above trials did not examine specifically the effects of RBC storage age at the end of expiration (28–42 days of storage). Patients in the perioperative phase who require massive transfusion protocol activation due to life threatening bleeding are often exposed to large volumes of RBCs at the end of expiration and as a result are exposed to RBCs with less efficacy and reduced safety. They are also transfused platelets stored at 22 °C, which have less hemostatic capacity than the platelets in whole blood stored at 4 °C [49].

The use of whole blood exposes the recipient to one donor compared to three for reconstituted whole blood with RBC, plasma, and platelet units, while providing a more concentrated and perhaps functional and safer product. Reduction of exposure to additional donors is important to the patient in the perioperative period since it reduces the risk of TTD's and alloimunization, which is important for patients that may eventually need an organ transplant.

Biologic risks of LR-CWB theoretically include thrombosis due to irreversible platelet activation, and immunologic consequences secondary to leukocytes. Experiments have previously concluded that platelets stored at 4 °C became activated based upon irreversible shape change. As a result there was concern that transfusion of whole blood stored at 4 °C would increase the risk of thrombosis. In trials of either whole blood or platelet units stored at 4 °C there has not been an increased risk of thrombotic events. Two trials that compared component blood products to whole blood in children requiring cardiac surgery did not reveal any increased risk of adverse events [17]. In a trial of adults randomized to receive platelet units at 4° or 22 °C there was also no increased risk of adverse events [30].

Recent studies suggest that cold storage of platelet containing blood products primes platelets for activation, but that these platelets still respond to inhibitory signals produced by the endothelium. Under static and flow conditions, platelets stored at 4 °C compared to the control group of freshly drawn un-stored platelets displayed similar ability to respond to control signals like nitric oxide and prostacyclin as measured by platelet aggregometry, adhesion to collagen in microfluidic flow cells and viscoelastic measures.

As suggested in studies of blood component products, residual WBC in WB may cause some degree of immunosuppression in the transfusion recipient, though this may also be mediated by cellular microparticles generated over time with storage. Any direct effect of whole blood on immune function has not been well studied. One trial examined inflammatory effects of whole blood compared to the use of RBCs and plasma units to prime cardiopulmonary by-pass circuits in children. It reported that measures of inflammation were similar between the two study groups except for an increase in lipopolysaccharide binding protein in the CWB group at 48 and 72 h [50]. Microparticle content is similar in whole blood stored at 4° vs. 22 °C. Therefore if microparticles have any clinical effect, the risks are potentially similar between CWB and platelet units stored at 22 °C. The clinical relevance of the effects of microparticles from all cell lines in blood products requires further study.

Viable WBC in blood products can cause transfusion-associated GVHD (TA-GVHD). TA-GVHD is a very rare and almost always fatal condition that occurs due to engraftment of donor WBCs into the recipient. The primary risk factor for TA-GVHD is transfusion of viable donor lymphocytes that share common antigens with host cells at either HLA Class I or II loci. Most cases of TA-GVHD have occurred when blood products were transfused within 10 days of collection. Leukoreduction may reduce, but not completely eliminate the risk of TA-GVHD; only gamma irradiation is licensed for inactivation of donor lymphocytes and TA-GVHD prevention [51]. Microchimerism is low level (<5 %) engraftment of donor lymphocytes or peripheral blood progenitor cells into the recipient. The clinical consequences of microchimerism are unknown but the condition may persist for decades. The risk of TA-GVHD and microchimerism is related to the presence of viable WBCs in vulnerable recipients, particularly those suffering from traumatic injury. These risks with whole blood are similar to those with LR-RBCs stored for less than 10–15 days.

Risks with WWB are increased risk of infection and immunogenic. Increased risk of transfusion-transmitted diseases occurs in austere settings when there is not adequate time or infrastructure to perform the standard testing for infectious agents. One study calculated the risk of TTD transmission in a military setting where US personnel were donors and were not screened with rapid tests for TTDs. In this setting there was a 1/800 rate of HCV transmission the recipients [52]. With over 10,000 units of WWB transfused to patients treated at US Military hospitals there have been one case of HTLV and one case of HCV transmission. To mitigate these risks, rapid screening tests for HCV, HBV, and HIV can be used to screen potential donors. Donors are also vaccinated against HBV, which would prevent HBV infection in the donor and reduce risks for future whole blood recipients. US Military donors are screened for HIV every 2 years and immediately prior to deployment. Registered donors in military walking blood banks are currently fully prescreened according to standard donor protocols and undergo repeat testing every 90 days during deployment.

Currently, platelet-sparing leukoreduction filters have not been implemented in military settings for WWB transfusion. Since WWB transfusion is generally an emergency procedure undertaken in the face of exsanguinating hemorrhage, the modest benefits of LR have been deemed insufficient to warrant the time delays in WB unit preparation, logistical requirements and cost when collection is required immediately prior to transfusion. Whether LR can or should be accomplished in the military context for whole blood that is collected and then stored is currently controversial. It is unknown if the potential risks of residual WBCs present in non-LR WB are different for warm whole blood that is transfused immediately compared to cold whole blood that is stored. There have actually been no studies that evaluate the immune consequences of white blood cells within WWB or stored CWB in recipients. The clinical relevance of leukoreduction is controversial due to conflicting reports regarding its effect on outcomes in critically ill populations. Aside from reducing febrile reactions leukoreduction has not consistently been associated with improved morbidity or mortality [53]. Furthermore, in these studies the blood products that were not leukoreduced were stored for a period of time. There may be a different biological effect of transfusing a blood product with normal amounts of WBCs in it that has been stored for weeks compared to one that is transfused immediately. The fresh WBCs in WFWB may or may not be harmful. It may even be helpful in patients who are in an immune deficient state early after traumatic injury or due to other critical illnesses. Additional research is needed in this area to more thoroughly understand the benefits of leukoreduction in whole blood that is transfused immediately compared to when it is stored.

The risk of both TTD transmission and immunologic challenges posed by viable donor WBCs (such as TA-GVHD) may be substantially mitigated by development of the Mirasol pathogen reduction technology for WB (Terumo BCT), which is based on the photochemical degradation of nucleic acids by riboflavin and UV light. Several pathogen reduction technologies prevent WBC replication and thus inactivate WBCs, although currently, the Mirasol technology is the only one demonstrated to be effective in WB [54]. The importance of WBC inactivation may be different for WWB compared to stored CWB and requires further study [55].

11.1.5.1 Logistical Benefits

According to a recent survey of 132 trauma centers, blood products are used for prehospital resuscitation by only 34 % of first responders [56]. A smaller minority is carrying both RBCs and plasma, and very few, if any, ever carry platelets, although multiple studies demonstrate that they are important for hemostasis. The use of blood products to resuscitate patients in the pre-hospital phase has the potential to reduce the large risk of death from hemorrhagic shock prior to hospital admission. Military data indicates that 90% of potentially survivable deaths occur from hemorrhagic shock [57]. Improved control of bleeding and early blood product use are required to reduce this toll. The logistical constraints of transporting RBCs, plasma, and platelets prehospital are considerable. In addition to the extra weight and complexity of transfusing product from multiple bags, intravenous or intraosseous catheters have a limited number of access ports. The use of whole blood stored at 4 °C stored for less than 10–14 days would dramatically reduce the logistical burden compared to the current approach, which too often provides the oldest component products in the inventory.

11.1.5.2 Costs and Benefits

Whole blood is very likely to be cost effective compared to the use of blood components for life-threatening bleeding. A formal cost-effectiveness analysis has not been performed, but when the following are considered it very well may be associated with reduced costs while at the very least maintaining equal efficacy if not improving it. A transition to the use of whole blood would include reduced costs associated with the fractionation of whole blood, including both equipment-related and staffing costs. If CWB was limited to 15 days this would virtually triple the inventory of platelet-containing products for patients who require them for severe bleeding. Contrary to the widely made assumption that the vast majority of platelet units, perhaps 80 %, are transfused prophylactically to thrombocytopenic cancer patients, the most recent survey of blood use in the US, conducted in 2011 by the Department of Health and Human Services, demonstrated that 25 % of platelet units are transfused for surgical or trauma patients, with another 12 % consumed in the ICU setting. By contrast, 34 % of platelets were transfused on Hematology-Oncology services. This clearly indicates that a substantial proportion of hospitalized patients require platelets for acute bleeding, and are therefore likely to require not only platelets but also RBCs and plasma – all of which could be conveniently supplied by WB. The improved hemostatic function of platelets in CWB may also reduce costs compared to the standard platelets stored at 22 °C. Reduced bleeding and donor exposure will improve safety and costs. The use of CWB instead of platelets at 22 °C would also eliminate the cost of bacterial testing required for products stored at room temperature. The additional costs associated with the use of CWB are the costs of performing Anti-A/B titers and the cost of lost revenue if the whole blood unit expires. While RBCs can be recovered from a whole blood unit stored at 4 °C at 15 days of storage, plasma and platelets from that unit are not suitable for further separation and storage.

Presuming that at the very least whole blood is just as efficacious as blood components the presumed net reduced costs associated with its use in patients with hemorrhagic shock are substantial and may in and of itself justify the incorporation of this product for this patient population. Formal cost-effectiveness studies are warranted to quantify the economic impact of transitioning to a CWB transfusion strategy for patients with hemorrhagic shock.

11.1.6 Plasma, Platelets, and RBCs in a 1:1:1 Unit Ratio

Since whole blood is not readily available in many hospitals except for some children's hospitals and in austere environments in the developing world, blood is often reconstituted into whole blood by providing plasma, platelets, and RBCs in a 1:1:1 unit ratio. The rationale for resuscitating patients with severe life threatening bleeding in the perioperative arena with plasma, platelets and RBCs in a 1:1:1 unit ratio is that it approximates whole blood functionality, providing a balanced therapeutic approach to both shock and coagulopathy. According to a recent survey of 132 US trauma centers, by policy in adults, 88 % of sites target high $\geq$ (1:2) plasma-to-red blood cell

(RBC) ratios. Likewise, 79 % of sites target high platelet-to-RBC ratios. To facilitate the early transfusion of plasma, 69 % of sites indicated that they have plasma immediately available for MTP activation. A similar survey of MTP policies in Children's hospitals revealed that 78 % of centers targeted a high plasma:RBC ratio in children with life-threatening bleeding, whereas 65 % of centers targeted a high ratio of platelets:RBCs. In children's hospitals 48 % of centers had plasma immediately available as thawed plasma. These survey data indicate that most adult and pediatric medical centers are attempting to emulate whole blood with components.

Data on the use of empiric ratios of blood products is predominantly in the trauma literature. A meta-analysis of retrospective studies that evaluated the effect of high plasma ratios on outcomes in trauma patients indicated a survival benefit with early and increased plasma administration, although the quality of the data was determined to be poor [58]. There has been less data published regarding high platelet to RBC ratios, although most indicate improved survival, similar to the plasma:RBC ratio data. In a retrospective analysis examining the association of plasma and platelet ratios to RBCs in patients with severe TBI, patients transfused a high ratio of platelets had more of an association with improved outcomes than a high plasma to RBC ratio [59]. The authors hypothesized that the platelets improved blood brain barrier function via oligodendrite precursor cells.

Recently, the Pragmatic Randomized Optimal Platelet and Plasma Ratios (PROPPR) trial, a large prospective randomized trial comparing a 1:1:1 ratio versus 1:1:2 in which the latter group received fewer platelets given later in resuscitation [60]. The study, while limited by the similarity of the treatment arms due to ethical considerations, provided higher quality evidence to support equal ratio component use and was still able to show a difference in secondary outcomes of reduced death from hemorrhage at 24 h and reduced time to hemostasis. Primary outcomes of mortality at 24 and 30 days were not different between the two study groups.

In summary, for patients in the perioperative period with hemorrhagic shock, the current component products may not be optimal, but until further research leads to changes in regulations and guidelines, transfusion of high ratios of plasma and RT-stored platelets to RBCs in massively bleeding patients is associated with better outcomes, to include reduced death from hemorrhage. Refrigerated stored whole blood is an underutilized potential source of platelets, plasma and RBCs that may be more efficacious, safe, and logistically feasible, all of which will positively affect the economics of health care provided.

References

1. Spinella PC, Holcomb JB (2009) Resuscitation and transfusion principles for traumatic hemorrhagic shock. Blood Rev 23(6):231–240
2. Gerhardt RT, Strandenes G, Cap AP, Rentas FJ, Glassberg E, Mott J et al (2013) Remote damage control resuscitation and the Solstrand conference: defining the need, the language, and a way forward. Transfusion 53(Suppl 1):9S–16S
3. Barnes A (1980) Transfusion of universal donor and uncrossmatched blood. Bibl Haematol 46:132–142

4. Carrico CJ, Canizaro PC, Shires GT (1976) Fluid resuscitation following injury: rationale for the use of balanced salt solutions. Crit Care Med 4(2):46–54
5. Cotton BA, Guy JS, Morris JA Jr, Abumrad NN (2006) The cellular, metabolic, and systemic consequences of aggressive fluid resuscitation strategies. Shock 26(2):115–121
6. Balogh Z, McKinley BA, Cocanour CS, Kozar RA, Valdivia A, Sailors RM et al (2003) Supranormal trauma resuscitation causes more cases of abdominal compartment syndrome. Arch Surg 138(6):637–642; discussion 42–43
7. Shoemaker WC (1976) Comparison of the relative effectiveness of whole blood transfusions and various types of fluid therapy in resuscitation. Crit Care Med 4(2):71–78
8. Makley AT, Goodman MD, Friend LA, Deters JS, Johannigman JA, Dorlac WC et al (2010) Resuscitation with fresh whole blood ameliorates the inflammatory response after hemorrhagic shock. J Trauma 68(2):305–311
9. Alam HB, Rhee P (2007) New developments in fluid resuscitation. Surg Clin North Am 87(1):55–72, vi
10. Johansson PI, Swiatek F, Jorgensen L, Jensen LP, Secher NH (2008) Intraoperative platelet and plasma improves survival in patients operated for a rAAA: a follow-up evaluation. Eur J Vasc Endovasc Surg Off J Eur Soc Vasc Surg 36(4):397–400
11. Johansson PI, Stensballe J, Rosenberg I, Hilslov TL, Jorgensen L, Secher NH (2007) Proactive administration of platelets and plasma for patients with a ruptured abdominal aortic aneurysm: evaluating a change in transfusion practice. Transfusion 47(4):593–598
12. Alexander JM, Sarode R, McIntire DD, Burner JD, Leveno KJ (2009) Whole blood in the management of hypovolemia due to obstetric hemorrhage. Obstet Gynecol 113(6): 1320–1326
13. Gutierrez MC, Goodnough LT, Druzin M, Butwick AJ (2012) Postpartum hemorrhage treated with a massive transfusion protocol at a tertiary obstetric center: a retrospective study. Int J Obstet Anesth 21(3):230–235
14. U.S. Department of Health and Human Services: Biologics Guidances 2015 [cited 8-31-2015]. Available from: http://www.fda.gov/BiologicsBloodVaccines/GuidanceComplianceRegulatoryInformation/Guidances/default.htm
15. Snyder EL, Whitley P, Kingsbury T, Miripol J, Tormey CA (2010) In vitro and in vivo evaluation of a whole blood platelet-sparing leukoreduction filtration system. Transfusion 50(10):2145–2151
16. Manno CS, Hedberg KW, Kim HC, Bunin GR, Nicolson S, Jobes D et al (1991) Comparison of the hemostatic effects of fresh whole blood, stored whole blood, and components after open heart surgery in children. Blood 77(5):930–936
17. Mou SS, Giroir BP, Molitor-Kirsch EA, Leonard SR, Nikaidoh H, Nizzi F et al (2004) Fresh whole blood versus reconstituted blood for pump priming in heart surgery in infants. N Engl J Med 351(16):1635–1644
18. Cotton BA, Podbielski J, Camp E, Welch T, del Junco D, Bai Y et al (2013) A randomized controlled pilot trial of modified whole blood versus component therapy in severely injured patients requiring large volume transfusions. Ann Surg 258(4):527–532; discussion 32–33
19. Golan M, Modan M, Lavee J, Martinowitz U, Savion N, Goor DA et al (1990) Transfusion of fresh whole blood stored (4 degrees C) for short period fails to improve platelet aggregation on extracellular matrix and clinical hemostasis after cardiopulmonary bypass. J Thorac Cardiovasc Surg 99(2):354–360
20. Nepstad I, Reikvam H, Strandenes G, Hess JR, Apelseth TO, Hervig TA (2014) Comparison of in vitro responses to fresh whole blood and reconstituted whole blood after collagen stimulation. Blood Transfus 12(1):50–55
21. Kornblith LZ, Howard BM, Cheung CK, Dayter Y, Pandey S, Busch MP et al (2014) The whole is greater than the sum of its parts: hemostatic profiles of whole blood variants. J Trauma Acute Care Surg 77(6):818–827
22. Jobes D, Wolfe Y, O'Neill D, Calder J, Jones L, Sesok-Pizzini D et al (2011) Toward a definition of "fresh" whole blood: an in vitro characterization of coagulation properties in refrigerated whole blood for transfusion. Transfusion 51(1):43–51

23. Pidcoke HF, McFaul SJ, Ramasubramanian AK, Parida BK, Mora AG, Fedyk CG et al (2013) Primary hemostatic capacity of whole blood: a comprehensive analysis of pathogen reduction and refrigeration effects over time. Transfusion 53(Suppl 1):137S–149S
24. Delgado KK, Pidcoke HF, Mora AG, Cap AP (2011) Platelet function in stored whole blood measured by a shear- and Von Willebrand factor-dependent methodology is retained during storage at 4 °C for up to 7 days. Transfusion 51(Suppl s3):65A
25. Babic AM, Josefsson EC, Bergmeier W, Wagner DD, Kaufman RM, Silberstein LE et al (2007) In vitro function and phagocytosis of galactosylated platelet concentrates after long-term refrigeration. Transfusion 47(3):442–451
26. Choi JW, Pai SH (2003) Influence of storage temperature on the responsiveness of human platelets to agonists. Ann Clin Lab Sci 33(1):79–85
27. Hornsey VS, Drummond O, McMillan L, Morrison A, Morrison L, MacGregor IR et al (2008) Cold storage of pooled, buffy-coat-derived, leucoreduced platelets in plasma. Vox Sang 95(1):26–32
28. Valeri CR (1976) Circulation and hemostatic effectiveness of platelets stored at 4 C or 22 C: studies in aspirin-treated normal volunteers. Transfusion 16(1):20–23
29. Valeri CR (1974) Hemostatic effectiveness of liquid-preserved and previously frozen human platelets. N Engl J Med 290(7):353–358
30. Becker GA, Tuccelli M, Kunicki T, Chalos MK, Aster RH (1973) Studies of platelet concentrates stored at 22 C nad 4 C. Transfusion 13(2):61–68
31. Spinella PC, Perkins JG, Grathwohl KW, Beekley AC, Holcomb JB (2009) Warm fresh whole blood is independently associated with improved survival for patients with combat-related traumatic injuries. J Trauma 66(4 Suppl):S69–S76
32. Perkins JG, Cap AP, Spinella PC, Shorr AF, Beekley AC, Grathwohl KW et al (2011) Comparison of platelet transfusion as fresh whole blood versus apheresis platelets for massively transfused combat trauma patients (CME). Transfusion 51(2):242–252
33. Ho KM, Leonard AD (2011) Lack of effect of unrefrigerated young whole blood transfusion on patient outcomes after massive transfusion in a civilian setting. Transfusion 51(8): 1669–1675
34. Lavee J, Martinowitz U, Mohr R, Goor DA, Golan M, Langsam J et al (1989) The effect of transfusion of fresh whole blood versus platelet concentrates after cardiac operations. A scanning electron microscope study of platelet aggregation on extracellular matrix. J Thorac Cardiovasc Surg 97(2):204–212
35. Mohr R, Martinowitz U, Lavee J, Amroch D, Ramot B, Goor DA (1988) The hemostatic effect of transfusing fresh whole blood versus platelet concentrates after cardiac operations. J Thorac Cardiovasc Surg 96(4):530–534
36. Spinella PC, Dressler A, Tucci M, Carroll CL, Rosen RS, Hume H et al (2010) Survey of transfusion policies at US and Canadian children's hospitals in 2008 and 2009. Transfusion 50(11):2328–2335
37. Strandenes G, Berseus O, Cap AP, Hervig T, Reade M, Prat N et al (2014) Low titer group O whole blood in emergency situations. Shock 41(Suppl 1):70–75
38. Hess JR, Thomas MJ (2003) Blood use in war and disaster: lessons from the past century. Transfusion 43(11):1622–1633
39. Kendrick D (1964) Blood program in World War II. Chapter XX: The blood, plasma, and related programs in the Korean War [cited 2-23-2015]. Available from: http://history.amedd.army.mil/booksdocs/wwii/blood/chapter20.htm
40. Neel MGS (1991) Vietnam studies: medical support of the US Army in Vietnam 1965–1970. Department of the Army [cited 2-23-2015]. Available from: http://history.amedd.army.mil/booksdocs/vietnam/medicalsupport/default.html
41. Crosby WH, Akeroyd JH (1954) Some immunohematologic results of large transfusions of group O blood in recipients of other blood groups; a study of battle casualties in Korea. Blood 9(2):103–116

42. Nessen SC, Eastridge BJ, Cronk D, Craig RM, Berseus O, Ellison R et al (2013) Fresh whole blood use by forward surgical teams in Afghanistan is associated with improved survival compared to component therapy without platelets. Transfusion 53(Suppl 1):107S–113S
43. Strandenes G, Austlid I, Apelseth TO, Hervig TA, Sommerfelt-Pettersen J, Herzig MC et al (2015) Coagulation function of stored whole blood is preserved for 14 days in austere conditions: a ROTEM feasibility study during a Norwegian antipiracy mission and comparison to equal ratio reconstituted blood. J Trauma Acute Care Surg 78(6 Suppl 1):S31–S38
44. Armand R, Hess JR (2003) Treating coagulopathy in trauma patients. Transfus Med Rev 17(3):223–231
45. Fergusson DA, Hebert P, Hogan DL, LeBel L, Rouvinez-Bouali N, Smyth JA et al (2012) Effect of fresh red blood cell transfusions on clinical outcomes in premature, very low-birthweight infants: the ARIPI randomized trial. JAMA 308(14):1443–1451
46. Steiner ME, Ness PM, Assmann SF, Triulzi DJ, Sloan SR, Delaney M et al (2015) Effects of red-cell storage duration on patients undergoing cardiac surgery. N Engl J Med 372(15):1419–1429
47. Lacroix J, Hebert PC, Fergusson DA, Tinmouth A, Cook DJ, Marshall JC et al (2015) Age of transfused blood in critically ill adults. N Engl J Med 372(15):1410–1418
48. Steiner ME, Stowell C (2009) Does red blood cell storage affect clinical outcome? When in doubt, do the experiment. Transfusion 49(7):1286–1290
49. Reddoch KM, Pidcoke HF, Montgomery RK, Fedyk CG, Aden JK, Ramasubramanian AK et al (2014) Hemostatic function of apheresis platelets stored at 4 degrees C and 22 degrees C. Shock 41(Suppl 1):54–61
50. Mou SS, Giroir BP, Molitor-Kirsch EA, Leonard SR, Nikaidoh H, Nizzi F et al (2004) Fresh whole blood versus reconstituted blood for pump priming in heart surgery in infants. N Engl J Med 351:1635–1644
51. Kopolovic I, Ostro J, Tsubota H, Lin Y, Cserti-Gazdewich CM, Messner HA et al (2015) A systematic review of transfusion-associated graft-versus-host disease. Blood 126(3):406–414
52. Spinella PC, Perkins JG, Grathwohl KW, Repine T, Beekley AC, Sebesta J et al (2007) Risks associated with fresh whole blood and red blood cell transfusions in a combat support hospital. Crit Care Med 35(11):2576–2581
53. Blajchman MA (2006) The clinical benefits of the leukoreduction of blood products. J Trauma 60(6 Suppl):S83–S90
54. Fast LD, Marschner S, Nevola M, Reddy HL, Goodrich RP (2011) Inactivation of leukocytes in whole blood treated with the Mirasol® System in comparison to gamma-irradiation. Transfusion 51(Suppl s3):41A
55. Grass JA, Wafa T, Reames A, Wages D, Corash L, Ferrara JL et al (1999) Prevention of transfusion-associated graft-versus-host disease by photochemical treatment. Blood 93(9):3140–3147
56. Camazine M, Hemmila M, Leonard JC, Jacobs R, Horst J, Kozar R et al (2015) Massive transfusion polices at trauma centers participating in the american college of surgeons trauma quality improvement program. J Trauma Acute Care Surg 78(6 Suppl 1):S48–S53
57. Eastridge BJ, Mabry RL, Seguin P, Cantrell J, Tops T, Uribe P et al (2012) Death on the battlefield (2001–2011): implications for the future of combat casualty care. J Trauma Acute Care Surg 73(6 Suppl 5):S431–S437
58. Murad MH, Stubbs JR, Gandhi MJ, Wang AT, Paul A, Erwin PJ et al (2010) The effect of plasma transfusion on morbidity and mortality: a systematic review and meta-analysis. Transfusion 50(6):1370–1383
59. Spinella PC, Wade CE, Blackbourne LH, Borgman MA, Zarzabal LA, Du F et al (2011) The association of blood component use ratios with the survival of massively transfused trauma patients with and without severe brain injury. J Trauma 71(2 Suppl 3):S343–S352
60. Holcomb JB, Tilley BC, Baraniuk S, Fox EE, Wade CE, Podbielski JM et al (2015) Transfusion of plasma, platelets, and red blood cells in a 1:1:1 vs a 1:1:2 ratio and mortality in patients with severe trauma: the PROPPR randomized clinical trial. JAMA 313(5):471–482

Chapter 12
Fluid Therapy in Trauma

M.F.M. James and D.P.D. Chappell

Abstract Fluid therapy in trauma has been a controversial issue since it was first introduced on a major scale during the First World War. Initially, fluid resuscitation concentrated on the use of blood and blood products supported by colloids and saline-based crystalloid solutions. In the 1960s, experience from battlefield resuscitations, together with laboratory research led to the concept of the "third space" and the use of massive crystalloid resuscitation. The consequence of this was initially improved survival, but also the development of pulmonary oedema and abdominal compartment syndromes. More recently, aggressive pre-hospital resuscitation has been questioned and more conservative, hypotensive resuscitation has been advocated prior to the patient being transported to hospital. The role of hypotensive resuscitation in neurologically injured patients remains controversial. Military field data has suggested that the early use of blood products in addition to red blood cell concentrates may improve survival in battle casualties, but this has not been confirmed in civilian trauma. However, the emphasis on resuscitation of coagulation has placed increasing emphasis on the use of procoagulants, such as tranexamic acid early in severe trauma and, in hospital, on the appropriate monitoring of coagulation. The use viscoelastic coagulation assessment has allowed more precise management of trauma-induced coagulopathy with possibly improved outcomes. It has also been recognised that crystalloid overload is a major component of post-resuscitation morbidity and several studies have emphasised that the combined use of appropriate fluids including limited crystalloids, colloids and appropriate blood products may result in improved outcomes.

Keywords Trauma • Fluid therapy • Haemorrhage • Blood transfusion • Coagulation • Renal glycocalyx • Resuscitation

M.F.M. James (✉)
Department of Anaesthesia, University of Cape Town, Cape Town 7700, South Africa
e-mail: mike.james@uct.ac.za

D.P.D. Chappell
Klinik für Anästhesiologie, Klinikum der Universität München, Munich 81664, Germany
e-mail: daniel.chappell@med.uni-muenchen.de

K. Stuart-Smith (ed.), *Perioperative Medicine – Current Controversies*,
DOI 10.1007/978-3-319-28821-5_12

Trauma of various kinds is the leading cause of death in humans under the age of 40 years, worldwide.

12.1 Physiological Principles in Trauma

The physiology of the response to trauma is well known, but it is useful to outline the basic principles in order to develop a sound understanding of the basis for volume therapy in trauma.

The basic principle of the physiological response is that of the so-called "fight or flight" response that evolved to enable an injured animal to maximise physical activity in the face of an immediate traumatic event.

Circulatory adaptations are aimed at the perfusion and supply of basic nutrients to the so-called "fight or flight organs". These organs include the heart, lungs, skeletal muscle, the brain and red blood cells. In order to maximise blood flow to these organs, the sympathetic nervous system, through the alpha-adrenergic receptors, causes intense vasoconstriction to all of the other organ systems not required for the immediate activity. Thus, there is a profound reduction in blood flow to the organs of the abdominal cavity (notably the gut, liver and kidneys) as well as to the skin. The beta-adrenergic system driven by adrenaline release from the adrenal gland results in myocardial stimulation and vasodilation of the coronary arteries, pulmonary vasculature, bronchial smooth muscle and skeletal muscle.

Simultaneously, a variety of metabolic responses occur aimed at optimising the available intravascular fluid volume and metabolic substrate for immediate physical activity. At the regional level, redistribution of intra-renal blood flow, secretion of antidiuretic hormone (ADH) and secretion of aldosterone all result in the maximal retention of sodium, minimal fluid volume excretion and retention of water in excess of sodium. The consequences are a reduction in urine volume, an increase in the concentration of any urine produced and a relatively small, but probably clinically important reduction in the concentration of plasma sodium with a loss of potassium.

At the metabolic level, inhibition of insulin production, together with the release of a variety of hormones that inhibit insulin release (glucocorticoids, glucagon, thyroid hormone and growth hormone) all result in a relative insulin resistance and an increase in plasma sugar concentration, maximising the available energy for immediate physical activity [1]. None of these adaptations can be shown to have beneficial effects on survival in the long term in badly injured subjects. Where these adaptive mechanisms fail to maintain adequate tissue perfusion in the face of trauma and volume loss, dire metabolic consequences ensue, resulting in tissue ischaemia and the metabolic consequences of cellular hypoxia.

Even more deleterious is the release of cytokines in response to injury. These substances, which are a group of low-molecular-weight proteins including the interleukins and the interferons are produced from activated leucocytes and fibroblasts as well as endothelial cells. Their function is primarily the mediation of local inflammatory responses, but with extensive injury, widespread release of cytokines predisposes to the development of systemic inflammation.

Critical to the understanding of modern concepts of volume therapy is the recognition of the importance of the endothelial glycocalyx and its response in trauma. The glycocalyx is a surface layer found on all eukaryotic cells that serves the purpose of interaction between cells and messenger substances. The endothelial glycocalyx is a negatively charged layer lining the vascular endothelial surface and is made up of proteoglycans and glycosaminoglycans. This layer occupies a substantial volume of the endovascular space and, across the entire organism, may comprise as much as one litre in volume. The layer has only recently been described as previous electron microscopic techniques were unable to visualise it. Various molecules make up the endothelial glycocalyx, predominantly membrane-bound syndecans and glypicans which carry the negative charges. The discovery of this layer has substantially altered the understanding of the dynamics of fluid flux across the vascular wall. Although the interaction between the fluid contents of blood vessels and the glycocalyx is complex, simplistically it is the endothelial glycocalyx rather than the vessel wall that forms a barrier to colloid transduction into the extracellular space. The functional endothelial surface layer consists of an interaction of plasma proteins with the chemical constituents of the glycocalyx and this combination forms the physiologically active vascular barrier. Nevertheless, an intact glycocalyx is critical for the functioning of infused colloids and damage to this surface membrane may reduce or even eliminate the benefits of colloidal infusions. In addition, the glycocalyx has a central position in mediating systemic anti-inflammatory processes by trapping and enveloping adhesion molecules. The glycocalyx is also critical in the transmission of shear forces to the endothelial cell and thus plays a vital role in the release of nitric oxide and local tissue vasodilatation. In trauma, the glycocalyx is readily damaged and shedding of the components of this layer occurs early and in proportion to the degree of injury. This can be detected through measurement of the plasma content of syndecans and glypicans. Johansson and colleagues measured glycocalyx constituents in the plasma of patients submitted to a level 1 trauma centre. Irrespective of the trauma severity, glycocalyx shedding was associated with increased mortality, inflammation, catecholamines and an impaired coagulation [2]. Rahbar et al. recently found that glycocalyx shedding following severe trauma was associated with increased vessel permeability, reduced plasma colloid osmotic pressure and reduced thrombin levels which impair coagulation [3].

The physiology of coagulation is significantly disturbed in trauma in a number of different ways and is also disrupted by a variety of resuscitation strategies. Trauma induced coagulopathy (TIC) has long been considered to be a secondary event related to consumption of coagulation factors and dilution of those factors by aggressive fluid administration, together with hypothermia and acidosis. When it occurs, it is a significant predictor of mortality [4] with an incidence ranging between 10 and 34 % [5]. However, more recently it has been suggested that trauma itself induces a significant derangement of the clotting process. Consumptive coagulopathy (DIC) has been advocated as a major mechanism for the development of TIC and the development of DIC has been associated with adverse outcomes including prolonged intensive care stay and multiple organ dysfunction syndromes [6]. The concept is that extensive tissue injury activates tissue factor throughout the body and that tissue factor pathway inhibitor activation is unable to control this excessive stimulus to coagulation leading to consumptive losses of fibrinogen and activation of fibrinolytic pathways. An alternative concept has been advocated more recently in which it is suggested that damage to the endothelial glycocalyx results in excess of production of activated protein C that not only inhibits the normal thrombin activation of the coagulation cascade but also inhibits plasminogen activator inhibitor, resulting in enhanced fibrinolysis [7, 8]. In an extensive review, Ward evaluated the role of severe tissue hypoxaemia and endothelial damage as drivers of the coagulopathic state [9]. In a recent study of TIC, two cohorts of moderately injured patients were studied, divided into those with and without early TIC (ETIC). Both cohorts showed evidence of increased thrombin activation with fibrin generation together with increased fibrinolysis, but the ETIC group and received greater volumes of crystalloid prior to admission to hospital and these authors concluded that decreasing the amount of prehospital crystalloid administration together with early administration of coagulation factors may prevent the development of TIC [10]. It thus appears that TIC is a complex entity that is related to numerous factors including the severity of injury, the duration and severity of the consequent shock state, the use of resuscitation solutions that do not contain coagulation factors as well as the more established contributors such as hypothermia and acidosis. Interestingly, acidosis may not be as severe an inhibitor of coagulation as has been previously thought since a recent study using thrombelastography (TEG) showed that only severe acidosis (pH = 7.15) impaired coagulation [11].

Neurological injury represents a specific form of trauma that carries its own considerations. Damage done at the time of injury is largely irreversible, but substantial secondary insults frequently occur that are potentially treatable or avoidable. Such secondary injuries can have serious implications for the long-term outcome of injured patients. The important consideration is to recall the Munro-Kelly doctrine regarding the inextricable relationship between pressure and volume in a confined space; of the three constituents in the intracranial cavity, neural tissue, CSF and blood, any increase in one must result in a compensatory decrease in the other (the so-called cerebral elastance) or a rise in intracranial pressure. Once intracranial pressure has started to rise, these compensatory mechanisms have been exceeded and any further increase in any of the intracranial contents will result in a steep increase in intracranial pressure. Brain tissue is particularly vulnerable to the consequences of trauma and the activation of inflammatory cascades. Cerebral ischaemia and inflammatory mediators result in neuronal and astrocytic swelling which together with vasogenic oedema contribute to brain swelling [12]. These secondary insults are influenced by changes in cerebral blood flow (hypo- and hyperperfusion), impairment of cerebrovascular autoregulation, cerebral metabolic dysfunction and inadequate cerebral oxygenation. Furthermore, excitotoxic cell damage and inflammation may lead to apoptotic and necrotic cell death [12]. The damaged brain is particularly sensitive to changes in plasma osmotic pressure and the management of neural injury requires an appreciation of the constituents of resuscitation fluids and their possible effects on osmotic swelling of neural tissues. Spinal cord trauma raises additional complications in the form of spinal shock resulting in disproportionate hypotension to the volume of fluid lost. In addition, there is some emerging evidence that high-level spinal-cord injury (above T6) may result in disordered cerebral blood flow regulation [13]. The early management phase following cerebral trauma aims to achieve haemodynamic stability and limit secondary insults (e.g. hypotension, hypoxia) prior to definitive management [14].

12.2 Trauma Resuscitation

The objective of volume therapy in trauma is the restoration of tissue perfusion through the replacement of lost volume and plasma components of shed blood. However, it is critical that the underlying physical principles inform the choice of volume in terms of both content of intravenous solutions and the quantities to be administered. Both of these choices have been highly controversial over the last 100 years and remain so today.

12.2.1 History

Blood transfusion for the management of trauma and haemorrhagic shock was first attempted as long ago as the seventeenth-century when disastrous attempts were made to transfuse humans with animal blood [15]. The first human-to-human transfusion is credited to James Blundell in 1818 who used a syringe with a two-way stopcock to transfuse blood into a woman with postpartum haemorrhage using her husband as the donor but the good outcome was somewhat fortuitous as there was no knowledge of the different blood groups at the time. Several scientists studied and attempted to classify different types of blood. Landsteiner in 1901 described blood groups A, B and later C on the basis of agglutination patterns. Working independently, Jansky (1907) in Poland described four blood groups which he labelled I, II, III and IV in order of their frequency. Confusingly, Moss (1910) in the United States described similar groups, but in the reverse order, with what is now known as group O classified as IV. This lead to potentially dangerous misunderstandings that were finally resolved at a meeting of the 1937 Congress of the International Society of Blood Transfusion held in Paris when the current ABO terminology was adopted [15].

The advent of the First World War led to substantial advances in blood transfusion for trauma victims as the severity of injuries rapidly demonstrated that something more substantial than saline was required. The discovery of sodium citrate as an anticoagulant allowed the preparation of stored whole blood that could be prepared in advance and held in reserve for the management of casualties. This strategy proved to be remarkably successful, in part due to the fact that the practitioners limited themselves to the use of Moss Type IV (modern blood group O) which had already been recognised as the "universal donor", [16] a practice that has been hailed as perhaps the most important medical advance of the war. Robertson further advocated the use of gelatin to replace lost plasma and the efficacy of this colloid in haemorrhagic shock was supported by Hogan [17]. Similar practices using saline, colloid and whole blood were adopted in the Second World War, but by this time albumin and freeze-dried plasma were available to expand the pool of blood-based volume therapy. An investigation during the second World War suggested that modified gelatin could be a very effective plasma substitute and subsequent developments led to the evolution of modern gelatin products during the 1950s [18].

Following the Second World War, a paradigm shift occurred in trauma resuscitation with greater emphasis on resuscitation of the extracellular fluid space. This philosophy, initiated by Wiggers, and subsequently given extensive support by Dillon and Shires [19] advocated the administration of three times the volume of crystalloid for every volume of shed blood. The objective at this time was full resuscitation before anaesthesia and surgery were attempted. Based on studies using a sulphated tracer to measure extracellular fluid volume, Shires developed the concept of the "third space"[20] in which it was hypothesised that tissue trauma led to swelling and oedema that resulted in sequestration of large volumes of fluid into non-functional, non-anatomic areas of the body and that this lost fluid needed to be replaced. This philosophy became the basis of fluid resuscitation during the Vietnam conflict and was widely adopted. The 3:1 ratio for crystalloids to shed blood was developed in the laboratory and became a strongly advocated strategy for resuscitation into the twenty-first-century despite a lack of randomised controlled trials to provide evidence for its validity in the clinical situation. Although survival from the initial injury improved and the incidence of acute renal failure diminished, these strategies led to the emergence of pulmonary fluid overload and ARDS [21]. In parallel with these high-crystalloid strategies, logistical issues led to the almost universal adoption of component therapy in place of the use of whole blood, although there is a lack of evidence demonstrating that such an approach is equivalent, let alone superior to a whole blood strategy.

In evaluating the consequence of high-volume crystalloid strategies, Cotton and colleagues commented "Early resuscitation of haemorrhagic shock with predominately saline-based regimens has been associated with cardiac dysfunction and lower cardiac output, as well as higher mortality. Numerous investigators have evaluated potential risk factors for developing abdominal compartment syndrome and have universally noted the excessive use of crystalloids as the primary determinant" [22]. In a study of conservative vs. liberal fluid strategies in patients with acute lung injury, there was no difference in mortality or non-pulmonary organ failures, but a significant reduction in the duration of the need for ventilatory support in the conservative group [23]. Surgical studies at this time demonstrated a significant reduction in perioperative complication rates, particularly those related to gut function, following the introduction of more conservative crystalloid administration strategies [24]. A review of the literature suggested that the third space concept was a fallacy that resulted from several methodological errors in the original studies [25].

Clear evidence of the dangers of high-volume resuscitation in trauma emanated from the study by Balogh and colleagues who attempted to achieve supra-normal oxygen delivery in trauma patients in an attempt to improve outcome. In order to achieve their targets, large volumes of crystalloid were required (>13 L in the first 24 h) and this resulted in an increased incidence of abdominal compartment syndrome, multiple organ failure and death compared to a standard resuscitation group who received crystalloid infusions in the order of 7 L [26].

12.2.2 *Timing of Resuscitation*

Timing of resuscitation is also controversial. During the First World War, the French physician, Santy, and his colleagues documented the critical temporal relationship between initial injury and treatment, thus defining the "Golden Hour" concept and its effect on mortality. This work demonstrated that soldiers treated appropriately within 1 h of being wounded sustained a mortality of 10 %, whereas those who had treatment rendered 8 h post-injury had mortality rates as high as 75 %. However, at the same time, surgeons Walter Cannon and John Fraser called for the limited use of fluids and blood during haemorrhage, indicating that "Injection of a fluid that will increase blood pressure has dangers in itself. Haemorrhage in a case of shock may not have occurred to a marked degree because the blood pressure has been too low and the flow too scant to overcome the obstacle offered by a clot. If the pressure is raised before the surgeon is ready to check any bleeding that may take place, blood that is sorely needed may be lost" [27]. These authors also stated as follows: "When a patient must wait for a considerable period, elevation of his systolic blood pressure to approximately 85 mmHg is all that is necessary and when profuse internal bleeding is present, it is wasteful of time and blood to attempt to get the patient's blood pressure up to normal." This principle was forgotten for many years but has recently been revisited. In 1994, Bickell and colleagues performed a randomised trial of 598 patients with penetrating torso injuries who presented with a pre-hospital systolic arterial pressure (SBP) ≤90 mmHg. An immediate resuscitation group received standard fluid resuscitation during hospital transfer whereas the delayed-resuscitation group received no fluid until they arrived in the operating room. The authors reported a small but significant 8 % reduction in mortality in the delayed group, but this was confined to patients with penetrating torso injuries [28]. A subsequent trial randomised patients in haemorrhagic shock to one of two protocols: SBP >100 mmHg (conventional) or a target SBP of 70 mmHg. Overall mortality was lower than previously reported and virtually identical between the two groups, which these authors attributed to better in-hospital management. The lack of differentiation between the two groups was ascribed to the heterogeneous nature the injuries, and the imprecision of SBP as a marker for tissue oxygen delivery [29].

These evolving ideas and strategies have resulted in the concept of damage control procedures, involving both damage control surgery and damage control resuscitation. The two concepts are inextricably linked. Current concepts of early resuscitation, particularly for penetrating trauma, involve a coordinated sequence starting with the prehospital phase where rapid hospital transport, hypotensive resuscitation for uncontrolled haemorrhage and maintenance of baseline tissue energy delivery followed by more definitive damage control resuscitation and surgery once the patient has arrived in hospital.

12.2.3 Prehospital Resuscitation

As indicated above, the use of pre-hospital fluid resuscitation remains controversial. In a large retrospective study, with rapid hospital transit times, initial SBP affected mortality, but fluid infusion had no influence [30]. An extensive review of 776,734 patients extracted from the TRAUMA Data Bank concluded that patients who received prehospital intravenous cannulation had an increased mortality [31]. However, this finding was widely criticised, predominantly on the basis that a causal relationship could not be established from the data available [32]. By contrast, a prospective study that collected data from 10 Level 1 trauma centres found that patients receiving prehospital intravenous fluid (up to 700 mL) had a decreased hospital mortality than those who received no fluid [33]. A recent retrospective review of the German Trauma Registry examined data from 948 matched pairs in which one group received <1500 mL prior to hospital admission and the other received higher than this resuscitation volume. The difference in shock on arrival at the hospital was not different between the groups but the high-volume group required significantly more blood and blood products and had a reduced ability to coagulate. Mortality was increased in the high-volume group [34]. The same group showed a similar trend in children that was not statistically significant [35].

Cotton et al., after an extensive literature search, concluded that there was insufficient evidence for or against the placement of an IV cannula in the prehospital setting, but that there was clear evidence that this attempt should not delay hospital transfer. Their consensus was that volume resuscitation should be limited to small boluses of fluid and that consideration might be given to hypertonic solutions, but there was insufficient evidence for a firm recommendation [36]. Given the lack of definitive evidence, the current status of prehospital resuscitation appears to favour hypotensive resuscitation prior to, and during hospital transfer targeting the presence of the radial artery pulse and adequate mentation. Various SBP targets between 70 and 90 mmHg have been proposed but lack definitive evidence.

The situation for patients with neurological injuries is even less clear, but generally higher SBP targets are advocated for patients with neurotrauma in order to maintain cerebral perfusion pressure. Current recommendations are unclear as to an acceptable approach for the patient with combined non-cranial and neurological injuries. While driving blood pressure to "normal" levels with large volume crystalloid infusions is clearly not appropriate, definitive data on the target blood pressure for hypotensive resuscitation are lacking, especially where trauma is complicated by head injury [37]. Harris et al. concluded that minimising secondary brain injury through optimal supportive care was critical to improving outcome in neurotrauma [38]. However, the definition of what is optimal remains unclear and definitive evidence for hypotensive resuscitation remains elusive.

The use of tranexamic acid (TXA) has been widely advocated after the publication of the CRASH-2 trial which demonstrated a small, but significant reduction in mortality in patients who received this anti-fibrinolytic agent [39]. However, it appears that maximal efficacy is only achieved in patients in whom this drug is administered in the first 1–3 h after injury and that later administration may be associated with harm. Where significant delay is likely to occur in transferring the patient to hospital, as is the case in many parts of the world, TXA should be regarded as a pre-hospital agent only [40].

12.2.4 In-hospital Resuscitation

The objective of limited-volume prehospital resuscitation is to minimise blood loss. Modern concepts of in-hospital care are also aimed at the same objective. Consequently, the concept of damage control resuscitation together with damage control surgery has evolved.

12.2.4.1 Damage Control

The aim is to minimise further blood loss through surgical procedures directed at haemorrhagic control together with resuscitation strategies aimed at optimising coagulation. In many ways, this has seen a return to the concepts of First World War resuscitation strategies with early use of blood coupled with mechanical control of bleeding. The currently advocated strategy involves early damage control surgery aimed at limiting and controlling further haemorrhage coupled with damage control resuscitation that targets patients likely to develop coagulopathy. Damage control surgery for severely injured patients is a process in which the initial operation is abbreviated to that required for control of bleeding and wound decontamination followed by ongoing resuscitation in the intensive-care unit [41]. Damage control resuscitation (DCR) is the overall guiding concept to emerge from recent military experience [42]. DCR, which includes permissive hypotension as first described by Cannon in 1918 [27], can be summarized as allowing the blood pressure to be lower than normal to promote thrombus formation while still providing enough perfusion to end organs to prevent further deterioration. DCR also advocates for the rapid control of surgical bleeding, prevention of acidosis, hypocalcaemia and hypothermia, and for the limitation of excessive crystalloid use to decrease haemodilution. It is important to note that the correction of hypocalcaemia has more to do with cardiac performance than coagulation as it has been shown that coagulation is unaffected until ionised calcium is <0.55 mmol/L [43].

Recent military experience has strongly argued in favour of attempting to recreate whole blood through the administration of packed red blood cells (RBCs), fresh frozen plasma (FFP) and platelets in a 1:1:1 unit ratio from the outset of resuscitation. In severely injured patients requiring massive transfusion there is evidence that this may contribute to a reduction in haemorrhagic mortality. The first study to explore this concept was by Borgman and colleagues performed a retrospective chart review of 246 patients in a combat support hospital who required massive transfusion (defined as $\geq$ 10 units RBCs in 24 h). They divided the groups into low FFP:RBC ratios (1:8) medium (1:2.5) and high (1:1.4). In these groups, overall mortality rates were 65 %, 34 % and 19 % ($p<0.001$) [44]. However, this was followed by a prospective study of 806 consecutive trauma patients of whom 250 received both RBCs and FFP. Analysis by stepwise regression controlling for significant variables found that the FFP:RBC ratio did not predict intensive care unit days, hospital days or mortality in patients requiring massive transfusion [45]. Riskin and colleagues studied 40 patients before and 37 patients after the introduction of a massive transfusion protocol (MTP) and found a reduction in mortality despite the fact that the FFP:RBC ratio was unchanged at 1:1.8. They concluded that early availability of blood and blood products appeared to decrease mortality significantly but that the FFP:RBC ratio had little impact [46]. Several other authors have supported the concept of a high FFP:RBC ratio (<1:2) [47, 48]. However, Ho et al. concluded that the current available evidence supporting higher plasma: erythrocyte resuscitation was inconclusive as a consequence of the prevalence of survivor bias in many of the observational studies [49]. This was supported by a time-based analysis that found no association between the FFP:RBC ratio and survival, suggesting that survival bias may have influenced earlier results [50]. In patients where no coagulation defect could be demonstrated, there was no evidence of benefit to a higher ratio and concern was expressed that the higher costs and higher risks associated with inappropriate administration of FFP required further study [51]. Current European Society of Anaesthesia guidelines recommend a restrictive transfusion strategy but do not support FFP for volume therapy.

A further consideration is that a 1:1:1 ratio of FFP: RBC: Platelets does not reconstitute the functionality of whole blood. The volume of anticoagulant added at the time of blood donation, together with the addition of suspending and nutrient solutions to RBCs, substantially diminishes the coagulation capacity of the final product. Platelet counts are significantly reduced and the "repackaged whole blood" has a haematocrit of around 29 % with around 65 % of the concentration of coagulation factors present in whole blood [52]. FFP is relatively deficient in fibrinogen, with a concentration of around 1 g/L compared to the normal value of 1.3 g/L. In cases of fibrinogen deficiency, cryoprecipitate is the better option as it has a concentration of fibrinogen of around 200 mg/15 mL or approximately 12 g/L. However, cryoprecipitate is a pooled product that does not undergo pathogen inactivation. Adverse events associated with this product include transmission of blood-borne diseases and transfusion-related acute lung injury. Consequently, a number of European countries have withdrawn the licensing of this product [53].

The level of fibrinogen at which treatment is required is by no means settled. Older recommendations suggested that a level of 1 g/L was appropriate, but more recent studies, particularly in obstetrics have shown an increased risk of bleeding with fibrinogen levels <2 g/L. Given the concerns regarding the safety and efficacy of plasma products to raise fibrinogen concentrations significantly, other haemostatic preparations have been considered, including fibrinogen concentrates and prothrombin complex concentrates. The use of these products has been reviewed recently [54].

The importance of mechanical control of haemorrhage is emphasised by a recent publication showing that despite aggressive haemostatic resuscitation, ROTEM parameters and lactate clearance did not improve until haemorrhage control was achieved [55].

12.2.4.2 Assessment of Coagulation

As discussed above, coagulation pathways are disrupted through a variety of mechanisms in severe trauma and attention to these disruptions is essential if improved survival is to be attained. It is, therefore, imperative to establish the presence of coagulopathy prior to embarking on aggressive coagulation resuscitation unless it is patently obvious from the outset that the patient will require massive transfusion. The conventional plasma-based coagulation tests, such as prothrombin time (PT), international normalised ratio (INR) and activated partial thromboplastin time (aPTT) were primarily designed to evaluate anticoagulation rather than coagulopathy. They are performed only on plasma, excluding cellular elements of blood and only provide a single snapshot of the whole process at an early stage of completion. They have almost no capability to detect clot lysis, an important component of TIC. The evidence that these tests are predictive for bleeding or have the potential to guide coagulation therapy, especially in trauma is poor [56].

Recent research has emphasised the value of point-of-care testing of coagulation such as thrombelastography (TEG®) and thrombelastometry (ROTEM®) as a means of providing rapid assessment of clot formation. The functional fibrinogen assay for the TEG and the FIBTEM assay on the ROTEM both provide a rapid assessment of the activity of fibrinogen. Early inspection of the traces on these devices can provide an indication of inadequate coagulation in as little as 10 min. These devices are particularly good at detecting clot lysis. Recent reports attest to the potential value of these technologies [57] but further evidence is required before they can be considered the standard of care [58].

12.2.5 Asanguinous Resuscitation Fluids

Whilst haemostatic resuscitation appears to be of value in the management of patients requiring massive transfusions, it must be recognised that a minority of trauma patients require damage control resuscitation. Only 25 % of trauma patients actually received blood transfusions and between 2 and 3 % of the civilian casualties and 7–8 % of military casualties require massive transfusion [59]. Thus, it is inappropriate to commence haemostatic resuscitation in all patients until either the presence of coagulopathy has been established or it is obvious that massive transfusion will be required. Given that current the recommendations for the management of the bleeding patient suggest that the haemoglobin should be maintained between 70 and 90 $g{\cdot}L^{-1}$ [60] most trauma patients with moderate bleeding will require minimal, if any, blood resuscitation. Although this recommendation is based on low-quality evidence, it is currently the accepted wisdom. For the majority of trauma patients, volume resuscitation should be based on standard principles using non-blood products as the primary resuscitating fluid with additional blood products as required.

12.2.5.1 Crystalloid Solutions

The most important evidence to emerge from trauma resuscitation data in the last 10 years is that crystalloid overload is uniformly harmful and should be avoided. Unless there are obvious sources of extracellular fluid loss (excessive peritoneal fluid or burns exudate) crystalloid administration should not greatly exceed the maximum daily crystalloid load of approximately 3 L per day in the average adult. Crystalloid remains the most widely used initial resuscitation fluid, but small volume resuscitation in the prehospital phase should be limited to 250 mL boluses and should not exceed 1 L in total volume. Progressive crystalloid resuscitation in hospital patients correlated with adverse outcomes other than mortality including increased incidence of adult respire distress syndrome, multiorgan failure and compartment syndromes [61]. Emergency department administration of crystalloid in excess of 1.5 L was an independent risk factor for mortality in a retrospective survey of 3137 patients. This effect was particularly marked in elderly patients [62]. Once the patient is in hospital, resuscitation should focus on fluids that replenish intravascular volume with limited crystalloid administration.

The electrolyte composition of crystalloid solutions may also be important. So-called physiological saline (0.9 % saline) is probably the most widely used crystalloid solution for trauma resuscitation. There appear to be two main reasons for this: the fact that saline is isotonic with plasma as compared with Ringer's lactate makes it an attractive resuscitation solution with broad applicability to trauma resuscitation whether or not head injury is present; the second reason is the absence of calcium which means that 0.9 % saline can be freely given with citrated blood products without triggering in-line coagulation of blood products. However, the downside is that the solution presents a considerable chloride load and these chloride loads have been associated with acidosis and adverse outcomes [63, 64]. However, definitive evidence of harm from 0.9 % saline for hard outcomes is lacking.

On the other hand, Ringer's lactate is also far from ideal as a resuscitation solution. Originally developed for the treatment of infantile diarrhoea and diabetes,

Hartman's solution (as it is also called) is substantially hypo-osmolar, with a sodium concentration of only 131 mmol/l and a calculated osmolarity of 276 mosm/l. This solution also contains a substantial concentration of calcium, depending on the precise details of the manufacture, ranging between 1.8 and 2.4 mmol/L. Since this is almost all ionised calcium, this creates significant problems if it is mixed with citrated solutions containing coagulation factors such as whole blood, fresh frozen plasma, cryoprecipitate or platelets. Even more importantly, since this is a non-ideal solution and all of the components are not completely ionised, the actual osmolality (the property that really matters) is substantially below that of plasma and is generally reported as 257 mosm/L [65]. Such a hypotonic solution may well contribute to worsening cerebral oedema in neurotrauma and this has been demonstrated in experimental animals [66] although there is little conclusive evidence in humans. Nevertheless, it is generally recommended that Ringer's lactate should not be used in resuscitation of head injured patients.

Other crystalloid solutions have been developed that are closer to the physiological constituents of plasma. However, the ongoing problem is balancing the missing anions that are provided by plasma proteins. This raises the debate about the suitability of various anions as replacements for bicarbonate, since bicarbonate is difficult to stabilise in solutions in stored in plastic containers. Lactate is hallowed by tradition and most older solutions contain a racemic mixture of this anion. However, there is evidence that the D-lactate stereoisomer is potentially harmful and lactate containing solutions should be composed exclusively of the L-lactate isomer. Several recently developed crystalloid solutions have replaced lactate with acetate as the primary metabolisable anion on the theoretical grounds that acetate is more rapidly metabolised than lactate and by a wider range of tissues. In the dialysis literature, acetate has been associated with cardiovascular depression, [67] but this has not been demonstrated when acetate-based solutions are used for resuscitation. There has been a recent report of severe lactic acidosis following sodium acetate resuscitation in a trauma patient, [68] but in a rat model of severe haemorrhagic shock acetate appeared superior to lactate or saline [69]. In an attempt to minimise the preponderance of a single anion, gluconate has been included in one solution, although good evidence of the safety of this anion is lacking. Gluconate is slowly metabolised and is likely to cause minimal change in acid–base balance making it potentially useful as an additive anion in balanced solutions [70]. Plasmalyte A (also known as Plasmalyte 148) is probably the best balanced crystalloid solution currently available and has been shown to produce less acid–base disturbances with better preservation of renal cortical tissue perfusion than saline in volunteers [71].

12.2.6 Colloid Solutions

There is a considerable variety of colloid solutions available, including various albumin preparations together with the non-human semi-synthetic colloids, gelatin, dextran and hydroxyethyl starch (HES). Although there is a substantial volume of literature on the crystalloid-colloid debate, the nature of trauma and the difficulties in conducting appropriate randomised, controlled trials in this area has meant that the quality of evidence is poor and there are few well-conducted trials demonstrating the relative advantages of each type of fluid. Several meta-analyses have failed to demonstrate any benefit of colloid solutions over crystalloid resuscitation in critically ill patients, [72, 73] but most of the literature is based on post-resuscitation intensive care studies and almost none of these studies include properly conducted randomised trials in acute trauma resuscitation.

12.2.6.1 Albumin

Albumin is a naturally-occurring plasma protein derived largely from human plasma. Unlike direct plasma products, albumin is sterilised and ultra-filtrated to diminish the risk of disease transmission, but contains no antibodies or coagulation factors. It is a mono-dispersed colloid with a molecular weight of around 69 kDa and, like the semi-synthetic colloids, requires no special storage facilities or blood grouping. Albumin preparations are generally available in isotonic (4–5 %) or hyperoncotic (20–25 %) formulations and mainly in relatively small volumes of around 100 mL or less.

Albumin has been associated with an increased risk of death in critically ill patients, although the data in trauma specifically was of poor quality and there were no randomised controlled trials [74]. This finding was disputed in a meta-analysis of patient survival which reported no significant excess mortality in surgery and trauma where albumin had been used (RR 1.12; 0.85–1.46) [75]. The subsequent SAFE study also refuted these concerns in a general ICU population, but there was a trend toward increased mortality in trauma, almost entirely accounted for by deaths in head-injured patients (see below) [76].

Overall, there appear to be limited data for or against the use of albumin in trauma resuscitation, with the major drawback of this product being its high cost and limited availability in most of the world.

In burns patients, there has been renewed interest in the addition of colloid to the early crystalloid strategy that is generally recommended in an effort to limit "fluid creep", and reduce the risk of pulmonary oedema and secondary abdominal compartment syndrome. A recent retrospective review examined the effect of adding albumin to the resuscitation fluid strategy on fluid input/output (I/O) ratios. The administration of albumin rapidly reduced hourly fluid requirements, restored normal I/O ratios and ameliorated fluid creep [77]. A study in paediatric burns confirmed these findings showing that the use of albumin restored a normal I/O ratio in paediatric patients, adding further weight to the inclusion of colloid in burns resuscitation strategies [78].

12.2.6.2 Gelatins

The gelatins are the oldest of the synthetic colloids available, and are derived from the degradation of bovine collagen with alkali. There are two types of gelatins available, the urea-linked form typified by Haemaccel® and the succinylated form as typified by Gelofusin®.

Gelatins have a negative charge which enhances intravascular retention, despite the molecular weight of only 30–35 kDa. They have an intense, but short lived, plasma volume expansion lasting for 1–3 h. The gelatins are fully cleared from the body through the kidney.

The gelatins exert little effect on coagulation, although they may impair clot strength, mainly on a dilutional basis. The succinylated gelatins may have a greater effect on the von Willebrand (vWF) factor than the urea-linked type, [79] but no coagulation disturbances are recognised and no volume limitation applies. The main adverse effect attributable to the gelatins is that of anaphylaxis. The risk is significantly greater for the urea-linked gelatins than the fluid gelatins, but both appear to have a higher risk of allergic reactions than other colloids.

12.2.6.3 Dextrans

The dextrans are probably the most effective volume-expanding solutions, but have numerous adverse effects. Dextran 40 10 % is substantially hyperoncotic and may lead to hyperosmotic renal failure [80]. Rarely, adverse reactions including the dextran syndrome (which consists of a combination of hypotension, ARDS, anaemia and coagulopathy) and anaphylaxis may occur. Dextrans interfere with blood cell aggregation which can interfere with cross-matching and also lead to impaired coagulation. They also enhance fibrinolysis [81]. Indeed, the anti-coagulant properties of the dextrans are of such significance that they have been recommended for post-operative vascular surgical management to minimise the risk of occlusive vascular events [82].

Recent research has suggested that the combination of hypertonic saline and dextran may have advantages in resuscitation, particularly in patients with brain injury [83]. However, the problems that the dextrans pose with cross matching together with the risk of anaphylaxis and bleeding have led to these products being rarely used in trauma resuscitation.

12.2.6.4 Hydroxyethyl Starch (HES)

Of the semi-synthetic colloids, HES has been by far the most widely investigated. A variety of different HES products exist, with considerable differences in their pharmacological properties. HES solutions are classified according to the *in vitro* molecular weight (MW) and molar substitution (MS) The current classification is as follows:

Molar substitution ratio:

MS	Generic name	Commercial products
0.7	Hetastarch	Hespan®, Plasmasteril®, Hextend®
0.6	Hexastarch	Elohes®
0.5	Pentastarch	HAES-Steril®, Pentaspan®, Hemohes®
0.4	Tetrastarch	Voluven®, Venofundin®, Tetraspan®

There has been considerable criticism of HES recently in the light of three "pragmatic" critical care trials in which HES solutions were associated with harm, notably renal complications. The first of these studies used a hyperoncotic 10 % HES in doses well in excess of those recommended by the manufacturer and showed a dose-related incidence of renal injury [84]. The second study (CHEST) was powered to detect a mortality difference between HES and saline and found no difference in mortality over nearly 7000 subjects. Although the HES group received more renal replacement therapy (RRT), the RIFLE criteria favoured the HES [85]. The third (6S) study found an increased incidence of RRT in septic patients in the HES 130/0.42 (potato starch) group together with an increased 90-day mortality [86]. None of these studies enrolled patients prior to initial resuscitation and are not pertinent to the consideration of trauma resuscitation. By contrast another intensive care study published recently in which colloid was compared to crystalloid for initial resuscitation found a similar mortality signal to that of the 6S study, but on this occasion in favour of colloid, with the predominant colloid being HES 130/0.4 [87].

The first study to demonstrate a survival benefit associated with the use of colloids in trauma was published recently. In this retrospective study of 1714 patients admitted to a level 1 trauma centre, the authors examined the outcomes in patients who did, or did not, receive a colloid (Hextend®) as part of the resuscitation strategy. Compared to the crystalloid-only standard of care treatment, the overall mortality analysed by univariate analysis was significantly lower in the colloid-treated patients (5.2 % vs. 8.9 %, $p=0.0035$). The results were slightly better when limited to patients with penetrating trauma ($p=0.0016$) and also when only severely injured patients with Injury Severity Scores (ISS) >26 were considered ($p=0.0014$). More starch-treated patients survived to reach intensive care and more blood and blood products were used in this group. However, coagulation measures (prothrombin time and partial thromboplastin time) were not different between the groups. Urine output and renal function were similar between the groups, but there were more early deaths in the standard of care group, which introduces the possibility of selection bias. Controlling for early deaths with multivariate analysis showed similar trends, but the data no longer reached statistical significance [88].

The first randomised, double-blind, controlled trial of crystalloid (0.9 % saline) versus isotonic colloid (HES 130/0.4 in saline) in trauma resuscitation has recently been published (FIRST trial). In this study of 109 patients, blunt and penetrating trauma were randomised separately. In the penetrating trauma group ($n=67$), lactate clearance was faster over the first 4 h of resuscitation and lactate and acid base balance were significantly better in the HES group on Day 1. The HES group had a zero incidence of renal injury (RIFLE criteria) compared to 17 % in patients treated with saline p=0.019). This difference persisted even after allowance for the early deaths in both groups. Maximum SOFA scores were significantly lower in the HES group. The ratio of colloid: crystalloid in the first 24 h was approximately 1:1.5, resulting in a reduction of 2.5 L of study fluid in the HES group. The use of blood and blood products was similar, with a non-significant trend to lower product use in the penetrating HES group. The blunt trauma analysis was severely hampered by the fact that patients in the blunt HES group were much more severely injured (median injury severity score 29.5 vs. 18; $P<0.01$). There was no significant difference in the use of study fluid between groups and the HES group required significantly more blood and blood products. Outcomes were similar in both blunt trauma groups in terms of renal function and organ recovery, with no differences in mortality despite substantially higher ISS in the blunt HES group [89].

In a retrospective analysis of penetrating trauma patients in whom the principles of damage control surgery and resuscitation were applied, outcomes were analysed by the volumes of fluid received in addition to the DCR fluids: RBCs, FFP and platelets. In the study of 307 patients, 43 % received <129 mL of crystalloid prior to surgery, while the comparator group received an average of 2757 mL in the emergency department. During subsequent intraoperative management, the restricted fluid group received significantly less crystalloid (3152 mL vs. 8931 mL) together with non-significantly higher volumes of PRBCs, FFP and platelets. The restricted fluid group had significantly lower overall mortality (OR 0.69; 95 % CI 0.37–0.91) [90]. A subsequent matched case–control study from the same group examined the influence of asanguineous fluids used in support of the DCR strategy in patients requiring ≥4 units PRBCs. In this study, mortality was significantly lower in patients who received low volumes (median 1.0 L) of colloid (Hextend®) than those who received crystalloid (median 3.6 L), and the mortality was directly related to the volumes of crystalloid administered. These authors concluded that DCR principles coupled with low-volume, primarily colloid, resuscitation improved survival in these patients [91].

The appropriate use of HES in surgery and trauma resuscitation has recently been reviewed [92].

12.2.7 Neurotrauma

As has been alluded to above, the issue of neuro trauma remains poorly resolved. In the prehospital phase, it is generally accepted that, in the absence of active bleeding, more aggressive resuscitation may be appropriate than in other forms of trauma. Where there is a conflict of interests in a bleeding patient who also has suffered neurological injury, the situation is much less clear, but most authorities would recommend a higher arterial pressure (SBP >100 mmHg) in the prehospital phase [37]. Once the patient is in hospital, similar principles apply, with damage control resuscitation being appropriate inpatients requiring massive transfusion. Aggressive management including resuscitation with blood products and hyperosmolar therapy in gunshot wounds of the brain was associated with significant improval in survival in a recent report [93]. In less seriously bleeding patients, fluid resuscitation should probably be minimised and there is an argument in favour of hypertonic solutions, although these have not been shown to be beneficial. On the basis of subgroup analysis of the SAFE study [76] albumin was suggested to be associated with a worse outcome in head injured patients, probably on the basis of increased intracranial pressure [94]. However, the albumin used in this study was substantially hypo-osmolar with an osmolality of around 260 mosm/L [65] quite sufficient to account for worse intracranial pressures in the albumin group. Other studies of the use of HES or albumin in head injured patients have not demonstrated similar increases in intracranial pressure [95]. A retrospective study of 93 patients with severe traumatic brain injury investigated the impact of a resuscitation protocol that incorporated the use of albumin. These authors reported an early achievement of negative fluid balance in these patients with well maintained plasma albumin concentrations. Over the first 10 days for which the patients were studied, colloids constituted 40–60 % of the total fluids given daily and fluid balance was assisted by the use of furosemide. Mortality at 28 days and 18 months was 11 % and 14 % respectively. The authors contrast this with the SAFE study. In the Scandinavian study, [96] mortality in the study group was substantially lower than that in the SAFE study, but these comparisons should be viewed with caution as the resuscitation strategies, particularly the management of intracranial pressure may well have been different between the study groups. However, a reappraisal of the use of albumin in brain injury has recently been advocated [97].

12.2.8 Burns

Fluid resuscitation formulas, such as the Parkland formula, have found wide application around the world and form the basis for calculation of fluid resuscitation regimes in burnt patients. There have been a number of recent papers examining the use of fluids in patients suffering burns, emphasising that most patients end up receiving crystalloids well in excess of the calculated needs. This fluid overload has been called "fluid creep" and associated with worsening of burn oedema, conversion of superficial into deep burns, and compartment syndromes [98].

A recent review of this topic commented that, while crystalloid fluid has been the mainstay of resuscitation for the better part of four decades, there has been a progressive but as yet unexplained recent trend toward provision of resuscitation volumes well in excess of those predicted by the formula. This increased fluid load has been associated with numerous oedema-related complications. This review emphasizes that, in burns involving >25 % body surface area capillary, capillary permeability is increased, not only in the damaged tissue, but also in non-burnt areas. It was proposed that correction of fluid creep will likely revolve around several strategies, which may include tighter control of titration, re-emergence of colloids and hypertonic salt solutions, and possibly the use of adjunctive markers of resuscitation other than urinary output [99]. In a subsequent retrospective review, the same author concluded that this trend had not improved and that, over an 8-year period, higher than recommended urine output had been attained as a result of increased crystalloid administration [100].

In one publication, the role of hyperoncotic 10 % HES 200/0.5, was examined in a pilot trial of 30 patients. Although the numbers are small, the study was abandoned as the hyperoncotic starch was associated with an apparent increased incidence of fatal outcome [101]. These data underline the view that there is virtually no place for hyperoncotic colloids as acute resuscitation solutions. However, a number of other studies and reviews suggest that the role of colloids in the resuscitation of patients suffering burns is being re-examined.

Another review concluded that starch colloids may limit burn oedema by ameliorating capillary leak and that improved endpoints for resuscitation may help to minimize the problems of over-resuscitation [98]. A further retrospective review examined the effect of adding albumin to the resuscitation fluid strategy on fluid input/output (I/O) ratios. The administration of albumin rapidly reduced hourly fluid requirements, restored normal I/O ratios and ameliorated fluid creep [102]. A study in paediatric burns confirmed these findings showing that the use of albumin restored a normal I/O ratio in paediatric patients, adding further weight to the inclusion of colloid in burns resuscitation strategies [103].

12.3 Conclusions

In recent years there has been a marked shift in the concepts governing trauma resuscitation, with modern views startlingly resembling the initial findings in the First World War. In patients with severe trauma requiring massive transfusions (<10 % of all trauma patients) aggressive damage-control resuscitation and surgery carries the prospect of reducing mortality particularly amongst battle casualties with penetrating injuries. In the civilian population, these data are less conclusive, but the principles are probably still valid. Prehospital resuscitation now focuses on minimal fluid interventions with low volumes of crystalloid as the first line fluid approach. Once the patient reaches hospital, more aggressive fluid resuscitation strategies may be appropriate. In patients requiring less than massive transfusion, substantial portions of the resuscitation volume should be administered with clear fluids. No more than 2–3 L of such fluid should be given as crystalloid, preferably as a balanced crystalloid that is isotonic with plasma. The use of colloids in penetrating trauma seems to be beneficial, but there is insufficient data at present in blunt trauma and burns to make soundly-based recommendations.

References

1. Desborough JP (2000) The stress response to trauma and surgery. Br J Anaesth 85(1):109–117
2. Johansson PI, Stensballe J, Rasmussen LS, Ostrowski SR (2011) A high admission syndecan-1 level, a marker of endothelial glycocalyx degradation, is associated with inflammation, protein C depletion, fibrinolysis, and increased mortality in trauma patients. Ann Surg 254(2):194–200. doi:10.1097/SLA.0b013e318226113d
3. Rahbar E, Cardenas JC, Baimukanova G, Usadi B, Bruhn R, Pati S, Ostrowski SR, Johansson PI, Holcomb JB, Wade CE (2015) Endothelial glycocalyx shedding and vascular permeability in severely injured trauma patients. J Transl Med 13:117. doi:10.1186/s12967-015-0481-5
4. Eddy VA, Morris JA Jr, Cullinane DC (2000) Hypothermia, coagulopathy, and acidosis. Surg Clin North Am 80(3):845–854
5. Brohi K, Cohen MJ, Ganter MT, Schultz MJ, Levi M, Mackersie RC, Pittet JF (2008) Acute coagulopathy of trauma: hypoperfusion induces systemic anticoagulation and hyperfibrinolysis. J Trauma 64(5):1211–1217
6. Gando S, Nanzaki S, Morimoto Y, Ishitani T, Kemmotsu O (2001) Tissue factor pathway inhibitor response does not correlate with tissue factor-induced disseminated intravascular coagulation and multiple organ dysfunction syndrome in trauma patients. Crit Care Med 29(2):262–266
7. Hess JR, Brohi K, Dutton RP, Hauser CJ, Holcomb JB, Kluger Y, Mackway-Jones K, Parr MJ, Rizoli SB, Yukioka T, Hoyt DB, Bouillon B (2008) The coagulopathy of trauma: a review of mechanisms. J Trauma 65(4):748–754
8. Noel P, Cashen S, Patel B (2013) Trauma-induced coagulopathy: from biology to therapy. Semin Hematol 50(3):259–269. doi:10.1053/j.seminhematol.2013.06.009
9. Ward KR (2013) The microcirculation: linking trauma and coagulopathy. Transfusion 53(Suppl 1):38S–47S. doi:10.1111/trf.12034
10. Shaz BH, Winkler AM, James AB, Hillyer CD, MacLeod JB (2011) Pathophysiology of early trauma-induced coagulopathy: emerging evidence for hemodilution and coagulation factor depletion. J Trauma 70(6):1401–1407. doi:10.1097/TA.0b013e31821266e0
11. Engstrom M, Schott U, Romner B, Reinstrup P (2006) Acidosis impairs the coagulation: a thromboelastographic study. J Trauma 61(3):624–628. doi:10.1097/01.ta.0000226739.30655.75
12. Werner C, Engelhard K (2007) Pathophysiology of traumatic brain injury. Br J Anaesth 99(1):4–9. doi:10.1093/bja/aem131
13. Phillips AA, Ainslie PN, Krassioukov AV, Warburton DE (2013) Regulation of cerebral blood flow after spinal cord injury. J Neurotrauma 30(18):1551–1563. doi:10.1089/neu.2013.2972
14. Kolias AG, Guilfoyle MR, Helmy A, Allanson J, Hutchinson PJ (2013) Traumatic brain injury in adults. Pract Neurol 13(4):228–235. doi:10.1136/practneurol-2012-000268
15. Giangrande PL (2000) The history of blood transfusion. Br J Haematol 110(4):758–767
16. Robertson OH (1918) A method of citrated blood transfusion. Br Med J 1(2991):477
17. Hogan JJ (1915) The intravenous use of colloidal (gelatin) solutions in shock. J Am Med Assoc LXIV 9:721–726. doi:10.1001/jama.1915.02570350015004
18. Saddler JM, Horsey PJ (1987) The new generation gelatins. Anaesthesia 42(9):998–1004. doi:10.1111/j.1365-2044.1987.tb05376.x
19. Shires GT, Canizaro PC (1973) Fluid resuscitation in the severely injured. Surg Clin North Am 53(6):1341–1366
20. Shires T, Williams J, Brown F (1961) Acute change in extracellular fluids associated with major surgical procedures. Ann Surg 154:803–810
21. Demling R (1980) The pathogenesis of respiratory failure after trauma and sepsis. Surg Clin North Am 60(6):1373
22. Cotton BA, Guy JS, Morris JA Jr, Abumrad NN (2006) The cellular, metabolic, and systemic consequences of aggressive fluid resuscitation strategies. Shock 26(2):115–121

23. Wiedemann HP, Wheeler AP, Bernard GR, Thompson BT, Hayden D, deBoisblanc B, Connors AF Jr, Hite RD, Harabin AL (2006) Comparison of two fluid-management strategies in acute lung injury. N Engl J Med 354(24):2564–2575. doi:10.1056/NEJMoa062200
24. Brandstrup B, Tonnesen H, Beier-Holgersen R, Hjortso E, Ording H, Lindorff-Larsen K, Rasmussen MS, Lanng C, Wallin L, Iversen LH, Gramkow CS, Okholm M, Blemmer T, Svendsen PE, Rottensten HH, Thage B, Riis J, Jeppesen IS, Teilum D, Christensen AM, Graungaard B, Pott F (2003) Effects of intravenous fluid restriction on postoperative complications: comparison of two perioperative fluid regimens: a randomized assessor-blinded multicenter trial. Ann Surg 238(5):641–648. doi:10.1097/01.sla.0000094387.50865.23
25. Brandstrup B, Svensen C, Engquist A (2006) Hemorrhage and operation cause a contraction of the extracellular space needing replacement–evidence and implications? A systematic review. Surgery 139(3):419–432
26. Balogh Z, McKinley BA, Cocanour CS, Kozar RA, Valdivia A, Sailors RM, Moore FA (2003) Supranormal trauma resuscitation causes more cases of abdominal compartment syndrome. Arch Surg 138(6):637–642
27. Cannon WB, Fraser J, Cowell EM (1918) The preventive treatment of wound shock. JAMA 70(9):618–621
28. Bickell WH, Wall MJ Jr, Pepe PE, Martin RR, Ginger VF, Allen MK, Mattox KL (1994) Immediate versus delayed fluid resuscitation for hypotensive patients with penetrating torso injuries [see comments]. N Engl J Med 331(17):1105–1109
29. Dutton RP, Mackenzie CF, Scalea TM (2002) Hypotensive resuscitation during active hemorrhage: impact on in-hospital mortality. J Trauma 52(6):1141–1146
30. Kaweski SM, Sise MJ, Virgilio RW (1990) The effect of prehospital fluids on survival in trauma patients. J Trauma 30(10):1215–1218; discussion 1218–1219
31. Haut ER, Kalish BT, Cotton BA, Efron DT, Haider AH, Stevens KA, Kieninger AN, Cornwell EE 3rd, Chang DC (2011) Prehospital intravenous fluid administration is associated with higher mortality in trauma patients: a National Trauma Data Bank analysis. Ann Surg 253(2):371–377. doi:10.1097/SLA.0b013e318207c24f
32. Rivkind AI, Ganor O, Glassberg E (2014) Prehospital intravenous fluid administration is associated with higher mortality in trauma patients. Ann Surg 259(2):e22. doi:10.1097/SLA.0b013e3182456b67
33. Hampton DA, Fabricant LJ, Differding J, Diggs B, Underwood S, De La Cruz D, Holcomb JB, Brasel KJ, Cohen MJ, Fox EE, Alarcon LH, Rahbar MH, Phelan HA, Bulger EM, Muskat P, Myers JG, del Junco DJ, Wade CE, Cotton BA, Schreiber MA (2013) Prehospital intravenous fluid is associated with increased survival in trauma patients. J Trauma Acute Care Surg 75(1 Suppl 1):S9–S15. doi:10.1097/TA.0b013e318290cd52
34. Hussmann B, Lefering R, Waydhas C, Touma A, Kauther MD, Ruchholtz S, Lendemans S (2013) Does increased prehospital replacement volume lead to a poor clinical course and an increased mortality? A matched-pair analysis of 1896 patients of the Trauma Registry of the German Society for Trauma Surgery who were managed by an emergency doctor at the accident site. Injury 44(5):611–617. doi:10.1016/j.injury.2012.02.004
35. Hussmann B, Lefering R, Kauther MD, Ruchholtz S, Moldzio P, Lendemans S (2012) Influence of prehospital volume replacement on outcome in polytraumatized children. Crit Care 16(5):R201. doi:10.1186/cc11809
36. Cotton BA, Jerome R, Collier BR, Khetarpal S, Holevar M, Tucker B, Kurek S, Mowery NT, Shah K, Bromberg W, Gunter OL, Riordan WP Jr (2009) Guidelines for prehospital fluid resuscitation in the injured patient. J Trauma 67(2):389–402. doi:10.1097/TA.0b013e3181a8b26f
37. Tobin JM, Dutton RP, Pittet JF, Sharma D (2014) Hypotensive resuscitation in a head-injured multi-trauma patient. J Crit Care 29(2):313.e1–5. doi:10.1016/j.jcrc.2013.11.017
38. Harris T, Davenport R, Hurst T, Hunt P, Fotheringham T, Jones J (2012) Improving outcome in severe trauma: what's new in ABC? Imaging, bleeding and brain injury. Postgrad Med J 88(1044):595–603. doi:10.1136/postgradmedj-2011-130285

39. Shakur H, Roberts I, Bautista R, Caballero J, Coats T, Dewan Y, El-Sayed H, Gogichaishvili T, Gupta S, Herrera J, Hunt B, Iribhogbe P, Izurieta M, Khamis H, Komolafe E, Marrero MA, Mejia-Mantilla J, Miranda J, Morales C, Olaomi O, Olldashi F, Perel P, Peto R, Ramana PV, Ravi RR, Yutthakasemsunt S (2010) Effects of tranexamic acid on death, vascular occlusive events, and blood transfusion in trauma patients with significant haemorrhage (CRASH-2): a randomised, placebo-controlled trial. Lancet 376(9734):23–32. doi:10.1016/S0140-6736(10)60835-5
40. Thurston B, Chowdhury S, Edu S, Nicol AJ, Navsaria PH (2015) Time since injury is the major factor in preventing tranexamic acid use in the trauma setting: an observational cohort study from a major trauma centre in a middle-income country. S Afr J Surg 53(1):13–18
41. Gruen RL, Brohi K, Schreiber M, Balogh ZJ, Pitt V, Narayan M, Maier RV (2012) Haemorrhage control in severely injured patients. Lancet 380(9847):1099–1108. doi:10.1016/S0140-6736(12)61224-0
42. Holcomb JB, Spinella PC (2010) Optimal use of blood in trauma patients. Biologicals 38(1):72–77
43. James MF, Roche AM (2004) Dose–response relationship between plasma ionized calcium concentration and thrombelastography. J Cardiothorac Vasc Anesth 18(5):581–586
44. Borgman MA, Spinella PC, Perkins JG, Grathwohl KW, Repine T, Beekley AC, Sebesta J, Jenkins D, Wade CE, Holcomb JB (2007) The ratio of blood products transfused affects mortality in patients receiving massive transfusions at a combat support hospital. J Trauma 63(4):805–813
45. Scalea TM, Bochicchio KM, Lumpkins K, Hess JR, Dutton R, Pyle A, Bochicchio GV (2008) Early aggressive use of fresh frozen plasma does not improve outcome in critically injured trauma patients. Ann Surg 248(4):578–584
46. Riskin DJ, Tsai TC, Riskin L, Hernandez-Boussard T, Purtill M, Maggio PM, Spain DA, Brundage SI (2009) Massive transfusion protocols: the role of aggressive resuscitation versus product ratio in mortality reduction. J Am Coll Surg 209(2):198–205
47. Campion EM, Pritts TA, Dorlac WC, Nguyen AQ, Fraley SM, Hanseman D, Robinson BR (2013) Implementation of a military-derived damage-control resuscitation strategy in a civilian trauma center decreases acute hypoxia in massively transfused patients. J Trauma Acute Care Surg 75(2 Suppl 2):S221–S227. doi:10.1097/TA.0b013e318299d59b
48. Brown JB, Cohen MJ, Minei JP, Maier RV, West MA, Billiar TR, Peitzman AB, Moore EE, Cushieri J, Sperry JL (2012) Debunking the survival bias myth: characterization of mortality during the initial 24 hours for patients requiring massive transfusion. J Trauma Acute Care Surg 73(2):358–364. doi:10.1097/TA.0b013e31825889ba; discussion 364
49. Ho AM, Dion PW, Yeung JH, Holcomb JB, Critchley LA, Ng CS, Karmakar MK, Cheung CW, Rainer TH (2012) Prevalence of survivor bias in observational studies on fresh frozen plasma: erythrocyte ratios in trauma requiring massive transfusion. Anesthesiology 116(3):716–728. doi:10.1097/ALN.0b013e318245c47b
50. Halmin M, Bostrom F, Brattstrom O, Lundahl J, Wikman A, Ostlund A, Edgren G (2013) Effect of plasma-to-RBC ratios in trauma patients: a cohort study with time-dependent data. Crit Care Med. doi:10.1097/CCM.0b013e31828a3214
51. Mitra B, Cameron PA, Gruen RL (2012) Aggressive fresh frozen plasma (FFP) with massive blood transfusion in the absence of acute traumatic coagulopathy. Injury 43(1):33–37. doi:10.1016/j.injury.2011.10.011
52. Cap A, Hunt BJ (2015) The pathogenesis of traumatic coagulopathy. Anaesthesia. 70(Suppl 1):96–101, e132–104. doi:10.1111/anae.12914
53. Nascimento B, Goodnough LT, Levy JH (2014) Cryoprecipitate therapy. Br J Anaesth 113(6):922–934. doi:10.1093/bja/aeu158
54. Gill R (2015) Practical management of major blood loss. Anaesthesia 70:54–e20. doi:10.1111/anae.12915
55. Khan S, Brohi K, Chana M, Raza I, Stanworth S, Gaarder C, Davenport R (2014) Hemostatic resuscitation is neither hemostatic nor resuscitative in trauma hemorrhage. J Trauma Acute Care Surg 76(3):561–567. doi:10.1097/TA.0000000000000146; discussion 567–568

56. Haas T, Fries D, Tanaka KA, Asmis L, Curry NS, Schochl H (2015) Usefulness of standard plasma coagulation tests in the management of perioperative coagulopathic bleeding: is there any evidence? Br J Anaesth 114(2):217–224. doi:10.1093/bja/aeu303
57. Schöchl H, Voelckel W, Schlimp CJ (2015) Management of traumatic haemorrhage – the European perspective. Anaesthesia 70:102–e137107. doi:10.1111/anae.12901
58. Whiting D, DiNardo JA (2014) TEG and ROTEM: technology and clinical applications. Am J Hematol 89(2):228–232. doi:10.1002/ajh.23599
59. Como JJ, Dutton RP, Scalea TM, Edelman BB, Hess JR (2004) Blood transfusion rates in the care of acute trauma. Transfusion 44(6):809–813. doi:10.1111/j.1537-2995.2004.03409.x
60. Kozek-Langenecker SA, Afshari A, Albaladejo P, Santullano CA, De Robertis E, Filipescu DC, Fries D, Gorlinger K, Haas T, Imberger G, Jacob M, Lance M, Llau J, Mallett S, Meier J, Rahe-Meyer N, Samama CM, Smith A, Solomon C, Van der Linden P, Wikkelso AJ, Wouters P, Wyffels P (2013) Management of severe perioperative bleeding: guidelines from the European Society of Anaesthesiology. Eur J Anaesthesiol 30(6):270–382. doi:10.1097/EJA.0b013e32835f4d5b
61. Kasotakis G, Sideris A, Yang Y, de Moya M, Alam H, King DR, Tompkins R, Velmahos G (2013) Aggressive early crystalloid resuscitation adversely affects outcomes in adult blunt trauma patients: an analysis of the Glue Grant database. J Trauma Acute Care Surg 74(5):1215–1221. doi:10.1097/TA.0b013e3182826e13; discussion 1221–1212
62. Ley EJ, Clond MA, Srour MK, Barnajian M, Mirocha J, Margulies DR, Salim A (2011) Emergency department crystalloid resuscitation of 1.5 L or more is associated with increased mortality in elderly and nonelderly trauma patients. J Trauma 70(2):398–400. doi:10.1097/TA.0b013e318208f99b
63. McCluskey SA, Karkouti K, Wijeysundera D, Minkovich L, Tait G, Beattie WS (2013) Hyperchloremia after noncardiac surgery is independently associated with increased morbidity and mortality: a propensity-matched cohort study. Anesth Analg 117(2):412–421. doi:10.1213/ANE.0b013e318293d81e
64. Shaw AD, Bagshaw SM, Goldstein SL, Scherer LA, Duan M, Schermer CR, Kellum JA (2012) Major complications, mortality, and resource utilization after open abdominal surgery: 0.9% saline compared to plasma-Lyte. Ann Surg 255(5):821–829. doi:10.1097/SLA.0b013e31825074f5
65. Van Aken HK, Kampmeier TG, Ertmer C, Westphal M (2012) Fluid resuscitation in patients with traumatic brain injury: what is a SAFE approach? Curr Opin Anaesthesiol 25(5):563–565. doi:10.1097/ACO.0b013e3283572274
66. Korosue K, Heros RC, Ogilvy CS, Hyodo A, Tu YK, Graichen R (1990) Comparison of crystalloids and colloids for hemodilution in a model of focal cerebral ischemia. J Neurosurg 73(4):576–584
67. Mehta BR, Fischer D, Ahmad M, Dubose TD Jr (1983) Effects of acetate and bicarbonate hemodialysis on cardiac function in chronic dialysis patients. Kidney Int 24(6)
68. McCague A, Bowman N, Wong DT (2012) Lactic acidosis after resuscitation with sodium acetate. J Surg Res 173(2):362–364
69. Rohrig R, Wegewitz C, Lendemans S, Petrat F, de Groot H (2014) Superiority of acetate compared with lactate in a rodent model of severe hemorrhagic shock. J Surg Res 186(1):338–345. doi:10.1016/j.jss.2013.09.005
70. Naylor JM, Forsyth GW (1986) The alkalinizing effects of metabolizable bases in the healthy calf. Can J Vet Res = Revue canadienne de recherche veterinaire 50(4):509–516
71. Chowdhury AH, Cox EF, Francis ST, Lobo DN (2012) A randomized, controlled, double-blind crossover study on the effects of 2-L infusions of 0.9% saline and plasma-lyte(R) 148 on renal blood flow velocity and renal cortical tissue perfusion in healthy volunteers. Ann Surg 256(1):18–24. doi:10.1097/SLA.0b013e318256be72
72. Perel P, Roberts I, Ker K (2013) Colloids versus crystalloids for fluid resuscitation in critically ill patients. Cochrane Database Syst Rev (2):CD000567. doi:10.1002/14651858.CD000567.pub6

73. Gattas DJ, Dan A, Myburgh J, Billot L, Lo S, Finfer S (2013) Fluid resuscitation with 6 % hydroxyethyl starch (130/0.4 and 130/0.42) in acutely ill patients: systematic review of effects on mortality and treatment with renal replacement therapy. Intensive Care Med 39(4):558–568. doi:10.1007/s00134-013-2840-0
74. Alderson P, Schierhout G, Roberts I, Bunn F (2000) Colloids versus crystalloids for fluid resuscitation in critically ill patients. Cochrane Database Syst Rev (2):CD000567
75. Wilkes MM, Navickis RJ (2001) Patient survival after human albumin administration. A meta-analysis of randomized, controlled trials. Ann Intern Med 135(3):149–164
76. Finfer S, Bellomo R, Boyce N, French J, Myburgh J, Norton R (2004) A comparison of albumin and saline for fluid resuscitation in the intensive care unit. N Engl J Med 350(22):2247–2256
77. Lawrence A, Faraklas I, Watkins H, Allen A, Cochran A, Morris S, Saffle J (2010) Colloid administration normalizes resuscitation ratio and ameliorates "fluid creep". J Burn Care Res 31(1):40–47
78. Faraklas I, Lam U, Cochran A, Stoddard G, Saffle J (2011) Colloid normalizes resuscitation ratio in pediatric burns. J Burn Care Res Off Publ Am Burn Assoc 32(1):91–97. doi:10.1097/BCR.0b013e318204b379
79. de Jonge E, Levi M (2001) Effects of different plasma substitutes on blood coagulation: a comparative review. Crit Care Med 29(6):1261–1267
80. Moran M, Kapsner C (1987) Acute renal failure associated with elevated plasma oncotic pressure. N Engl J Med 317(3):150–153
81. Jones CI, Payne DA, Hayes PD, Naylor AR, Bell PR, Thompson MM, Goodall AH (2008) The antithrombotic effect of dextran-40 in man is due to enhanced fibrinolysis in vivo. J Vasc Surg 48(3):715–722
82. Abir F, Barkhordarian S, Sumpio BE (2004) Efficacy of dextran solutions in vascular surgery. Vasc Endovasc Surg 38(6):483–491
83. Baker AJ, Rhind SG, Morrison LJ, Black S, Crnko NT, Shek PN, Rizoli SB (2009) Resuscitation with hypertonic saline-dextran reduces serum biomarker levels and correlates with outcome in severe traumatic brain injury patients. J Neurotrauma 26(8):1227–1240. doi:10.1089/neu.2008.0868
84. Brunkhorst FM, Engel C, Bloos F, Meier-Hellmann A, Ragaller M, Weiler N, Moerer O, Gruendling M, Oppert M, Grond S, Olthoff D, Jaschinski U, John S, Rossaint R, Welte T, Schaefer M, Kern P, Kuhnt E, Kiehntopf M, Hartog C, Natanson C, Loeffler M, Reinhart K (2008) Intensive insulin therapy and pentastarch resuscitation in severe sepsis. N Engl J Med 358(2):125–139
85. Myburgh JA, Finfer S, Billot L, Investigators C (2013) Hydroxyethyl starch or saline in intensive care. N Engl J Med 368(8):775. doi:10.1056/NEJMc1215977
86. Perner A, Haase N, Guttormsen AB, Tenhunen J, Klemenzson G, Aneman A, Madsen KR, Moller MH, Elkjaer JM, Poulsen LM, Bendtsen A, Winding R, Steensen M, Berezowicz P, Soe-Jensen P, Bestle M, Strand K, Wiis J, White JO, Thornberg KJ, Quist L, Nielsen J, Andersen LH, Holst LB, Thormar K, Kjaeldgaard AL, Fabritius ML, Mondrup F, Pott FC, Moller TP, Winkel P, Wetterslev J (2012) Hydroxyethyl starch 130/0.42 versus Ringer's acetate in severe sepsis. N Engl J Med 367(2):124–134. doi:10.1056/NEJMoa1204242
87. Annane D, Siami S, Jaber S, Martin C, Elatrous S, Declere AD, Preiser JC, Outin H, Troche G, Charpentier C, Trouillet JL, Kimmoun A, Forceville X, Darmon M, Lesur O, Reignier J, Abroug F, Berger P, Clec'h C, Cousson J, Thibault L, Chevret S, Investigators C (2013) Effects of fluid resuscitation with colloids vs crystalloids on mortality in critically ill patients presenting with hypovolemic shock: the CRISTAL randomized trial. JAMA 310(17):1809–1817. doi:10.1001/jama.2013.280502
88. Ogilvie MP, Pereira BM, McKenney MG, McMahon PJ, Manning RJ, Namias N, Livingstone AS, Schulman CI, Proctor KG (2010) First report on safety and efficacy of hetastarch solution for initial fluid resuscitation at a level 1 trauma center. J Am Coll Surg 210(5):870–872
89. James MF, Michell WL, Joubert IA, Nicol AJ, Navsaria PH, Gillespie RS (2011) Resuscitation with hydroxyethyl starch improves renal function and lactate clearance in penetrating trauma in a randomized controlled study: the FIRST trial (Fluids in Resuscitation of Severe Trauma). Br J Anaesth 107(5):693–702. doi:10.1093/bja/aer229

90. Duke MD, Guidry C, Guice J, Stuke L, Marr AB, Hunt JP, Meade P, McSwain NE Jr, Duchesne JC (2012) Restrictive fluid resuscitation in combination with damage control resuscitation: time for adaptation. J Trauma Acute Care Surg 73(3):674–678. doi:10.1097/TA.0b013e318265ce1f
91. Guidry C, Gleeson E, Simms ER, Stuke L, Meade P, McSwain NE Jr, Duchesne JC (2013) Initial assessment on the impact of crystalloids versus colloids during damage control resuscitation. J Surg Res 185(1):294–299. doi:10.1016/j.jss.2013.05.086
92. Weiskopf RB, James MF (2015) Update of use of hydroxyethyl starches in surgery and trauma. J Trauma Acute Care Surg 78(6 Suppl 1):S54–S59. doi:10.1097/TA.0000000000000636
93. Joseph B, Aziz H, Pandit V, Kulvatunyou N, O'Keeffe T, Wynne J, Tang A, Friese RS, Rhee P (2014) Improving survival rates after civilian gunshot wounds to the brain. J Am Coll Surg 218(1):58–65. doi:10.1016/j.jamcollsurg.2013.08.018
94. Cooper DJ, Myburgh J, Heritier S, Finfer S, Bellomo R, Billot L, Murray L, Vallance S (2013) Albumin resuscitation for traumatic brain injury: is intracranial hypertension the cause of increased mortality? J Neurotrauma 30(7):512–518. doi:10.1089/neu.2012.2573
95. Neff TA, Doelberg M, Jungheinrich C, Sauerland A, Spahn DR, Stocker R (2003) Repetitive large-dose infusion of the novel hydroxyethyl starch 130/0.4 in patients with severe head injury. Anesth Analg 96(5):1453–1459
96. Rodling Wahlstrom M, Olivecrona M, Nystrom F, Koskinen LO, Naredi S (2009) Fluid therapy and the use of albumin in the treatment of severe traumatic brain injury. Acta Anaesthesiol Scand 53(1):18–25. doi:10.1111/j.1399-6576.2008.01798.x
97. Statler KD (2011) Albumin and brain injury: time for a new approach? Crit Care Med 39(1):217–218. doi:10.1097/CCM.0b013e318202e408
98. Tricklebank S (2009) Modern trends in fluid therapy for burns. Burns 35(6):757–767
99. Cartotto R (2009) Fluid resuscitation of the thermally injured patient. Clin Plast Surg 36(4):569–581. doi:10.1016/j.cps.2009.05.002
100. Cartotto R, Zhou A (2010) Fluid creep: the pendulum hasn't swung back yet! J Burn Care Res Off Publ Am Burn Assoc 31(4):551–558. doi:10.1097/BCR.0b013e3181e4d732
101. Bechir M, Puhan MA, Neff SB, Guggenheim M, Wedler V, Stover JF, Stocker R, Neff TA (2010) Early fluid resuscitation with hyperoncotic hydroxyethyl starch 200/0.5 (10%) in severe burn injury. Crit Care 14(3):R123
102. Azzopardi EA, McWilliams B, Iyer S, Whitaker IS (2009) Fluid resuscitation in adults with severe burns at risk of secondary abdominal compartment syndrome–an evidence based systematic review. Burns 35(7):911–920
103. Faraklas I, Cochran A, Saffle J (2012) Review of a fluid resuscitation protocol: "fluid creep" is not due to nursing error. J Burn Care Res Off Publ Am Burn Assoc 33(1):74–83. doi:10.1097/BCR.0b013e318234d949

Chapter 13
Role of Multimodal Monitoring (MMM) in the Perioperative Period: Improving Outcomes in High Risk Surgical Patients

D. Green

Abstract The early 1980s saw radical advances in perioperative monitoring. In a relatively short period of time, the "finger on the pulse", manual measurement of blood pressure using Riva Rocci cuff, aneroid mercury column and listening for Korotkov sounds with a stethoscope and (if you were lucky) a primitive ECG machine were replaced from the early to mid 1980s with pulse oximetry, automated non-invasive BP, end tidal carbon dioxide and anaesthetic agent monitoring and sophisticated ECG technology. This was a major transition in monitoring technology but, since then, progress has been halted. From the late 1990s technology has become widely available to measure the extent of anaesthesia induced cortical suppression using EEG monitoring technology (e.g. BIS, Medtronic, USA), cerebral and tissue oxygenation using the Invos cerebral oximeter (Medtronic, USA) and various flow monitors using Doppler and arterial pulse waveform analysis from the radial or femoral artery lines. In 2011 NICE (National Institute for Health and Care Excellence, UK) guidelines were introduced advocating the use of flow monitoring and then in 2012 "depth of anaesthesia" in high-risk surgical patients. Although widely available this technology is still only used in a small percentage of patients in the UK. Recent evidence indicates that the introduction of flow monitoring alongside existing monitoring does not radically improve outcome in terms of reduction in complications and has no effect on 30 day and in-hospital mortality, which remains at around 5 % in the elderly, high risk surgical population. This is probably due to a fundamental misunderstanding of circulatory physiology which has led to protocols and interventions using flow monitoring which are inappropriate. Often they result in massive fluid and sodium overload. This is particularly a problem in the highest risk patients. The rapid rise in the number of such patients has I believe necessitated a radical re-evaluation of circulatory physiology under anaesthesia and its implication for management. Using the *combination* of these new monitoring technologies (flow, depth of anaesthesia and cerebral and tissue oximetry) in a multimodal monitoring strategy provides maximum benefit by reducing complications,

D. Green
King's College Hospital NHS Foundation Trust, King's Health Partners,
Denmark Hill, London SE5 9RS, UK
e-mail: davidgreen1@nhs.net

K. Stuart-Smith (ed.), *Perioperative Medicine – Current Controversies*,
DOI 10.1007/978-3-319-28821-5_13

the necessity for high dependency care and mortality. This combined strategy and its rationale is the subject of this chapter.

Keywords Monitoring • Cardiac output • Depth of anaesthesia • Cerebral oxygenation • Circulation during anaesthesia • Crystalloids • Colloids • Outcome following surgery

13.1 Introduction

Why do we need to change our current perioperative monitoring strategy when it is clear from recent analyses that death in the *immediate* perioperative period is incredibly low? Anaesthesia is completely safe isn't it? In a survey linked to the NAP 5 (National Audit Project) study of the Royal College of Anaesthetists in the United Kingdom on accidental awareness during general anaesthesia (AAGA) investigators made several pertinent comments [1]:

- "The low mortality rate of GA (0.06 % or one in 1718) during surgery is notable".
- "Many patients are scared of anaesthesia and this figure can only be reassuring for them".
- "During the period of time when they are cared for by anaesthetists, the risk of death is low indeed".
- "This low mortality rate is in marked contrast to the report by EUSOS [2] of an overall **4 %** or one in 25 mortality rate for in patient elective major surgery"

The implication then is that anaesthesia is completely safe and with a mortality rate of one in 2000 one might expect little to be gained from a new multimodal monitoring (MMM) strategy. However, there is a very important caveat which arises from the following statement made in the paper.

"These differences (in immediate and intermediate mortality) highlight the potential impact advances in perioperative care by anaesthetists, surgeons and intensivists might have on overall mortality rates after surgery".

Patients thus die very rarely (0.05 %) on the operating table or in the first 1 or 2 days post-operatively but up to 4 % may die during a hospital admission [2]. It is thus for anaesthetists to recognise that deficiencies in intraoperative and immediate perioperative management may start the "time bomb ticking" for complications leading to mortality occurring later on during hospital admission. These deaths cannot simply be accounted entirely by surgical complications. The challenge we have today is to translate the very low immediate perioperative mortality indicated by the NAP5 study data to the whole perioperative period.

13.2 The Increasingly High-Risk Surgical Population

A recent review highlighted the type of patients and surgery who constitute a high-risk population [3]. The elderly patient with one or more comorbidities of the cardiovascular, respiratory or renal systems [4] is now the largest category of high risk patient. The increasing burden of obesity and diabetes must also be taken into account and may affect a younger population. It is estimated that one million of the estimated three million operations carried out in the UK in 2013 were on patients over 65 [1]. In a study in the USA the elderly underwent surgery four times more often than the young and in a recent Australian and New Zealand study, 5 % of patients over 70 died within 30 days of surgery [5]. The "Reason" study investigators commented that "strategies are needed to reduce complications and mortality in older surgical patients". Given the 4 % mortality in the EUSOS study we might conclude that strategies are required for younger and fitter patients too [2]! Strategies and protocols need to be guided by monitors not "recipe books" and we need to individualise patient care [6]. Anaesthetists must recognise that deficiencies in intraoperative management are deleterious for long-term patient outcome.

13.3 Monitoring Technology

The late 70s and early 80s saw a dramatic advance in monitoring technology in the perioperative period. The "finger on the pulse" was replaced by the pulse oximeter and it also became possible to routinely monitor end tidal carbon dioxide and inhalation agent concentration. In addition, manual intermittent blood pressure using a conventional arm cuff or oscillotonometer was replaced by automated non-invasive BP machines. ECG monitoring advanced with the appearance of memory scopes and automated analysis of ST segment depression. This represented an enormous advance in a very short period of time. However, since from around 1984 until 2011 there have been no major advances in routine perioperative monitoring. Indeed, assessment of patient status still relies solely on finger sPO_2, blood pressure monitoring and the presence of ischaemia indicated on a bipolar, three lead ECG system not really designed to pick this up. In addition, in the UK even if more sophisticated monitoring was available it was usually only placed in the operating room and not available in the induction room, still beloved by 80 % of anaesthetists [1]. Most hospitals in Europe and the USA do not have a rigorous protocol of haemodynamic management even for high risk patients [7].

13.4 The Implications of Poor Blood Pressure Control During Anaesthesia

Most anaesthetists will try and maintain mean arterial pressure (MAP) within about 20 % of the pre-induction value. What is the evidence that low MAP is harmful? A recent large retrospective study from the Cleveland clinic found that "a mean arterial pressure less than 55 mmHg predicted adverse cardiac and renal related outcomes; however, the limitations of retrospective analyses must be considered. Notably, any amount of time at a MAP less than 55 mmHg was associated with adverse outcomes [8]". No room for complacency! They also go on to say that "further research is required to determine whether interventions to prevent and rapidly treat intraoperative hypotension ameliorate the risk of acute kidney injury and myocardial injury in patients having non-cardiac surgery". Although this seems a reasonable statement it may lead to further problems as correction of MAP by vasoconstrictors may result in a decrease in flow and oxygen delivery, further compromising oxygenation [9]. So, it is crucial that we know what causes the fall in MAP on induction and during maintenance of anaesthesia in moderate and high-risk patients in order to make appropriate interventions (see later).

13.5 Interventions Used by Anaesthetists to Maintain MAP During Anaesthesia

Most anaesthetists will use fluids and/or vasoactive drugs to correct changes in MAP without consideration of the "depth of anaesthesia" or degree of cortical suppression induced by the anaesthetics used.

13.5.1 Using Fluids

As far as fluids are concerned current practice throughout the world is still influenced by the ground breaking paper from Shires et al from Dallas published in 1961 in the Annals of Surgery [10]. The investigators used a triple indicator dilution technique which suggested that there was a profound loss of functional extracellular fluid during major surgery, later nicknamed "third space", which could amount during surgery to up to 15 ml.kg^{-1} body weight hr^{-1}. Given that some patients might therefore need 1 L or more per hour of extracellular fluid equivalent replacement they proposed the introduction of a lactated version of Ringer's solution, a balanced salt solution which contains a more physiological concentration of chloride in contradistinction to so-called "normal saline", 0.9 % NaCl. It literally opened the floodgates as anaesthetists could point to excessive "third space loss" as the cause of intra-operative hypotension in major surgery and were given *carte blanche* to use excessive amounts of fluid such as lactated Ringer's (Hartmann's) in trying to reverse the fall in blood pressure.

It is clear that the original authors very quickly became concerned that excessive fluids were now been given rather than the fluid restrictive practice which had existed prior to the paper of 1961. Moore and Shires in 1967 stated [11] "the objective of (perioperative) care is restoration to normal physiology and normal function of organs, with a normal blood volume and functional body water and electrolytes. This can never be achieved by inundation (with fluids)". It is clear that giving massive volumes of balanced salt solutions will not achieve anything resembling normal physiology! The dangers of excessive fluids and volume overload with high CVP have recently been re-iterated by Marik. e.g. pulmonary oedema and increased extra-vascular lung water, impaired oxygenation, altered pulmonary and chest wall mechanics, increased work of breathing, myocardial oedema etc. The list is long [12].

Interestingly, recent papers and reviews have confirmed that “there is no convincing evidence supporting the existence of the non-anatomical third space loss neither in haemorrhagic shock nor in surgery of any kind” [13–15]. This highlights another problem. Third space loss **does** exist in conditions such as septic shock, anaphylaxis and burn injury and it is important to distinguish fluid and flow requirements in elective surgery in contradistinction to haemodynamic changes in the intensive care unit where sepsis may well be a serious problem. Reviews of fluid management often mix up these two completely different scenarios and thus cause unnecessary confusion.

It is thus disappointing to see a large scale prospective trial of *restrictive* versus *liberal* fluids during surgery, the *Relief* study, which will be running until November 2018 where large volumes of fluid are given in the liberal group. Intervention patients may receive 4 days Na^+ requirement in a 2-h procedure! In 2016 patients are still receiving too much fluid and Na^+ in the perioperative period with implications for poor outcome (see later).

13.5.2 Using Vasoactive Agents

It is easy to restore MAP to “normal” using vasoactive agents such as metaraminol, phenylephrine or even noradrenaline [16]. However, MAP may then be maintained at the expense of cardiac output as shown in Fig. 13.1 which demonstrates that reliance on MAP alone may be associated with profound falls in cardiac output (CO) and oxygen delivery (DO_2) and subsequent build-up of oxygen debt (vide infra). It is crucial to know the *cause* of the fall in MAP before using vasoactive compounds. This may be a real problem when the flow monitor does not accurately track changes in flow with changes in vasomotor tone [17]. This may lead in some cases to a fall sense of security.

13.6 Cardiac Output Changes Following Induction and Maintenance of Anaesthesia

13.6.1 Introduction

Most anaesthetists opine that the fall in MAP following induction is mainly due to peripheral vasodilation with a fall in systemic vascular resistance (SVR) but little change in CO and oxygen delivery DO_2 and incorrectly assume that a simple vasoconstrictor such as metaraminol will both correct MAP whilst maintaining flow

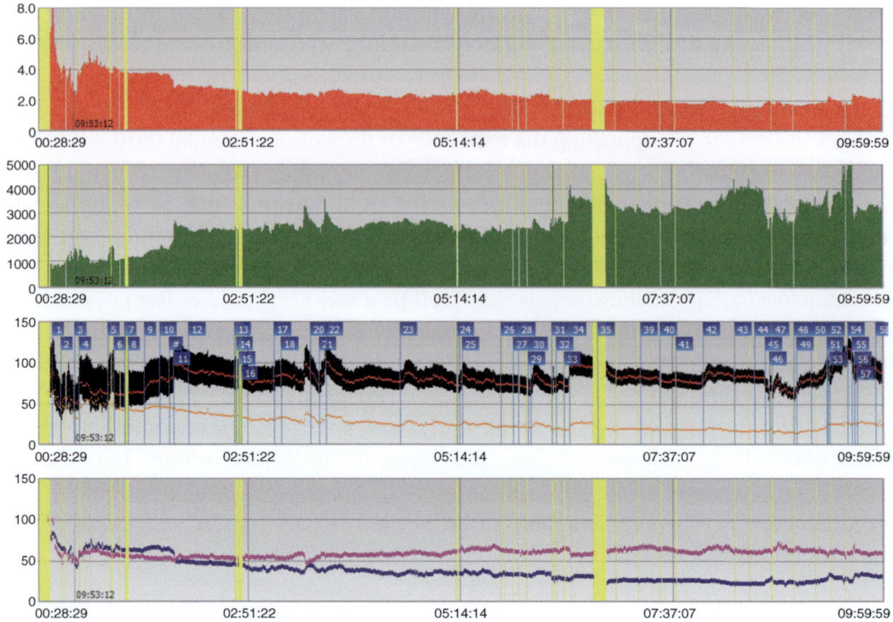

Fig. 13.1 LiDCOviewPro Case analysis using LiDCOview pro software (www.lidco.com). The top horizontal trace in *red* shows nominal cardiac output (nCO). The next trace in *green* shows systemic vascular resistance (SVR). The third trace *black* hatching is blood pressure (BP) from an arterial line, mean pressure (MAP) is indicated by the *red line*. The *bottom trace* shows stroke volume (SV) in *blue* and heart rate (HR) in *purple*. The numbers on the BP segment indicate interventions. 1 indicates induction of anaesthesia and 2 is intubation. Various vasoactive compounds were given during the 10 h. procedure but it can be seen that nCO is maintained at only 30 % of starting value (2.2 vs. 6.6 lpm) for the majority of the operation. However, MAP and HR are maintained at normal values and would not suggest any problem. The massive reduction in oxygen delivery (DO2) and build-up of oxygen debt thus occurred without any indication from conventional monitoring of BP and HR

(see above). The first measurement of CO during surgery using the dye dilution technique was carried out in 1960 and showed that the CO dropped by 20 % after induction [18]. However, it was only following the advent of continuous CO monitoring technology, particularly using analysis of the radial artery waveform (see later) that it was possible to follow the rapid changes of pressure **and** flow following induction of anaesthesia. Kamenik and Petrun, using the LiDCO rapid (LiDCO, UK), were able to demonstrate that the fall in MAP following BIS (Medtronic, USA) targeted induction using either propofol or etomidate was mainly due to a fall in CO and not due to a fall in SVR [19]. The fall may be as much as 40 % in elderly, high-risk patients. Knowledge of this potentially dramatic fall is essential to allow appropriate intraoperative fluid and vasoactive drug management of these patients.

13.6.2 Why Does Cardiac Output Fall on Induction and During Maintenance of Anaesthesia?

To understand what happens to the circulation during anaesthesia induction we need to go back more than 50 years to a series of papers by Professor Sharpey-Schafer at St Thomas's Hospital Medical School in London [20–22]. He studied **venous tone** in patients and volunteers who were healthy, anaemic, hypertensive, following Valsalva manoeuvre, ganglion blocking drugs and blood donation. His comment that "it is clear that the usual clinical practice of following such cases (hypotension with ganglion blocking drugs, then being introduced for the treatment of hypertension [23]) by the measurement of arterial pressure by the antiquated cuff method can be of limited value". He appreciated that the changes in MAP caused by ganglion blocking drugs especially when the patient was raised to the upright position were occurring too quickly to be picked up by conventional methods of measuring MAP in use at the time, such as Korotkoff sounds and the mercury manometer. Interestingly, hypotensive anaesthesia had been introduced into clinical practice by Enderby and colleagues at East Grinstead, Sussex, UK to allow more accurate plastic surgery to be carried out on airmen who had received facial burns during the Second World War. Enderby also noted the restrictions imposed on accurately measuring MAP during these cases and introduced the mechanical von Recklinghausen and Boulitte oscillotonometers into clinical anaesthetic practice in the UK. These he found more accurate at low pressures than the conventional blood pressure cuff [24]. This also allowed remote measurement of blood pressure and removed the necessity of placing the stethoscope over the brachial artery. It is noteworthy that modern automatic oscillo(tono)meters are now the most common way of measuring blood pressure during anaesthesia. Sharpey-Schafer's conclusion was that "it appears that the **veins and not the arteries** are the major factor in causing hypotension with ganglion blocking drugs when assuming the upright position". Is the same true with anaesthetic drugs? Isolated hepatic portal veins and aorta taken from the rat were used to investigate the direct action of propofol. It caused a dose related decrease of tone in both types of blood vessel but concentrations required to produce this effect on veins were significantly lower than those required to produce similar changes in the isolated artery preparation. The authors concluded that this direct action may contribute towards the hypotensive effects of propofol [25]. Venodilation with propofol resulted in an increase in venous capacitance in dogs and there was no change in SVR to account for the change in MAP. They concluded that anaesthesia with propofol is accompanied by decreased CO secondary to venodilation and reduction in preload. Cardiac output and MAP are preserved as long as preload is maintained [26]. It was also noted by Sharpey-Schafer that underlying venous tone is markedly increased in patients with heart failure and anaemia, particularly pertinent to the high-risk and elderly patient population. It is therefore reasonable to conclude that these patients are more susceptible to the venodilation induced by propofol and probably other anaesthetics.

Venodilation is thus a key feature of anaesthesia which has been very little recognised [27]. Following venodilation the fall in MAP is due not only to falling preload, stroke volume (SV) and CO with a consequent reduction in DO_2 but also due to a shift of volume out of the arterial tree into the dilated venous compartment [27]. The question is whether we should maintain venous capacitance with fluids (as in the dog study above) or by administration of a venoconstrictor. Our experience with the venoconstrictor, phenylephrine, the effect monitored by the LiDCOrapid (see later) commenced in low-dose pre-induction can maintain venous tone, cardiac output and MAP without unnecessary fluid administration to maintain venous capacitance [28].

Crucially important is the fact that measuring CO with whatever technology **after** induction of anaesthesia may greatly underestimate the true resting CO and therefore any intervention which is supposedly designed to "optimise stroke volume" (whatever that means) and maintain DO_2 will be flawed and may result in unnecessary excess fluid and Na^+ administration to the detriment of outcome in high-risk patients [12].

13.7 Recommendation to Add Flow Monitoring

Although flow monitoring technology had been around for many years mainly using the pulmonary artery thermodilution catheter (PAC) this technology came to be considered too invasive for routine use and some studies seemed to question its usefulness in high-risk surgery [29]. In addition, it only provided intermittent values for flow. Doppler technology, providing continuous measurement of flow velocity, from which was calculated stroke volume and CO, was first employed in perioperative management in the 1990s [30] but was popularised by the introduction of the Deltex oesophageal Doppler in the early 2000s (Deltex, UK). Three major trials published between 2002 and 2006 suggested benefits to higher risk abdominal surgery patients by modest reductions in hospital length of stay [31–33]. However, adoption of this technology had been extremely limited and it was left to the National Institute of Health and Care Excellence (NICE) in the UK to recommend this technology for a distinct group of high-risk surgical patients undergoing particular surgical operations in 2011 [34]. Even then, the guidance was questioned and argued over in editorials [35, 36]. In the intervening period many other continuous flow monitoring technologies had been developed by various companies, usually using analysis of the arterial waveform either from the descending aorta via a catheter placed in the femoral artery (PiCCO, Pulsion, Germany) or from the radial artery (LiDCO, UK and Flotrac, Edwards Lab, USA). Both PiCCO and LiDCO technologies allowed for accurate calibration either using a trans-pulmonary thermodilution technique as with PiCCO or a lithium dilution technique with LiDCO. It was not surprising that these companies protested that a single flow monitoring technology had been advocated by NICE and thus in 2012 the advice was extended to any flow monitoring technology which in the European Union had a CE mark. (It is important to point out that the CE mark does not guarantee that the device measures flow accurately only that it is been manufactured to a certain standard).

It is now possible using relatively non-invasive technology to measure flow accurately and continuously throughout the perioperative period.

13.8 How Did the Addition of Flow Monitoring Affect Anaesthetic Practice and Patient Outcome?

13.8.1 Protocols and Interventions

Although flow monitoring was advocated in high-risk patients by NICE it was not clear what protocol would or should be used by anaesthetists. The principle of goal directed therapy (GDT), see later, had been introduced to try and achieve population-based DO_2 targets and the figure of 600 ml of oxygen delivery per m^{-2} Body Surface Area (BSA) is often quoted. However, although this target is usually used in the immediate postoperative period to reduce repay oxygen debt and reduce complications [37] it is less frequently used during high-risk surgery [38] as will be discussed in the next section.

13.8.2 Flow Time Corrected (FTc) and Stroke Volume "Optimisation" Strategies

The original Doppler studies suggested that flow time, corrected for heart rate, (Ftc) maintained at about 350 msec seemed to indicate that patients were optimally filled [31]. This important parameter was discarded following the introduction of SV optimisation or maximisation strategies (see Fig. 13.2). This was based on the opinion that poor outcome was a result of inadequate tissue perfusion and giving additional fluids to improve perfusion would result in a better outcome. Thus if the patient was fluid responsive it was assumed that they needed fluids. Thus 200–300 ml. boluses of colloid (3–4 $ml.kg^{-1}$) are given over a period of less than 5 min and repeated until additional boluses no longer produced a greater than 10 % increase in stroke volume. (A 10 % threshold is chosen as a 300 ml. fluid challenge will produce haemodilution and around a 5 % reduction in oxygen carrying capacity and this would need to be offset by at least a 5 % increase in SV for the fluid challenge to increase DO_2). It is clear that just because the patient may *respond* to fluids with an increase in SV does not necessarily mean that they need fluids (see also next section). In addition, since the Doppler is placed post induction none of the studies appreciated the extent of the fall in CO and DO_2 post induction.

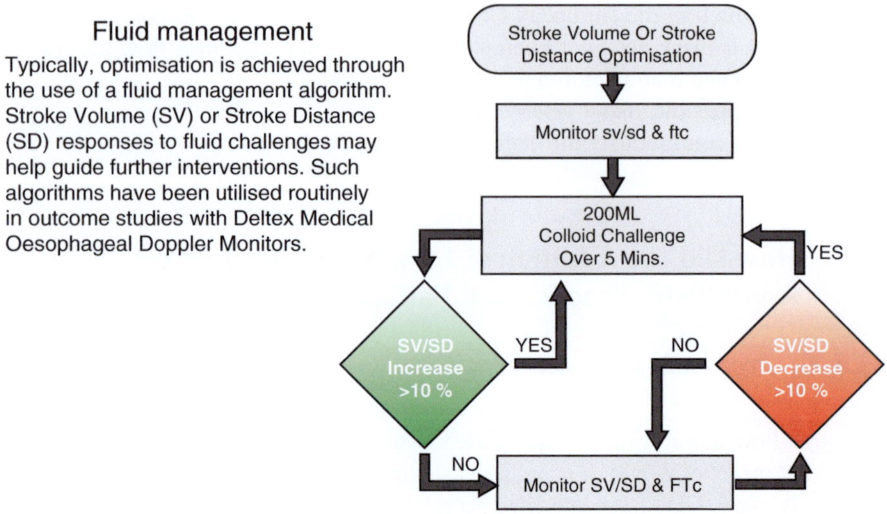

Fig. 13.2 Fluid management using the Oesophageal Doppler algorithm suggested by the Deltex Medical Company, Chichester, UK (Reproduced with permission from Bidd et al. [60])

13.8.3 Systolic Pressure and Stroke Volume Variation as Predictors of Fluid Responsiveness

Another parameter available from monitors showing the dynamic pressure waveform from an arterial line is pulse pressure variation (PPV) [39]. Fluid depleted patients show high values of PPV and this can be used as an indication to give fluids. In addition, if flow is calculated from the pressure waveform (see above) stroke volume variation (SVV) can also be used as an indication of fluid responsiveness. These parameters are only available following intubation and institution of intermittent positive pressure ventilation (IPPV) using a tidal volume of at least 8 $ml.kg^{-1}$ in a patient with a closed chest and in sinus rhythm. A SVV of between 5 and 10 % suggested adequate filling whereas a figure above 13 % suggested fluid depletion [39, 40]. As will be seen later excessive quantities of unnecessary fluid and Na^+ have deleterious implications for both the peripheral circulation and the cardiovascular system.

13.9 Does the Current Practice of Adding Flow Monitoring to Existing Monitoring Actually Improve Outcome?

Treating individual patients using population-based estimates of DO_2, "maximising" or "optimising" SV and trying to achieve certain targets of PPV or SVV may not produce the benefits expected. In a randomised controlled trial of intraoperative goal directed fluid therapy in aerobically fit and unfit patients having major colorectal surgery it was concluded that "SV optimisation conferred no additional benefit and in aerobically fit patients had a detrimental effect on the primary outcome in terms of increased patient length of stay". This study used a SV maximisation strategy and also included replacement of "third space loss" at 10 $ml.kg^{-1}.h^{-1}$[41]. Another study from Brazil using a calibrated LiDCO plus monitor tried to achieve GDT targets of 600 $ml.m^{-2}$ BSA DO_2 using fluids and inotropes (dobutamine) in a high-risk surgical population randomised to two baseline fluid groups i.e. a restrictive group received 4 $ml.kg^{-1}.hr^{-1}$ of lactated Ringer's per hour versus 12 in the standard group. Although DO_2 targets were better achieved in the standard group the outcome was better in the restricted group [38]. It is also clear that SV maximisation using the Doppler or LiDCO may not produce any additional benefit over a simple restrictive fluid strategy [41].

13.10 Fundamental Misconceptions of Circulatory Physiology During Anaesthesia

I have already considered the importance of distinguishing between venodilation and vasodilation and implications for management. Other factors have come to light recently which will have profound implications for fluid management during major surgery in high risk patients. These are discussed in more detail in a recent review [27].

13.10.1 The Endothelial Glycocalyx Layer

The clinical implications and physiological importance of the endothelial glycocalyx layer (EGL) have been pointed out in three recent reviews and have altered our understanding and management of fluids during surgery [42–44]. The identification of the EGL, a fragile, sieve like barrier between the plasma and the endothelial cells, requires a fundamental re-evaluation of fluid transfer into and out of the interstitial space from arteriolar to venular end as proposed by the Starling equation. It now appears that gross fluid transfer is much less than previously suggested. Fluid crossing the EGL and into the interstitium returns to the circulation predominantly by the lymphatics. The EGL is the true effective barrier between the circulation and the interstitial fluid. As it is fragile it is easily damaged by ischaemia, sepsis, burn injury, anaphylaxis and pertinent to this discussion, hypervolaemia. Excess fluids and Na^+ to replace non-existent "third space loss" may not only damage the EGL directly by compression and distortion but also lead to an increase in atrial natriuretic peptide (ANP) which itself damages the layer. Albumin appears to be protective and sits within the sieve like membrane partially preventing excess fluid loss. Excess fluid administration in the perioperative period leads to a fall in albumin which again damages the EGL. The implication is that "third space loss" may actually be *precipitated* by excess fluids rather than they being a replacement for third space loss. *Maintaining* pre-induction values of CO and DO_2 (e.g. by using colloids/blood to replace obvious losses to maintain circulating volume) rather than trying to *maximise* them to nominal population based targets may be the more appropriate strategy. This will also help to minimise unnecessary intraoperative fluids and Na^+ which may be deleterious in the high risk patient.

13.10.2 Cardiac Output and the Peripheral Circulation

In experiments in the 1950s Guyton showed that the need of the body for oxygen is the real regulator of cardiac output [45]. In other words, CO is determined by the metabolic needs of the tissues and not the other way round. In general, tissues (except skin and kidney) regulate their own flow according to their metabolic demands. Summation of these tissue flows determines venous return and venous tone determines how much of that venous return actually gets back to the heart as preload. Venous tone is decreased by anaesthesia and small doses of venoconstrictors prevent or substantially reduce the venodilation (see earlier). The heart plays a permissive role and does not regulate its own output, so it is not appropriate to use a SV optimisation or maximisation strategy or try to achieve a perioperative *increase* in global blood flow as an intervention strategy in the elective patient. It seems much more appropriate to just maintain pre-induction CO and DO_2.

13.11 Does The Use of Flow Monitoring and Maximising Stroke Volume and Oxygen Delivery Improve Outcome and Reduce Mortality?

This is clearly a very important question as if our current strategies are not improving outcome to the extent we desire then we must consider a new strategy based on a clearer understanding of circulatory physiology, as indicated above [27].

In a recent Cochrane collaboration review entitled "Perioperative increase in global blood flow to explicit defined goals and outcomes after surgery", the investigators noted that using this strategy there was no decrease in mortality and length of stay was decreased on average by only 1 day [46]. There were reductions in complications which accounted for the reduction in length of stay. However, the investigators concluded "the balance of current evidence does not support widespread implementation of this approach to reduce mortality but does suggest complications and duration of hospital stay are reduced". This was hardly a ringing endorsement of this strategy and worryingly there was no mortality reduction. In addition, the latest large-scale prospective randomised trial of the effect of perioperative, CO guided haemodynamic therapy on outcomes following major gastrointestinal surgery, the so-called *Optimise* trial, did not reduce a composite outcome of complications and 30 day mortality [47]. A recent editorial by two authors of the Challand et al. study [48] mentioned earlier concluded that SV optimisation/maximisation using the algorithm shown in Fig. 13.2 provides "no marginal benefit in aerobically fit patients having elective surgery within a contemporary enhanced recovery pathway" [49].

13.12 Why Does Flow Monitoring Using Current Strategies Not Produce the Expected Benefits, Especially Reduction in Mortality?

Earlier reviews of outcome using flow monitoring technology appeared to suggest that mortality was reduced in high risk patients [50] especially in studies where mortality in the control group exceeded 10 %. However, the introduction of the enhanced recovery after surgery program (ERAS) has meant that overall surgical care has improved in the last 10 years or so [51]. This would likely produce an improvement in outcome and a reduction in mortality independent of any effect of flow monitoring. Any benefit expected from flow monitoring is also likely to be reduced due to misunderstanding of circulatory physiology, the role of the heart in the circulation and third space loss alluded to above. Patients may receive too much fluid and Na^+ and this may be deleterious in these patients and offset any benefits from improved care such as ERAS.

13.13 Conclusion on Goal Directed Therapy and Outcome

It is clear that there has not been as much benefit as expected in particular there has been no effect on mortality reduction versus current practice and mortality remains unacceptably high. What is the way forward?

13.14 The Role of Individualised Haemodynamic Therapy

In a review of perioperative haemodynamic therapy the authors stated that "once an individualised approach will be identified, the terms of liberal, restrictive and supranormal values could eventually be replaced by adequate haemodynamic support that fits every patient's own needs" [6]. The question is, can we identify a better individualised approach to further improve outcomes and in particular reduce mortality in this high risk patient group?

The goal of twenty-first-century individualised intraoperative management of the high-risk surgical patient should be directed towards avoiding the build-up of oxygen debt. This can only be achieved by using multi-modal monitoring (MMM) to assess and maintain DO_2 using a reliable flow monitor from pre-induction in the elective patient and then maintained as close to the pre-induction level as possible during surgery and into the immediate postoperative period.

DO_2 is the product of arterial oxygen content (CaO_2) multiplied by cardiac output.

CaO_2 is the product of Hb concentration in g.L^{-1} multiplied by arterial oxygen saturation (sPO_2) as a percentage divided by 100 multiplied by 1.34, giving the number of mls of oxygen carried by 1 g of Hb when fully saturated. If "normal" CO is 5 l.min^{-1} then DO_2 is:

$$\mathbf{(150 * 98 / 100 * 1.34) * 5 = 1000\,ml.min^{-1}}$$

in the average 70 kg patient.

So, clearly we need to track CaO_2 as well as CO throughout the perioperative period to be able to maintain DO_2. In patients who are experiencing blood loss, frequent measurements of Hb are clearly necessary. To be able to maintain CO and DO_2 during surgery, but at the same time prevent excessive or inadequate levels of anaesthesia, we need to have some monitor to assess depth of anaesthesia. The latter term is better expressed as adequate cortical suppression [52, 53], maintained at an appropriate level in response to the intensity of surgery throughout the procedure. Since our main strategy is to maintain DO_2 and tissue oxygenation, a rational monitor to use alongside flow monitoring and depth of anaesthesia monitoring is a cerebral/tissue oximeter (see later) [54]. These three components of flow, depth of anaesthesia and cerebral/tissue oxygenation form the three strands of the MMM strategy.

13.15 Oxygen Debt and Why It Is Bad for High-Risk Patients

The concept of build-up of oxygen debt and contribution to poor outcome in high-risk surgical patients was first evaluated by Shoemaker et al in the late 80s and early 90s [55–58]. He measured oxygen consumption (VO_2) in his high-risk patients at pre-induction and then throughout anaesthesia using a PAC. The difference between VO_2 intra-operatively (less 15 % which Shoemaker considered was the likely reduction in VO_2 due to the effects of anaesthesia) and that obtained from the pre-induction reading constituted oxygen debt and clearly the greater the difference and the longer it went on for, the greater the oxygen debt would be at the end of the procedure.

Shoemaker's assumption was that patients experience *shock* during high-risk surgery: that is they build up oxygen debt inexorably during surgery and this build-up may continue into the post-operative period. Crucially he discovered that those patients with limited oxygen debt and who repay the debt quickly will recover without complications [56]. However, he noted that the debt could continue to accumulate in the post-operative period and if it reached a critical value of about 30 $L.m^{-2}$ then the patient inevitably went on to suffer complications and died. This led to the introduction of GDT in the **post**-operative period using inotropes and fluids to try and repay this debt as quickly as possible in order to try and reduce complications and mortality. It is clear that this strategy worked to a certain degree [37]. However, is it obligatory that oxygen debt should build up at all during the intraoperative period? It is rare that VO_2 is measured during anaesthesia and in Shoemaker's studies required the invasive PAC. So, how do we know if oxygen debt is building up or not during a procedure? Looking at Shoemaker's data there was a clear relationship between decline in DO_2, build-up of oxygen debt and poor outcome. In other words, if DO_2 was maintained during surgery (as it was in a few patients) very little oxygen debt built up.

The implication of this in elective, elderly, high risk patients is that we can avoid significant build-up of oxygen debt if we can maintain pre-induction CO, and DO_2 during surgery using a MMM strategy. Lack of oxygen debt at the end of the procedure means that strategies of GDT intended to increase DO_2 by fluids and inotropes in HDU/ICU and repay debt are no longer needed

13.16 New Monitoring Technology and Its Use in the Operating Room

It is clear from the above that there have been very little advances in routine intraoperative monitoring in anaesthesia in the last 30 years since the introduction of the pulse oximeter. The addition of flow monitoring does not appear to have had much benefit as far as reduction in mortality is concerned, and it is still around 5 % in the elderly surgical population [5]. In 2010 we proposed that "although there is now evidence to suggest that the use of individual new monitors can influence outcome, it will only be their combination that will radically improve the perioperative management on outcome of high-risk surgical patients" [59]. In other words, many new monitors had become available such as a variety of depth of anaesthesia monitors, cerebral/tissue oximeters and flow monitors but to get maximum benefit it would be best for these to be used together in a systematic way. After all, trying to manage a hypotensive patient during anaesthesia without recognising the fact the patient is very sensitive to anaesthetics and is excessively deeply anaesthetised may lead to unnecessary interventions of excess fluids and vasoactive drugs to correct hypotension, when simply reducing the amount of anaesthetic may be all that's necessary. Optimising "depth of anaesthesia" is best achieved by the use of specific monitors.

In 2013, an *Opinion* article in *Perioperative Medicine* made the case for this new multi-modal monitoring strategy to improve outcome in high-risk patients [60].

13.17 Cerebral and Tissue Oxygen Monitoring as Part of the Multi-modal Monitoring (MMM) Strategy

13.17.1 Introduction

"The proper management of brain oxygenation should be one of the principal endpoints of all anaesthesia procedures yet the brain remains one of the least monitored organs during clinical anaesthesiology" [61]. Near infrared technology using a reflectance technique can be used to measure cerebral tissue oxygenation to a depth of about 2.5 cm below the measuring electrodes which are usually placed on the forehead. Regional brain oxygen saturation (rSO_2) reflects overall *tissue* rather than *arterial* oxygen saturation as with the pulse oximeter. Cerebral rSO_2 is venous weighted with a characteristic value in healthy patients of around about 70 % and is similar to jugular venous oxygen saturation. A detailed consideration of this technology is outside the scope of this chapter so the reader is referred to excellent reviews [54, 61]. The wavelengths used relate to the maximum absorption of haemoglobin in both oxygenated and de-oxygenated state. The first clinically available cerebral oximeter came in the late 90s (Invos, Somanetics, Medtronics, USA) and since that time additional monitors from Casmed, USA and Nonin, USA have become clinically available. Although the later oximeters use additional wavelengths and claim to measure *absolute* cerebral oxygenation, confounders in the underlying technology will limit their ability to do this in all clinical situations. This is due to variations in anatomy, path length of the photons through varying depths and consistency of cerebral and extra cerebral tissues and the presence of oedema. It is generally considered that these monitors work best as *trend* monitors rather than absolute monitors [61]. Thus the importance of getting a pre-induction value in elective patients submitted to general anaesthesia.

Apart from those neurological conditions where brain oxygenation is obviously suspect, for example carotid endarterectomy [62], many operations on the elderly may lead to reductions in rSO_2 [63] and poor outcome [64]. The problem is to define suitable interventions if rSO_2 diminishes by more than about 10 % from starting value. A MMM strategy which maintains flow and DO_2 almost always insures that rSO_2 will remain at or above the pre-induction level [65]. One of the commonest reasons for desaturation in major non-cardiac surgery is inadvertent hyperventilation of the elderly patient. It is essential that arterial carbon dioxide is maintained at or marginally above the normal level. Another reason is that cerebral blood flow to maintain oxygenation cannot increase sufficiently in the elderly to maintain oxygenation if the oxygen content falls too low. This usually only occurs when haemoglobin concentration drops to below around 80 $g.L^{-1}$ and is virtually never seen if the haemoglobin is above 100 $g.L^{-1}$ thus unnecessary transfusion based simply on cerebral oxygenation values rather than the usual clinical criteria is rarely necessary.

Consideration of the role of cerebral oximetry in cardiac surgery where its clinical applicability is greatest is outside the scope of this chapter. However, the jury is out on its true role [66] but several studies have indicated possible benefits [67, 68].

My own experience with this monitoring technology amounts to nearly 1000 cases. These mainly include high-risk cases with an average ASA class of three or above, long duration where the build-up of oxygen debt will potentially be a problem and in the elderly population (average age 68) where oxygen debt is likely to cause most problems.

Figure 13.3a show patient haemodynamic data where basic parameters, including flow, were maintained at normal levels, despite a period of 1.5 L of blood loss. This occurred in a patient who was already modestly anaemic with a haemoglobin of approximately 100 $g.L^{-1}$. Figure 13.3b shows the cerebral oxygenation trace where during the period of blood loss and increasing anaemia saturation fell by about 25 % and only recovered when oxygen content was increased by the use of blood transfusion and the Hb level raised to 90 $g.L^{-1}$.

13.17.2 Using RSO_2 as a Surrogate of Adequacy of Oxygen Consumption and Delivery in the Intraoperative Period

A recent review covered the whole concept of tissue oxygenation measurement including cerebral and muscle oxygenation. The reviewers concluded that "measuring and obtaining adequate tissue oxygenation may prevent post-operative complications and may thus be cost-effective" [54]. The cerebral/tissue oximeter is thus a good monitor to detect supply/demand imbalances in cerebral/tissue oxygenation, especially in the elderly. In general, correction of desaturation seems to reduce complications versus controls and the important point is that conventional monitoring, even including flow, may not necessarily indicate problems. It is a key component of the MMM strategy.

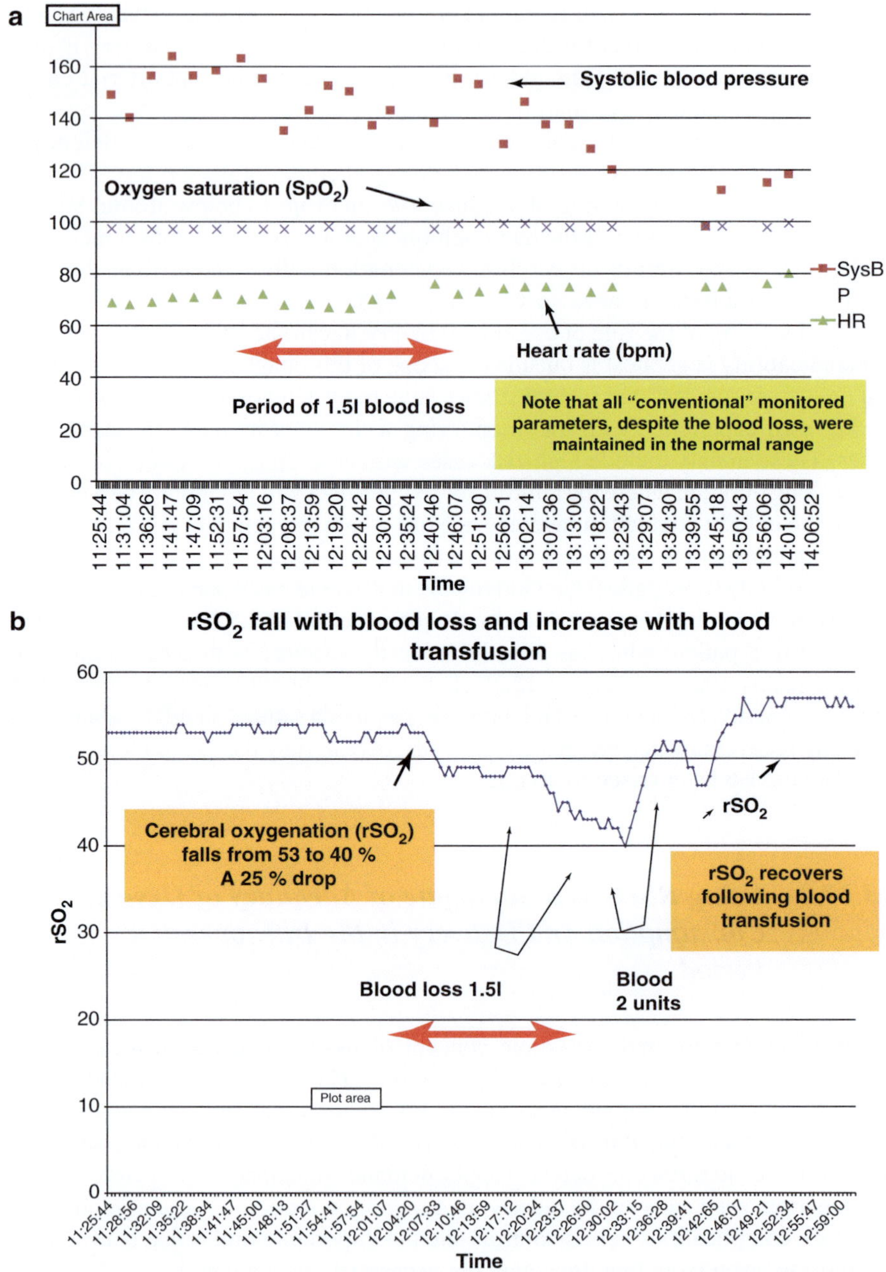

Fig. 13.3 (**a**) Shows that conventional parameters, systolic blood pressure, oxygen saturation and heart rate were maintained in the normal range with crystalloid and colloids infusion during a period of 1.5 l of blood loss in a high risk patient undergoing nephrectomy. All seems well. (**b**) This shows that a simultaneous time trace (**a**) of cerebral oxygenation (rSO_2) indicates a profound fall of 25 % despite maintenance of "conventional" parameters shown in (**a**). This could result in cerebral hypoxaemia and is due to excessive haemodilution and a fall in Hb to around 70 $g.L^{-1}$. Transfusion of two units of blood restored blood oxygen carrying capacity and thus cerebral oxygen delivery and the rSO_2 recovers

13.18 Depth of Anaesthesia Monitoring as Part of the MMM Strategy

13.18.1 Introduction

Although more than 50 million cases of anaesthesia and deep sedation have been monitored using depth of anaesthesia technology, mainly bispectral index (BIS, Medtronic, USA) uptake has been patchy. In a recent survey in the UK only around 2 % of anaesthetists use this technology [1] whereas in some countries it is nearly 70 % or more. However, it is now recommended by NICE in the UK for high-risk patients although the support for this move by anaesthetists in the UK is controversial [69, 70]!

13.18.2 What Do Depth of Anaesthesia Monitors Such as BIS Really Measure?

Considerable confusion surrounds the use of this monitoring technology. In a recent review of the impact of monitoring the brain with processed EEG (pEEG) and outcome several statements were made which reflect misunderstandings of what the monitor is measuring [71]. For example:

- *There should be a proportional relationship between increasing ET concentrations of inhalation agents and a BIS number.* This might be so in volunteers but is not so in the face of different amounts of surgical stimulation and degree of analgesia. After all, if the BIS is increasing, the ET concentration of anaesthetic will be increased by the practitioner. This will lead to the false assumption that there is no relationship between BIS and ET concentration (i.e. BIS *should go down* if ET concentration *goes up*)
- *There should be a "switch" value of BIS which reliably predicts the transition from unconsciousness to consciousness.* This is not the case but it does not invalidate the use of the monitor.
- *Using a BIS monitor should necessarily mean a reduction in anaesthetic agent usage.* Since most patients are too deeply anaesthetised if the BIS number is to be believed this is likely to be the case, but it is not necessarily so if some patients are too lightly anaesthetised. Most studies show a reduction in anaesthesia usage with BIS as, in general, most patients otherwise get too much anaesthetic.
- *BIS provides a reliable indication of degree of "cortical suppression" even in the face of multiple anaesthetic agents.* This is probably not the case as different anaesthetics affect cerebral metabolism, the EEG and thus degree of cortical suppression differently.

Many anaesthetists appear to believe that since BIS purports to be a monitor of "consciousness" and as we don't know what consciousness really is, it means that it simply doesn't or can't work and is therefore useless! However, once we appreciate what the BIS number means, the monitor becomes crucial in the MMM strategy.

Several studies have shown EEG and BIS correlations with brain glucose metabolic rate during anaesthesia in volunteers. There is a clear link between cerebral metabolism, the effect on the EEG and the number produced by the BIS monitor. As the amount of anaesthetic with halothane, isoflurane or propofol was increased the EEG changed, the BIS number decreased and this was paralleled by a reduction in uptake of brain glucose, indicating decreased cerebral metabolic rate of between 30 and 50 % depending on the anaesthetic at the time of unresponsiveness [52, 53]. Providing cortical suppression is adequate and the BIS number is around 40–50 and maintained at that level in spite of surgical stimulation, there is a negligible risk that the patient will have any explicit recall in the post-operative period. However, this does not mean that one can therefore simply avoid explicit recall by giving every patient a "deep anaesthetic". NICE recognised that some patients are considered at high risk of excessively deep levels of anaesthesia and include older patients, patients with a high body mass index and patients with poor cardiovascular function [70]. Since the dawn of anaesthesia in 1846 it has been recognised that excessive depth of anaesthesia will lead to cardiovascular depression, and possibly poor outcome.

13.18.3 Why Monitoring Depth of Anaesthesia Is So Important with High Risk, Elderly Patients

Figure 13.4 indicates why monitoring the effect of the anaesthetic on cortical suppression is important for high-risk patients and may not be as important for fit and healthy patients. Looking along the x-axis from left to right, increasing anaesthetic concentration will inevitably lead to cardiovascular depression and decreased DO_2. This obviously has implications for build- up of oxygen debt, complications in the post-operative period and poor outcome. It also is likely to produce neuronal toxicity as the concentration of anaesthetic increases [72]. On the other hand, moving from right to left along the x-axis, too little anaesthetic obviously gives one a risk of patient explicit recall and also we may lose the benefits of anaesthetic neuronal protection. So, there is a balance to be struck. The red curve indicates a low risk patient where the target anaesthetic concentration is rather broad and the implications of misjudging the depth of anaesthesia (as indicated by the shallowness of the curve) means that not too much harm will come to the patient.

This is not the same situation in the elderly or high-risk patient in whom it is usually assumed that the amount of anaesthetic required is less and too light anaesthesia will lead to the patient waking up and moving around whereas a slight increase to deepen the anaesthetic may result in profound cardiovascular depression. All anaesthetists have observed this in their practice! However, 10 years experience with the BIS monitor in high-risk patients, especially the elderly, have shown that some older patients require much more anaesthetic than others and indeed there is up to a fourfold variation in propofol anaesthesia requirement [65]. This variation cannot be predicted by simple clinical signs alone.

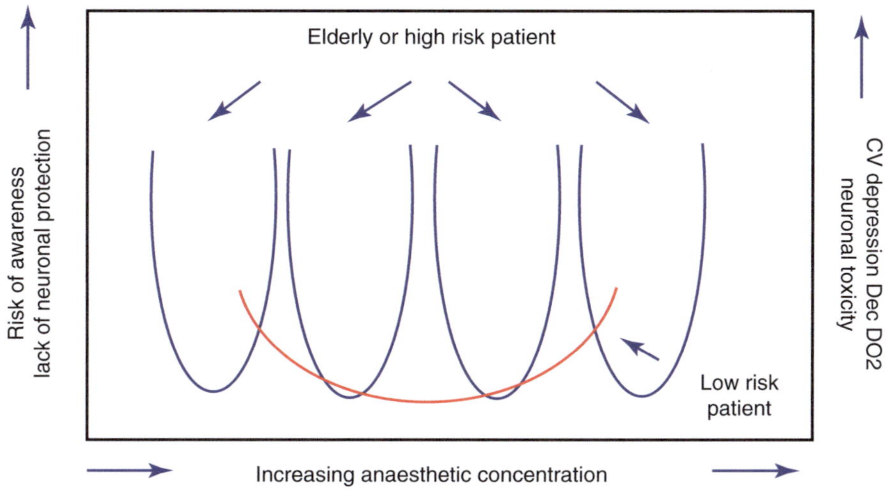

Fig. 13.4 The x axis indicates that increasing anaesthetic concentration from low to high has an effect on the risk of awareness and lack of neuronal protection if the concentration is too low (*left* y axis) or on the other hand excessive cardiovascular depression and risk of neuronal toxicity in the concentration is too high (*right* y axis). The low-risk patient is indicated by the red line which suggests that there is a wide margin of safety when choosing a suitable anaesthetic concentration and the risk of harm is minimal. This is indicated by the shallowness of the curve. On the other hand, with an elderly or high-risk patient, the implications for getting it wrong as indicated by the blue curves are much greater (*steeper curves*) and there is a much narrower margin of safety. Conventional wisdom suggests that most elderly or high-risk patients need very little anaesthesia as indicated by the left-hand blue curve. However, there is a wide spectrum of anaesthesia requirements in the elderly or high-risk patient which means that they may require even more anaesthetic than the low-risk patient. Use of a cortical suppression monitor allows us to adjust for these differences and administer the correct amount of anaesthesia for the individual high risk patient (Reproduced with permission from Bidd et al. [60])

13.18.4 Evidence that Excessive Depth of Anaesthesia is Harmful

Evidence from the large-scale randomised "CODA" trial showed that BIS guided anaesthesia decreases post-operative delirium and cognitive decline (POCD). This good outcome was achieved as the intervention group average BIS was 53 versus 36 in the control group. For every 1000 elderly patients, using BIS guided monitoring would prevent 83 cases of delirium and 23 cases of POCD. Since delirium is associated with poor immediate outcome and POCD is also associated with increased mortality and increased reliance on social services in the later post-operative period [73] then this result shows the benefit of using BIS technology [74]. However, a similar large trial from Berlin was unable to show a difference in POCD but a significant reduction in delirium. Unfortunately, the investigators were unable to sustain a difference in mean BIS readings between the two groups. For an intervention to be effective one must be able to show a difference between the endpoint i.e. BIS reading obtained using technology versus that obtained without it [75]. It was also interesting to note that there was a 5 % mortality in both groups of high-risk, elderly patients in this large study.

13.18.5 Use of Combined BIS and Cerebral Oximetry to Reduce the Incidence of POCD

In a cohort study and nested randomised controlled trial of high risk elderly patients undergoing major abdominal and orthopaedic surgery, using combined technology of BIS and cerebral oximetry as interventions, investigators were able to show a profound reduction in the intervention group in POCD [76]. Of interest was the fact that maintaining the BIS in the "normal" range (40–60) i.e. not too deep and not too light led to significantly lower levels of S100 B, an indicator of neuronal damage, in the intervention group.

13.19 Should We Use a MMM Strategy in High-Risk Patients?

One of the highest risk patient categories that are dealt with on a routine basis are those elderly patients who suffer a proximal femoral fracture. Management and monitoring guidelines were produced by the Association of Anaesthetists of Great Britain and Ireland in 2012 for dealing with this high-risk group where post-operative mortality in some centres approaches 25 %. They suggest that monitoring of cardiac output, depth of anaesthesia and cerebral oxygenation should be considered alongside a basic monitoring setup in this high-risk group [77].

13.20 Using a Multimodal Monitoring (MMM) Strategy in Clinical Practice

13.20.1 Getting Baseline Readings Pre-induction

It would be rare for anaesthetists to begin anaesthetising a patient in the twenty-first century without having routine monitoring in place. This would include baseline oxygen saturation (SpO2), ECG and non-invasive BP. Having taken baseline readings anaesthesia would then commence. Using a MMM strategy it is also essential to have baseline readings of the three major parameters measured i.e. SV and CO from the flow monitor, baseline indication of cortical suppression (BIS) and cerebral oxygenation. This means that in major cases a radial artery line is placed prior to induction and connected up to a reliable CO monitor such as the LiDCO rapid. LiDCO technology of monitoring pulse power rather than pulse contour is in my opinion the most accurate from the pressure signal derived from the radial artery. The PulseCo algorithm has remained unchanged since 2000. If a femoral artery catheter is preferred, then PiCCO or EV1000 (Edwards Lab., USA) technology can be used. If a central venous line is considered necessary for the procedure, remembering that central venous pressure has no relevance in managing fluid therapy as shown by recent meta-analyses [12, 78, 79], then the PiCCO or EV 1000 can be calibrated by trans-pulmonary thermodilution. However, in elective cases a baseline nominal cardiac output (nCO) using the LiDCO algorithm scaled up by using the height, weight and age of the patient is sufficient. Together with baseline Hb and SpO_2 the nominal DO_2 prior to induction can be calculated easily. It is this value that should be maintained within about 10–15 % of baseline throughout the procedure to avoid significant build-up of oxygen debt. However, it is somewhat surprising that the majority of trials of fluid management using flow monitoring in recent years have not considered this baseline value as important. The potential problems of using a SV maximisation or GDT approach have already been considered.

13.20.2 What Are the Benefits of Measuring Cardiac Output and Oxygen Delivery From Pre-induction?

Modern perioperative care should be individualised and in the elective patient the individualisation involves maintenance of pre-induction haemodynamic values. This is well indicated in Fig. 13.5, which shows CO monitoring commenced pre-induction, and induction occurring at mark 2. Mark 3 indicates that the fall in CO post induction reached nearly 50 %. If CO monitoring had commenced at this time it would have massively underestimated the starting value. Since in this case CO monitoring began pre-induction and was then maintained at near the pre-induction level, a significant build-up of oxygen debt was avoided. Obtaining and maintaining the pre-induction CO and then DO_2 level throughout the procedure is a key part of the MMM strategy.

Comparison of fluid requirement during major surgery using the MMM strategy has shown marked reductions in fluid requirement in comparison to the three major trials mentioned earlier on which the Doppler recommendations were made. For example, in high risk vascular surgery, fluid requirement per hour was only about 5 $ml.kg^{-1}.hr^{-1}$ versus about 20–30 ml. per $kg.hr^{-1}$ in the trials mentioned [65].

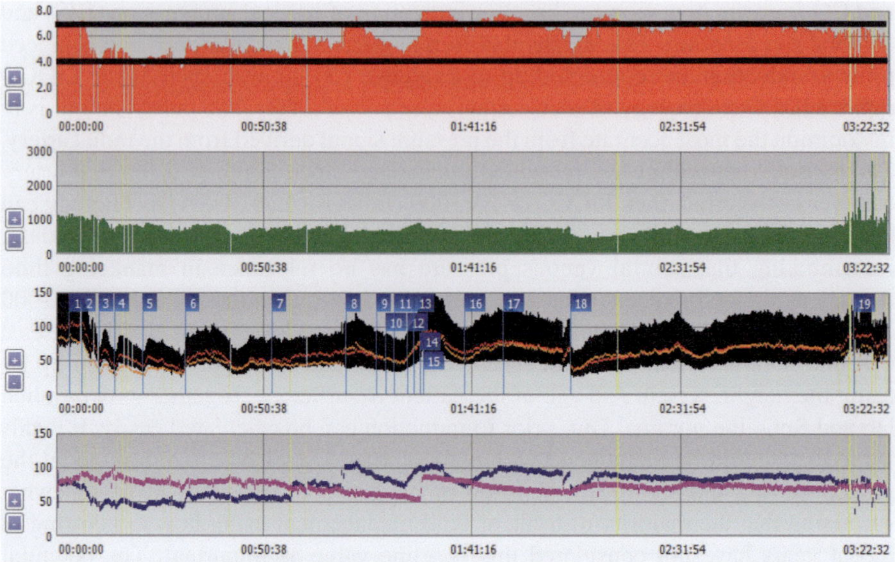

Fig. 13.5 This LiDCOviewpro trace shows the value of measuring cardiac output from pre-induction. As in Fig. 13.1, the top horizontal trace in red a indicates cardiac output. The third horizontal trace indicates blood pressure. 2 indicates induction with propofol and 3 intubation. Pre-induction nominal cardiac output was 7 and post induction around 4 lpm. If cardiac output monitoring had commenced at 3 and the latter value had been used as a baseline then one can see that the difference between actual cardiac output achieved versus the post induction baseline would have been considerable and would have resulted in a buildup of oxygen debt due to reduced oxygen delivery (horizontal *black lines* indicate pre-induction and post-induction cardiac output values)

13.21 Does the MMM Strategy Make a Difference to Outcome in High-Risk Surgery

In an observational case series in 120 very high-risk patients undergoing major peripheral vascular surgery, predicted 30-day mortality from similar series and also from calculation of the V-possum scores was estimated to be between 5 and 10 %. However, the 30 day mortality using the MMM strategy was only 0.8 %. In addition, nominal CO and rSO_2 were maintained at 87 % or above the baseline value. The major amputation rate at 1 year was only 2 % and only 8 % of patients were admitted to the high dependency or intensive care units [65]. This suggests benefits using the MMM strategy.

13.22 The MMM Strategy in Practice. Using the Pre Vent™ Protocol and Decision Matrix

This protocol is now routinely employed in these high-risk patients and is outlined in Fig. 13.6. Management strategy during the case is indicated by a four quadrant box with the "normal" box (Green) in the top right-hand corner. This is where MAP and CO are within 10–15 % of pre-induction values. Suitable strategies are employed for the amber boxes where either MAP or CO are below these limits (see below). In all cases, it is absolutely essential to pay attention to the degree of cortical suppression as indicated by the depth of anaesthesia monitor. For example, treating a low MAP or CO by administration of fluids or vasoactive agents is not attempted if the BIS reading is around 20–30. Unless it is in an emergency, it is much more rational to lighten the anaesthesia first before commencing fluid and vasoactive drug treatment.

If both CO and MAP are satisfactory (Green Box) then no intervention is necessary and the only additional fluids given are those for basic maintenance such as 1–2 $ml.kg^{-1}.hr^{-1}$ of 0.18 % NaCl in 4 % glucose.

If CO is at target but MAP is lower than required (Lower Amber box) then increasing the dose or commencing infusion of phenylephrine (0.25–0.5 $ug.kg^{-1}.hr^{-1}$) is all that is usually required. If on the other hand, the MAP is within acceptable limits but the CO is below target (upper Amber box) then SVV, if available, is used to guide therapy. If it is greater than 10–13 % then a fluid challenge is given using 4–5 $ml.kg^{-1}$ of a colloid solution followed by phenylephrine infusion of 0.25–0.5 $ug.kg^{-1}.hr^{-1}$ if fluid does not produce the desired response. If on the other hand, the SVV is low then there is no point in administering extra fluid and ephedrine then phenylephrine would be employed.

If CO and MAP are both below target (Red box) then both need to be restored to the pre-induction value. Here again, SVV, if available, is useful. If it is high, phenylephrine is started or the infusion increased to induce venoconstriction, reduce stroke volume variation and also raise MAP and cardiac output. Additional colloid fluid challenges can also be given. If on the other hand the SVV is low then additional fluids and phenylephrine are usually ineffective in increasing CO. In this case ephedrine is employed and in the most severe cases inotropes will be started such as dobutamine or noradrenaline. However, this is rarely necessary. The great advantage of MMM is that you can see if the results of intervention are beneficial.

This strategy, especially if the venoconstrictor phenylephrine is commenced pre-induction, makes management in the perioperative period much simpler and consistent. It also avoids excessive amounts of fluid and Na^+ which might otherwise damage the EGL, reduce albumin concentration and actually precipitate third space loss.

Clearly an observational case series is not sufficient evidence for widespread implementation of this strategy. However, unlike other management strategies, it does adhere to basic physiological principles [27].

It is crucial that future randomised controlled trials using flow monitoring should use the approach mentioned above with efforts made to have appropriate levels of cortical suppression and maintain pre-induction values of CO and DO_2 to within 10–15 % of baseline. The technique of SV maximisation and GDT to a population-based target of 600 $ml.m^{-2}$ BSA should be abandoned. In addition, future trials should recognise that third space loss does not occur in elective major surgery and thus patients should only receive minimal maintenance fluid in addition to replacement of obvious volume losses, a point recognised in the *Optimise* trial protocol [47].

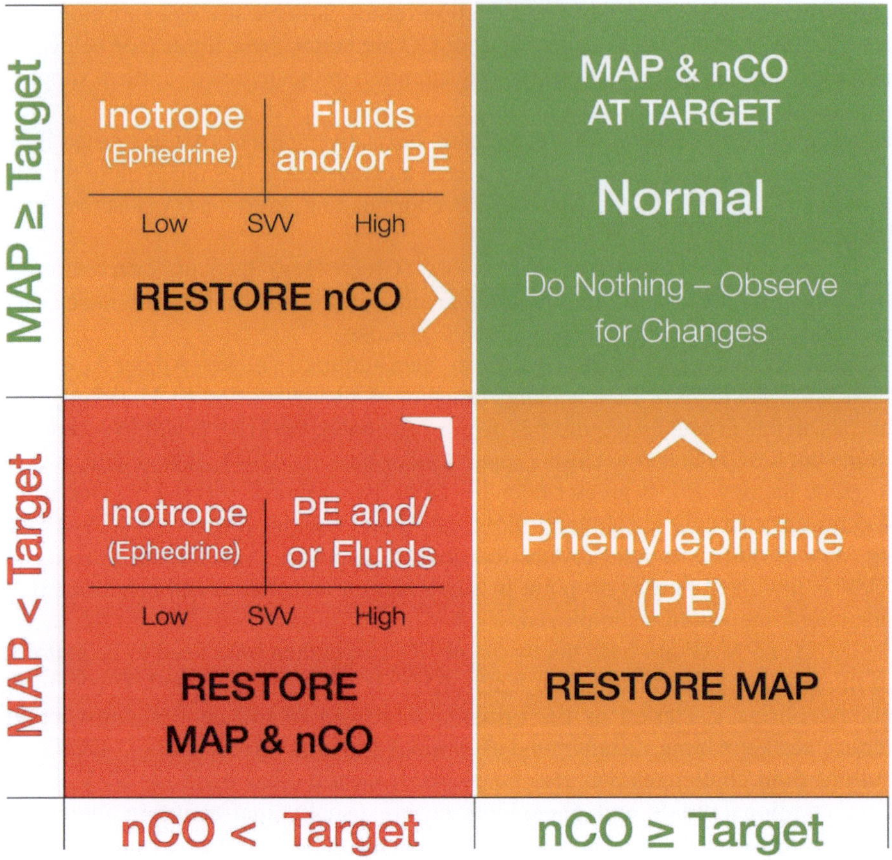

Fig. 13.6 The Pre Vent™ protocol for intraoperative management of the high risk patient. For explanation, see text. Copyright LiDCO PLC London, UK (Reproduced with permission from LiDCO's)

13.23 Summary

MMM has advantages for the anaesthetist, the patient and the healthcare system. It pinpoints very clearly the physiological changes associated with anaesthesia and surgery from pre-induction to post anaesthesia care unit. It allows intervention strategies such as Pre-Vent to be more focused and physiologically appropriate and allows a more rational approach to intraoperative haemodynamic management. As we get greater insight into intraoperative physiological change in our patients we can individualise management using strategies designed and centred on minimising the build-up of oxygen debt and thus this should improve outcome. The burgeoning of the elderly high-risk patient population mandates a new perioperative anaesthetic management strategy alongside ERAS. It would appear that MMM reduces the requirement for expensive HDU and ICU facilities and may decrease hospital length of stay and therefore reduce the cost of perioperative care by reducing patient morbidity and mortality

13.24 The Future

It must become routine practice to use a MMM strategy to try to maintain perioperative oxygen delivery to minimise build-up of oxygen debt as a key for successful outcomes in high-risk surgical patients. This is achievable now with the monitoring technology available. The recent arrival of finger-based non-invasive and continuous blood pressure monitoring [80] with provision for converting the waveform into flow means that in the future all high risk patients could benefit from this strategy.

References

1. Sury MR, Palmer JH, Cook TM, Pandit JJ (2014) The state of UK anaesthesia: a survey of National Health Service activity in 2013. Br J Anaesth 113:575–584
2. Pearse RM, Moreno RP, Bauer P, Pelosi P, Metnitz P, Spies C, Vallet B, Vincent JL, Hoeft A, Rhodes A (2012) Mortality after surgery in Europe: a 7 day cohort study. Lancet 380:1059–1065
3. Lees N, Hamilton M, Rhodes A (2009) Clinical review: goal-directed therapy in high risk surgical patients. Crit Care 13:231
4. Boddaert J, Raux M, Khiami F, Riou B (2014) Perioperative management of elderly patients with hip fracture. Anesthesiology 121:1336–1341
5. Story DA, Leslie K, Myles PS, Fink M, Poustie SJ, Forbes A, Yap S, Beavis V, Kerridge R (2010) Complications and mortality in older surgical patients in Australia and New Zealand (the REASON study): a multicentre, prospective, observational study*. Anaesthesia 65:1022–1030
6. Kirov MY, Kuzkov VV, Molnar Z (2010) Perioperative haemodynamic therapy. Curr Opin Crit Care 16:384–392
7. Cannesson M, Pestel G, Ricks C, Hoeft A, Perel A (2011) Hemodynamic monitoring and management in patients undergoing high risk surgery: a survey among North American and European anesthesiologists. Crit Care 15:R197
8. Walsh M, Devereaux PJ, Garg AX, Kurz A, Turan A, Rodseth RN, Cywinski J, Thabane L, Sessler DI (2013) Relationship between intraoperative mean arterial pressure and clinical outcomes after noncardiac surgery: toward an empirical definition of hypotension. Anesthesiology 119:507–515
9. Meng L, Cannesson M, Alexander BS, Yu Z, Kain ZN, Cerussi AE, Tromberg BJ, Mantulin WW (2011) Effect of phenylephrine and ephedrine bolus treatment on cerebral oxygenation in anaesthetized patients. Br J Anaesth 107:209–217
10. Shires T, Williams J, Brown F (1961) Acute change in extracellular fluids associated with major surgical procedures. Ann Surg 154:803–810
11. Moore FD, Shires G (1967) Moderation. Ann Surg 166:300–301
12. Marik PE (2014) Iatrogenic salt water drowning and the hazards of a high central venous pressure. Ann Intens care 4:21
13. Brandstrup B, Svensen C, Engquist A (2006) Hemorrhage and operation cause a contraction of the extracellular space needing replacement--evidence and implications? A systematic review. Surgery 139:419–432
14. Chappell D, Jacob M, Hofmann-Kiefer K, Conzen P, Rehm M (2008) A rational approach to perioperative fluid management. Anesthesiology 109:723–740
15. Jacob M, Chappell D, Rehm M (2009) The 'third space'--fact or fiction? Best practice & research. Clini Anaes 23:145–157

16. Wuethrich PY, Burkhard FC, Thalmann GN, Stueber F, Studer UE (2014) Restrictive deferred hydration combined with preemptive norepinephrine infusion during radical cystectomy reduces postoperative complications and hospitalization time: a randomized clinical trial. Anesthesiology 120(2):365–377
17. Slagt C, Malagon I, Groeneveld AB (2014) Systematic review of uncalibrated arterial pressure waveform analysis to determine cardiac output and stroke volume variation. Br J Anaesth 112:626–637
18. Heilbrunn A, Allbritten FF Jr (1960) Cardiac output during the following surgical operations. Ann Surg 152:197–210
19. Moller Petrun A, Kamenik M (2013) Bispectral index-guided induction of general anaesthesia in patients undergoing major abdominal surgery using propofol or etomidate: a double-blind, randomized, clinical trial. Br J Anaesth 110:388–396
20. Sharpey-Schafer EP (1961) Venous tone. Br Med J 2:1589–1595
21. Sharpey-Schafer EP (1963) Venous tone: effects of reflex changes, humoral agents and exercise. Br Med Bull 19:145–148
22. Sharpey-Schafer EP, Ginsburg J (1962) Humoral agents and venous tone. Effects of catecholamines, 5-hydroxytryptamine, histamine, and nitritex. Lancet 2:1337–1340
23. Paton WD, Zaimis EJ (1949) The pharmacological actions of polymethylene bistrimethylammonium salts. Br J Pharmacol Chemother 4:381–400
24. Enderby GE (1975) Guest editorial. Some observations on the practice of deliberate hypotension. Br J Anaesth 47:743–744
25. Bentley GN, Gent JP, Goodchild CS (1989) Vascular effects of propofol: smooth muscle relaxation in isolated veins and arteries. J Pharm Pharmacol 41:797–798
26. Goodchild CS, Serrao JM (1989) Cardiovascular effects of propofol in the anaesthetized dog. Br J Anaesth 63:87–92
27. Wolff CB, Green DW (2014) Clarification of the circulatory patho-physiology of anaesthesia – implications for high-risk surgical patients. Int J Surg 12:1348–1356
28. Green DW (2015) Cardiac output decrease and propofol: what is the mechanism? Br J Anaesth 114:163–164
29. Sandham JD, Hull RD, Brant RF, Knox L, Pineo GF, Doig CJ, Laporta DP, Viner S, Passerini L, Devitt H, Kirby A, Jacka M (2003) A randomized, controlled trial of the use of pulmonary-artery catheters in high-risk surgical patients. N Engl J Med 348:5–14
30. Stein MS, Barratt SM, Purcell GJ (1991) Intraoperative assessment of the Lawrence 3000 Doppler cardiac output monitor. Anaesth Intensive Care 19:251–255
31. Gan TJ, Soppitt A, Maroof M, el-Moalem H, Robertson KM, Moretti E, Dwane P, Glass PS (2002) Goal-directed intraoperative fluid administration reduces length of hospital stay after major surgery. Anesthesiology 97:820–826
32. Wakeling HG, McFall MR, Jenkins CS, Woods WG, Miles WF, Barclay GR, Fleming SC (2005) Intraoperative oesophageal Doppler guided fluid management shortens postoperative hospital stay after major bowel surgery. Br J Anaesth 95:634–642
33. Noblett SE, Snowden CP, Shenton BK, Horgan AF (2006) Randomized clinical trial assessing the effect of Doppler-optimized fluid management on outcome after elective colorectal resection. Br J Surg 93:1069–1076
34. National Institute for Health and Clinical Excellence (2011) Medical technologies guidance MTG3: CardioQODM oesophageal doppler monitor. http://www.nice.org.uk/MTG3
35. Ghosh S, Arthur B, Klein AA (2011) NICE guidance on CardioQ(TM) oesophageal Doppler monitoring. Anaesthesia 66:1081–1083
36. Singer M (2012) Oesophageal Doppler monitoring: a not-so-NICE editorial. Anaesthesia 67:428–430; author reply 431–432
37. Pearse R, Dawson D, Fawcett J, Rhodes A, Grounds RM, Bennett ED (2005) Early goal-directed therapy after major surgery reduces complications and duration of hospital stay, a randomised, controlled trial [ISRCTN38797445]. Crit Care 9:R687–R693
38. Lobo SM, Ronchi LS, Oliveira NE, Brandao PG, Froes A, Cunrath GS, Nishiyama KG, Netinho JG, Lobo FR (2011) Restrictive strategy of intraoperative fluid maintenance during optimization of oxygen delivery decreases major complications after high-risk surgery. Crit Care 15:R226

39. Cannesson M, Le Manach Y, Hofer CK, Goarin JP, Lehot JJ, Vallet B, Tavernier B (2011) Assessing the diagnostic accuracy of pulse pressure variations for the prediction of fluid responsiveness: a "Gray Zone" approach. Anesthesiology 115:231–241
40. Willars C, Dada A, Hughes T, Green D (2012) Functional haemodynamic monitoring: the value of SVV as measured by the LiDCORapid in predicting fluid responsiveness in high risk vascular surgical patients. Int J Surg 10:148–152
41. Brandstrup B, Svendsen PE, Rasmussen M, Belhage B, Rodt SA, Hansen B, Moller DR, Lundbech LB, Andersen N, Berg V, Thomassen N, Andersen ST, Simonsen L (2012) Which goal for fluid therapy during colorectal surgery is followed by the best outcome: near-maximal stroke volume or zero fluid balance? Br J Anaesth 109:191–199
42. Levick JR, Michel CC (2010) Microvascular fluid exchange and the revised Starling principle. Cardiovasc Res 87:198–210
43. Woodcock TE, Woodcock TM (2012) Revised Starling equation and the glycocalyx model of transvascular fluid exchange: an improved paradigm for prescribing intravenous fluid therapy. Br J Anaesth 108:384–394
44. Alphonsus CS, Rodseth RN (2014) The endothelial glycocalyx: a review of the vascular barrier. Anaesthesia 69:777–784
45. Wolff CB (2007) Normal cardiac output, oxygen delivery and oxygen extraction. Adv Exp Med Biol 599:169–182
46. Grocott MP, Dushianthan A, Hamilton MA, Mythen MG, Harrison D, Rowan K (2013) Perioperative increase in global blood flow to explicit defined goals and outcomes after surgery: a Cochrane Systematic Review. Br J Anaesth 111:535–548
47. Pearse RM, Harrison DA, MacDonald N, Gillies MA, Blunt M, Ackland G, Grocott MP, Ahern A, Griggs K, Scott R, Hinds C, Rowan K (2014) Effect of a perioperative, cardiac output-guided hemodynamic therapy algorithm on outcomes following major gastrointestinal surgery: a randomized clinical trial and systematic review. JAMA 311:2181–2190
48. Challand C, Struthers R, Sneyd JR, Erasmus PD, Mellor N, Hosie KB, Minto G (2012) Randomized controlled trial of intraoperative goal-directed fluid therapy in aerobically fit and unfit patients having major colorectal surgery. Br J Anaesth 108:53–62
49. Minto G, Struthers R (2014) Stroke volume optimisation: is the fairy tale over? Anaesthesia 69:291–296
50. Hamilton MA, Cecconi M, Rhodes A (2011) A systematic review and meta-analysis on the use of preemptive hemodynamic intervention to improve postoperative outcomes in moderate and high-risk surgical patients. Anesth Analg 112:1392–1402
51. Kehlet H, Wilmore DW (2008) Evidence-based surgical care and the evolution of fast-track surgery. Ann Surg 248:189–198
52. Alkire MT (1998) Quantitative EEG correlations with brain glucose metabolic rate during anesthesia in volunteers. Anesthesiology 89:323–333
53. Alkire MT, Pomfrett CJ, Haier RJ, Gianzero MV, Chan CM, Jacobsen BP, Fallon JH (1999) Functional brain imaging during anesthesia in humans: effects of halothane on global and regional cerebral glucose metabolism. Anesthesiology 90:701–709
54. Scheeren TW, Schober P, Schwarte LA (2012) Monitoring tissue oxygenation by near infrared spectroscopy (NIRS): background and current applications. J Clin Monit Comput 26:279–287
55. Shoemaker WC, Appel PL, Kram HB (1988) Tissue oxygen debt as a determinant of lethal and nonlethal postoperative organ failure. Crit Care Med 16:1117–1120
56. Shoemaker WC, Appel PL, Kram HB (1992) Role of oxygen debt in the development of organ failure sepsis, and death in high-risk surgical patients. Chest 102:208–215
57. Shoemaker WC, Appel PL, Kram HB (1993) Hemodynamic and oxygen transport responses in survivors and nonsurvivors of high-risk surgery. Crit Care Med 21:977–990
58. Shoemaker WC, Thangathurai D, Wo CC, Kuchta K, Canas M, Sullivan MJ, Farlo J, Roffey P, Zellman V, Katz RL (1999) Intraoperative evaluation of tissue perfusion in high-risk patients by invasive and noninvasive hemodynamic monitoring. Crit Care Med 27:2147–2152
59. Green D, Paklet L (2010) Latest developments in peri-operative monitoring of the high-risk major surgery patient. Int J Surg 8:90–99
60. Bidd H, Tann A, Green DW (2013) Using bispectral index and cerebral oximetry to guide hemodynamic therapy in high-risk surgical patients. Perioperative Med 2:1–9

61. Murkin JM, Arango M (2009) Near-infrared spectroscopy as an index of brain and tissue oxygenation. Br J Anaesth 103(Suppl 1):i3–i13
62. Ritter JC, Green D, Slim H, Tiwari A, Brown J, Rashid H (2011) The role of cerebral oximetry in combination with awake testing in patients undergoing carotid endarterectomy under local anaesthesia. Eur J Vasc Endovasc Surg 41:599–605
63. Green DW (2007) A retrospective study of changes in cerebral oxygenation using a cerebral oximeter in older patients undergoing prolonged major abdominal surgery. Eur J Anaesthesiol 24:230–234
64. Casati A, Fanelli G, Pietropaoli P, Proietti R, Tufano R, Danelli G, Fierro G, De Cosmo G, Servillo G (2005) Continuous monitoring of cerebral oxygen saturation in elderly patients undergoing major abdominal surgery minimizes brain exposure to potential hypoxia. Anesth Analg 101:740–747
65. Green D, Bidd H, Rashid H (2014) Multimodal intraoperative monitoring: an observational case series in high risk patients undergoing major peripheral vascular surgery. Int J Surg 12:231–236
66. Zheng F, Sheinberg R, Yee MS, Ono M, Zheng Y, Hogue CW (2013) Cerebral near-infrared spectroscopy monitoring and neurologic outcomes in adult cardiac surgery patients: a systematic review. Anesth Analg 116:663–676
67. Murkin JM, Adams SJ, Novick RJ, Quantz M, Bainbridge D, Iglesias I, Cleland A, Schaefer B, Irwin B, Fox S (2007) Monitoring brain oxygen saturation during coronary bypass surgery: a randomized, prospective study. Anesth Analg 104:51–58
68. Slater JP, Guarino T, Stack J, Vinod K, Bustami RT, Brown JM 3rd, Rodriguez AL, Magovern CJ, Zaubler T, Freundlich K, Parr GV (2009) Cerebral oxygen desaturation predicts cognitive decline and longer hospital stay after cardiac surgery. Ann Thorac Surg 87:36–44; discussion 44–45
69. Pandit JJ, Cook TM (2013) I. National Institute for Clinical Excellence guidance on measuring depth of anaesthesia: limitations of EEG-based technology. Br J Anaesth 110:325–328
70. Shepherd J, Jones J, Frampton G, Bryant J, Baxter L, Cooper K (2012) Depth of anaesthesia monitoring (E-Entropy, Bispectral Index and Narcotrend) Health Technology Assessment
71. Escallier KE, Nadelson MR, Zhou D, Avidan MS (2014) Monitoring the brain: processed electroencephalogram and peri-operative outcomes. Anaesthesia 69(8):899–910
72. Jevtovic-Todorovic V, Absalom AR, Blomgren K, Brambrink A, Crosby G, Culley DJ, Fiskum G, Giffard RG, Herold KF, Loepke AW, Ma D, Orser BA, Planel E, Slikker W Jr, Soriano SG, Stratmann G, Vutskits L, Xie Z, Hemmings HC Jr (2013) Anaesthetic neurotoxicity and neuroplasticity: an expert group report and statement based on the BJA Salzburg Seminar. Br J Anaesth 111:143–151
73. Steinmetz J, Christensen KB, Lund T, Lohse N, Rasmussen LS (2009) Long-term consequences of postoperative cognitive dysfunction. Anesthesiology 110:548–555
74. Chan MT, Cheng BC, Lee TM, Gin T (2013) BIS-guided anesthesia decreases postoperative delirium and cognitive decline. J Neurosurg Anesthesiol 25:33–42
75. Radtke FM, Franck M, Lendner J, Kruger S, Wernecke KD, Spies CD (2013) Monitoring depth of anaesthesia in a randomized trial decreases the rate of postoperative delirium but not postoperative cognitive dysfunction. Br J Anaesth 110(Suppl 1):i98–i105
76. Ballard C, Jones E, Gauge N, Aarsland D, Nilsen OB, Saxby BK, Lowery D, Corbett A, Wesnes K, Katsaiti E, Arden J, Amaoko D, Prophet N, Purushothaman B, Green D (2012) Optimised anaesthesia to reduce Post Operative Cognitive Decline (POCD) in older Patients undergoing elective surgery, a randomised controlled trial. PLoS One 7:e37410
77. Griffiths R, Alper J, Beckingsale A, Goldhill D, Heyburn G, Holloway J, Leaper E, Parker M, Ridgway S, White S, Wiese M, Wilson I (2012) Management of proximal femoral fractures 2011: Association of Anaesthetists of Great Britain and Ireland. Anaesthesia 67:85–98
78. Marik PE, Baram M, Vahid B (2008) Does central venous pressure predict fluid responsiveness? A systematic review of the literature and the tale of seven mares. Chest 134:172–178
79. Marik PE, Cavallazzi R (2013) Does the central venous pressure predict fluid responsiveness? An updated meta-analysis and a plea for some common sense*. Crit Care Med 41:1774–1781
80. Saugel B, Reuter DA (2014) III. Are we ready for the age of non-invasive haemodynamic monitoring? Br J Anaesth 113:340–343

Chapter 14
Can Perioperative Interventions During Cancer Surgery Affect Recurrence or Metastasis?

Dalia Abdelrahman and Donal J. Buggy

Abstract Cancer continues to be a key cause of morbidity and mortality worldwide and its overall incidence continues to increase. Anaesthetists are increasingly faced with the challenge of managing cancer patients, either for surgical resection to debulk or excise the primary tumour, surgical emergencies in patients on chemotherapy, or for the analgesic management of acute or chronic pain. Metastatic recurrence is usually the cause of death in cancer patients. There is some experimental data suggesting that surgery, some general anaesthetics, acute postoperative pain and opioid analgesics may accelerate tumour growth and potentially increase the risk of metastatic recurrence. It has been hypothesized that awareness of perioperative factors promoting metastasis could enable designing anaesthetic and analgesic techniques that might attenuate these factors and reduce recurrence or metastasis. Evaluation of emerging literature indicates a signal from cell culture models of cancer cell growth supporting aspects of this hypothesis. Retrospective clinical studies of the association between various anaesthetic techniques, including regional anaesthesia, have yielded conflicting results. Currently, the most compelling evidence is for a direct effect of amide local anaesthetics on cancer cell metastatic potential, inhibiting migration by both a sodium channel and intracellular signaling mechanisms. Only prospective, randomized, clinical trials can prove a cause-and-effect link between anaesthetic technique and cancer. Although a number are ongoing, they are necessitate prolonged patient follow-up and large patient numbers, but their results will inform best practice in the perioperative care of cancer patients.

Keywords Anaesthesia, general • Anaesthesia regional • Local anaesthetics • Cancer, metastasis • Cancer, recurrence

---

D. Abdelrahman, MRCSI, EDRA, FCAI
D.J. Buggy, MD, MSc, DME, FRCPI, FCAI, FRCA (✉)
Department of Anaesthesia, Mater University Hospital, School of Medicine & Medical Science, University College Dublin, Dublin, Ireland

Outcomes Research, Cleveland Clinic, Ohio, USA
e-mail: donal.buggy@ucd.ie

K. Stuart-Smith (ed.), *Perioperative Medicine – Current Controversies*,
DOI 10.1007/978-3-319-28821-5_14

14.1 Introduction

Cancer is among the leading causes of morbidity and mortality worldwide, with approximately 14 million new cases and 8.2 million cancer related deaths in 2012 [1]. Most mortality from cancer is attributable to recurrence or metastatic spread to other organs, rather than direct effects of the primary tumour itself. The mechanism of metastasis remains the subject of ongoing research, and is beyond the scope of the present review, but the "seed and soil" hypothesis outlines a plausible mechanism by which tumour cells may be dispersed and thrive to become the focus of subsequent metastatic disease [2].

In 2006 it was postulated that anaesthetic and perioperative interventions during primary cancer surgery might attenuate the risk of recurrence or metastasis [15] This chapter will summarise the rationale by which different anaesthetic and analgesic techniques and other perioperative interventions might influence the risk of metastasis either in a detrimental or beneficial manner, and the evidence base underpinning it.

14.2 Understanding Cancer Metastasis

Carcinogenesis is caused by mutation, whereby normal cells are reprogrammed to undergo uncontrolled cell division, upsetting the balance between cell proliferation and programmed death (apoptosis). Due to high metabolic activity, cancer cells must establish their own vascular supply, a process termed angiogenesis. This requires the release of pro-angiogenic factors such as prostaglandin (PGE_2) and vascular endothelial growth factor (VEGF) from the tumour. Some cells separate from the primary tumour and invade the basement membrane of lymphatics and blood vessels. From here they are passively transported to distal organs and seed in a capillary bed, where the tumour cells can establish their own blood supply and continue to grow forming distant metastasis.

In order for the tumour cells to form an established metastatic site, the host immune elimination mechanisms must be overcome and the cells must adapt to an adverse growth environment [3].

14.3 The Immune System, Cancer and Anaesthesia

Natural Killer cells (NK) and T-cell lymphocytes represent the primary immune response against neoplasm development. NK cells are large lymphocytes that are able to recognise and lyse tumour cells with prior sensitization. A reduction in NK activity is found to be associated with increased numbers of circulating tumour cells and metastasis in animal models and in breast, colon and prostate cancer patients [4, 5].

In a translational study, serum from women randomized to receive either a propofol-paravertebral or standard volatile agent-opioid analgesic anaesthetic technique in women undergoing surgery for breast cancer was incubated with NK cells and breast cancer cells, the former anaesthetic technique was associated with preserved NK cytotoxicity and increased cancer cell apoptosis. As this observed difference could only be attributed to anaesthetic technique, the implication is that the propofol-paravertebral anaesthetic technique creates a serum milieu that facilitates immune activity against cancer cells [6]. In a separate prospective study of non-small cell lung cancer patients, NK cell dysfunction was evident whether regional anaesthesia was used or not, highlighting the independent depressive effect of cancer and surgery on innate immunity.

T-cell lymphocytes are part of the adaptive immune system, which requires sensitization by means of tumour-specific antigen. Infiltration of tumour by T cell lymphocytes is associated with favourable prognosis in colorectal cancer, as reported in the analysis of tumour specimens of a large cohort of patients with colorectal cancer, who were followed up for 15 years [7].

In contrast, some immune cells facilitate tumour progression. Tumour associated macrophages (TAM), were found to aid tumour extravasation and growth in animal models of breast cancer [8]. Myeloid-derived suppressor cells (MDSC), which accumulate in lymph nodes and bone marrow as well as tumour sites and inhibits both innate and adaptive immunity as shown in human and experimental animal models [9].

Sevoflurane and isoflurane (but not isoflurane) are implicated in T lymphocyte apoptosis in vitro. These volatile anaesthetic agents increase mitochondrial membrane permeability, which indirectly favours tumour dissemination by weakening cellular Immunity [10].

14.4 Influence of Surgery on Tumour Progression

Surgery remains the primary treatment of many localised solid neoplasms. However, manipulation of the tumour during surgery may accelerate tumour dissemination as a result of dispersal of malignant cells via the vascular and lymphatic systems to distant organs [11]. Although the vast majority are handled by the body's cellular immunity, some may develop the conditions necessary to facilitate cellular development locally to the point of becoming self-sustaining metastatic lesion [3]. Secondly, the surgical stress response, involving both neuroendocrine and cytokine elements [12], depresses cellular immunity transiently during this critical period. Studies have demonstrated NK cell dysfunction following surgery [13], furthermore, the inflammatory state triggered by the surgical stress leads to activation of several pro inflammatory cytokines, importantly, pro-angiogenic factors such as vascular endothelial growth factor (VEGF) stimulated by prostaglandins PGE2 [14].

14.5 Anaesthetic Agents and Their Impact on Cancer Recurrence

14.5.1 Intravenous Agents

Propofol was found to inhibit the migration of breast cancer cells in vitro, via an inhibitory action on neuroepithelial cell transforming geneNET1 [16]. In another cell culture model, Propofol down-regulates hypoxia-inducible factor 1-alpha (HIFα), which promotes neoangeogensis essential for tumour growth [17].

In a rat inoculation model of cancer, rats were inoculated with a cancer cell known to metastasize to the lung. The animals were subjected to surgery (e.g. a laparotomy) under different forms of anaesthesia using propofol, ketamine, thiopental or halothane. Animals were followed up for NK cells activity and they were scarified weeks later to assess for lung metastasis. Propofol has been shown to preserve NK cell activity, in contrast to the other agents who were associated with decreased NK cells, with ketamine being the most deleterious; decreased NK cytotoxic activity was associated with increased metastasis and progression of cancer in this model [5].

A translational study which took serum from women randomised to receive either a Propofol- paravertebral anaesthetic technique for breast cancer surgery found that this serum caused more cancer cell apoptosis and contained less of the angiogenesis factor VEGF compared with the serum from patients who had received standard balanced anaesthesia for their breast cancer surgery, consisting of sevoflurane vapour and opioid analgesia [18, 19].

Besides the immunomodulatory effects on NK cells [5], ketamine is also found to inhibit maturation of dendritic cells which influence T-cell lymphocytes activation [20].

Taken together, the evidence from largely in vitro studies suggests that propofol has at least a neutral or perhaps a potentially beneficial effect on molecular factors effecting cancer cell development, whereas the opposite signal is emerging from similar studies on ketamine. There are insufficient data on other, less frequently used IV induction agents to comment.

14.5.2 Inhalation Agents

Inhalation agents been shown to aid tumour progression in cell culture models. Many have a direct effect on promoting growth and progression of breast cancer cells in vitro. This has been demonstrated with sevoflurane [21]. Similarly, exposure of colon cancer cells to isoflurane has been shown to protect them against apoptosis, the underlying mechanism involves over expression of cancer cells to caveolin-1, a structural molecule of the cellular wall, suggesting potential resistance to chemotherapeutic agents [22]. Isoflurane was also found to enhance renal cell migration by a Hypoxia Inducible Factor alpha (HIF alpha) mechanism [23]. Sevoflurane also activates (HIFα) [17, 24]. Recently published studies demonstrated expansion of glioma cells when exposed to sevoflurane and isoflurane, associated with increased proliferation and angiogenesis of ovarian cancer cell cultures [24, 25].

The noble gas xenon, which has clinically significant anaesthetic properties, has been shown to inhibit migration of breast adenocarcinoma cells in vitro to a greater extent than sevoflurane [26].

By contrast, nitrous oxide does not enhance cancer cell migration or proliferation, despite the fact that it interferes with DNA synthesis. Secondary analysis of two randomised controlled trials that were designed originally to examine the cardiovascular effect of N_2O, demonstrate no difference in mortality in the cancer subgroup who received N_2O, compared to patients who did not receive it [27, 28].

Taken together, the balance of evidence seems to indicate the potential of volatile agents to adversely affect metastasis, the safety of nitrous oxide, and the potential of xenon to deliver a beneficial effect, subject to future clinical evaluation. While these long-term clinical trials are in progress, clinicians should continue to choose anaesthetic drugs and techniques based on their assessment of individual patients' co-morbidities.

14.5.3 Opioids

Opioids are commonly used to provide intraoperative analgesia and as an adjunct of anaesthesia, in addition to post-operative pain management following surgery, and palliation of chronic cancer pain. Although known to alter the immune system at both cellular and humeral level, the evidence for their influence on the cancer process is somewhat inconsistent, but the balance of current evidence suggests that opioids may have an adverse effect. In terms of the immune system, NK cell dysfunction is demonstrated in an animal inoculation model of cancer injected with fentanyl [29], with the extent of metastasis being proportional to the dose given and inversely proportional to the degree of NK cell activity. In another animal study, morphine reduced macrophage toll-like receptors activity [30], an effect that is believed to be mediated by μ-opioid receptors MOR. Similarly, opioid-sparing techniques for postoperative analgesia of breast cancer patients resulted in preservation of cytotoxic effect of NK cells [6]. A retrospective analysis of patients with non-small cell lung cancer, suggests an association between increased doses of opioids during the initial postoperative period and a higher recurrence rate within 5 years [31]. These findings could be linked to another study where analysis of tissue samples obtained from lung cancer patients, showed significantly increased expression of μ-opioid receptors (MOR) in cancer tissues compared with adjacent control ($P=0.0242$); furthermore, a twofold increase in MOR expression in patients with metastatic disease was observed compared with non-metastatic ones ($P=0.0013$) [32].

On the other hand, chronic high dose morphine promotes apoptosis in some cancer cell lines in vitro [33].

The endogenous opiate (B-endorphin) in contrast appears to have anti-cancer properties; β-endorphin neuron transplantation reduces the incidence of cancer in rat models of breast cancer [34]. Although not fully understood, attenuation of the surgical stress response by this opioid and increased peripheral NK cell activity are plausible mechanisms of this effect [35].

On balance therefore, there appears to be a signal suggesting that opioids may adversely affect the processes of resisting cancer. However, there is insufficient evidence to date to justify a change of practice. Patients for whom opioids are indicated should continue to receive them, at the discretion of their anaesthetist and the multidisciplinary team.

14.6 Non-steroidal Anti-inflammatory Drugs (NSAIDs)

Non-steroidal anti-inflammatory drugs interrupt the synthesis of prostaglandins from arachidonic acid by inhibition of the cyclooxygenase enzyme COX-2. Prostaglandins are pro-inflammatory cytokines that facilitate malignant cell migration and invasion as evident from several studies of different epithelial tumours [36–38]. A laboratory study of human breast cancer cells linked over-expression of COX-2 to increased level of pro-urokinase plasminogen activator, resulting in an increased potential for motility and invasiveness [39].

Alternative studies suggest that use of NSAIDs reduces risk of certain cancers, A meta-analysis of eight randomised controlled trials (RCTs) of the use of aspirin for prevention of vascular events, found benefit of daily use of aspirin in prevention of some common cancers, especially adenocarcinomas [40]. Similarly, a review of 91 epidemiological studies concluded that regular use of NSAIDs, principally ibuprofen and aspirin, is associated with reduction of common malignancies including GI tract, breast, lung, prostate and ovaries [41]. The use of intraoperative ketorolac and diclofenac for conservative breast cancer surgery was associated with improved disease-free survival (DFS) in a retrospective analysis of a cohort of 720 patients [42].

The sum of currently available evidence therefore is suggestive of a beneficial role of NSAIDs in cancer protection; further specifically designed, prospective, randomized studies are warranted before any recommendations can be made about changes in practice.

14.7 Steroids

Dexamethasone is used routinely as an antiemetic during anaesthesia. However, its potential long term effect on cancer progression was highlighted by a recent retrospective study with 4 years median follow up, where patients undergoing elective surgery received dexamethasone or placebo. While there was no difference in disease–free survival, a significant increase in metastatic recurrence was observed in patients receiving dexamethasone [43]. Again, this association requires cause-and-effect evaluation in a randomized controlled trial.

14.8 Local Anaesthetics

Several studies attribute anti-proliferative properties among cancer cells to amide local anaesthetics. A number of mechanisms underpinning this effect have been elucidated. Voltage gated sodium channels (VGSC), which are the site of action of local anaesthetics, are also expressed by a number of tumour cells, VGSC are thought to create a hyper-excitable state that is implicated in the process of metastasis. Therefore, it is plausible to suggest that inhibition of VGSC should lead to tumour suppression [44]. The amide local anaesthetic ropivacaine has been shown in vitro to inhibit metastatic colon cancer cells by deactivating VGSC [45].

Local anaesthetics also exhibit anti-inflammatory activity, and inhibition of various pro-inflammatory molecules was tested in cell culture models. Lidocaine inhibits epidermal growth factor activity at receptor level, thereby; suppressing the growth of cancer cells in vitro [46].

Inhibition of lung cancer cell migration by lidocaine and ropivacaine was associated with the inhibition of tumour necrosis factor-α and intercellular adhesion molecule-1 phosphorylation in a cell culture model of colon cancer cells [47].

An alternative and interesting mode of action of local anaesthetics is modulation of gene expression via demethylation of deoxyribonucleic acid (DNA). Two cell culture studies examined this effect and found that most amide, but not ester local anaesthetics, interfered with cancer cell DNA in this way [46, 48], in these experiments, breast cancer cell lines were incubated with different local anaesthetics, and lidocaine and ropivacaine but not bupivacaine demonstrated the inhibitory effect on breast cancer cells [49]. Moreover, this property was shown to improve the cytotoxic effect of different chemotherapeutic agents like 5-aza-2′-deoxycytidine (DAC) [49] and cisplatin on breast cancer cells in vitro [48].

14.9 Regional Anaesthesia and Cancer Recurrence

There are a number of plausible theories to explain the influence of regional anaesthesia and analgesia on recurrence or metastasis. Amide local anaesthetics have demonstrated an anti-neoplastic effect. This is mediated by both a sodium channel effect and a non-sodium channel mechanism [45]. The inhibition of the sodium channel-mediated effect on colon cancer cells has been shown to inhibit cancer cell migration and invasion in vitro. The latter non-sodium channel effect is inhibition of cancer cell intracellular Src signaling, which inhibits its migratory and invasive ability [47].

Regional anaesthesia as a clinical technique attenuates the surgical stress response, which preserves immune function from the transient inhibition associated with the surgical stress response perioperatively. Effective regional anaesthesia also enables sparing of perioperative use of both volatile agents and opioids, perhaps facilitating the host's resistance to cancer progression [50]. Therefore, it is plausible to suggest that use of a regional anaesthesia technique during cancer surgery, alone or in combination with GA, might alter the balance or pro- and anti-metastatic forces acting on cancer cells in the perioperative period in a manner, which might prevent or reduce risk of metastasis. While initial retrospective clinical analysis suggested an association between use of regional anaesthesia and improved time to recurrence, subsequently published analyses have given conflicting results in a variety of tumour types.

In one retrospective clinical study on prostate cancer, the use of local anaesthetic and epidural analgesia led to 57 % lower recurrence rate after radical prostatectomy, in comparison with general anaesthetic and morphine [51]. Another study however, showed no such benefit of combined general anaesthesia with epidural analgesia for the same type of patients [52]. Retrospective analysis of breast cancer patients who underwent mastectomy and axillary clearance suggested that the use of paravertebral analgesia was associated with reduced recurrence [15]. Although this apparent benefit has also been reported in number of similar retrospective studies including cancers of the lung, liver, ovaries, and larynx [53–55], there are also a significant number of contradictory studies, which showed no specific benefit of anaesthetic technique. A meta-analysis of a large cohort of over 42,000 patients with colorectal cancer showed that perioperative use of epidural anaesthesia was associated with improved survival at 5 years but no difference in cancer recurrence rate was observed [56].

With regards to the immuno-modulatory effect of anaesthesia, a meta-analysis showed that anaesthetic technique does not appear to significantly affect postoperative NK T-lymphocyte function [57]. In another study in patients undergoing breast cancer surgery, although regional anaesthesia techniques was associated with attenuation of surgical stress response as measured by lower concentration of serum cortisol and glucose, it didn't affect the concentration of pro-angiogenic factors such as vascular endothelial growth factor VEGF or prostaglandin E2 [58].

Possible contributing factors for the conflicting evidence are: heterogeneity of patients' groups, the different grades and stages of tumours and the impact of other therapies and complications during the course of treatment.

14.10 Fractional Inspired Oxygen and Cancer Progression

Whether high FiO_2 during anaesthesia reduces surgical site infection has been controversial for over a decade now. Current evidence suggests that higher FiO_2 does not reduce surgical site infection or pulmonary complications [59]. At molecular level, higher fraction of inspired oxygen is associated with generation of reactive oxygen species which have potential to damage the DNA, in addition high FiO_2 induces angiogenesis and erythropoietin synthesis which is a potent growth factor, all of these effects could contribute towards cancer progression [60]. The recent large PROXI trial, designed to evaluate surgical site infection as the primary outcome, was re-analysed, in a follow-up study of this previous randomized trial, to determine if there was an association between the different FiO_2 patients had been randomized to, and cancer outcomes. The results were unsettling: Higher (FiO_2) of 80 % compared to 30 % was associated with increased 5 year mortality, and reduced cancer free survival [59], although cancer recurrence rates were similar. This study does not, of course, prove that high inspired O_2 during surgery drives bad outcomes, but rather suggests that in the absence of benefit in terms of SSI, caution should be exercised in deciding on the optimal FiO_2 for patients unless higher FiO_2 is indicated for adequate oxygenation itself. Again, randomized, controlled trials are the only technique to prove whether FiO_2 affects longer-term outcomes.

In summary, while there appears to be a signal from some cell culture models and some animal models of cancer, supported by approximately half of the published retrospective analyses, and a few translational studies of small numbers of patients in randomized trials, none of this amounts to convincing evidence demanding a change of practice. The results of prospective, randomised clinical trials are needed to prove a cause and effect link between anaesthetic-analgesic technique and cancer outcomes While these trials are difficult to undertake and will require many years to complete, they are fully justified, because even a small improvement effect of anaesthetic technique during cancer surgery on recurrence or metastasis would have an immensely favourable impact for cancer patients, both economically (because widespread changes in anaesthetic practice to deliver such benefits could be accomplished quickly without additional healthcare budget expenditure), and also in terms of negligible adverse effects for patients.

References

1. World cancer reports 2014. http://www.who.int/mediacentre/factsheets/fs297/en/
2. Fidler IJ (2003) The pathogenesis of cancer metastasis: the "seed and soil" hypothesis revisited. Nat Rev Cancer 3:453–458
3. Chu GC, Chung LW (2014) RANK-mediated signaling network and cancer metastasis. Cancer Metastasis Rev 33(2–3):497–509
4. Santos MF, Mannam VKR, Craft BS, Puneky LV, Sheehan NT, Lewis RE, Cruse JM (2014) Comparative analysis of innate immune system function in metastatic breast, Colorectal, and prostate cancer patients with circulating tumour cells. Exp Mol Pathol 96(3):367–374
5. Melamed R, Bar-Yosef S, Shakhar G, Shakhar K, Ben-Eliyahu S (2003) Suppression of natural killer cell activity and promotion of tumour metastasis by ketamine, thiopental, and halothane, but not by propofol: mediating mechanisms and prophylactic measures. Anesth Analg 97(5):1331–1339
6. Buckley A, McQuaid S, Johnson P, Buggy DJ (2014) Effect of anaesthetic technique on the natural killer cell anti-tumour activity of serum from women undergoing breast cancer surgery: a pilot study. Br J Anaesth 113(suppl 1):i56–i62. Oxford University Press

7. Pages F, Galon J, Dieu-Nosjen M-C, Tartour E, Sautes-Fridman C, Fridman W-H (2010) Immune infiltration in human tumors: a prognostic factor that should not be ignored. Oncogene 29:1093–1102
8. Qian B, Deng Y, Im JH, Muschel RJ, Zou Y, Lang RA, Pollard JW (2009) A distinct macrophage population mediates metastatic breast cancer cell extravasation, establishment and growth. PLoS One 4(8):e6562
9. Ostrand-Rosenberg S, Sinha P (2009) Myeloid-derived suppressor cells: linking inflammation and cancer. J Immunol 182(8):4499–4506
10. Loop T, Dovi-Akue D, Frick M, Roesslein M, Egger L, Humar M, Hoetzel A, Schmidt R, Borner C, Pahl HL, Geiger KK, Pannen BH (2005) Volatile anesthetics induce caspase-dependent, mitochondria-mediated apoptosis in human T-lymphocytes in vitro. Anesthesiology 102(6):1147–1157
11. Neeman E, Ben-Eliyahu S (2013) Surgery and stress promote cancer metastasis: new outlooks on perioperative mediating mechanisms and immune involvement. Brain Behav Immun 30 Suppl:S32–S40
12. Buggy DJ, Smith G (1999) Epidural anaesthesia and analgesia, better outcome after major surgery? Growing evidence suggests so. BMJ 319(7209):530–531
13. Goldfarb Y, Ben-Eliyahu S (2006) Surgery as a risk factor for breast cancer recurrence and metastasis: mediating mechanisms and clinical prophylactic approaches. Breast Dis 26:99–114
14. Carmeliet P (2005) Angiogenesis in life, disease and medicine. Nature 438(70707):932–936
15. Exadaktylos AK, Buggy DJ, Moriarty DC, Mascha E, Sessler DI (2006) Can anesthetic technique for primary breast cancer surgery affect recurrence or metastasis? Anesthesiology 105(4):660–664
16. Ecimovic P, Murray D, Doran P, Buggy DJ (2014) Propofol and bupivacaine in breast cancer cell function in vitro – role of the NET1 gene. Anticancer Res 34(3):1321–1331
17. Huang H, Benzonana LL, Zhao H, Watts HR, Perry NJS, Bevan C, Brown R, Ma D (2014) Prostate cancer cell malignancy via modulation of HIF-1α pathway with isoflurane and propofol alone and in combination. Br J Cancer 111:1338–1349
18. Jaura AI, Flood G, Gallagher HC, Buggy DJ (2014) Differential effects of serum from patients administered distinct anaesthetic techniques on apoptosis in breast cancer cells in vitro: a pilot study. Br J Anaesth 113(Suppl1):i63–i67. I. Oxford University Press
19. Looney M, Doran P, Buggy DJ (2010) Effect of anesthetic technique on serum vascular endothelial growth factor C and transforming growth factor β in women undergoing anesthesia and surgery for breast cancer. Anesthesiology 113(5):1118–1125
20. Ohta N, Ohashi Y, Fujino Y (2009) Ketamine inhibits maturation of bone marrow-derived dendritic cells and priming of the Th1-type immune response. Anesth Analg 109:793–800
21. Ecimovic P, McHugh B, Murray D, Doran P, Buggy DJ (2013) Effects of sevoflurane on breast cancer cell function in vitro. Anticancer Res 33(10):4255–4260
22. Kawaraguchi Y, Horikawa YT, Murphy AN (2011) Volatile anesthetics protect cancer cells against tumor necrosis factor-related apoptosis-inducing ligand-induced apoptosis via caveolins. Anesthesiology 115(3):499–508
23. Benzonana LL, Perry N, Watt HR, Yang B, Perry IA, Coombes C, Takata M, Ma D (2013) Isoflurane, a commonly used volatile anesthetic enhances renal cancer growth and malignant potential via the hypoxia-inducible factor cellular signaling pathway in vitro. Anesthesiolog 119:593–605
24. Shi QY, Zhang SJ, Liu L, Chen QS, Yu LN, Zhang FJ, Yan M (2015) Sevoflurane promotes the expansion of glioma stem cells through activation of hypoxia-inducible factors in vitro. Br J Anaesth 114(5):715–716
25. Luo X, Zhao H, Hennah L, Ning J, Liu J, Tu H, Ma D (2015) Impact of isoflurane on malignant capability of ovarian cancer in vitro. Br J Anaesth 114(5):831–839
26. Ash SA, Valchev GI, Looney M, Ni Mhathuna A, Crowley PD, Gallagher HC, Buggy DJ (2014) Xenon decreases cell migration and secretion of a pro-angiogenesis factor in breast adenocarcinoma cells: comparison with Sevoflurane. Br J Anaesth 113(suppl 1):i14–i21

27. Myles PS, Leslie K, Chan MTV, Forbes A, Peyton PJ, Paech MJ, Beattie WS, Sessler DI, Devereaux PJ, Silbert B, Schricker T, Wallace S (2014) The safety of addition of nitrous oxide to general anaesthesia in at-risk patients having major non-cardiac surgery (ENIGMA-II): a randomised, single-blind trial. Lancet 384(9952):1446–1454
28. Fleischmann E, Marschalek C, Schlemitz K (2009) Nitrous oxide may not increase the risk of cancer recurrence after colorectal surgery: a follow-up of a randomized controlled trial. BMC Anesthesiol 9:1
29. Forget P, Collet V, Lavand'homme P, De Kock M (2010) Does analgesia and condition influence immunity after surgery? Effects of fentanyl, ketamine and clonidine on natural killer activity at different ages. Eur J Anaesthesiol 27(3):233–240
30. Franchi S, Moretti S, Castelli M et al (2012) Mu opioid receptor activation modulates toll like receptor 4 in murine macrophages. Behav Immun 26(3):480–488
31. Maher DP, Wong W, White PF, McKenna R, Roster H, Shamloo B, Louy C, Wender R, Yumul R, Zhang V (2014) Association of increased postoperative opioid administration with non-small-cell lung cancer recurrence: a retrospective analysis. Br J Anaesth 113(suppl 1):i4–i13
32. Singleton PA, Mirzapoiazova T, Hasina R, Salgia R, Moss J (2014) Increased μ-opioid receptor expression in metastatic lung cancer. Br J Anaesth 113(suppl 1):i4–i1356
33. Lin X, Wang YJ, Li Q, Hou YY, Hong MH, Cao YL, Chi ZO, Liu JG (2009) Chronic high-dose morphine treatment promotes SH-SY5Y cell apoptosis via c-Jun N-terminal kinase-mediated activation of mitochondria-dependent pathway. FEBS J 276(7):2022–2036
34. Sarkar DK, Zhang C, Murugan S, Dokur M, Boyadjieva NI, Ortiguela M, Reuhl KR, Mojtehedzadeh S (2011) Transplantation of β endorphin neurons into the hypothalamus promotes immune function and restricts the growth and metastasis of mammary carcinoma. Cancer Res 71(19):6282–6291
35. Sarkar DK, Murugan S, Zhang C, Boyadjieva N (2012) Regulation of cancer progression by β endorphin neuron. Cancer Res 72(4):836–840
36. Soslow RA, Dannenberg AJ, Rush D, Woerner BM, Khan KM, Masferrer J, Koki AT (2000) COX-2 is expressed in human pulmonary, colonic, and mammary tumors. Cancer 89(12):2637–2645
37. Cheng I, Liu X, Plummer SJ, Krumroy LM, Casey G, Whitte JS (2007) COX2 genetic variation, NSAIDs, and advanced prostate cancer risk. Br J Cancer 97:557–561
38. Brown JR, DuBois RN (2005) COX-2: a molecular target for colorectal cancer prevention. J Clin Oncol 23(12):2840–2855
39. Singh B, Berry JA, Shoher A, Ramakrishnan V, Lucci A (2005) COX-2 overexpression increases motility and invasion of breast cancer cells. Int J Oncol 26(5):1393–1399
40. Rothwell PM, Fowkes F, Belch J, Ogawa H, Warlow CP, Meade TW (2011) Effect of daily aspirin on long-term risk of death due to cancer: analysis of individual patient data from randomised trials. Lancet 377(9759):31–41
41. Harris RE, Beebe-Donk J, Doss H, Doss DB (2005) Aspirin, ibuprofen, and other non-steroidal anti-inflammatory drugs in cancer prevention: a critical review of non-selective COX-2 blockade. Oncol Rep 13(4):559–583 (25)
42. Forget P, Bentin C, Machiels J-P, Berliere M, Coulie PG, De Kock M (2014) Intraoperative use of ketorolac or diclofenac is associated with improved disease-free survival and overall survival in conservative breast cancer surgery. Br J Anaesth 113(suppl 1):i82–i87
43. Singh PP, Lemanu DP, Taylor M, Hill AG (2014) Association between preoperative glucocorticoids and long-term survival and cancer recurrence after colectomy: follow-up analysis of a previous randomized controlled trial. Br J Anaesth 113(suppl 1):i4–i13
44. Fraser SP, Foo I, Djamgoz M (2014) Anaesthetic use in cancer surgery and disease recurrence: role of voltage-gated sodium channels. Br J Anaesth 113(6):899–902
45. Baptista-Hon DT, Robertson FM, Robertson GB, Owen SJ, Rogers GW, Lydon EL, Lee NH, Hales TG (2014) Potent inhibition by ropivacaine of metastatic colon cancer SW620 cell invasion and NaV1.5 channel function. Br J Anaesth 113(suppl 1):i39–i48
46. Lirk P, Berger R, Hollmann MW et al (2012) Lidocaine time- and dose-dependently demethylates deoxyribonucleic acid in breast cancer cell lines in vitro. Br J Anaesth 109(2):200–207

47. Piegeler T, Votta-Velis EG, Liu G, Place AT, Schwartz DE, Beck-Schimmer B, Minshall RD, Borgeat A (2012) Antimetastatic potential of amide-linked local anesthetics: inhibition of lung adenocarcinoma cell migration and inflammatory Src signaling independent of sodium channel blockade. Anesthesiology 117(3):548–559
48. Li K, Yang J, Han X (2014) Lidocaine sensitizes the cytotoxicity of cisplatin in breast cancer cells via up-regulation of RARβ2 and RASSF1A demethylation. Int J Mol Sci 15(12):23519–23536
49. Lirk P, Hollmann MW, Fleischer M, Weber NC, Fiegl H (2014) Lidocaine and ropivacaine, but not bupivacaine, demethylate deoxyribonucleic acid in breast cancer cells in vitro. Br J Anaesth 113(suppl 1):i32–i38
50. Ash SA, Buggy DJ (2013) Does regional anaesthesia and analgesia or opioid analgesia influence recurrence after primary cancer surgery? An update of available evidence. Best Pract Res Clin Anaesthesiol 27(4):441–456
51. Biki B, Mascha E, Moriarty DC, Fitzpatrick JM, Sessler DI, Buggy DJ (2008) Anesthetic technique for radical prostatectomy surgery affects cancer recurrence: a retrospective analysis. Anesthesiology 109:180–187
52. Wuethrich PY, Thalmann GN, Studer UE, Burkhard FC (2013) Epidural analgesia during open radical prostatectomy does not improve long-term cancer-related outcome: a retrospective study in patients with advanced prostate cancer. PLoS One 8(8):e72873
53. Merquiol F, Montelimard A-S, Nourissat A, Molliex S, Zufferey PJ (2013) Cervical epidural anesthesia is associated with increased cancer-free survival in laryngeal and hypopharyngeal cancer surgery: a retrospective propensity-matched analysis. Reg Anesth Pain Med 38(5):398–402
54. Lai R, Peng Z, Chen D, Wang X, Xing W, Zeng W, Chen M (2012) The effects of anesthetic technique on cancer recurrence in percutaneous radiofrequency ablation of small hepatocellular carcinoma. Anesth Analg 114(2):290–296
55. de Oliveira GS Jr, Ahmad S, Schink JC, Singh DK, Fitzgerald PC, McCarthy RJ (2011) Intraoperative neuraxial anesthesia but not postoperative neuraxial analgesia is associated with increased relapse-free survival in ovarian cancer patients after primary cytoreductive surgery. Reg Anesth Pain Med 36(3):271–277
56. Cummings KC 3rd, Xu F, Cummings LC, Cooper GS (2012) A comparison of epidural analgesia and traditional pain management effects on survival and cancer recurrence after colectomy, a population-based study. Anesthesiology 116(4):797–806
57. Conrick-Martin I, Kell MR, Buggy DJ (2012) Meta-analysis of the effect of central neuraxial regional anesthesia compared with general anesthesia on postoperative natural killer T lymphocyte function. J Clin Anesth 24(1):3–7
58. O'Riain SC, Buggy DJ, Kerin MJ, Watson RW, Moriarty DC (2005) Inhibition of the stress response to breast cancer surgery by regional anesthesia and analgesia does not affect vascular endothelial growth factor and prostaglandin E2. Anesth Analg 100(1):244–249
59. Togioka B, Galvagno S, Sumida S, Murphy J, Ouanes JP, Wu C (2012) The role of perioperative high inspired oxygen therapy in reducing surgical site infection: a meta-analysis. Anesth Analg 114(2):334–342
60. Cooke MS, Evans MD, Dizdaroglu M, Lunec J (2003) Oxidative DNA damage: mechanisms, mutation, and disease. FASEB J 17(10):1195–1214

Chapter 15
Transverse Abdominis Plane Block: Evolution and Current Understanding

Ruth-Aoibheann O'Leary and John G. McDonnell

Abstract Regional blockade of the anterior abdominal wall can contribute significantly to postoperative analgesia for a wide range of abdominal and pelvic procedures, as part of a multimodal analgesic regime. Abdominal wall blocks are not new in the field of anaesthesia but they have gained in popularity recently with the widespread adoption of ultrasound guided regional anaesthesia. In this chapter we describe the evolution of the transverse abdominis plane block, with reference to its anatomy, sonoanatomy and the progression from landmark to ultrasound-guided techniques. The quadratus lumborum block developed as a consequence of the move from anterior to posterior in transverse abdominis plane blocks and the approaches used for this are described. We detail the evidence for the analgesic efficacy of these blocks compared to other modes of analgesia and also discuss areas of ongoing interest in the field. Finally, we describe its use in special patient populations as well as reported complications.

Keywords Abdominal wall block • Transversus abdominis plane block • Quadratus lumborum block • Regional • Analgesia

R.-A. O'Leary, FCAI, FJFICMI
Department of Anaesthesia and Intensive Care Medicine, Galway University Hospitals, Galway, Ireland
e-mail: R.OLEARY1@nuigalway.ie

J.G. McDonnell, MD, FCARCSI (✉)
Department of Anaesthesia and Intensive Care Medicine, Galway University Hospitals, Galway, Ireland

Department of Anaesthesia, Clinical Sciences Institute, National University of Ireland, Galway, Ireland

Clinical Research Facility, Galway University Hospitals, Galway, Ireland

Breastcheck West, Galway University Hospital, Galway, Ireland
e-mail: johngmcdonnell@gmail.com; bonsanaesthesia@gmail.com

K. Stuart-Smith (ed.), *Perioperative Medicine – Current Controversies*,
DOI 10.1007/978-3-319-28821-5_15

15.1 Introduction

Since its description by Rafi in 2001 [1], the transversus abdominis plane (TAP) block has been used successfully as part of a multimodal analgesic regime for a range of abdominal operations. The original TAP block used a "double pop" landmark based technique, with the needle inserted through the apex of the Triangle of Petit. It was initially described as a regional abdominal field infiltration (RAFI) block, prior to being renamed as the TAP block. Multiple distribution studies, both cadaveric and clinical, have shown a wider pattern of local anaesthetic distribution than originally perceived, with extension being observed into the thoracic paravertebral space, which might explain the extent of its efficacy [2].

The TAP block has evolved with the use of ultrasound. Ultrasound guided anterior abdominal TAP blocks have been shown to have a markedly different distribution both to those performed with the landmark technique and to posterior approaches, and in turn ultrasound has led to a further refinement of the block [2–4]. In the originally described ultrasound-guided TAP block, the three muscular layers of the abdominal wall are identified. These are followed posterolaterally to approximately the midaxillary line, superior to the iliac crest where local anaesthetic may be deposited under direct vision into the plane between the internal oblique and the transversus abdominis muscles [5].

Dr. Rafa Blanco described an alternative ultrasound-guided posterior approach to the TAP block in 2007, the quadratus lumborum (QL) block [6]. This produces an analgesic effect similar to the original landmark based TAP block, with spread to the paravertebral space. Although the TAP and QL blocks remain underutilized compared to central neuraxial blockade [7], they have been described as part of a multimodal analgesic regime in a wide variety of procedures from open prostatectomies to renal transplants [8–10]. The QL blocks are the subject of ongoing research by Drs Blanco and Borglum [11, 12].

15.2 Anatomy

15.2.1 Transverse Abdominis Plane Block

The anterior abdominal wall spans the area between the costal margin and pelvic bones. It is bounded superiorly by the costal margin and xiphoid process, inferiorly by the inguinal ligament and pelvic bones and laterally by the mid-axillary line [13]. The sensory innervation of the anterolateral abdominal wall derives from the anterior rami of the lower six thoracic and upper two lumbar sensory nerves [14]. These rami form the intercostal and subcostal nerves (T6–T12) and the iliohypogastric and ilioinguinal nerves (L1–L2). T6–T10 supply the skin superior to and including the umbilicus, while T11, 12 and the iliohypogastric and ilioinguinal nerves supple the skin inferior to the umbilicus. These nerves pass between the internal oblique and transversus abdominis muscles and form an interconnecting 'TAP plexus' [14]. Before this plexus forms, the nerves can reliably be found in close proximity to the triangle of Petit. Loukas et al. describe the anatomy of the triangle of Petit [15]. The external oblique muscle forms the medial edge of the triangle. The lateral edge is formed by latissimus dorsi, while the inferior edge is the iliac crest. The transversalis fascia is found within the TAP in 82 % of people, although its size and location is somewhat variable [15]. The triangle lies posterior to the mid-axillary line, just above the iliac crest [16]. Its size varies considerable from <8 to >12 cm^2 [15].

15.2.2 Quadratus Lumborum Block

The quadratus lumborum block involves deposition of local anaesthetic in the area of the quadratus lumborum muscle. This lies anterolateral to the body of L4, which is used as a landmark. The quadratus lumborum and psoas muscles have their embryonic origin in the thoracic cage and are thus closely associated with the thoracic paravertebral space. As a result, the QL block can block central visceral pain conduction as well as blocking the nerves supplying the anterior abdominal wall, as described in the TAP block. To date there have been a number of different approaches to the QL block described in the literature. Dr. Blanco's original approach describes deposition of local anaesthetic solution at the anterolateral aspect of the quadratus lumborum muscle, and this was refined further by Borglum et al. They have described a number of different approaches. In the transmuscular approach the needle passes from lateral to medial through the body of quadratus lumborum to lie in the space between the psoas and quadratus lumborum muscles. This technique is facilitated by using the 'Shamrock' sign as described by Borglum [17]. Psoas major (anterior), erector spinae (posterior) and quadratus lumborum form three leaves of the shamrock and the transverse process of L4 forms the stalk. Differences in approach have been described but the sonographic anatomy remains the same.

15.2.3 Distribution Studies

Contrast studies using computed tomography and magnetic resonance imaging have furthered our understanding of the spread of local anaesthetic following TAP blocks. A cadaveric study, injecting dye into the TAP using landmark technique, showed distribution of dye from the iliac crest to the costal margin [3]. Following on from this, healthy volunteers underwent injection into the TAP using landmark technique. CT and MRI scans then confirmed the spread of dye. However, Carney [2] used serial MRIs to compare the spread of local anaesthetic from four blocks: landmark TAP, ultrasound-guided midaxillary (lateral) TAP, ultrasound-guided subcostal TAP and ultrasound-guided posterior approach to the Triangle of Petit [6]. The ultrasound-guided midaxillary and subcostal TAP blocks resulted in anterior spread within the transversus abdominis plane. However, the landmark TAP block showed significant posterior spread to the paravertebral space. This, most likely, explains the difference in analgesia noted between the different approaches. Similarly, the posterior ultrasound-guided approach to the triangle of Petit, which has evolved into the quadratus lumborum block, resulted in paravertebral, epidural and lymphatic contrast enhancement from T5 to T10. Again, this accounts for its extensive and predictable analgesic effect.

15.3 Techniques

15.3.1 TAP Block

15.3.1.1 Landmark Technique

In the initial descriptions of the TAP block Rafi used surface anatomic landmarks to locate the triangle of Petit [1]. The patient is positioned in the supine position. The triangle is felt as an inward dip just on top of the iliac crest. This is most easily located by running a finger posteriorly along the iliac crest from the anterior superior iliac spine (ASIS). Posterior to this dip the edge of latissimus dorsi can be palpated. This represents the lateral edge of the triangle so the needle insertion point is anterior to this point. A cutting needle can be used to break skin and thereafter a blunt tipped needle is used. The needle is slowly advanced until a definite pop is felt. At this point the needle lies in the transversus abdominis plane. After careful aspiration to exclude vascular puncture, the local anaesthetic solution can be injected. For incisions crossing the midline, the TAP block can be repeated on the contralateral side, with attention being paid to the overall dose of local anaesthetic administered.

15.3.1.2 Most Popular Ultrasound Techniques

The use of ultrasound has prompted a move away from the triangle of Petit with the ability to visualise the muscle layers now of paramount importance. With the patient in the supine position, the ultrasound probe is placed in transverse orientation on the anterior abdominal wall. Once the three muscle layers are identified, the probe is moved laterally to lie in the midaxillary line superior to the iliac crest. The needle passes from medial to lateral under direct vision until it lies in the correct plane (Fig. 15.1). This can be confirmed by visualisation of local anaesthetic spread in the transversus abdominis plane [5].

Hebbard and Borglum described subcostal TAP blocks, which can be used when the aim is to provide analgesia to the supra-umbilical abdominal wall. The probe is placed at an oblique angle along the subcostal margin. The needle passes inferolaterally from a point near the xiphoid process and local anaesthetic can be deposited in the TAP along the costal margin [18, 19].

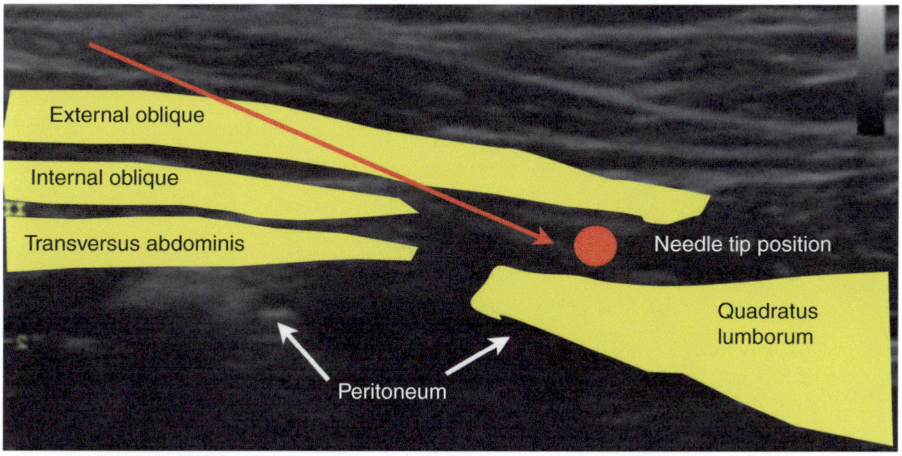

Fig. 15.1 Ultrasound-guided TAP block. *Red arrow* needle path, *red dot* plane for deposition of local anaesthetic. Note muscle layers and proximity of peritoneum

The introduction of ultrasound to widespread clinical practice of regional anaesthetic techniques caused dramatic differences in analgesic efficacy to present depending on the needle insertion and endpoints in the transversus abdominis plane. The differences demonstrated caused renewed interest in exploration of the posterior aspect of the lateral abdominal wall and in particular the lumbar triangle of Petit, its surrounding muscle groups and the pattern of spread of the local anaesthetic agents from differing approaches to the TAP block. The concept of the TAP block subsequently evolved, with the description of the various approaches to the quadratus lumborum blocks.

15.3.2 Quadratus Lumborum Block

Unlike the TAP block, for which both a landmark and ultrasound guided technique have been described, the QL block is unusual in that it can only be performed under ultrasound guidance. A number of different ultrasound-guided approaches to the QL block have evolved over the last decade (Fig. 15.2). In the original description by Dr. Blanco, the patient is positioned in the lateral position and the ultrasound probe is placed above the iliac crest. A curved array probe is usually advised due to the depth required. The three abdominal muscle layers are identified. The probe is then moved posterolaterally until the transversus abdominis muscle becomes aponeurotic. This aponeurosis is followed until the quadratus lumborum muscle comes into view. This is approached from medial to lateral using an in-plane technique. After passing through the external and internal oblique muscles the local anaesthetic can be deposited on the anterolateral surface of QL in close proximity to the lateral aspect of the TAP [6]. This approach is dubbed QL1. Borglum modified this technique, passing the needle from lateral to medial to deposit local anaesthetic

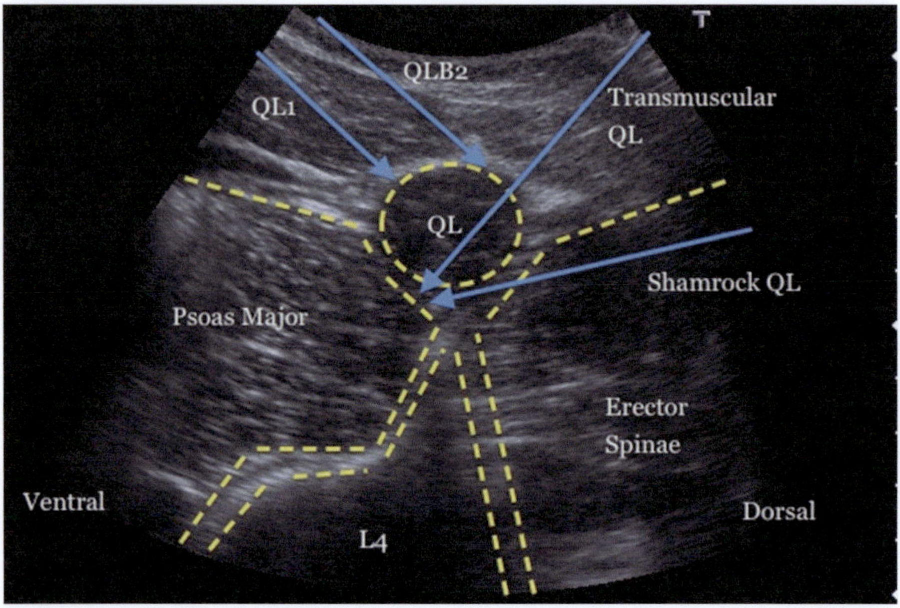

Fig. 15.2 Ultrasound-guided QL block approaches. See text for details

solution on the anterolateral aspect of the QL, dubbed transmuscular QL [12]. They further refined their approach by describing the Shamrock sign, to facilitate identification of the relevant structures. In this technique, the needle is passed in-plane, from posterior to anterior mimicking the landmark based approach to the psoas compartment block, through the body of QL to lie in the space between psoas and QL. As this required passing from the posterior to anterior and occasionally significantly deep views to be obtained to observe the erector spinae muscles, they modified the needle insertion point to be just medial to the probe position on the over the mid-axillary line [12]. This approach is referred to as the Shamrock QL. Blanco has subsequently revised his technique, dubbed QLB 2. He advises a more superficial approach, depositing local anaesthetic posterior or superficial to QL (Fig. 15.3). The needle passes only through subcutaneous tissue and the latissimus dorsi muscle. This approach provides a greater degree of paravertebral spread with improved safety margin [11].

15.4 Analgesia

Since the original description of the TAP block, there have been a plethora of studies detailing its efficacy. These have described its use in a wide variety of abdominal and pelvic procedures and have looked at outcomes such as reduction in pain scores, opioid consumption or opioid-related side effects. Essentially, when the block is

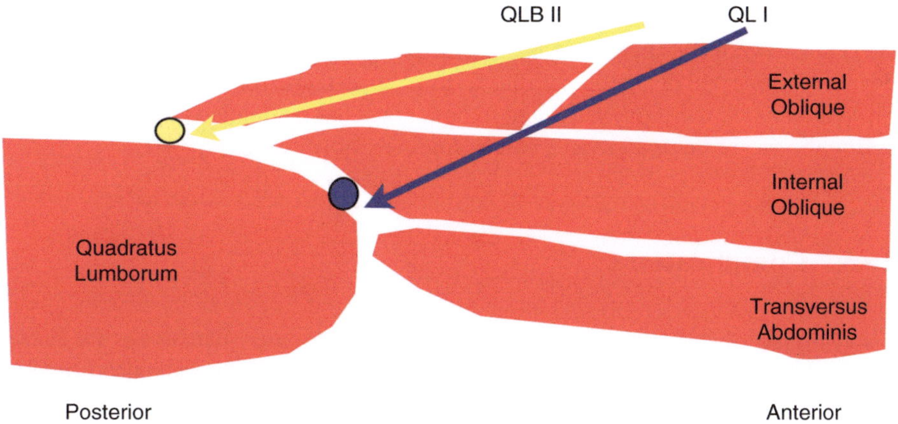

Fig. 15.3 Anterior approaches to QL block. See text for details

used in the correct manner for an appropriate surgery, there will be analgesic benefit to the patient. When compared to placebo, the landmark TAP block shows significant analgesic benefit and opioid-sparing effect in patients, both adult and paediatric, undergoing abdominal and pelvic surgery [20–22]. This effect is not seen in lower segment caesarean sections when compared to intrathecal opioid [23]. Interestingly, recent reviews suggest that the posterior, landmark, TAP block provides more prolonged analgesia than the mid-axillary ultrasound guided approach [24, 25]. The reason for this is likely to be extension of the local anaesthetic into the paravertebral space. This paravertebral extension does not occur in the most commonly used ultrasound guided TAP blocks, with the probe positioned in the mid-axillary line or the subcostal areas. The realisation that posterior approaches improve efficacy has led to increased interest in the posterior ultrasound based TAP and the quadratus lumborum blocks. These blocks have demonstrated spread to the paravertebral space.

TAP blocks have also been compared to other regional analgesic techniques. When compared to epidural analgesia, patients with TAP blocks consumed significantly more opiate [26, 27]. TAP blocks have been compared with ilioinguinal and iliohypogastric nerve blocks for inguinal hernia repair surgery and have been shown to provide superior analgesia [28]. Continuous TAP infusions have been described using the ultrasound approach but they have not been compared to wound infiltration catheters, sited during surgery. As both techniques affect the abdominal wall it is probable that their effects will be similar. Abdominal wall blocks do not and should not be expected to confer similar analgesia to neuraxial blocks. However, they reduce opioid requirements when used as part of a multimodal approach.

There has been little research into the pharmacokinetics and pharmacodynamics of the local anaesthetics used in TAP and QL blocks. As these blocks have a centrally mediated effect, the appropriate volume seems to be important in ensuring spread to the paravertebral space. Doses of 0.3–0.6 ml/kg of local anaesthetic in low concentrations have been described but the exact volume needed to produce analgesia is unknown [29]. There have been some small studies looking at the addition of dexamethasone and dexmedetomidine to local anaesthetic for these blocks [30, 31]. Prolonged block duration was found in both cases but randomised trials are needed before the use of these adjuvants can be routinely advocated.

15.5 Complications

Clinically significant complications of transversus abdominis plan blocks are thankfully rare. However, with the introduction of more technically advanced blocks, such as the QL block, there is potential for an increase in complications, particularly with inexperienced sonographers. The TAP block uses high volumes of local anaesthetic and as a result there is a risk of toxic quantities of the agent being absorbed. Griffiths et al. measured the peak plasma concentration of ropivicaine after bilateral TAP blocks using ropivicaine 3 mg/kg [32]. They found a mean peak plasma ropivicaine level of 2.2 mcg/ml, which is above the minimum toxic plasma levels. There were no clinical sequelae associated with this potentially toxic level.

Interestingly, Griffiths proceeded to look at plasma ropivicaine levels in pregnant women undergoing bilateral TAP blocks for Caesarean sections [33]. Although peak plasma levels were in fact lower in this population, a number of women complained of symptoms of local anaesthetic toxicity. The mechanism for this toxicity was felt to be systemic absorption as opposed to intravascular injection, due to a 15-min delay between injection and onset of symptoms. Changes in cardiac output and blood flow during pregnancy may make these patients more susceptible to toxicity. This is further suggested by case reports of seizures following bilateral TAP blocks in this population. The patients received a similar dose of ropivicaine and their symptoms were rapidly reversed with lipid emulsion, supporting a diagnosis of local anaesthetic toxicity [34]. These reports should, however, alert one to the potential risks. The dose of local anaesthetic should be closely monitored and, in the case of these large field-like blocks, lower concentrations are advisable.

Patients with liver disease are similarly at increased risk of systemic absorption and toxicity. Landy et al. describe a case of local anaesthetic toxicity in a patient with chronic liver disease [35]. They hypothesized that the toxicity, which occurred after administering 3 ml/kg of ropivicaine, was due to a combination of reduced protein binding of local anaesthetic and increased abdominal wall vascularity from portal hypertension promoting systemic absorption.

Parturients and patients with chronic liver disease have a similar risk profile due to altered blood flow and protein binding. Caution is needed when performing local anaesthetic blocks in these populations. Similarly, as these blocks require a high volume of anaesthetic, paediatric and underweight patients are at particular risk of inadvertent overdose. Sakai et al. describe local anaesthetic toxicity following bilateral TAP blocks in a 40 kg patient who received 150 mg of ropivicaine (3.75 mg/kg). Although peak plasma levels of ropivicaine were not measured, it is reasonable to assume that they were significantly higher than the minimum toxic plasma level [36].

Damage to intraperitoneal organs has been described following both landmark and ultrasound guided TAP blocks. Farooq et al. described a liver laceration following landmark based TAP block, resulting in gross liver enlargement [37]. Similarly, Lancaster described a liver laceration after ultrasound TAP block for repair of an inguinal hernia. Poor needle visualization was felt to be the primary factor in this complication [38]. Although no reports of spleen or kidney injury have been reported, it is reasonable to assume that they are also at risk during TAP placement. Transient femoral nerve block is another complication, which can rarely be seen with TAP blocks [39]. Although transient, this may lead to delayed mobilisation and discharge, and should be considered before undertaking this block.

15.6 Conclusion

Transverse abdominis plane blocks have, over the last 14 years, become established as a valuable part of the multimodal analgesic armamentarium for a wide range of abdominal and pelvic procedures. Advances in our understanding of the neuroanatomy of the abdominal wall and the distribution of local anaesthetic have led to the evolution of this block culminating in the description of the quadratus lumborum block. Recent reports have emphasised the need for caution as these techniques are technically challenging, but these blocks remain safe and well-tolerated in the management of postoperative pain.

References

1. Rafi AN (2001) Abdominal field block: a new approach via the lumbar triangle. Anaesthesia 56:1024–1026
2. Carney J, Finnerty O, Rauf J et al (2011) Studies on the spread of local anaesthetic solution in transversus abdominis plane blocks. Anaesthesia 66:1023–1030
3. McDonnell JG, O'Donnell BD, Farrell T et al (2007) Transversus abdominis plane block: a cadaveric and radiological evaluation. Reg Anesth Pain Med 32:399–404
4. Tran TM, Ivanusic JJ, Hebbard P, Barrington M (2009) Determination of spread of injectate after ultrasound-guided transversus abdominis plane block: a cadaveric study. Br J Anaesth 102:123–127
5. Hebbard P, Fujiwara Y, Shibata Y, Royse C (2007) Ultrasound-guided transversus abdominis plane (TAP) block. Anaesth Intensive Care 35(4):616–617
6. Blanco R (2007) TAP block under ultrasound guidance: the description of a 'no pops technique'. Reg Anesth Pain Med 32(Suppl 1):130
7. Kearns RJ, Young SJ (2011) Transversus abdominis plane blocks; a national survey of techniques used by UK obstetric anaesthetists. Int J Obstet Anesth 20(1):103–104
8. O'Donnell BD, McDonnell JG, McShane AJ (2006) The transversus abdominis plane (TAP) block in open retropubic prostatectomy. Reg Anesth Pain Med 31(1):91
9. Jankovic ZB, Pollard SG, Nachiappan MM (2009) Continuous transversus abdominis plane block for renal transplant recipients. Anesth Analg 109(5):1710–1711
10. Bharti N, Kumar P, Bala I, Gupta V (2011) The efficacy of a novel approach to transversus abdominis plane block for postoperative analgesia after colorectal surgery. Anesth Analg 112(6):1504–1508

11. Blanco R (2014) Optimal point of injection: the quadratus lumborum type I and II blocks. Anaesthesia. Available via: http://www.respond2articles.com/ANA/forums/post/1550.aspx. Accessed 23 Sept 2014
12. Borglum J, Moriggl B, Jensen K et al (2013) Ultrasound-guided transmuscular quadratus lumborum blockade. Br J Anaesth. Available via: http://www.bjaoxfordjournals.org. Accessed 22 Sept 2014
13. Moore K, Dalley A (1999) Clinically oriented anatomy, 4th edn. Lippincott Williams and Watkins, Philadelphia
14. Rozen WM, Tran TM, Ashton MW, Barrington MJ, Ivanusic JJ, Taylor GI (2008) Refining the course of the thoracolumbar nerves: a new understanding of the innervation of the anterior abdominal wall. Clin Anat 21:325–333
15. Loukas M, Tubbs RS, El-Sedfy A et al (2007) The clinical anatomy of the Triangle of Petit. Hernia 11:441–444
16. Jankovic ZB, duFeu FM, McConnell P (2009) An anatomical study of the transversus abdominis plane block: location of the lumbar Triangle of Petit and adjacent nerves. Anesth Analg 109:981–985
17. Sauter AR, Ullensvang K, Bendtsen TF, Børglum J (2013) The "Shamrock Method" a new and promising technique for ultrasound guided lumbar plexus block. Br J Anaesth. Available via: http://bja.oxfordjournals.org/forum/topic/brjana_el%3b1. Accessed 11 Mar 2013
18. Hebbard P (2008) Subcostal transversus abdominis plane block under ultrasound guidance. Anesth Analg 106(2):674–675
19. Borglum J, Maschmann C, Belhage B, Jensen K (2011) Ultrasound-guided bilateral dual transversus abdominis plane block: a new four-point approach. Acta Anaesthesiol Scand 55(6):658–663
20. Carney J, McDonnell JG, Ochana A, Bhinder R, Laffey JG (2008) The transversus abdominis plane block provides effective postoperative analgesia in patients undergoing total abdominal hysterectomy. Anesth Analg 107:2056–2060
21. Carney J, Finnerty O, Rauf J, Curley G, McDonnell JG, Laffey JG (2010) Ipsilateral transversus abdominis plane block provides effective analgesia after appendectomy in children: a randomized controlled trial. Anesth Analg 111:998–1003
22. Peterse PL, Mathiesen O, Torup H, Dahl JB (2010) The transversus abdominis plane block: a valuable option for postoperative analgesia? A topical review. Acta Anaesthesiol Scand 54:529–535
23. Costello JF, Moore AR, Wieczorek PM, Macarthur AJ, Balki M, Carvalho JC (2009) The transversus abdominis plane block, when used as part of a multimodal regimen inclusive of intrathecal morphine, does not improve analgesia after cesarean delivery. Reg Anesth Pain Med 34(6):586–589
24. Abdallah FW, Laffey JG, Halpern SH, Brull R (2013) Duration of analgesic effectiveness after the posterior and lateral transversus abdominis plane block techniques for transverse lower abdominal incisions: a meta-analysis. Br J Anaesth 111(5):721–735
25. Abdallah FW, Chan VW, Brull R (2012) Transversus abdominis plane block: a systematic review. Reg Anesth Pain Med 37:193–209
26. Kadam VR, Moran JL (2011) Epidural infusions versus transversus abdominis plane (TAP) block infusions: retrospective study. J Anesth 25(5):786–787
27. Niraj G, Kelkar A, Jeyapalan I et al (2011) Comparison of analgesic efficacy of subcostal transversus abdominis plane blocks with epidural analgesia following upper abdominal surgery. Anaesthesia 66(6):465–471
28. Aveline C, Le Hetet H, Le Roux A et al (2011) Comparison between ultrasound-guided transversus abdominis plane and conventional ilioinguinal/iliohypogastric nerve blocks for day-case open inguinal hernia repair. Br J Anaesth 106(3):380–386
29. Forero M, Heikkila A, Paul JE, Cheng J, Thabane L (2015) Lumbar transversus abdominis plane block: the role of local anesthetic volume and concentration – a pilot, prospective, randomised, controlled trial. Pilot Feasibil Stud 1:10

30. Ammar A, Mahmoud K (2012) Effect of adding dexamethasone to bupivacaine on transversus abdominis plane block for abdominal hysterectomy: a prospective randomized controlled trial. Saudi J Anaesth 6(3):229–233
31. Almarakbi W, Kaki A (2014) Addition of dexmedetomidine to bupivacaine in transversus abdominis plane block potentiates post-operative pain relief among abdominal hysterectomy patients: a prospective randomized controlled trial. Saudi J Anaesth 8(2):161–166
32. Griffiths JD, Barron FA, Grant S et al (2010) Plasma ropivacaine concentrations after ultrasound-guided transversus abdominis plane block. Br J Anaesth 105(6):853–856
33. Griffiths JD, Le NV, Grant S et al (2013) Symptomatic local anaesthetic toxicity and plasma ropivacaine concentrations after transversus abdominis plane block for Caesarean section. Br J Anaesth 110(6):996–1000
34. Weiss E, Jolly C, Dumoulin JL et al (2014) Convulsions in 2 patients after bilateral ultrasound-guided transversus abdominis plane analgesia. Reg Anesth Pain Med 39(3):248–251
35. Landy C, Gagnon N, Boulladn P et al (2012) Seizures associated with local anaesthetic intoxication. Br J Anaesth 109:463–464
36. Sakai T, Manabe W, Kamitani T et al (2010) Ropivicaine-induced late-onset systemic toxicity after transversus abdominis plane block under general anesthesia: successful reversal with 20 % lipid emulsion. Masui 59:1502–1505
37. Farooq M, Carey M (2008) A case of liver trauma with a blunt regional anesthesia needle while performing transversus abdominis plane block. Reg Anesth Pain Med 33(3):273–275
38. Lancaster P, Chadwick M (2010) Liver trauma secondary to ultrasound-guided transversus abdominis plane block. Br J Anaesth 104:509–510
39. Walker G (2010) Transversus abdominis plane block: a note of caution. Br J Anaesth 104(2):265

Chapter 16
Future Ultrasound Technologies for the Perioperative Physician

Graeme McLeod

Abstract Commercial ultrasound systems lack sufficient resolution to differentiate exactly between tissue planes – a 10 MHz transducer has a resolution to only 300 μm. Incorporation of reflectors has improved needle echogenicity but visibility is still impaired out-of-plane to the transducer and with increasing depths. Active needles using electro-magnetic guidance or peizo-electric vibration promise better needle direction. Optical spectroscopy needles needles are able to differentiate between blood, lipid and protein and will be available in practice soon. However, a fundamental physical problem remains – the greater the depth and the higher the transducer frequency, the greater the attenuation of energy, and the poorer tissue visibility. High frequency ultrasound using frequencies between 30 MHz and 50 MHz provide resolution to between 100 μm and 60 μm respectively. Micro-ultrasound is already being used to visualize and measure the properties of tissue in laboratories. Miniature ultrasound transducers integrated at the tip of needles will enable clinicians to image regions such as the epidural space or perineural spaces and identify tissue morphology in colour using features such as strain and shear wave elastography.

Keywords Ultrasonography • Needle • Nerve • Micro-ultrasound • Elastography • Colour Doppler

16.1 Introduction

The origins of ultrasound can be traced back to measurements of the harmonics of string instruments by Pythagoras in the sixth century BC, acoustic studies of water conducted by Leonardo da Vinci in 1490 and navigational experiments on bats

G. McLeod
Ninewells Hospital, NHS Tayside, Dundee, UK

Institute of Academic Anaesthesia, Medical Research Institute, School of Medicine, University of Dundee, Dundee DD1 9SY, UK
e-mail: g.a.mcleod@dundee.ac.uk

K. Stuart-Smith (ed.), *Perioperative Medicine – Current Controversies*,
DOI 10.1007/978-3-319-28821-5_16

carried out by the Italian biologist Lazzaro Spallanzani in 1790. More recently, discovery of the peizoelectric properties of crystal by Pierre and Jacques Curie – the generation of electricity secondary to relative displacement (otherwise termed strain) within a crystal – enabled the development of echolocation of submarines by Paul Langevin in 1917. The first ultrasound studies on animal tissue were first conducted by George Ludwig at the Bethesda Naval Medical Research Institute, Maryland in 1949, followed shortly after by the application of ultrasound to the diagnosis of heart disease in 1953 by Edler and Hertz at Lund University, Sweden. At the same time in Glasgow, Brown, Donald and MacVicar introduced ultrasound for obstetric examination. Since then ultrasound has expanded to almost all clinical specialities, and compared to CT and MRI technologies, is easy to use, safe, portable, and inexpensive.

The aim of this chapter is to discuss recent innovation in clinical ultrasound technology. In order to understand the scientific rationale behind recent developments, it is important to understand the current limitations of clinical imaging and the fundamental physical mechanisms underpinning the interaction between ultrasound and tissue. The approach to the chapter will be to describe the clinical limitations of transducers, needles and ultrasound images, and for each, describe technological innovations that have been tested in phantoms, animals or humans that have the potential to be translatable to perioperative patient care.

16.2 Needle Visibility

Ultrasound use is now more widespread, and is used by anaesthetists, radiologists and surgeons, not only as a diagnostic tool, but as an interventional, whereby needles are placed within anatomical spaces and drugs such as local anaesthetics injected, biopsies taken or fluid or pus aspirated.

Optimal needle and tip visibility is best when needles are inserted in-plane, parallel to the longitudinal array of elements in the ultrasound transducer [1]. In practice, injection at acute angles up to 30° to the transducer face provides good needle tip and shaft visibility. However, with increased angulation needle tip visibility deteriorates; within a soft embalmed cadaver model, only 30 % of tips of traditional smooth shafted Tuohy needles [2] were visible over a range of angles between 30° and 75° (Fig. 16.1). Even at low angulation, poor coordination of hand and transducer movement may inadvertently place needles across the transducer midline, limiting visibility. Experienced anaesthetists sometimes insert needles out-of-plane whereby a needle is placed at the middle of the transducer and inserted across the plane of the ultrasound beam. Detection of the needle tip out-of-plane is technically more challenging as visibility relies on identification of the tip by a single ultrasound element rather than a row of elements when inserted in-plane. Even then, unless the needle tip has a characteristic ultrasonic signature, reflection from a portion of the shaft crossing the single ultrasound beam may be misinterpreted as the needle tip, rendering any intervention inaccurate and potentially unsafe.

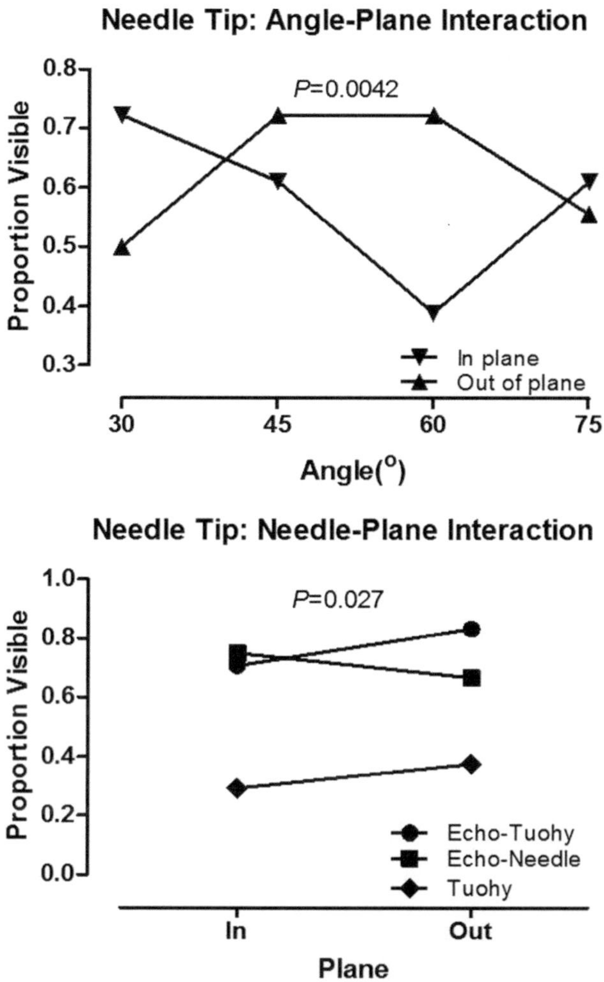

Fig. 16.1 Needle tip: Angle-plane and needle-plane interactions. The *upper* image shows the proportion of visible needle tips on the *y-axis* relative to the angle of needle insertion on the *x-axis* for in-plane and out-of-plane orientations. Insertion of needles at wider angles in-plane and more shallow angles out-of-plane reduced tip visibility. The *lower* image shows the proportion of visible needle tips on the *y-axis* relative to in-plane and out-of-plane orientations. Both textured needles are more visible than the smooth surfaced Tuohy needle

In practice, when using the out-of-plane technique, experienced anaesthetists rely on visibility of tissue distortion secondary to needle travel and hydrolocation, 0.5–1 ml incremental boluses of local anaesthetic [3], rather than needle tip and shaft visibility, in order to position needle tips accurately. Low volumes of local anaesthetic are readily seen on the B-Mode image as a spreading anechogenic area and may be useful in separating anatomical planes as practionners carefully "dissect" down to the target.

16.2.1 Echogenic Needles

In order to improve ultrasonic needle visibility, manufacturers indented the surface of needles with angular depressions. Studies of single injection and Tuohy needles in Thiel embalmed human cadavers [2, 4] and patients have shown improved visibility of needle tip and needle shaft for both in-plane (Fig. 16.1) out-of-plane needle insertion. In a study conducted on the soft embalmed Thiel cadaver [2] the median (IQR) proportions of visible needle tips were 83 % (67–83 %) for the echogenic Tuohy needle, 75 % (67–83 %) for the echogenic single-shot needle and 33 % (33–46 %) for the smooth-surfaced Tuohy needle ($p<0.001$). Median (IQR) needle shaft visibility (between 1 and 5) was rated as 4.0 (3.5–4.7) for the echogenic Tuohy needle, 4.0 (3.8–4.5) for the echogenic single-shot needle and 3.0 (2.4–3.3) for the smooth-surfaced Tuohy ($p=0.015$).

Improved overall visibility can be explained by the circumferential placement of reflectors, in contrast to standard needles which scatter beams randomly and transmit them poorly back to the transducer [5]. Clinical experience confirms cadaver findings, although pain associated with needle movement can be problematic, and poor visibility persists when using high angles in-plane and narrow angles out-of-plane are used.

16.2.2 Needle Guidance System

The development of needle guidance systems has been driven by the need for accurate cancer biopsy [6] and diagnosis, particularly in liver and prostate cancer. For surgical procedures, including laparoscopy, ultrasound based systems using electromagnetic and/or robotic guidance [7, 8] have been shown to be more accurate than freehand targeting of lesions.

Perioperative interventional ultrasound procedures are also complex. The 3D alignment of the needle and position of the needle tip relative to the target is difficult to calculate, and using, hydrolocation, the incremental injection of small aliquots of local anaesthetic, tissue anatomy becomes distorted and the target moves. With these difficulties in mind, a needle guidance system (Sonix GPS, Ultrasonix, Vancouver, Canada) has been developed. The equipment consists of a transmitter and two sensors – at the needle tip and in the transducer – allowing real-time tracking in three dimensions. Before conducting the block, the sensor needle is positioned either in-plane or out-of-plane to the ultrasound transducer at an angle aiming for the target. From this information a white trajectory is plotted on the screen. During needle insertion, guidance is optimized with the aid of coloured orientation bars (where green represents good orientation and red represents poor orientation) and a schematic diagram of needle transducer alignment (Fig. 16.2). The advantages of this system are that the optimal needle orientation is decided before insertion, potentially reducing the number of needle passes, the needle tip is seen irrespective of plane or angle, and that it is applicable to linear and curvilinear transducers. Two clinical

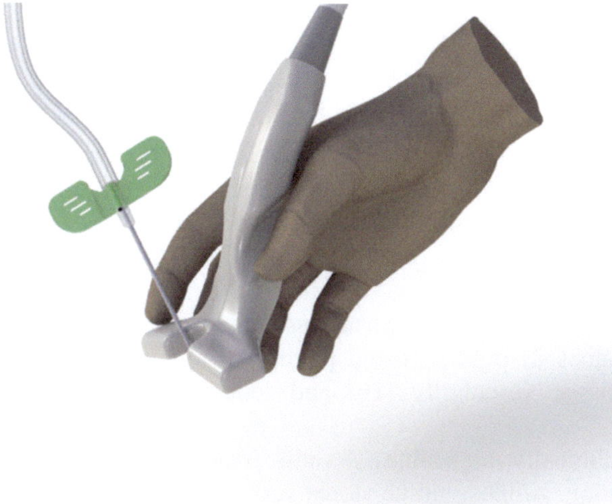

Fig. 16.2 Real-time 3D tracking system (Image courtesy of Sonix GPS, Ultrasonix, Vancouver, Canada) The *white* trajectory plotted on the screen, aims for the target pre-block. During needle insertion, guidance is optimized with the aid of coloured orientation bars (where *green* represents good orientation and *red* represents poor orientation) and a schematic diagram of needle transducer alignment

studies have reported on performance of spinal anaesthesia using electromagnetic guidance. Successful spinal anaesthesia was achieved in between 70 and 90 % of patients [9, 10], but on the first pass of the needle only in between 35 and 57 % of patients. These results show the potential utility of a GPS system, but one must bear in mind that these results were obtained in patients with BMI <35 who had no history of back surgery or spinal deformation – a low risk group! It is crucial, when introducing new technology to the market place in any speciality, that efficacy and side effects are investigated on a predefined population that encapsulates the population of patients with most to gain and/or at high risk of full or partial technical failure. Comparison between low risk and high risk groups is called stratification and offers a more precise evaluation of efficacy and side effects [11].

The GPS system was also evaluated in 26 medical students who conducted 30 simulated ultrasound nerve blocks on a pork simulator [12]. Students were randomised into GPS tracking or control groups, and block success or failure recorded. The GPS group had fewer attempts and conducted blocks in a shorter length of time compared to those using standard B-Mode ultrasound. It would appear that the GPS system may have benefits to training novice practionners.

16.2.3 Optical Needles

Knowledge of real-time tissue properties during needle movement could potentially improve the accuracy of regional anaesthesia and cancer biopsy. The optical characteristics of tissues, measured using spectroscopy, have been investigated as a means of differentiating between different tissues and may indicate the position of the needle tip. Integrated optical fibres transmit near infra red light from the tip of the needle or introducer. Spectroscopic analysis of reflected light estimates the scattering coefficient and the absorption coefficient of tissue such as haemoglobin, oxygenated haemoglobin, water and adipose tissue. Each tissue has a characteristic narrow absorption peak; oxyhaemoglobin absorbs light in the visible spectrum, whereas lipid and water absorb light in the near-infrared spectrum.

Experiments conducted in pigs have identified the ligamentum flavum and epidural space [13] and transitions from muscle to nerve in the axillary region [14]. Proximity to peripheral nerves was associated with higher lipid and lower hemoglobin values in pigs [14] and patients [15]. Future benefits may include detection of tissue vascularity [16] and reduction in accidental intravascular injection.

Although clinically appealing, several issues need to be resolved before commercialization. Adipose tissue stores vary between patients and with age, sex and BMI; spectral change from muscle to nerve is more acute than adipose tissue to nerve; and false positive results may arise from blood accumulation at the needle tip due to repeated injections. Spectral analysis may also be altered by transducer compression, reducing the distance between the skin and nerve target, and needle angulation. Future clinical trials will establish diagnostic accuracy by measuring the sensitivity and specificity of the test compared to traditional loss of resistance techniques.

16.2.4 Optical Coherence Tomography

Optical coherence tomography (OCT) is a laser based imaging technology already used clinically to image retinal disease. Resolution is remarkable down to between 10 and 30 μm. However, backscatter and absorption of the laser beam limit imaging to the first 2–3 mm of tissue. Incorporation of OCT into an epidural needle [17–19] has the potential to Image anatomy in deep spaces with very high resolution, delineate anatomical boundaries and reconstruct images in 3-D.

16.2.5 Vibrating Needles

Vibration of needles at ultra-high frequencies has been investigated as a means of visualising interventional needle. Actuation of peizo-electric crystals up to 20,000 Hz has recently demonstrated full visibility of needles using colour Doppler ultrasound (Fig. 16.3). Initial cadaver work have shown improved visualisation along the length of needle and at the needle tip at any angle, tissue depth and plane relative to the ultrasound transducer [20]. Added advantages include reduced force of tissue penetration and less deflection.

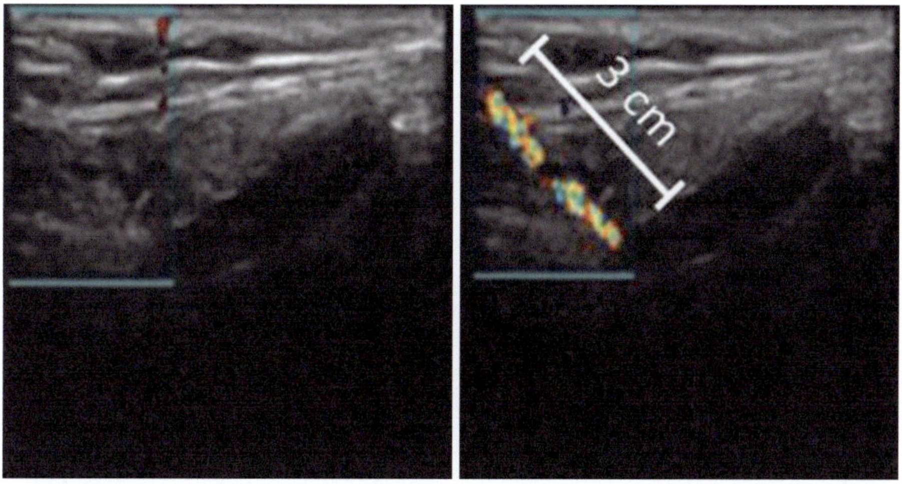

Fig. 16.3 Vibrating needle. *Left hand* image shows B-Mode image with echogenic needle poorly visible. Activation of the vibrating needle shows the same needle in colour visible using colour doppler imaging (Image courtesy of Dr M Sadiq PhD)

16.3 Transducer

16.3.1 Beam Steering

In order to improve image resolution, the ultrasound beam may be steered electronically by firing individual elements [21]. As the furthermost elements have further to travel, they are fired earlier while maintaining a constant distance between each line (Fig. 16.4). A common application of beam steering is spatial compounding whereby several beams are fired at several small angles then combined to form a smooth image. When imaging needles in-plane to the ultrasound beam, increasing the angle of needle insertion results in greater echo scatter, reducing needle visibility. Beam steering [22] increases visibility of needles and reduces artifacts by changing the relative angle of the needle to the beam.

Transducer design is limited by the configuration of ultrasound elements and anatomy. For example, transducers with a smaller footprint or diameter are often useful when visualizing needle insertion underneath the clavicle. Traditional linear or curvilinear arrays, although used regularly by radiologists for diagnostic work, dictate the orientation and angulation of needles during perioperative interventional work. There is a clear need for a clinical interventional transducer that can accommodate full angulation and manipulation of the needle, enhance visibility of needles and tissue interactions and minimize the distance from skin to target nerve, reducing tissue trauma.

In order to accommodate the needs of the anaesthetist, the design of the ultrasound transducer has been adapted in various ways. Designs incorporating a gap within the

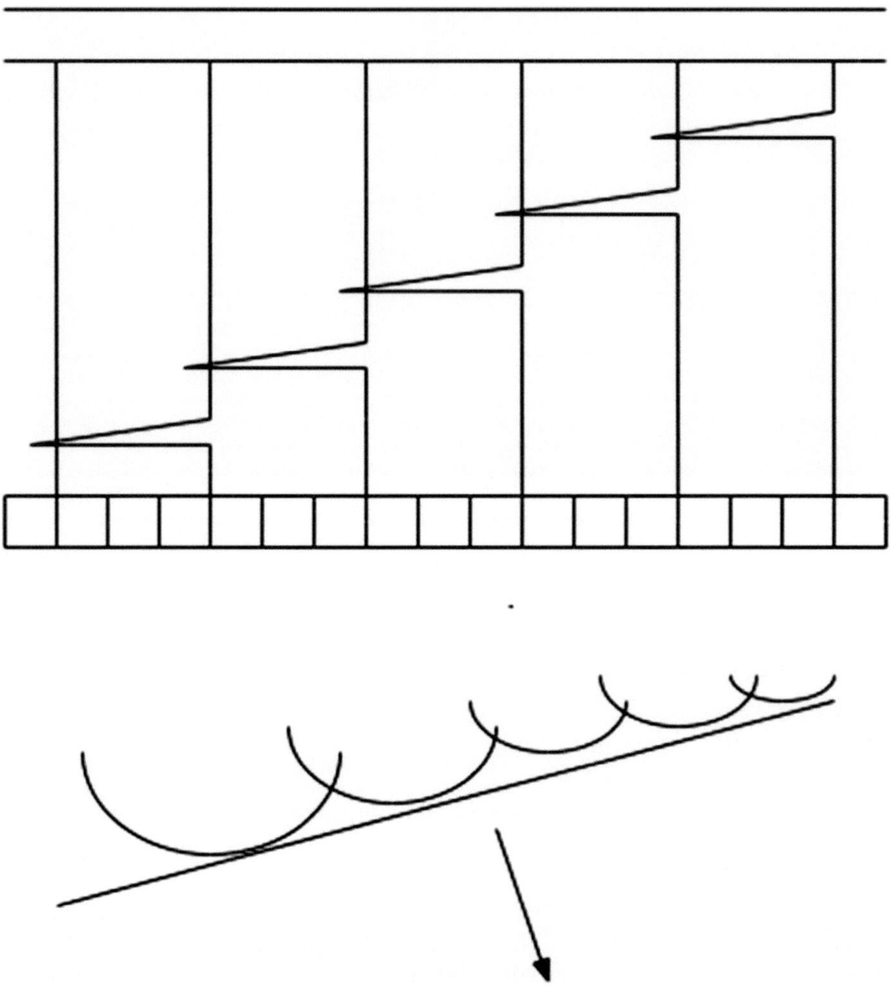

Fig. 16.4 Electronic beam steering. Programmed variation in firing of individual or groups of ultrasound elements steers the direction of ultrasound beams and improves imaging. A common example of use is transthoracic 3D echocardiography

middle of a linear array [23] or between two linear arrays have been patented [24, 25]. The principal advantage of such adaptations is that the needle may be inserted directly over the target and the distance from skin to nerve reduced. However, two major disadvantages arise: the central gap invariably masks visibility of the nerve, and the smaller the size of the gap, the more difficult it is to manipulate the needle.

An alternative design has been patented which comprises two linear arrays angled between 120° and 140° to each other to form a "V" shape [26]. Several advantages arise from this configuration (Fig. 16.5). The transducer may conform to the shape of body surfaces, the angulation being specific for each anatomical region;

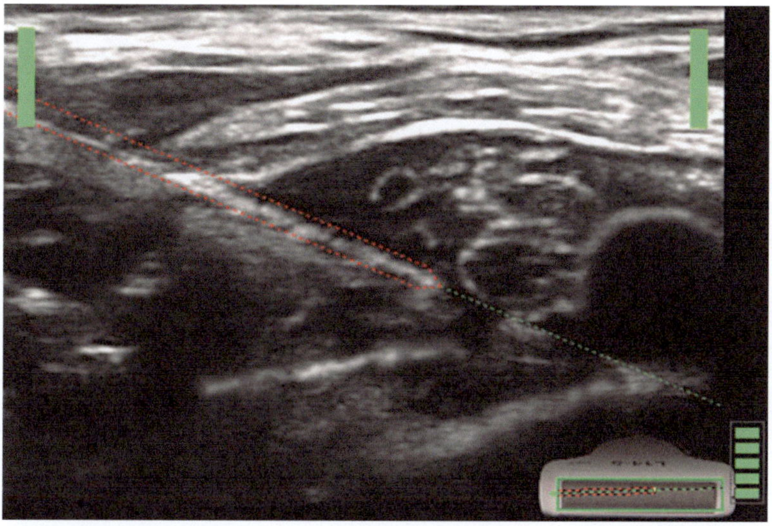

Fig. 16.5 New bi-array ultrasound transducer. Two linear arrays angled 130° to each other. A central gap allows needle manipulation; Incorporation of beam steering within the arrays reduces angle between needle and transducer to 45°, and improves needle visibility

a central gap allows needle manipulation; and needle visibility is improved because the relative angle of the needle inserted perpendicular to the transducer is reduced from 90° to between 60° and 70°. Incorporation of beam steering within the arrays further reduces the angle to 45° with good visibility of the needle. Overlapping arrays also reduce the superficial gap, allowing needle visibility. Potential applications beyond UGRA include breast and axillary biopsy, and joint or varicose vein injection under direct vision.

16.3.2 Three/Four Dimensional Ultrasound

Current 2-D imaging during regional anaesthesia has several limitations: 3-D anatomy must be inferred from 2-D knowledge and experience; once the transducer has moved, areas of interest are difficult to relocate; measurement of 2-D areas of interest do not necessarily correlate with 3-D volumes of interest; and needle visibility is only possible when the needle lies within the narrow line of the beam.

Two types of transducer are used in the creation of 3D images over time, otherwise termed 4D ultrasound [27]. Mechanical 3D transducers generate a series of 2-D images by sweeping a linear transducer fan-like around a pivot over the volume of interest. However, this mechanism has an in-built flaw. When acquiring images from greater depths, increased physical separation occurs between images and resolution degrades.

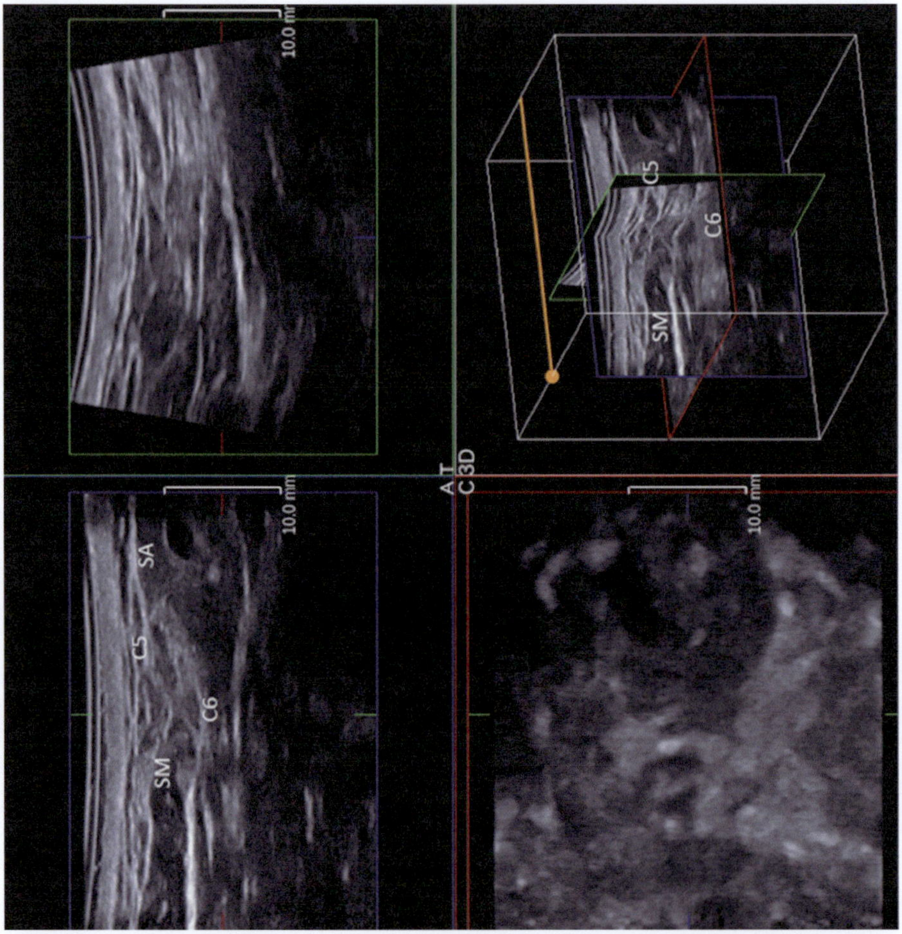

Fig. 16.6 3D image of male volunteers neck. The multiplanar image (*white square*) is constructed from a sagittal plane (*blue square*) and a coronal plane (*green square*). In the sagittal plane the interscalene groove and C5 and C6 nerve roots are visible, along with the scalenus medius (*SM*) and scalenus anterior (*SA*) muscles. The nerve roots are more difficult to see in the coronal plane. The multiplanar view offers little extra information and the rendered view (*red square*) is uninterpretable

New 2D matrix transducers with several thousand elements in a square configuration show better resolution. Using this technology, a stationary 2D phased array sweeps out a pyramid-like volume, and displays a set of multiple planes in real time. With 4D ultrasound the transducer is held still and real-time visualization of the needle and local anaesthetic spread is possible in multiple planes.

Three dimensional images are visualized using two techniques. The first, multiplanar reformatting, presents images as a 3D crossed plane (Fig. 16.6). With crossed planes, flat 2D planes are visualized in three orthogonal or perpendicular planes and each plane moved relative to the others in order to select the optimal image (Fig. 16.6, bottom right image).

The second, volume rendering, is a process which projects 3D data onto a 2D surface by using a technique termed ray-casting in order to optimize the projection effect, degree of translucency and surface shadowing of the image (Fig. 16.6, bottom left image). Rendering works best in situations where fluid provides a contrast to tissue planes such as foetal imaging and cardiac 4D transthoracic imaging. A good example in anaesthesia is the preoperative real-time visualization of heart valve and ventricular wall motion [28].

Few studies have investigated the role of 3D imaging in regional anaesthesia. In separate volunteer studies, the 3D anatomy of the brachial plexus [29] and paravertebral region [30] has been described, whereas, clinically, 3D ultrasound has been used to place popliteal [31] and infraclavicular catheters [32]. However, transducer selection was limited to low frequency devices designed to visualize deep structures for non-anaesthetic application such as breast imaging. In a study of 16 patients, the ligamentum flavum/dura complex was visualized in all except 2 patients. Loss of resistance and epidural catheter insertion were successful in all patients but epidural local anaesthetic spread was only seen in 5 out of 16 patients. However, the authors [33] state that "for the best results, good 2D image scanning should be achieved", indicating a new technology that superficially appears helpful, is still dependant on current B-Mode ultrasound technology, and moreover subject to the same biological and technical constraints, Thus, one would anticipate the poor visibility of 2D B-Mode ultrasound through the acoustic window in an obese patient would translate to poor 4D visibility.

16.4 Micro-ultrasound

As ultrasound waves pass through tissue, acoustic intensity diminishes as energy is attenuated within tissue. The greater the distance travelled and the higher the transducer frequency the more energy is lost. Attenuation of ultrasound energy occurs through absorption, scattering, reflection or refraction of ultrasound. Reflection and refraction of ultrasound waves occurs at the interface of tissues with different acoustic impedance, the product of density (σ) and speed of sound of acoustic waves (c). The greater the relative difference in impedance, the greater the reflection and the less the transmission of ultrasound waves [34]. The brightness of the image on the screen is dependant on the amount of acoustic energy returning back to the transducer; surfaces such as periosteum, tendons, pleura are hyperechoic because they reflect more energy, whereas interscalene nerve roots, blood and fluids are hypoechoic because they reflect less energy. Thus, in current clinical practice, linear array high transducer frequencies between 8 and 15 MHz are recommended order to image superficial anatomy whereas to image deeper structures convex low frequency 2–6 MHz transducers are used, albeit with less resolution.

In basic science, there is a need to scan small animal models in the fields of developmental biology, genomics and cancer in order to characterize anatomy. Both CT and MRI provide high quality images but are expensive and cumbersome. In order to observe changes in tissue configuration, a high resolution of 20–100 μm is

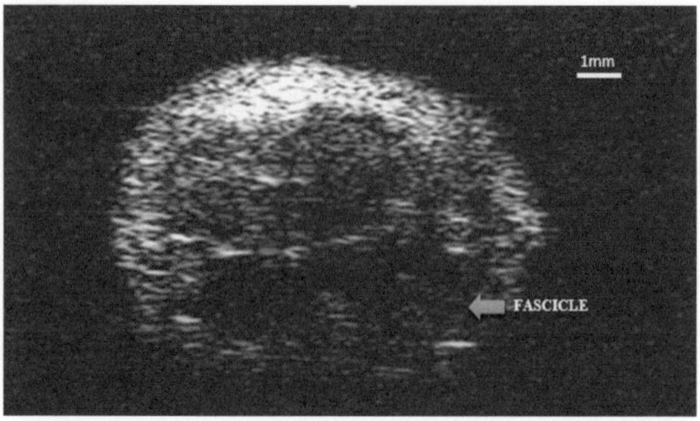

Fig. 16.7 Microultrasound image of cadaver median nerve. Fascicles seen as poorly echogenic round areas surrounded by brighter interfascicular or subepineural tissue

needed, corresponding to ultrasound frequencies between 150 and 30 MHz. Mechanical single element transducers with a capacity to generate frequencies up to 100 MHz have traditionally been used, but are now being replaced by linear transducer arrays with greater functionality. As well as investigating the molecular basis of disease, micro-ultrasound will, in future, be used clinically to characterize the ultrasound signatures of organs such as breast, prostate and skin in order to detect and treat pre-cancerous tissue changes.

Application of micro-ultrasound to UGRA may not be immediately obvious until one recognizes that, although the widespread application of B-Mode ultrasound to UGRA has improved the efficacy of nerve block, the incidence of transient neuropraxia has remained unchanged compared to peripheral nerve stimulation techniques [35]. As described above, the greater the distance between transducer and target, the greater the energy attenuation and the poorer image resolution. Experimentation with different shaped transducers using a wide range of frequencies is unlikely to resolve this clinical dilemma as obesity levels are rising within Western economies. One solution is to develop active ultrasound needles with peizo-electric crystals at the tip in order to more accurately identify relevant anatomy and target local anaesthetic injection. For clinical application we would propose using a combination of an active needle with a traditional ultrasound transducer and two images. The needle would be guided to, for example, within 1–1.5 cm of the target nerve, then the micro-image activated, nerve and connective tissue identified and the needle tip guided precisely to its destination. Two studies have been published on the application of microultrasound in regional anaesthesia. Scanning fresh cadaver nerves using 30 MHz frequencies, laboratory images [36] were able to delineate the epineurium, the size of the nerve, the number and size of fascicles, as well as visualise needle insertion and intraneural tissue trauma. Ultrasound images showed fascicles down to 0.5 mm diameter (Fig. 16.7), and splitting or rotation of fascicles when a needle was inserted into a nerve. In the second study [33], a single crystal 40 MHz ultrasound transducer was placed within an 18 g Tuohy needle in order to locate the epidural space of experimental pigs. All spaces were successfully identified and catherised using the paramedian approach in five pigs. Injected contrast confirmed anatomical spread.

16.5 Elastography

Technologies which describe or measure tissue stiffness or elasticity are described under the umbrella term elastography, suggested by Ophir et al. in 1991 [37]. Many forms of elastography exist but common to all are two factors – a means of applying stress to tissue and a means of measurement. Stress may be applied mechanically using transducer compression [38] or internally using an ultrasonic radiation force [39] while measurement may be achieved using magnetic resonance imaging (MRI) or ultrasound. The latter is popular because it has a high spatial resolution, works in real time, and uses conventional ultrasound equipment. Elastography is increasingly used for diagnosis of breast [40, 41] and prostate [42] cancer because it provides a colour image representative of tissue stiffness, a parameter analogous to clinical palpation.

Awareness of the limitations of B-mode ultrasound stimulated the development of new technology able to differentiate tissue on the basis of its physical properties. Palpation is ostensibly an assessment of tissue elasticity – a material's resistance to deformation, and described mathematically as the application of force over a specified area (stress) with resultant displacement of tissue (strain). Stress (σ) is defined as Force per unit Area with the units kPa or Nm^{-2} whereas strain(ϵ) may be regarded mathematically as $\epsilon = (L_f - L_0)/L_0$ where L_f is the length after and L_0 is the length before tissue displacement. Thus strain is a unitless entity [43].

Tissue elasticity [44] may be regarded as longitudinal, transverse and volume and described mathematically by Young's (E), Shear (G) and Bulk (K) moduli [43]. The fourth important concept is Poissons's ratio (σ), the ratio of lateral contraction to longitudinal extension. All four terms are mathematically interrelated.

The mechanics of soft tissues are very complex and in order to measure their elastic properties various assumptions are made. Tissues are regarded as homogeneous, isotropic, incompressible solids that obey Hooke's law (i.e. low values of strain are directly proportional to stress), have a density ρ of 1000 $kg.m^{-3}$ and waveform velocity equivalent to the speed of sound, 1540 $m.s^{-1}$. For incompressible tissue, Poisson's ratio is accepted as 0.5 and the relationship between Young's modulus and Shear modulus is fixed according to the equation: $E = 3.G$.

The practical implications of elasticity measurement are that bulk modulus tends to remain similar fairly similar throughout tissue whereas shear modulus may vary tenfold.

16.5.1 Strain Elastography

The elastic properties of tissue may be visualised using the qualitative method of strain elastography. Raw radiofrequency data (otherwise termed ultrasound A-lines), collected before and after transducer compression, is split into "windows" and compared using cross-correlation methods. The difference between waveforms equates to tissue displacement or shift. The rate of change in tissue displacement as a function of the distance from the transducer, is computed and displayed as a colour elastogram.

Low strain or tissue displacement in response to transducer pressure is visualised as a blue background colour whereas injection of local anaesthetic is displayed as the yellow, red and orange colours, indicative of greater strain due to formation of a hydrospace. An example of strain elastography in the same patient before and during infraclavicular nerve block is seen in Fig. 16.8. The elastogram shows a coloured "U" shaped strain pattern secondary to fluid spread between the subclavian artery and posterior cord of the brachial plexus. Circumferential flow around nerves and blood vessels [45], inadvertent spread and, in the cadaver, intraneural and intravascular placement are also seen [46].

Several factors account for the differing size and shape of strain patterns during the conduct of regional anaesthesia. The 2D area of the strain pattern does not equate to the 2D area of the hydrospace because injection into an enclosed sheath displaces both fluid and tissue. Therefore the edges of the tissue will be coloured depending on their relative strain or displacement.

The site of regional block determines the size and pattern of tissue strain. Increased spread of fluid around the femoral or popliteal nerves is associated with greater strain than around the interscalene nerves [47]. Both stress and strain decrease with tissue depth as structures beyond 4 cm are poorly compressed and are difficult to see. Unfortunately, needles, even when moving, are indistinct with a wide strain pattern. The role of elastography in regional anaesthesia is yet to be established and studies are ongoing to determine its role in the recognition of accidental intraneural injection.

16.5.2 Shear Wave Elastography

Developments in elastography have progressed from investigation of axial strain to exploration of shear forces, which move at right angles to an applied force. Ultrasound beams are transmitted at high speed and focused in a cone-like shape, transferring energy to tissues and inducing transverse shear waves, not unlike the ripples of water generated by a thrown stone. The velocity of shear waves varies between 1 and 10 m.s^{-1}, crossing an ultrasound image plane of 3 cm width in 3–30 ms. As a tracking frequency of 5–20 Hz is needed to detect shear waves, a dedicated ultrasound machine (Supersonic Imagine, Aix-en-Provence, France) is needed.

The speed of a shear wave is directly related to the elasticity or stiffness of tissue and can be quantified in kilopascals and mapped as bands of tissue elasticity. In contrast to strain elastography, stiffer tissues are coded in red and softer tissues in blue. The elasticity of a solid is a function of its density and longitudinal ultrasound wave speed according to the formula:

$$K = \rho . c_s^2$$

where K = shear modulus, Young's modulus = 3. K,
c_s = shearwave speed and ρ = tissue density

The advantages of shear wave elastography are that no manual compression is required, and that data is repeatable and reproducible. However, application of the transducer requires a light touch as any pressure will show as a surface tissue strain. Shear waves cannot pass through fluid pockets, and thus injectate appears as dark shadows, albeit accurately mapping the 2D size of the hydrospace. Diagnosis of hard lesions breast cancer has a high sensitivity and specificity [41]. Application of shear wave elastography to regional anaesthesia has successfully differentiated nerve from surrounding muscle and interstitial tissue [48]. Young's modulus in Thiel embalmed human cadavers [49] and volunteers [48] show a consistent four-fold difference in tissue elasticity between nerve and tissue, illustrated in Fig. 16.9 clinical studies quantifying the relative stiffness and validating the cross-sectional area of peripheral nerves are in progress.

16.5.3 Acoustic Force Radiation Impulse

Acoustic force radiation impulse (*AFRI*) imaging is based on the principle that distinct tissue types react differently to mechanical force [50]. For example stiff tissue is barely displaced in contrast to the greater movement of elastic tissue. Acoustic force radiation impulse imaging generates a short burst of acoustic energy in order to move tissue a distance of 1–10 μ within a region of excitation (ROE).

The distribution of the radiation force field is determined by both the acoustic energy and the acoustic properties of the tissue. In soft tissues, where the majority of attenuation results from absorption, the following equation can be used to determine radiation force magnitude:

$$F = \frac{2 \propto I}{c}$$

where the intensity of the force is determined by the local acoustic intensity (I), acoustic attenuation (α) and speed of sound (c). Transient shear waves are generated that propagate away from the ROE. By measuring the speed of these waves, the local value of the shear modulus can be estimated. Higher shear wave speeds and smaller displacements are associated with stiffer tissues, and slower shear wave speeds and larger displacements occur with more compliant tissues.

AFRI imaging has spatial resolution comparable to that of B-mode, and provides greater contrast in both cadaver models and patients for brachial plexus and sciatic blocks [51]. However, real-time ARFI imaging, encounters significant challenges with heating of the tissue and transducer.

Many peripheral nerves targeted for regional anesthetic injection (e.g., the brachial plexus and femoral nerves) are adjacent to arteries that can introduce significant physiologic motion when tracking ARFI-induced displacements [51]. To minimize such motion artifacts, B-mode and AFRI image acquisitions are ECG-triggered during *in-vivo* imaging. Needles appear in AFRI images as areas with

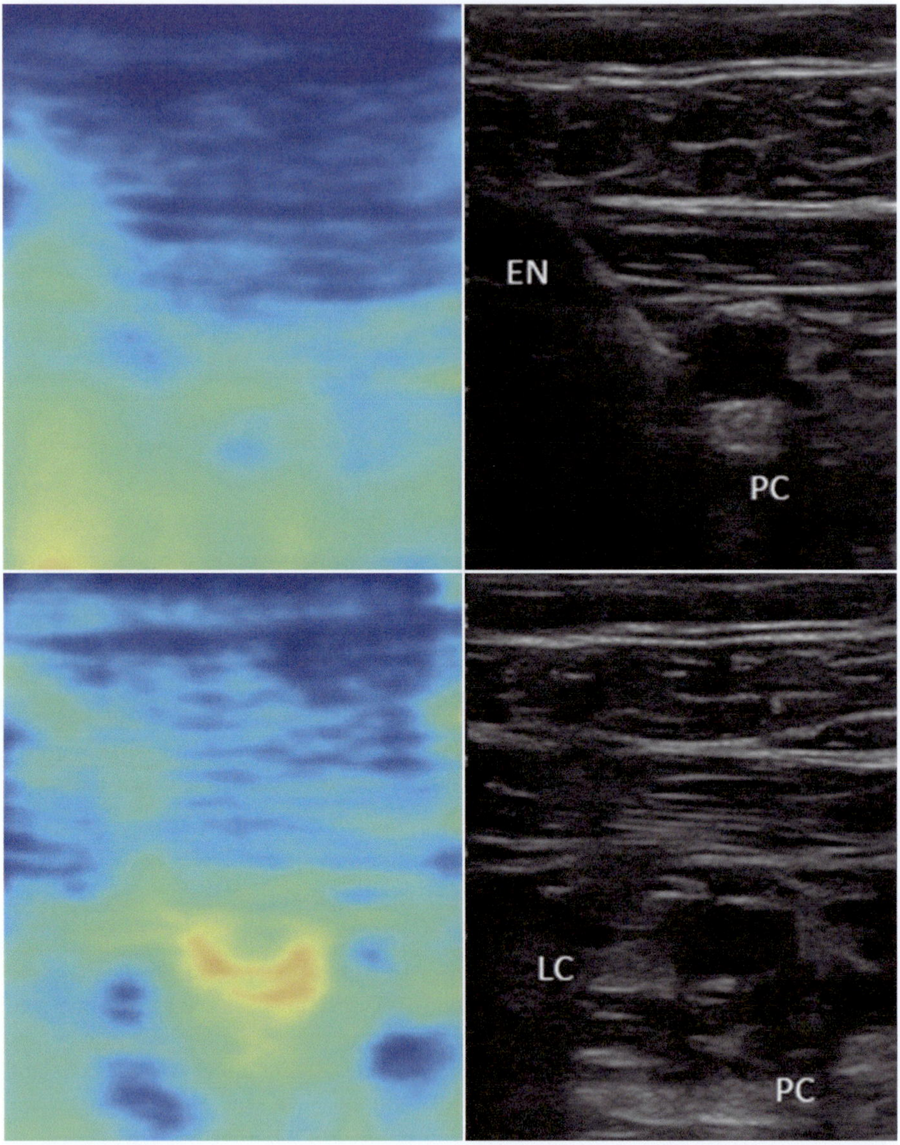

Fig. 16.8 B-Mode ultrasound image of infraclavicular block with single injection echogenic needle (*EN*) guided in-plane towards junction of subclavian artery and posterior cord (*PC*) of brachial plexus. Coloured area shows tissue strain in a classical U-shaped pattern between subclavian artery and posterior cord postero-inferior to lateral cord (*LC*)

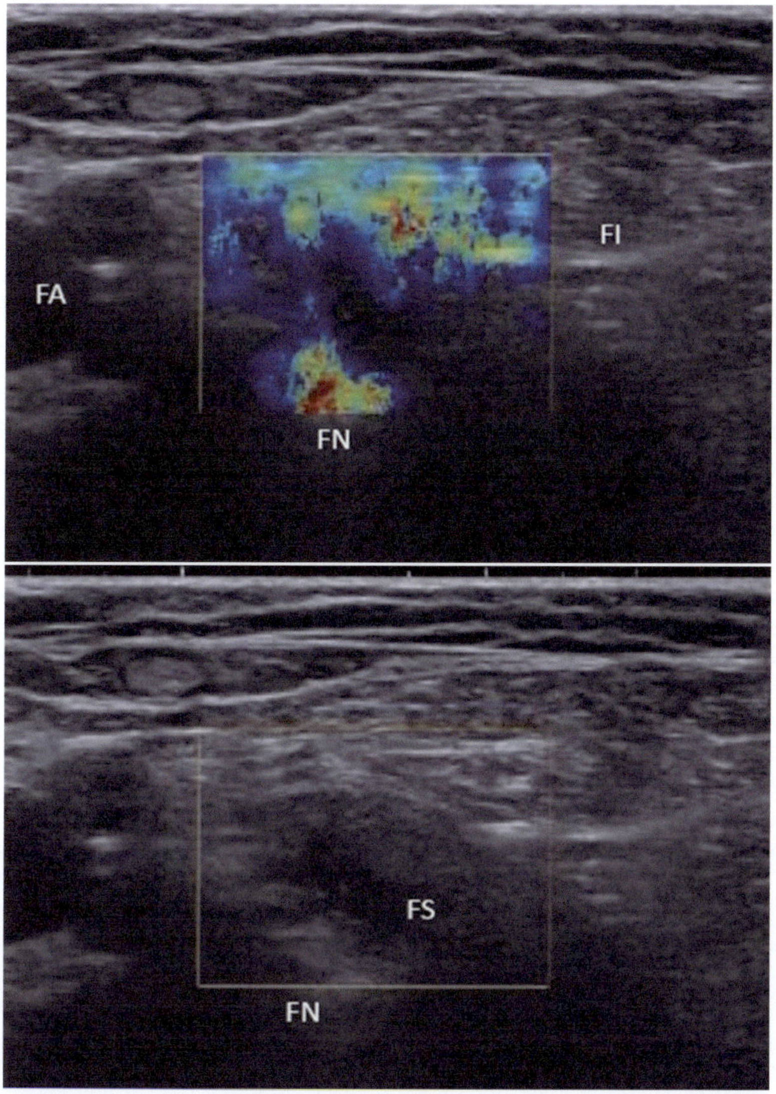

Fig. 16.9 B-Mode (*lower image*) and B-Mode overlaid with shear wave elastogram (*upper image*) in patient receiving a left sided femoral nerve block. Note the coloured areas of higher Young's modulus representing the femoral nerve (*FN*) and the fascia iliaca (*FI*). Between them is an anechogenic area of local anaesthetic similar in size in both images, delineating local anaesthetic fluid spread (*FS*). (*FA*) femoral artery

lower displacement than their surroundings. AFRI images overlaid on matched B-mode images hold promise for improved needle visibility in many clinical applications.

16.6 Conclusion

The future of ultrasound holds great promise. Ultrasound technology providing resolution to at least 20 μm is available in bioengineering laboratories, and is already being used to visualize and measure the properties of tissue. For surgery, it is envisioned that electromagnetic (EM) tracking will be fused with other modalities such as CT or MRI [6–8]. Already a commercial high-end ultrasound system is able to link CT images of prostate cancer to ultrasound images for needle biopsy. For patients undergoing ultrasound guided regional anaesthesia, benefits will be derived in some part from design and production of anaesthesia specific ultrasound transducers, software modifications enabling clear visibility of non-echogenic needles in colour and from advances in ultrasound guided cancer biopsy. Rather than attempt to develop ultrasound in order to visualize deep structures, research is ongoing designing high frequency needles which incorporate ultrasound elements at their tip. Miniature ultrasound transducers integrated within needles will enable clinicians to image regions such as the epidural space or perineural spaces and identify tissue morphology from within the tissue region of interest.

Examples of recent innovation include: 4D ultrasound imaging using electronic square arrays with beam steering, echogenic needles and electro-magnetic needle guidance systems. However, new technology offers much more in the future: novel ergonomic ultrasound transducer design; highly echogenic needles visible at any angle or depth; active needles incorporating ultrasound peizo-crystals at the tip operating above 30 MHz able to visualize distinct anatomical features both inside and outwith nerves; and colour images of tissue strain or tissue stiffness classified under the umbrella term of elastography.

Statement Dr Graeme McLeod is Consultant & Honorary Professor in the Institute of Academic Anaesthesia, Ninewells Hospital & Medical School, Dundee. Dr McLeod has 20 % share of patent WO2013030556 A1, 2013/CA2847015 A1, 2013 and 5 % share of patent GB 1304798.0, 2013.

References

1. Wiesmann T, Borntrager A, Neff M, Wulf H, Steinfeldt T (2012) Needle visibility in different tissue models for ultrasound-guided regional anaesthesia. Acta Anaesthesiol Scand 56:1152–1155
2. Munirama S, Joy J, Columb M et al (2015) A randomised, single-blind technical study comparing the ultrasonic visibility of smooth-surfaced and textured needles in a soft embalmed cadaver model. Anaesthesia 70:537–542

3. Moayeri N, Krediet AC, Welleweerd JC, Bleys RL, Groen GJ (2012) Early ultrasonographic detection of low-volume intraneural injection. Br J Anaesth 109:432–438
4. Guo S, Schwab A, McLeod G et al (2012) Echogenic regional anaesthesia needles: a comparison study in Thiel cadavers. Ultrasound Med Biol 38:702–707
5. Edgcombe H, Hocking G (2010) Sonographic identification of needle tip by specialists and novices: a blinded comparison of 5 regional block needles in fresh human cadavers. Reg Anesth Pain Med 35:207–211
6. Sonn GA, Natarajan S, Margolis DJ et al (2013) Targeted biopsy in the detection of prostate cancer using an office based magnetic resonance ultrasound fusion device. J Urol 189:86–91
7. Amalou H, Wood BJ (2012) Multimodality fusion with MRI, CT, and ultrasound contrast for ablation of renal cell carcinoma. Case Rep Urol 2012:390912
8. Appelbaum L, Solbiati L, Sosna J, Nissenbaum Y, Greenbaum N, Goldberg SN (2013) Evaluation of an electromagnetic image-fusion navigation system for biopsy of small lesions: assessment of accuracy in an in vivo swine model. Acad Radiol 20:209–217
9. Brinkmann S, Tang R, Sawka A, Vaghadia H (2013) Single-operator real-time ultrasound-guided spinal injection using SonixGPS: a case series. Can J Anaesth J Can Anesth 60:896–901
10. Niazi AU, Chin KJ, Jin R, Chan VW (2014) Real-time ultrasound-guided spinal anesthesia using the SonixGPS ultrasound guidance system: a feasibility study. Acta Anaesthesiol Scand 58:875–881
11. Munirama S, McLeod G (2015) 'Stratified' approach to individualized anaesthetic care. Br J Anaesth 114:543–545
12. McVicar J, Niazi AU, Murgatroyd H, Chin KJ, Chan VW (2015) Novice performance of ultrasound-guided needling skills: effect of a needle guidance system. Reg Anesth Pain Med 40:150–153
13. Desjardins AE, Hendriks BH, van der Voort M et al (2011) Epidural needle with embedded optical fibers for spectroscopic differentiation of tissue: ex vivo feasibility study. Biomed Opt Express 2:1452–1461
14. Brynolf M, Sommer M, Desjardins AE et al (2011) Optical detection of the brachial plexus for peripheral nerve blocks: an in vivo swine study. Reg Anesth Pain Med 36:350–357
15. Balthasar A, Desjardins AE, van der Voort M et al (2012) Optical detection of peripheral nerves: an in vivo human study. Reg Anesth Pain Med 37:277–282
16. Soto-Astorga RP, West S, Putnis S, Hebden JC, Desjardins AE (2013) Epidural catheter with integrated light guides for spectroscopic tissue characterization. Biomed Opt Express 4:2619–2628
17. Tang Q, Liang C-P, Wu K, Sandler A, Chen Y (2014) Real-time epidural anesthesia guidance using optical coherence tomography needle probe. Quant Imaging Med Surg. 2015; 5:118–24
18. Scolaro L, Lorenser D, McLaughlin RA, Quirk BC, Kirk RW, Sampson DD (2012) High-sensitivity anastigmatic imaging needle for optical coherence tomography. Opt Lett 37:5247–5249
19. Tang Q, Liang CP, Wu K, Sandler A, Chen Y (2015) Real-time epidural anesthesia guidance using optical coherence tomography needle probe. Quant Imaging Med Surg 5:118–124
20. Sadiq M, Cochran S, Huang Z, Corner G, McLeod G, Carena P (2014) Medical apparatus and its visualisation. Patent No. WO2014140556 A1. UK
21. Powers J, Kremkau F (2011) Medical ultrasound systems. Interface Focus 1:477–489
22. Wiesmann T, Borntrager A, Zoremba M, Neff M, Wulf H, Steinfeldt T (2013) Compound imaging technology and echogenic needle design: effects on needle visibility and tissue imaging. Reg Anesth Pain Med 38:452–455
23. Hossack JA (2002) Intra-operative diagnostic ultrasound multiple-array transducer probe and optional surgical tool US. Patent NO.: US 6,423,002 B1
24. Watson M, Corner G, Kirk K, Cochran A, Lines D, Rajagopal S (2008) Ultrasound Waveguide. Patent No. WO2005122903A1. UK
25. Cochran S, Habeshaw R, McLeod G, Corner G, Marsh J (2013) Ultrasound probe. Patent No. WO2013030556 A1. UK

26. Cochran S, Habeshaw R, Mcleod G, Corner G, Marsh J (2013) Ultrasound probe. In: Dundee Uo (ed). Patent No. WO2013030556 A1. UK
27. Fenster A, Parraga G, Bax J (2011) Three-dimensional ultrasound scanning. Interface Focus 1:503–519
28. Lynes H, Griffiths R (2013) Focused transthoracic echocardiography in hip fracture surgery. Anaesthesia 68:206–207
29. Cash CJ, Sardesai AM, Berman LH et al (2005) Spatial mapping of the brachial plexus using three-dimensional ultrasound. Br J Radiol 78:1086–1094
30. Karmakar MK, Li X, Li J, Hadzic A (2012) Volumetric three-dimensional ultrasound imaging of the anatomy relevant for thoracic paravertebral block. Anesth Analg 115:1246–1250
31. Feinglass NG, Clendenen SR, Torp KD, Wang RD, Castello R, Greengrass RA (2007) Real-time three-dimensional ultrasound for continuous popliteal blockade: a case report and image description. Anesth Analg 105:272–274
32. Clendenen SR, Robards CB, Clendenen NJ, Freidenstein JE, Greengrass RA (2010) Real-time 3-dimensional ultrasound-assisted infraclavicular brachial plexus catheter placement: implications of a new technology. Anesthesiol Res Pract 2010: 208025 doi: 10.1155/2010/208025
33. Voloshin AG (2015) Four-dimensional ultrasound guidance during epidural anaesthesia. J Ultrasound 18:135–142
34. Hoskins P, Martin K, Thrush A (2010) Diagnostic ultrasound: physics and equipment. Cambridge University Press, Cambridge
35. Munirama S, McLeod G (2013) Systematic review and meta-analysis. Anaesthesia. 2015 Sep;70(9):1084–91
36. Chandra A, Felts P, Eisma R, Munirama S, Demore C, McLeod G (2014) Visualizing nerve trauma caused by needle penetration during regional anaesthesia. Br J Anaesth 113:318
37. Ophir J, Cespedes I, Ponnekanti H, Yazdi Y, Li X (1991) Elastography: a quantitative method for imaging the elasticity of biological tissues. Ultrason Imaging 13:111–134
38. Treece G, Lindop J, Chen L, Housden J, Prager R, Gee A (2011) Real-time quasi-static ultrasound elastography. Interface Focus 1:540–552
39. Palmeri ML, Nightingale KR (2011) Acoustic radiation force-based elasticity imaging methods. Interface Focus 1:553–564
40. Evans A, Whelehan P, Thomson K et al (2010) Quantitative shear wave ultrasound elastography: initial experience in solid breast masses. Breast Cancer Res BCR 12:R104
41. Evans A, Whelehan P, Thomson K et al (2012) Differentiating benign from malignant solid breast masses: value of shear wave elastography according to lesion stiffness combined with greyscale ultrasound according to BI-RADS classification. Br J Cancer 107:224–229
42. Barr RG, Memo R, Schaub CR (2012) Shear wave ultrasound elastography of the prostate: initial results. Ultrasound Q 28:13–20
43. Wells PN, Liang HD (2011) Medical ultrasound: imaging of soft tissue strain and elasticity. J R Soc Interface R Soc 8:1521–1549
44. Wells PN, Liang HD, Young TP (2011) Ultrasonic imaging technologies in perspective. J Med Eng Technol 35:289–299
45. Munirama S, Joyce J, Eisma R et al (2012) A comparison of intraneural and extraneural shear modulus in the Thiel-embalmed human cadaver. British Journal of Anaesthesia 2012;108:712–713.
46. Munirama S, Schwab A (2012) Application of sonoelastography to regional anaesthesia: a descriptive study with the Thiel embalmed cadaver model. Ultrasound 20:41–48
47. Munirama S, Satapathy AR, Schwab A et al (2012) Translation of sonoelastography from Thiel cadaver to patients for peripheral nerve blocks. Anaesthesia 67:721–728
48. Munirama S, Joy J, Eisma R, Corner G, Cochran S, McLeod G (2013) Shear wave elastography: novel technology for ultrasound-guided regional anesthesia. Anesthesiology 119:698
49. Munirama S, Joyce J, Eisma R et al (2012) Comparison of intraneural and extraneural shear modulus in the Thiel-embalmed human cadaver. Br J Anaesth 108:712–713
50. Palmeri ML, Nightingale KR (2011) What challenges must be overcome before ultrasound elasticity imaging is ready for the clinic? Imaging Med 3:433–444
51. Palmeri ML, Dahl JJ, MacLeod DB, Grant SA, Nightingale KR (2009) On the feasibility of imaging peripheral nerves using acoustic radiation force impulse imaging. Ultrason Imaging 31:172–182

Chapter 17
Lung Ultrasound in Anaesthesia and Critical Care Medicine

David Canty, Kavi Haji, André Denault, and Alistair Royse

Abstract Lung ultrasound (respiratory or thoracic ultrasound) has traditionally been used for evaluation and guidance of drainage of pleural effusion, where it has been shown to reduce iatrogenic injuries from intercostal catheter insertion into adjacent organs. Only recently has its use in bedside diagnosis of respiratory disease become popular. Lung ultrasound is more accurate than chest radiography and approaches the accuracy of computed tomography in diagnosis of pleural effusion, pneumothorax, pulmonary oedema, consolidation and collapse, abscess, emphysema, and even pulmonary embolus. Improved accuracy and speed of diagnosis may also reduce the need for chest radiography and CT, reducing exposure of patients and staff to ionising radiation and the requirement to transport the critically ill. This useful technique is becoming incorporated into clinical practice and training in emergency and critical care medicine, where it has been shown to be effective in rapid diagnosis of the cause of respiratory distress. Additional potential uses for in anaesthetic practice include preoperative assessment in patients with dyspnoea, and rapid assessment of respiratory failure that may occur intra or postoperatively, for example from pneumothorax, acute pulmonary oedema and massive atelectasis or consolidation. Rapid bedside exclusion of pneumothorax may be useful after insertion of central line or paravertebral catheter, or prior to or during mechanical ventilation. There is also an emerging role of lung ultrasound in guidance of

---

D. Canty, MBBS, PhD, FANZCA, PGDipEcho (✉) • A. Royse, MBBS, MD, FCSANZ, FRACS
Department of Surgery, Ultrasound Education Group, University of Melbourne, Parkville, VIC 3050, Australia
e-mail: dcanty@unimelb.edu.au

K. Haji, FCICM, FACEM, PGDipPeriopEcho
Intensive Care Unit, Frankston Hospital, Frankston, VIC 3199, Australia

A. Denault, MD, PhD, FRCPC ABIM-CCM, FASE
The Montreal Heart Institute, Montreal, QC, Canada

Centre Hospitalier de l'Université de Montréal Hospitals, Montreal, QC, Canada

K. Stuart-Smith (ed.), *Perioperative Medicine – Current Controversies*,
DOI 10.1007/978-3-319-28821-5_17

endotracheal and subglottic airway management. Lung ultrasound is relatively easy to learn using portable ultrasound machines and can be integrated into routine clinical 'ultrasound-assisted examination'.

Keywords Ultrasonography • Lung • Point-of-Care Systems • Anaesthesia

17.1 Introduction

It has long been recognised that the location and volume of pleural effusions are relatively accurately assessed with surface ultrasound. More accurate quantification and localisation results in more precise and safer insertion of intercostal catheters. Ultrasound provides very specific location of pleural effusion. Despite evidence that ultrasound guided intercostal catheter insertion leads to reduced incidence of serious organ injury compared with blind placement [1], uptake has been slow [2]. Furthermore, point-of-care lung ultrasound is increasingly used [3] to non-invasively improve speed and accuracy in bedside diagnosis of respiratory pathology [4]. Lung ultrasound is more accurate than chest radiography and approaches the accuracy of computed tomography (CT) in the bedside diagnosis of pleural effusion, pneumothorax, pulmonary oedema, consolidation and collapse, abscess, emphysema, and even pulmonary embolus [5]. Improved accuracy and speed of diagnosis may also reduce the need for chest radiography and CT, reducing exposure of patients and staff to ionising radiation [6] and the requirement to transport the critically ill.

With the widespread availability of portable or personal "pocket" ultrasound devices, it is expected that point of care ultrasound will dramatically alter clinical practice, which has already being observed with transthoracic echocardiography [7]. Table 17.1 outlines the important features of point of care ultrasound. Ideally, lung ultrasound should be integrated with transthoracic echocardiography and into routine clinical examination forming "ultrasound assisted examination" [8].

With a practical focus, this chapter will cover how to perform lung ultrasound, diagnose respiratory pathology and how to use ultrasound to guide invasive thoracic procedures including drainage of pleural effusions and advanced airway management.

Table 17.1 Point of care use of lung ultrasound

The treating physician performs the scan
Scan at the time of clinical assessment is in real time
Scanning is focused on the clinical question at the time; without necessarily completing a comprehensive study in time critical circumstances
Scanning without altering patient position (e.g. for guidance of pleural catheter insertion for effusion)
Scanning whilst immediate changes to management are made to observe change
Repeated interval scanning to gauge response to management changes
Integration with other ultrasound examinations such as echocardiography or abdominal ultrasound

17.2 Lung Ultrasound Examination

The lung ultrasound examination is quite different to transthoracic echocardiography (TTE) because normal lung contains air, which prevents ultrasound transmission. Fortunately, the presence of lung pathology tends to be visible with ultrasound due to the accumulation of fluid within the lung and in the pleural space, which is permeable to ultrasound. As the lungs are superficial the practical skill required for lung ultrasound is easy to learn and most lung pathology is interpreted by *pattern recognition* rather than by quantifiable measurements. This makes integration into real time clinical practice simple and efficient.

In clinical practice however, lung ultrasound is often not an isolated examination; but rather an extension of another study such as TTE or Focused Assessment of Sonography in Trauma (FAST), or simply a focused study to exclude a specific clinical diagnosis. For example, in a patient with respiratory failure, where there is urgency, the ultrasound examination could start at the upper anterior chest for pneumothorax and pulmonary oedema, and then posteriorly at the lung bases, for consolidation and effusion. If a pneumothorax is found, then a contact or lung point may be searched for in order to roughly estimate size; and if a pulmonary embolus or cardiac failure is suspected then TTE and ultrasound of the venous system of the legs, could follow. TTE is complementary to lung ultrasound and should occur at the same time, resulting in cardiorespiratory ultrasound or ultrasound-assisted examination of the heart and lungs.

A description of a "standard" lung examination will follow but scanning technique will vary according to clinical urgency and whether the patient is supine or sitting.

17.2.1 Probe Selection

Probe selection depends on the likely pathology, obesity, physician preference and availability of equipment (Fig. 17.1). Pleural disease is usually superficial and therefore a linear high frequency probe may be more suitable (a and b), which

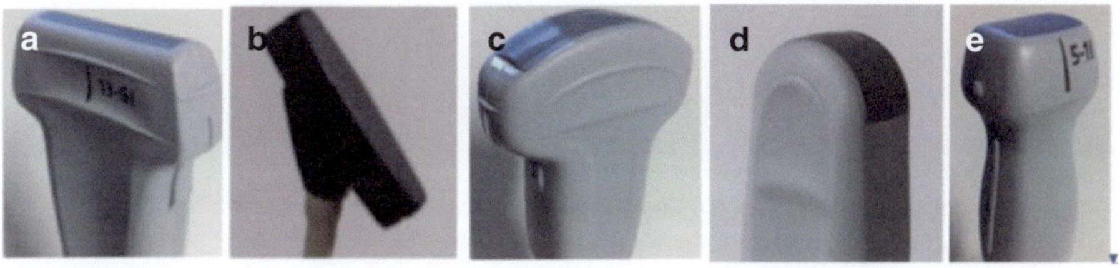

Fig. 17.1 Probes used for lung ultrasound (Reproduced with permission. © University of Melbourne). (**a**) Linear probe with handle, (**b**) linear probe without handle, (**c**) curvilinear (abdominal) probe, (**d**) microconvex probe, and (**e**) transthoracic (TTE) probe

would provide the best resolution for diagnosis or exclusion of pneumothorax and alveolar interstitial syndrome (described in Sect. 17.3). A linear probe without a handle (b), or curved probe head such as with a curvilinear (c) or microconvex (d) probe facilitates reaching posteriorly to the lung bases in patients who are unable to sit up. This is where pleural effusions, atelectasis, consolidation, empyema and lung abscess are usually found. The curvilinear probe has the largest footprint, providing the largest field of view and is probably the best probe for assessing the diaphragm. The microconvex probe (d) fits well in the intercostal space but has less depth penetration, however is less useful for sonography of other organs. Transthoracic probes (e) have similar penetration but less pleural resolution than curvilinear probes, however are more versatile as they can be used to image the heart and are a good general purpose probe for cardiothoracic assessment.

17.2.2 Ultrasound Anatomy of the Lung

Some authors have described scanning regions of the lung, with as many as 28 scanning regions, or zones, on each side [5]. However, these do not correspond to the anatomic regions of the lung. As it is feasible to scan the entire chest surface in under 5 min, we propose three scanning zones on each side, which correspond more closely to the anatomy (Table 17.2 and Fig. 17.2).

Table 17.2 Lung ultrasound zones

Lung ultrasound zone	Anatomy
Anterior (ANT)	Anterior and lateral surface of upper lobes
Posterior upper (PU)	Posterior surface of upper lobes
Posterior lower (PL)	Posterior and lateral surface of lower lobes

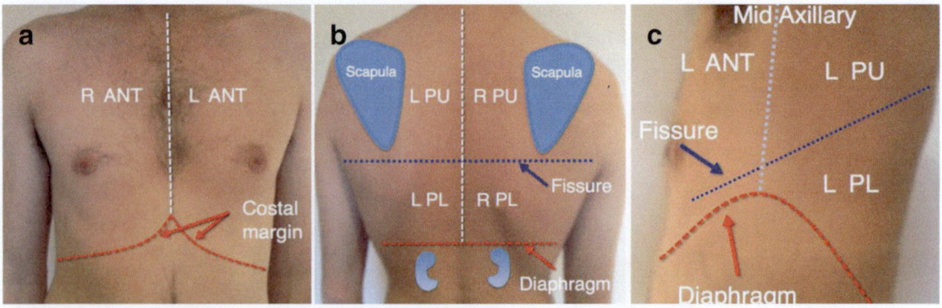

Fig. 17.2 Anterior (**a**), Posterior (**b**) and Lateral (**c**) chest surface landmarks (Reproduced with permission. © University of Melbourne). *R ANT* right anterior zone, *L ANT* left anterior zone, *LPU* left posterior upper zone, *RPU* right posterior upper zone, LPL left posterior lower zone, *RPL* right posterior lower zone

17.2.2.1 Anterior

The anterior zone, ANT (Fig. 17.2a), is defined as the whole of the anterior chest from the clavicle above, from the sternum medially to the mid-axillary line laterally and the costal margin inferiorly. The anterior zone is best considered to reflect the anterior aspect of each upper lobe, as the right middle lobe and lingular lobe can be considered as part of the upper lobe, as they are small and medially placed, with very little if any apposition to the chest wall. Additionally, the fissure that separates the upper and middle lobes is usually incomplete on the right and absent on the left. The upper lobe may be scanned anteriorly and then laterally to the upper aspect in the axilla, where it is arbitrarily separated from the posterior zone at the mid-axillary line. The lung rarely is seen below the seventh intercostal space, where the pleural reflections and diaphragm continue inferiorly to their lowermost costal attachments. The liver is seen on the patient's right, and the stomach and intestine are seen on the left. The anterior aspect of the lower lobe is not generally seen from the front, as it lies deep to the upper lobe tissue, which contains air, preventing transmission of ultrasound.

17.2.2.2 Posterior

The posterior zones (b) represent the posterior surface of the upper and lower lobes and include the posterior upper (PU) and posterior lower (PL) zones. A line joining the tips of the scapulae, approximating the fissure separating the upper and lower lobes, arbitrarily separates these. However this fissure varies considerably in position, for example from atelectasis (lung collapse), raised intra-abdominal pressure, hepatomegaly and diaphragmatic paralysis. The scanning window of the PU zone is narrow due to the scapula covering much of the posterolateral aspect of the chest. The lower lobe is wedge shaped and is accessed at the PL zone and around laterally and anteriorly to the mid-axillary line. The lower border of the lower lobe is just above the kidney, which is usually easily identified with ultrasound.

17.2.2.3 Lateral

The lateral aspect (c) is not considered to be a separate scanning zone since in this region the upper and lower lobes are continuous with the anterior or posterior imaging windows. The lateral border of the anterior and posterior zones is the mid-axillary line. The oblique fissure divides the upper and lower lobes, which can be estimated with a line from the tip of the scapula to the costal margin at the mid axillary line. However this has little clinical relevance. The upper lobe is scanned using the anterior zone and around laterally to the upper lateral area, and the lower lobe is scanned from the posterior lower window around laterally to the lower lateral area.

17.2.3 Lung Ultrasound Examination

Refer to Figs. 17.3 and 17.4 and Video 17.2.

- Probe selection depends on the pathology expected and personal preference.
- The vertical orientation of the probe (marker inferiorly for radiology convention or superiorly for cardiology convention) is not particularly important since the images are usually simple to interpret. The images should be labeled if the images are intended to be interpreted by another.
- The sequence need not differ from the user's "usual" examination sequence of the chest when using a stethoscope.
- The time taken to complete a comprehensive lung ultrasound examination, including any additional measurements, should be approximately 5–10 min.
- The lung examination is more conveniently performed with the patient sitting up, however the method for examination in the supine patient is also described.

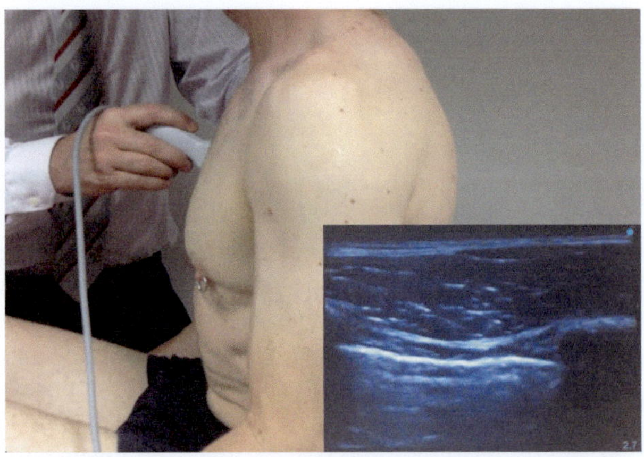

Fig. 17.3 Lung ultrasound examination with the patient in the sitting position (Reproduced with permission. © University of Melbourne)

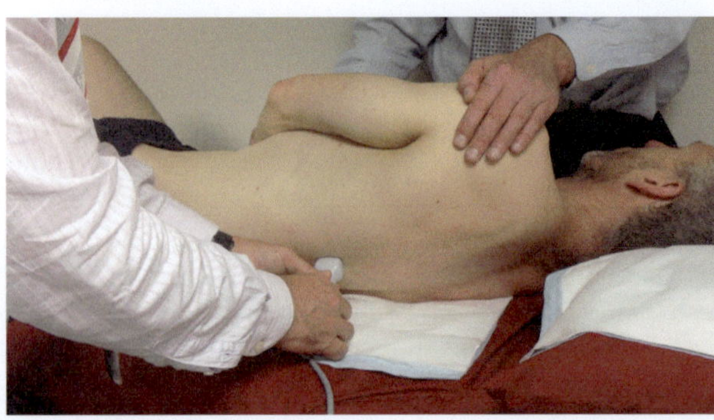

Fig. 17.4 Lung ultrasound examination with the patient in the supine position (Reproduced with permission. © University of Melbourne)

17.2.3.1 Anterior

1. Place the probe vertically oriented (sagittal plane), perpendicular to the ribs, below the midpoint of the clavicle (Fig. 17.3 and Video 17.1).
2. Adjust the depth so that the pleural line is in the middle of the screen.
3. In the supine position, pneumothorax is most commonly identified with ultrasound in the anterior zones. Search for pneumothorax and alveolar interstitial syndrome (described in Sect. 17.3.2 and 17.3.7) over the whole anterior zone, inferiorly until the costal margin is reached (liver on the right, stomach/intestine or spleen on the left) and laterally into to the upper axilla. This examines the anterolateral surface of the upper lobe.
4. If a pneumothorax is suspected select the higher frequency linear array probe for better resolution, with the patient supine to avoid displacement of air to the apex of the chest, which occurs in the erect patient.
5. Repeat on the other side before proceeding to posterior.

17.2.3.2 Posterior

6. Start with placing the probe vertically and high in the PU zone and scan the entire PU and PL zones. Identify the kidney to ensure that the whole of the lower lobe has been scanned. Continue laterally and anteriorly into the lower axilla. This examines the posterolateral surface of the lower lobe.
7. In the supine patient, the posterior zones are better exposed by log-rolling the patient 30–45° with anterior displacement of the arm to move it out of the way. It is important not to excessively abduct the arm as this rotates the scapula toward the spine and reduces the size of the PU window. Improved exposure may be obtained by placing a pillow under the shoulder or using a curvilinear, microconvex or linear probe without a handle.
8. A large pleural effusion or significant consolidation may provide a window to the entire posterior lower chest, enabling imaging of the whole lower lobe without log-rolling the patient. Imaging of the posterior upper zone (PU) may be obtained by placing the probe in the axilla angling superiorly and posteriorly.

Video 17.1. Lung ultrasound examination – sitting position (Reproduced with permission. © University of Melbourne. https://s3.amazonaws.com/iTU/LU+chapter/sitting+lungscan.mp4).

Video 17.2. Lung ultrasound examination – supine position (Reproduced with permission. © University of Melbourne. https://s3.amazonaws.com/iTU/LU+chapter/Supine+lungscan.mp4).

17.2.4 Reporting

Many of the principles for reporting lung ultrasound are similar to echocardiography. However due to the lack of anatomic landmarks, if the images are to be reviewed by someone who did not perform the ultrasound, or if lung pathology is tracked over time, then it is important to label the images. Although the liver and spleen are recognisable landmarks for the right and lower lobe, they can still be confused with each other.

Figure 17.5 shown below is a sample report form that is used in the University of Melbourne Graduate Certificate of Clinical Ultrasound and Advanced iHeartScan™ course. It is useful to record the patient position during the examination and whether they are mechanically ventilated and position of pleural drains and chest wall dressings.

17.2.4.1 iLungScan™ iPhone Reporting App

The free App can be downloaded from the iTunes store at https://itunes.apple.com/au/app/ilungscan/id962741429?mt=8. This app provides for demographic and diagnosis classification as well as the option to upload representative images or videos (Fig. 17.6).

17.3 Lung Ultrasound Pathology

17.3.1 Normal Appearance

The pleural surfaces are superficial and usually reached with a high frequency linear probe (8–12 MHz), which provides the best resolution, however lower frequency probes are usually sufficient. Placing the probe perpendicular to the chest wall and the ribs and sliding the probe superiorly and inferiorly will enable visualisation of two rib shadows that are intervened by horizontally placed intercostal space (Fig. 17.7). Below the intercostal muscles is a hyperechoic horizontal line, which represents the adherent parietal and visceral pleura, the pleural line. In the normal lung during respiration the two layers of pleura slide past one another resulting in subtle movement on the ultrasound display at the pleural line and is termed **lung sliding sign** (Fig. 17.8 and Video 17.4). Sliding sign has been described as the appearance of "crawling ants". The aerated lung below the pleura appears as a grey speculated shadow, which changes appearance during respiration quite similar to television 'white noise'. A normal appearance may include the occasional short vertical line seen to extend inferiorly from the pleural line, which move and

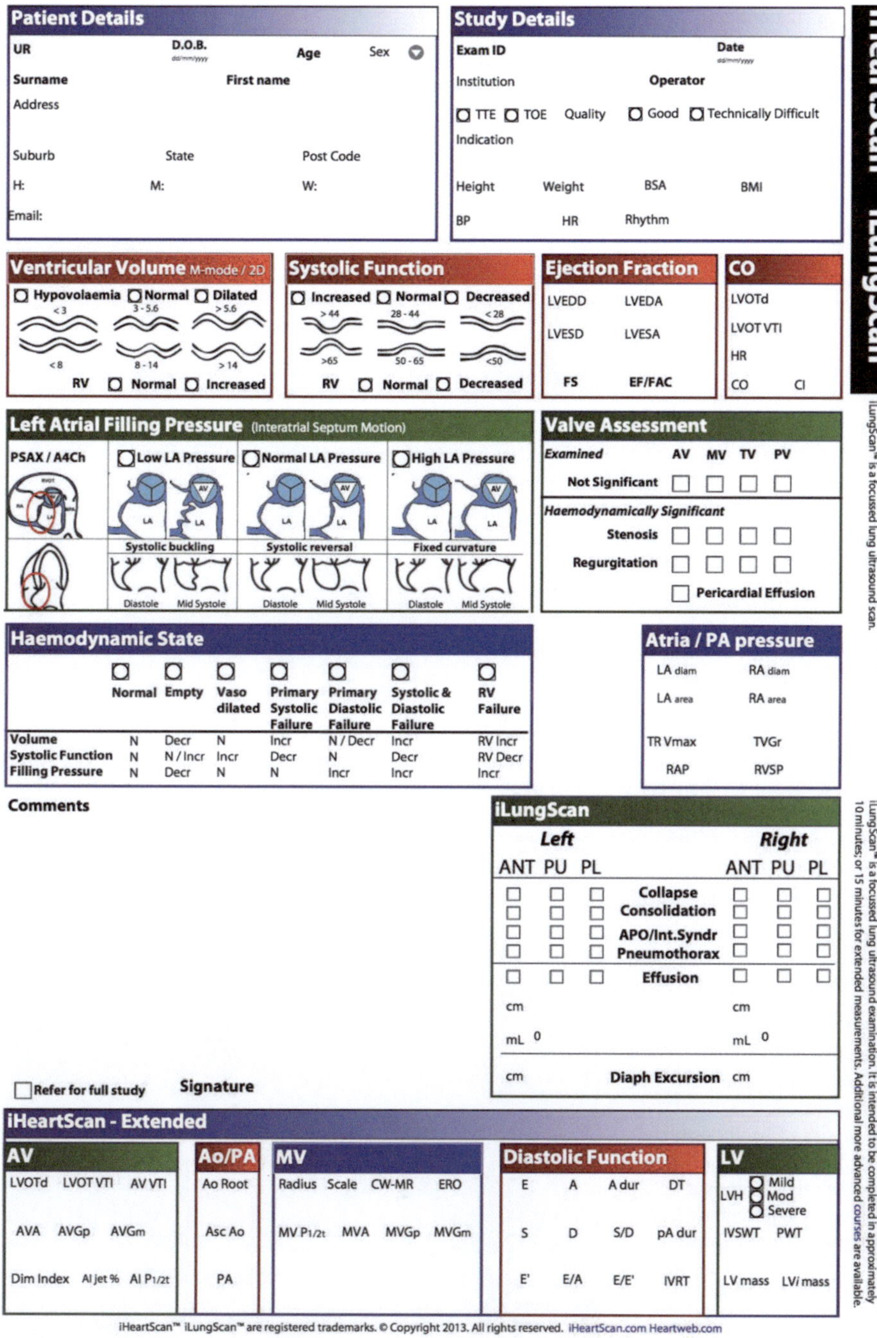

iHeartScan™ iLungScan™

iHeartScan™ Haemodynamic Echocardiography Assessment in Real Time
iLungScan™ is a focused lung ultrasound scan.

Patient Details

UR D.O.B. dd/mm/yyyy Age Sex
Surname First name
Address
Suburb State Post Code
H: M: W:
Email:

Study Details

Exam ID Date dd/mm/yyyy
Institution Operator
☐ TTE ☐ TOE Quality ☐ Good ☐ Technically Difficult
Indication
Height Weight BSA BMI
BP HR Rhythm

Ventricular Volume M-mode / 2D

☐ Hypovolaemia ☐ Normal ☐ Dilated
<3 3 - 5.6 >5.6
<8 8 - 14 >14
RV ☐ Normal ☐ Increased

Systolic Function

☐ Increased ☐ Normal ☐ Decreased
>44 28 - 44 <28
>65 50 - 65 <50
RV ☐ Normal ☐ Decreased

Ejection Fraction

LVEDD LVEDA
LVESD LVESA
FS EF/FAC

CO

LVOTd
LVOT VTI
HR
CO CI

Left Atrial Filling Pressure (Interatrial Septum Motion)

PSAX / A4Ch
☐ Low LA Pressure — Systolic buckling — Diastole, Mid Systole
☐ Normal LA Pressure — Systolic reversal — Diastole, Mid Systole
☐ High LA Pressure — Fixed curvature — Diastole, Mid Systole

Valve Assessment

Examined	AV	MV	TV	PV
Not Significant	☐	☐	☐	☐
Haemodynamically Significant				
Stenosis	☐	☐	☐	☐
Regurgitation	☐	☐	☐	☐

☐ Pericardial Effusion

Haemodynamic State

	☐ Normal	☐ Empty	☐ Vaso dilated	☐ Primary Systolic Failure	☐ Primary Diastolic Failure	☐ Systolic & Diastolic Failure	☐ RV Failure
Volume	N	Decr	N	Incr	N / Decr	Incr	RV Incr
Systolic Function	N	N / Incr	Incr	Decr	N	Decr	RV Decr
Filling Pressure	N	Decr	N	N	Incr	Incr	Incr

Atria / PA pressure

LA diam RA diam
LA area RA area
TR Vmax TVGr
RAP RVSP

Comments

☐ Refer for full study **Signature**

iLungScan

Left ANT	Left PU	Left PL		Right ANT	Right PU	Right PL
☐	☐	☐	Collapse	☐	☐	☐
☐	☐	☐	Consolidation	☐	☐	☐
☐	☐	☐	APO/Int.Syndr	☐	☐	☐
☐	☐	☐	Pneumothorax	☐	☐	☐
☐	☐	☐	Effusion	☐	☐	☐
cm				cm		
mL 0				mL 0		
cm			Diaph Excursion	cm		

iHeartScan - Extended

AV
LVOTd LVOT VTI AV VTI
AVA AVGp AVGm
Dim Index AI jet % AI P1/2t

Ao/PA
Ao Root
Asc Ao
PA

MV
Radius Scale CW-MR ERO
MV P1/2t MVA MVGp MVGm

Diastolic Function
E A A dur DT
S D S/D pA dur
E' E/A E/E' IVRT

LV
LVH ☐ Mild ☐ Mod ☐ Severe
IVSWT PWT
LV mass LV/mass

iHeartScan™ is a focused transthoracic echocardiography study that is qualitative and quantitative. iLungScan™ is a focused lung ultrasound examination. It is intended to be completed in approximately 10 minutes; or 15 minutes for extended measurements. Additional more advanced courses are available.

Fig. 17.5 Focused transthoracic echocardiography and lung ultrasound report form (Reproduced with permission. © University of Melbourne). *ANT* anterior – includes the upper lateral region (axilla), *PU* posterior upper, *PL* posterior lower – includes the lower lateral region (to the diaphragm) Download link: https://s3.amazonaws.com/iTU/LU+chapter/Appendix+TTE+and+lung+ultrasound+report+form.jpg

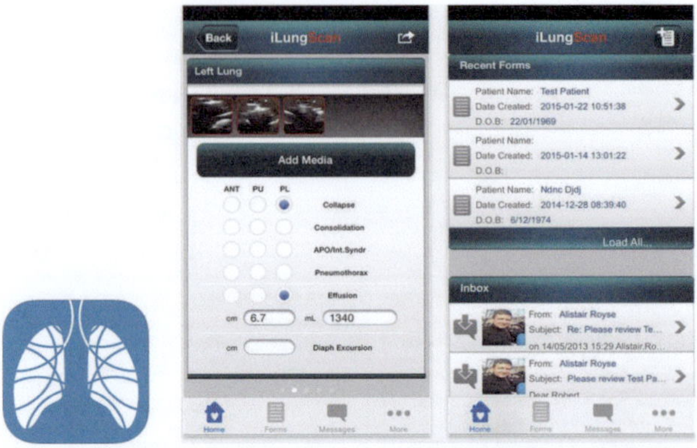

Fig. 17.6 iLungScan reporting app (Reproduced with permission. © University of Melbourne)

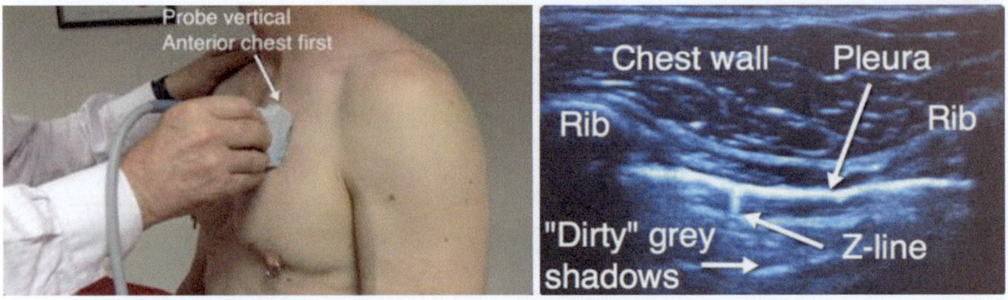

Fig. 17.7 Standard lung ultrasound appearance (Reproduced with permission. © University of Melbourne). With the linear probe placed longitudinally anteriorly below the clavicle, an interspace is demonstrated on the right with two hyperechoic shadows (ribs) with an adjoining hyperechoic line (pleural line). Above the pleural line is the chest wall (intercostal muscles and superficial tissues) and below is the lung parenchyma (dirty grey shadows). During respiration, the normal appearance is lung sliding sign, which is the parietal and visceral pleura sliding along each other from left to right, creating a shimmering or movement of small vertical lines (z-lines)

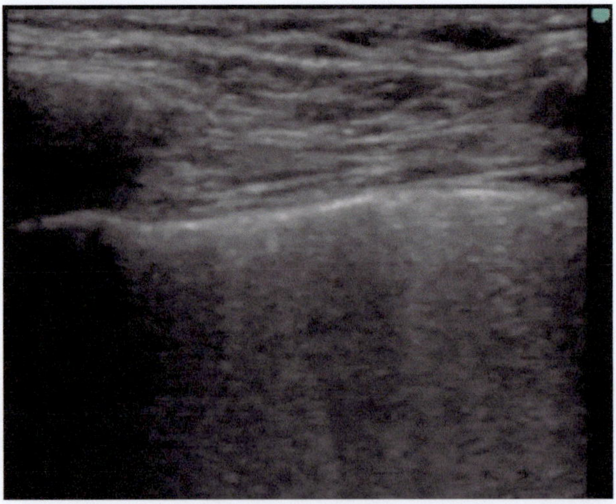

Fig. 17.8 Normal lung appearance using ultrasound (Reproduced with permission. © University of Melbourne). The pleural line is demonstrated here as a horizontal hyperechoic line in-between the intercostal muscles above and the lung below. Several z-lines are seen as short vertical lines originating at the pleural line extending down into the grey lung zone. These are normal findings

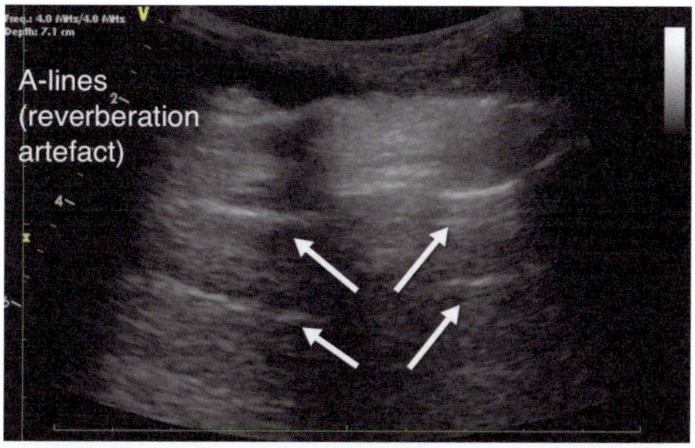

Fig. 17.9 A-lines (Reproduced with permission. © University of Melbourne). A-lines are horizontal copies of the pleural line reflection (reverberation artefacts) at multiples of the distance between the skin and pleura line indicated by the *white arrows*. These are normal findings but are accentuated in pneumothorax

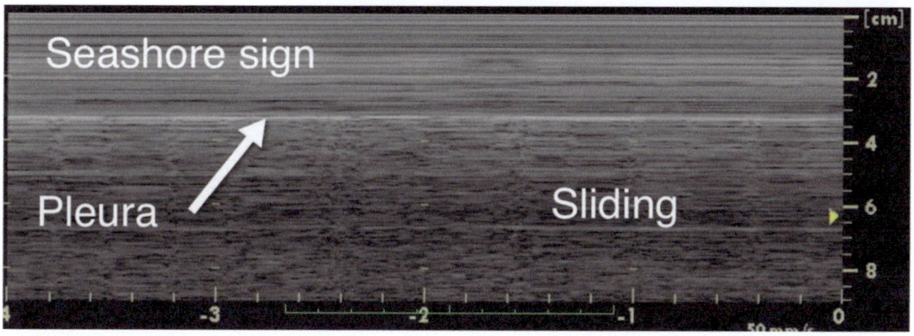

Fig. 17.10 M-mode appearance of lung sliding sign, *seashore sign (*Reproduced with permission. © University of Melbourne)

disappear with respiration, which are referred to as **Z-lines**. As the pleural line is highly reflective it is often duplicated below as reverberation artefacts that are equally spaced – A-lines (Fig. 17.9).

Video 17.4. Normal lung sliding sign (Reproduced with permission. © University of Melbourne. https://s3.amazonaws.com/iTU/LU+chapter/Lung+sliding.mp4).

17.3.1.1 M-Mode

M-mode displays motion at a single point over time and is able to illustrate changes in 2-D ultrasound appearance with movement from respiration over time. This enables the display of the characteristic appearance of lung sliding. With the probe placed in a horizontal position along the intercostal space, and the M-mode cursor placed through the interspace, lung sliding has a characteristic appearance containing two distinct layers (Fig. 17.10). On the upper zone of the ultrasound sector, a series of parallel stationary horizontal lines represent the superficial tissues and intercostal muscles, which usually remain still during respiration. This appearance has been likened to oceanic waves lapping onto a beach. Below the horizontal waves

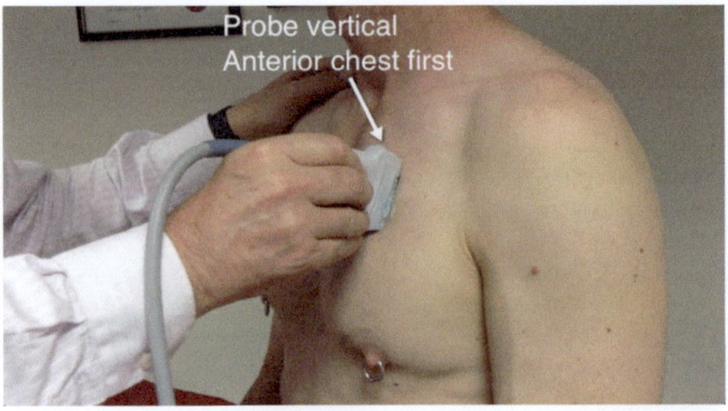

Fig. 17.11 Pneumothorax examination (Reproduced with permission. © University of Melbourne)

is a bright horizontal line representing the pleural line. Below the pleural line has a grey pixilated appearance, similar to sand on a beach, and the overall appearance of lung sliding has been termed **seashore sign**. The pixilated appearance is caused by the varying ultrasound reflections of the moving lung during respiration as the ultrasound beam intermittently crosses fluid containing structures.

17.3.2 Pneumothorax

Rapid diagnosis or exclusion of pneumothorax with ultrasound is a valuable skill for the anaesthetist and critical care physician. The detection of pneumothorax is simply based on the inability to see lung sliding sign, caused by air between the pleura (Compare lung sliding seen in Video 17.4 and no-lung sliding in Video 17.5). As air normally collects anteriorly in the supine patient, the probe should be initially placed vertically in the anterior zone below the clavicle in the mid-clavicular line and then moved inferiorly and laterally over the entire anterior zone (Fig. 17.11 and Videos 17.5 and 17.6). It is important to keep the probe motionless, as movements of the probe may be misinterpreted as lung sliding.

Video 17.5. Pneumothorax (no lung sliding) (Reproduced with permission. © University of Melbourne. https://s3.amazonaws.com/iTU/LU+chapter/No+lung+sliding.mp4).

Video 17.6. Pneumothorax examination. Reproduced with permission. © University of Melbourne. https://s3.amazonaws.com/iTU/LU+chapter/Pneumothorax+examination.mp4.

Identification of lung sliding excludes the presence of pneumothorax in the scanned region with a sensitivity and specificity close to 100 % [4]. The presence of Z-lines or B-lines (described in the Sect. 17.3.7) also excludes pneumothorax because these signs rely on apposition of the two layers of pleura. Unfortunately, lack of lung sliding may be caused by other pleural or parenchymal pathology that prevents movement of the pleura and hence lung ultrasound is better at ruling pneumothorax out rather than ruling it in. Conditions that also cause lack of lung sliding include pleural adhesions, bullous disease and severe atelectasis or hypoventilation

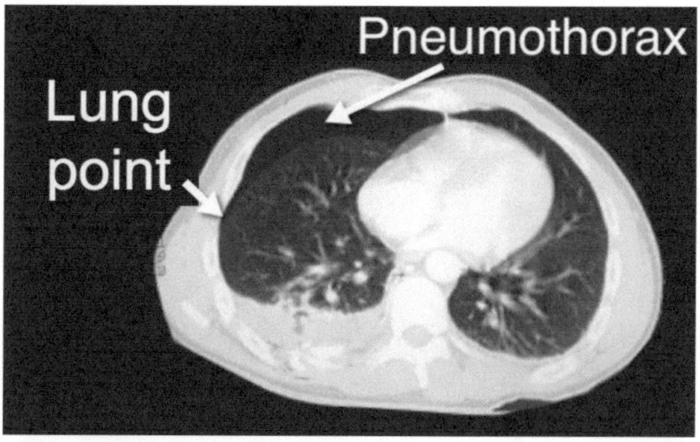

Fig. 17.12 Lung point (Reproduced with permission. © University of Melbourne)

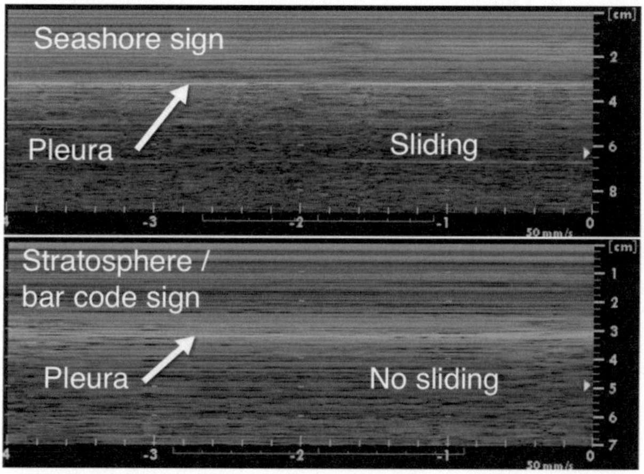

Fig. 17.13 Seashore sign and barcode sign (Reproduced with permission. © University of Melbourne). M-mode of the lung demonstrating the difference in appearance of lung sliding (*above*) with no lung sliding (*below*) as seen with pneumothorax

for example from endotracheal tube malposition or sputum obstruction, and interstitial diseases such as acute respiratory distress syndrome (ARDS) and pulmonary fibrosis. The ultrasound feature specific to pneumothorax is identification of a point where lung sliding abuts an area of no lung sliding, which identifies the point that the pleural surfaces regain contact, the contact point or **lung point** (Fig. 17.12).

Lack of lung sliding also has a characteristic appearance with M-mode (Fig. 17.13). The horizontal bars or waves extend all the way from the top to the bottom of the field without the area of pixelation due to lack of perceived lung movement, resulting in horizontal waves but no area of sand. This appearance has been termed **stratosphere** sign and **bar code** sign [9]. During M-Mode scanning with the linear probe in a horizontal position, if respiration results in the lung point moving from out of ultrasound field into the ultrasound field, a characteristic vertical linear demarcation appears, separating an area of lung sliding (seashore sign) from no long sliding (barcode sign) as shown in Fig. 17.14. This linear border represents the lung point or contact point, where the pneumothorax ends and the two layers of pleura contact each other and lung sliding recurs. Identification of

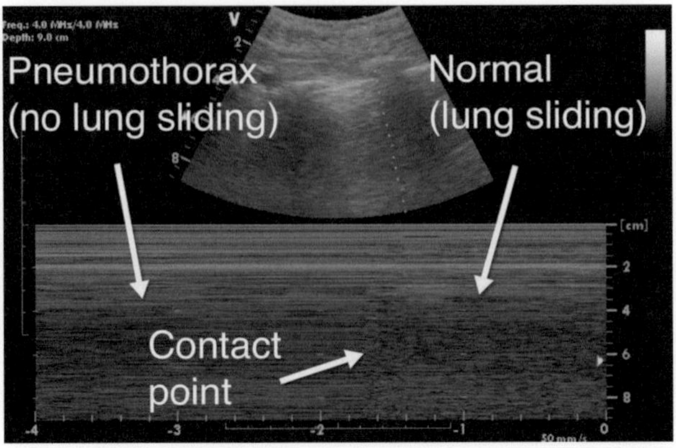

Fig. 17.14 Lung/Contact point of pneumothorax demonstrated with M-mode (Reproduced with permission. © University of Melbourne). M-mode of the lung demonstrating the lung or contact point, which is the point where the pneumothorax ends (no lung sliding) and the two pleural layers become re-apposed (lung sliding). The lung point appears as a vertical line in between these two regions

this lung point is specific for pneumothorax, and may enable quantification of the size of a pneumothorax and allow tracking of the size of the pneumothorax over time, aiding in the decision whether or not to drain the pneumothorax. For example a pneumothorax with lung point located at the anterior axillary line would be smaller than one located at the posterior axillary line or not located at all. However this is not a very reliable indication of size as the position of the lung point is dependent on other factors, such as diaphragm position, patient posture and lung pathology such as collapse, consolidation and pleural adhesions. The lung point is usually sufficiently detected with the use of 2-D ultrasound, but M-mode enables documentation of the lung point, which can be printed and stored in the patient's medical file.

When there is lack of lung sliding from complete absence of ventilation, pneumothorax may be still be excluded by observing a **lung pulse** [9]. When the probe is kept stationary, the movement of the heart may transmit oscillatory movements in the adjacent lung, resulting in very short bursts of lung sliding in time with the pulse (Fig. 17.15 and Video 17.7). Even though there is no ventilation, and no movement of the lung would occur with respiration, small movement of the lung is observed due to the cardiac motion displacing the lung synchronous with the electrocardiogram.

Video 17.7. Lung pulse (Reproduced with permission. © University of Melbourne. https://s3.amazonaws.com/iTU/LU+chapter/LungPulse.mp4).

Surgical emphysema (occasionally occurring with a pneumothorax) will not only impair imaging of the pleura; but in minor cases B-lines may arise from air bubbles within the chest wall tissues (superficial to the pleura), which are referred to as E-lines [9].

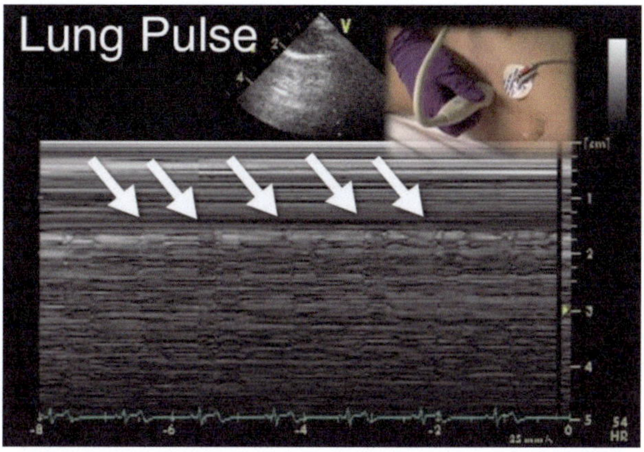

Fig. 17.15 Lung pulse (Reproduced with permission. © University of Melbourne)

17.3.2.1 Summary

The sonographic signs of pneumothorax include [5]:

- Presence of lung point(s)
- Absence of lung sliding
- Absence of B-lines and Z-lines
- Absence of lung pulse

Lung ultrasound is more accurate than supine anterior chest x-ray in ruling out and in pneumothorax [5].

17.3.3 Pleural Effusion

The typical sonographic appearance of pleural effusion is an anechoic (black) area between the parietal and visceral pleura (Fig. 17.16), which usually changes size with respiratory movements. It is easy to detect with lung ultrasound because fluid is an effective transmitter of ultrasound, which has the added advantage of enabling direct visualisation of underlying lung pathology such as atelectasis and consolidation. The heart and aorta may also be visualised through an effusion. For detection of effusion, lung ultrasound is more accurate than supine chest X-ray [2] and is as accurate as computed tomography [2, 10]. For lung opacities found on chest X-ray, lung ultrasound is more accurate in distinguishing between effusion and consolidation [11]. Pleural effusions are commonly seen in patients with cardiac failure, malignancy and after cardiac or thoracic surgery.

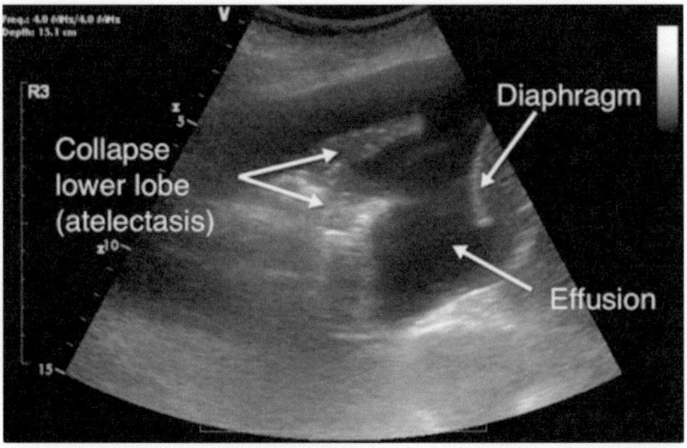

Fig. 17.16 Pleural effusion (Reproduced with permission. © University of Melbourne). A pleural effusion is shown here as a large hypo echoic area beneath the chest wall (*above*) and above the diaphragm (to the right). Within the effusion is a complex hypo echoic structure, which represents the lower lobe of the lung that contains atelectasis (collapse)

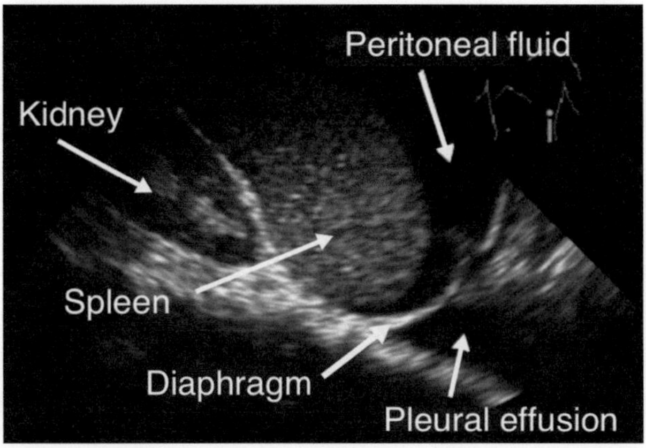

Fig. 17.17 Peritoneal fluid (Reproduced with permission. © University of Melbourne. The key landmarks are the diaphragm, kidney and spleen. Peritoneal fluid is seen here in-between the spleen and diaphragm

Video 17.8. Pleural effusion (Reproduced with permission. © University of Melbourne. https://s3.amazonaws.com/iTU/LU+chapter/Pleural+effusion+2.mov).

A low frequency probe with a large footprint such as a curvilinear probe or TTE probe enables sufficient penetration, which is typically 5–8 cm but may need to be increased in large effusions. Mostly effusions are not loculated and will collect in the dependent zone of the chest. Therefore effusions are mostly detected in the posterior lower zone, whether the patient is supine or erect. It is important to routinely identify the diaphragm (Sect. 17.3.10.3) to avoid confusion of pleural fluid and peritoneal fluid (Fig. 17.17). Very rarely, there may be confusion between a vascular structure and free fluid, and the use of color flow or pulsed wave Doppler will easily identify the nature of the vessel.

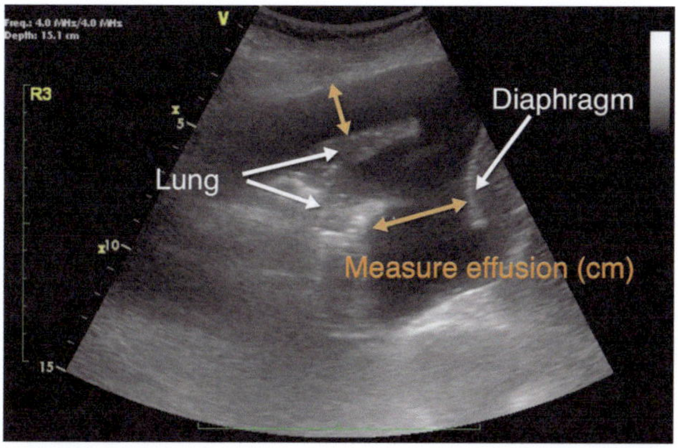

Fig. 17.18 Estimating the volume a pleural effusion. (Reproduced with permission. © University of Melbourne). Estimation of pleural fluid volume by measurement of the maximal pleural distance may be performed either between the diaphragm and the inferior surface of the lung or between the chest wall and the lateral surface of the adjacent lung

17.3.3.1 Quantification of Pleural Effusion with Ultrasound

A variety of methods have been described to estimate the volume of pleural effusion with ultrasound. The easiest technique is to image the effusion in the vertical plane using the maximum interpleural distance of the effusion (Fig. 17.18) [12]. The probe is positioned as posteriorly as possible in the supine patient, or at the lateral edge of the erector spinae muscle in the erect patient. The maximum distance of the effusion is measured perpendicularly between either the diaphragm and the visceral pleura (inferiorly) or the parietal and visceral pleura (laterally) as shown in the Figure below. The volume, in mL, is calculated by multiplying the maximum interpleural distance, in cm, by 200. For example the maximum interpleural distance of 6 cm corresponds to an effusion volume of 6 cm × 200 = 1,200 mL.

$$\text{Pleural effusion Volume}(\text{mL}) = \text{Interpleural Distance}(\text{cm}) \times 200$$

The volume calculated by this method tends to over estimate by approximately 10 %. Ultrasound-guided pleural catheter insertion is described in Sect. 17.4 below.

A serous or haemoserous pleural effusion that has no reflective (echogenic) material contained within is referred to as "simple" fluid and appears black (anechoic). Effusions containing particulate material or fine tissue strands or septae are referred to as "complex" fluid. A common example of a complex effusion is haematoma, which may develop fibrinous septations (Fig. 17.19). Mobile particulate matter or septations are usually due to infection and may represent empyema (covered in Sect. 17.3.5). However, atelectic lung may appear to move within an associated effusion, which has been referred to as jellyfish sign (Fig. 17.20 and Video 17.9). Whilst ultrasound may provide a clue to the composition of pleural fluid, the definitive test is thoraco-centesis (covered in Sect. 17.4).

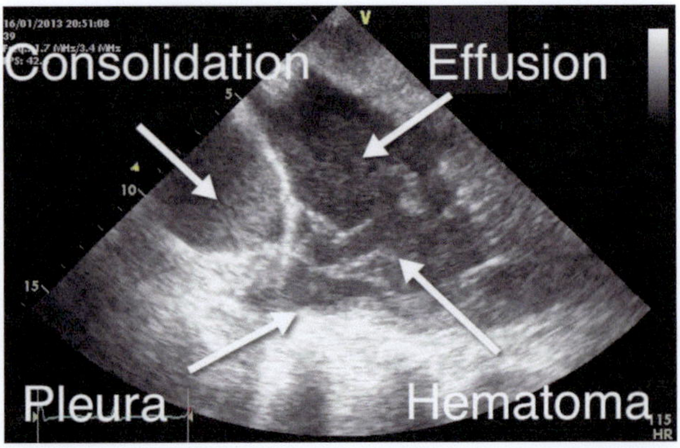

Fig. 17.19 Haemothorax (Reproduced with permission. © University of Melbourne)

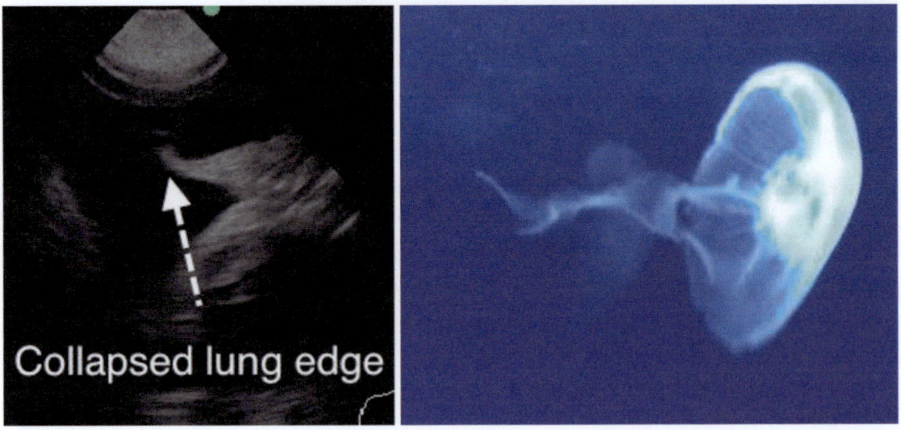

Fig. 17.20 Jellyfish sign (Reproduced with permission. © University of Melbourne). A pleural effusion is shown here as a large hypo echoic area beneath the chest wall. Within the effusion is a tongue-shaped strand of atelectic lung shown by the *arrow*, which is seen to float to and fro (in Video 17.9) as the fluid moves with respiration, which has been described as jellyfish sign

Video 17.9. Jellyfish sign of atelectasis mobile within an effusion (Reproduced with permission. © University of Melbourne. https://s3.amazonaws.com/iTU/LU+chapter/JellyfishSign.mp4).

17.3.4 Consolidation and Atelectasis

17.3.4.1 Consolidation

Lung consolidation is easily seen with ultrasound because the lung air spaces have become filled with fluid respectively, which allows the transmission of ultrasound. Consolidation appears with ultrasound grey and 'tissue like', similar to liver tissue (hepatisation) as the homogenous patches of lung separated by blood vessels that

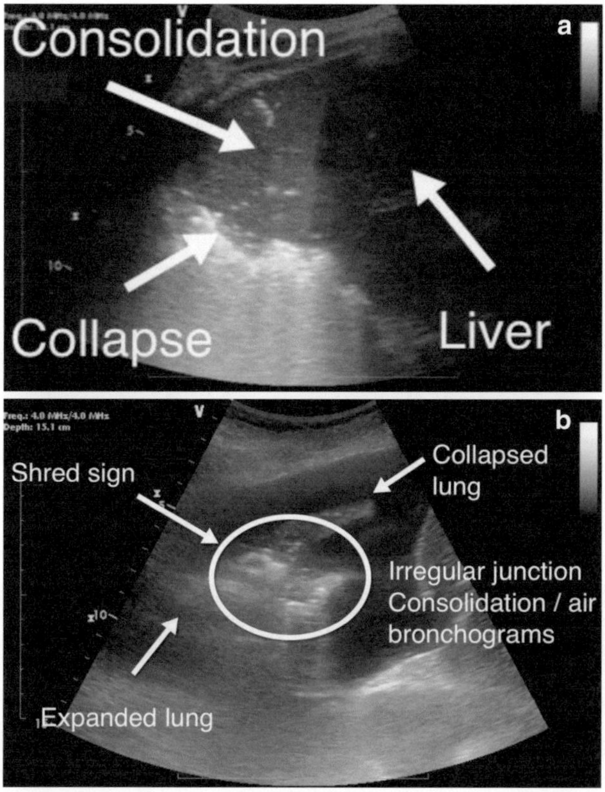

Fig. 17.21 (**a**) Lung consolidation and atelectasis (Reproduced with permission. © University of Melbourne). Lung ultrasound demonstrating the features of consolidation (*above*), collapse/atelectasis (*below*) and liver (*right*). (**b**) Linear hyper echoic lines adjacent to the areas of collapse are caused by trapped air within the lung (*shred sign*)

have the appearance of septae (Fig. 17.21). However, unlike the ultrasound appearance of liver, consolidated lung may contain cartilage of the bronchi and bronchioles, seen as hyperechoic dots and streaks. Consolidation can have a variety of causes including infection, cancer, contusion and pulmonary embolism. Differentiation of these pathologies with ultrasound may be aided by the quality of the deep margins of consolidation, presence of air and fluid and vascular pattern within the consolidation [13, 14]. Lung ultrasound is better than chest x-ray in diagnosis and distinguishing different causes of consolidation in mechanically ventilated patients [2] and in patients presenting with pleuritic pain [15], and is comparable with chest X-ray in a variety of clinical settings [16, 17].

17.3.4.2 Atelectasis

Atelectasis, or lung collapse, has a similar appearance to consolidation, which often appear together. Linear hyperechoic lines originating from an even greater hyperechoic core, which are caused by trapped air within the collapsed lung (Fig. 17.21a).

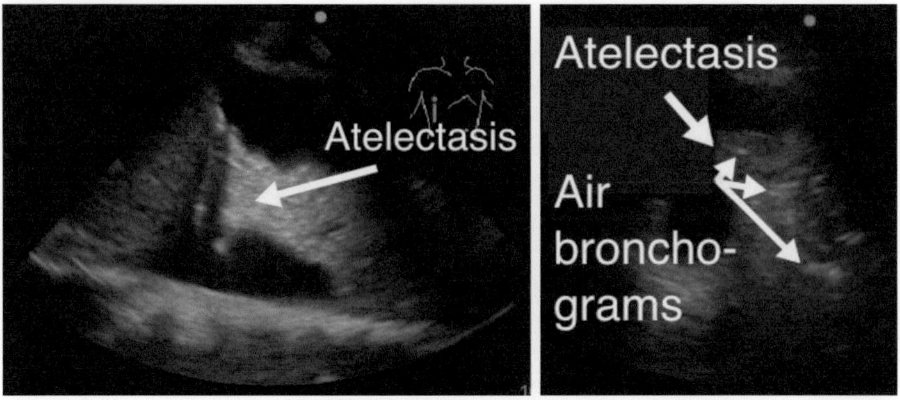

Fig. 17.22 Atelectasis (Reproduced with permission. © University of Melbourne). Ultrasound appearance of atelectasis is shown in these two examples – grey tissue with volume loss. The example on the right contains hyper echoic regions that are caused by trapped air (air bronchograms)

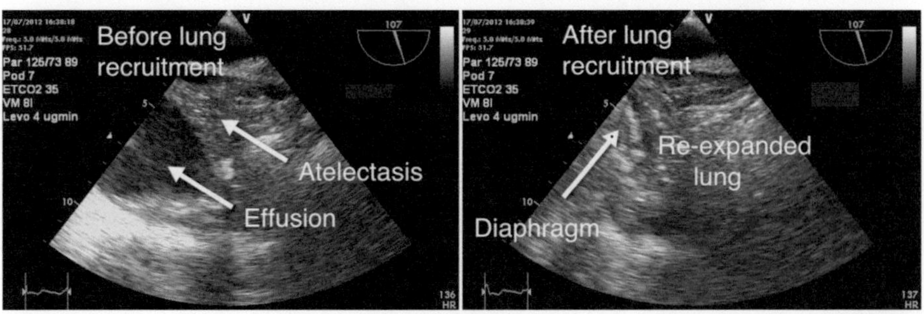

Fig. 17.23 Lung recruitment of atelectasis (Reproduced with permission. © University of Melbourne). Left lower lobe lung examination using transoesophageal echocardiography before and after a recruitment maneuver. Note the loss of pleural fluid and increase in lung volume after a recruitment maneuver. This indicates dynamic atelectasis

The most striking feature of atelectasis is loss of lung volume (Fig. 17.22). When this appears adjacent to an area of consolidation it has been termed **shred sign**, as shown in Fig. 17.21b [9]. The other key ultrasound feature that distinguishes atelectasis from consolidation is that atelectasis almost always results in *loss of lung volume* because the air spaces are squashed flat, whereas in consolidation there is less volume loss because the air spaces are filled with fluid and inflammatory material. However the definitive test that distinguishes atelectasis from consolidation is demonstration with ultrasound of re-expansion of collapsed lung units after a recruitment maneuver (Fig. 17.23). The repeated use of lung ultrasound can be used to assess the progress of lung collapse or consolidation with treatment. Lung sliding may be reduced with consolidation and atelectasis due to reduced lung expansion during respiration.

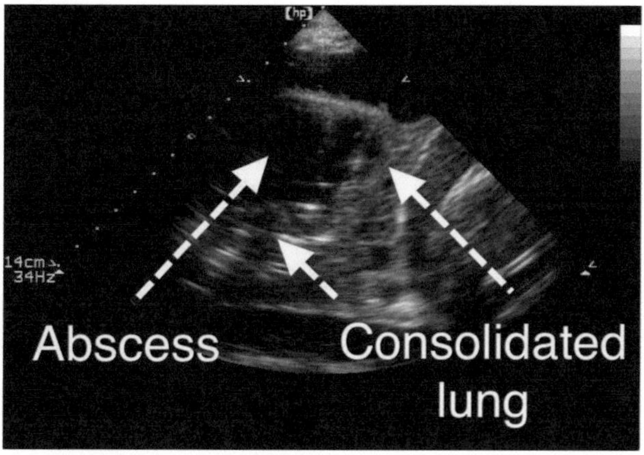

Fig. 17.24 Lung abscess (Reproduced with permission. © University of Melbourne). The abscess cavity in this example is surrounded by collapsed and consolidated lung and an effusion

17.3.5 Lung Abscess

Lung abscess may appear as an anechoic cavitation with irregular opacities or septae within (Fig. 17.24 and Video 17.10). There are usually areas of surrounding consolidation, atelectasis, effusion and pleural thickening. The fluid is often "complex" as a result of being filled with purulent and fibrinous or necrotic material. Empyema has a similar appearance but is restricted to the pleural space, rather than occurring within the lung, as with lung abscess.

Video 17.10. Lung abscess (https://s3.amazonaws.com/iTU/LU+chapter/Lung+abscess.mp4).

17.3.6 Pulmonary Embolus

The ability to diagnose pulmonary embolus with point of care ultrasound is desirable as it may avoid transporting a critically unwell patient to the radiology department for conventional investigation such as CT or angiography.

Pulmonary embolism appears as a localized or wedge shaped consolidation when it abuts the pleura. Mathis reported a sensitivity of 74 % and specificity of 95 %, with a positive predictive value of 95 % in 353 patients examined in intensive care who had a suspected pulmonary embolus [18]. If the embolus is haemodynamically significant echocardiography may show a dilated and poorly functioning right ventricle (Fig. 17.25 and Video 17.11) with raised estimated pulmonary artery systolic pressure. Mobile or fixed thromboemboli are only occasionally seen in the cardiac chambers or pulmonary arteries. Diagnosis of a deep venous thrombosis (Fig. 17.26 and Video 17.12) increases the accuracy of ultrasound diagnosis of pulmonary embolus [4]. However sub-clinical pulmonary embolism can still be present in the presence of a normal echocardiographic exam. If suspected, other modalities such as CT scan should be used [19].

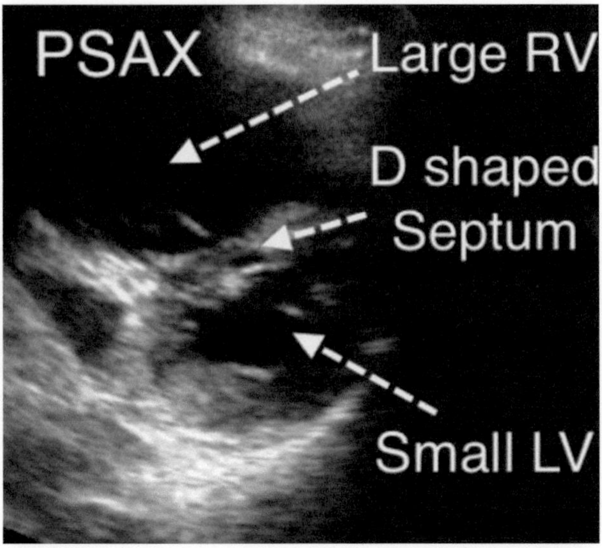

Fig. 17.25 Echocardiographic features of pulmonary embolus (Reproduced with permission. © University of Melbourne)

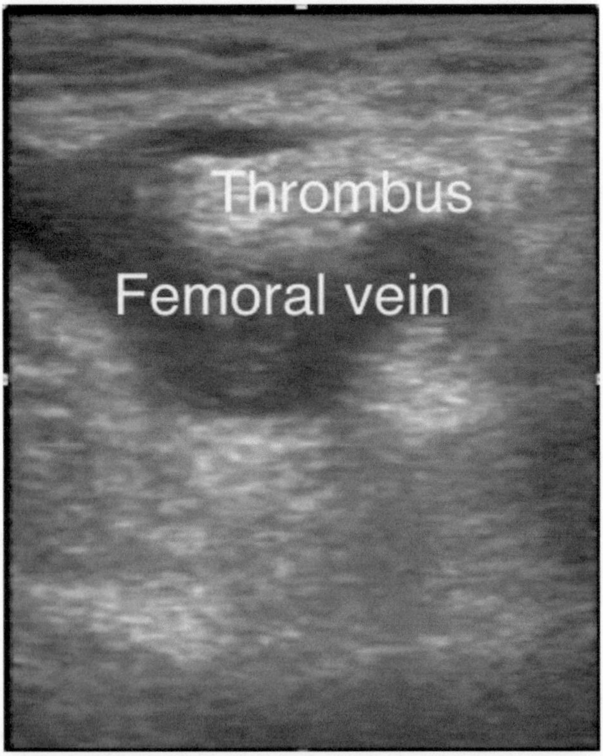

Fig. 17.26 Deep venous thrombosis (Reproduced with permission. © University of Melbourne)

Video 17.11. Echocardiographic features of pulmonary embolus (Reproduced with permission. © University of Melbourne. https://s3.amazonaws.com/iTU/LU+chapter/PE.mov).

Video 17.12. Deep venous thrombosis (Reproduced with permission. © University of Melbourne. https://s3.amazonaws.com/iTU/LU+chapter/DVT+(Converted).mov).

17.3.7 Alveolar Interstitial Syndrome

The ability to rapidly, non-invasively and accurately detect acute pulmonary oedema is a major advancement in clinical assessment in both critical and acute care medicine. Along with detection of pneumothorax, this has been responsible for the increasing popularity of lung ultrasound.

Alveolar interstitial syndrome (AIS) is a group of conditions that result in a characteristic lung ultrasound appearance of multiple **B-lines** (Fig. 17.27). These conditions include pulmonary oedema, interstitial pneumonia or pneumonitis, and diffuse parenchymal lung disease such as acute respiratory distress syndrome (ARDS) and pulmonary fibrosis [5]. Fortunately lung ultrasound can be used to separate these disorders.

The key ultrasound feature of AIS is the B-line, previously termed "comet tails" or "lung rockets". B-lines are hyper echoic, vertical, laser-like lines that extend without fading from the pleural line all the way to the bottom of the image, which

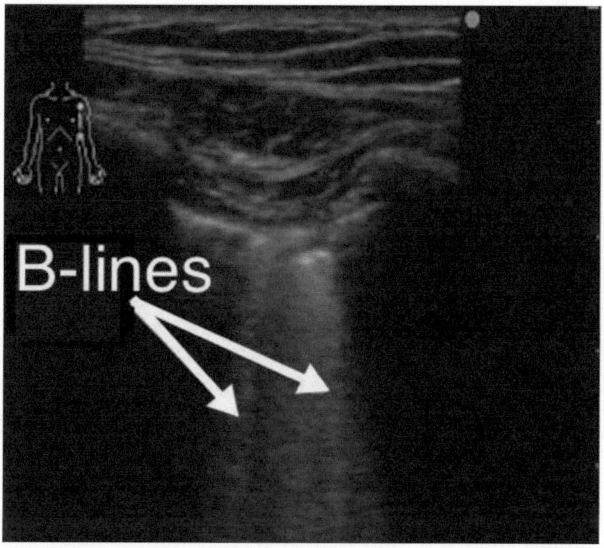

Fig. 17.27 B-lines, comet tails, lung rockets (Reproduced with permission. © University of Melbourne). B-lines are hyper echoic, vertical, laser-like lines that extend without fading from the pleural line all the way to the bottom of the image, which move synchronously with respiration

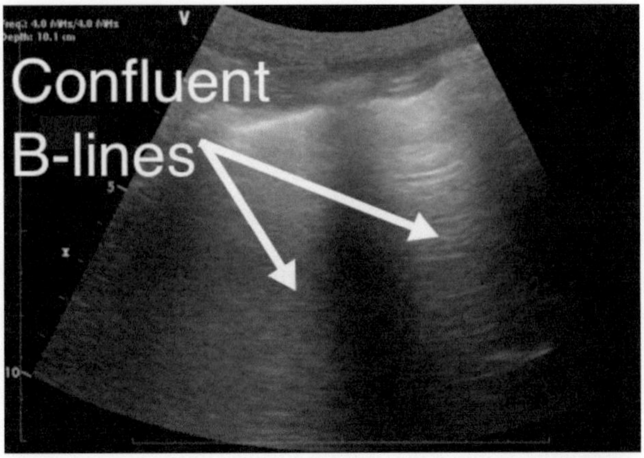

Fig. 17.28 Confluent B-lines in acute pulmonary oedema (Reproduced with permission. © University of Melbourne). *Dense* B-lines may result in confluence with a resulting 'snow storm' appearance

move synchronously with respiration [5], as shown in Fig. 17.27 and Video 17.13. B- lines are reverberation artefacts formed by reverberation of ultrasound within fluid within the pleural aspect of the lung, probably within the interlobular septae, however the exact mechanism is not known at the time of writing this review. B-lines are frequently seen in normal lungs but only 1 or 2 within a single interspace and normally only in the lower zones. Three or more B-lines observed in a single interspace defines AIS. Z-lines are normal findings and are distinguished from B-lines by their shallow depth, and fading with depth and they do not obscure A-lines, which are reverberation copies of the pleural line. Focal B-lines may be present in normal lung, pneumonia and pneumonitis, atelectasis, pulmonary contusion and infarction, pleural disease and neoplasia [5].

Video 17.13. B-lines, comet tails, lung rockets (Reproduced with permission. © University of Melbourne. https://s3.amazonaws.com/iTU/LU+chapter/B-lines.mp4).

17.3.7.1 Pulmonary Oedema

Lung ultrasound is superior to chest X-ray in diagnosis of pulmonary oedema [20]. Multiple B-lines are seen in both cardiogenic and non-cardiogenic pulmonary oedema but diffuse B-lines allows bedside distinction between cardiogenic versus a respiratory cause of acute dyspnoea [5]. Confidence in diagnosis may be improved with echocardiography. A key feature that separates pulmonary oedema from the other conditions causing AIS is that in oedema the B-lines are typically less than 3 mm apart. The B-lines are usually bilateral and in adjacent areas, without areas of "skipping" and are usually more prominent in the posterior lower zones, where associated pleural effusions are commonly seen. Lung sliding is normally preserved, as the pleura are not normally affected. As pulmonary oedema worsens, the density of B-lines increases and they may become confluent, appearing as a 'snow storm' (Fig. 17.28). In acute decompensated heart failure, the severity of oedema is proportional to the number of lung regions that contain B-lines [15, 21]. B-lines appear and disappear in response to fluid loading and removal respectively and repeated interval

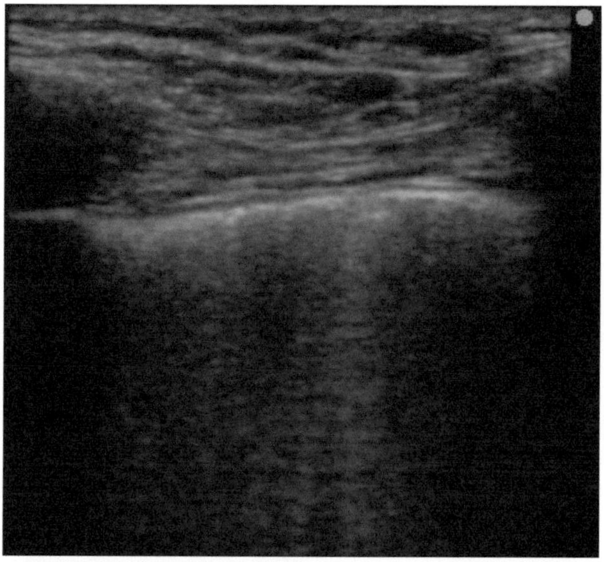

Fig. 17.29 Pulmonary fibrosis (Reproduced with permission. © University of Melbourne). Lung ultrasound features of pulmonary fibrosis include pleural thickening and B-lines with differing appearance

lung scanning can be used to determine clinical progress and response to treatments. Pulmonary fibrosis and cardiac disease may coexist, particularly in smokers, and hence the ultrasound appearance alone may not distinguish between these disorders and the cardiac findings and clinical context are important.

17.3.7.2 Interstitial Lung Disease

B-lines are seen in a variety of other conditions as similar reverberation of ultrasound occurs within fibrosis and other tissue, including tumour and infection. Lung infiltrations causing B-lines occur in pulmonary fibrosis, ARDS, interstitial pneumonia, Wegener's granulomatosis and lung carcinomatosis (e.g. Kaposi sarcoma), as shown in the Figures 17.29, 17.30 and 17.31, and Videos 17.14, 17.15, 17.16, and 17.17. Artefacts similar to B-lines are seen in atelectasis (shred sign), which are easily distinguished from B-lines in that they do not originate from both parietal and visceral pleura. B-lines are commonly seen with transoesophageal echocardiography of the descending thoracic aorta (Fig. 17.32). Features that are common to these disorders that separate them from the ultrasound appearance of pulmonary oedema include:

- Inflammatory pathologies often have pleural adhesions appearing as pleural line abnormalities (e.g. thickening), which may prevent lung sliding.
- B-lines are >3 mm apart, less widespread (diffuse) and are often asymmetrical and non homogenous (differing appearance)
- "Spared areas" of normal parenchyma
- Other focal lung pathology such as sub-pleural consolidation and tumor

Video 17.14. Pulmonary fibrosis (Reproduced with permission. © University of Melbourne. https://s3.amazonaws.com/iTU/LU+chapter/Pulmonary+fibrosis.mp4).

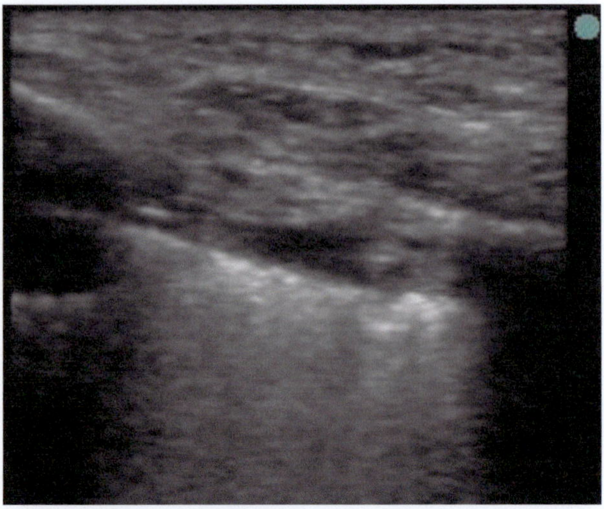

Fig. 17.30 Acute respiratory distress syndrome (Reproduced with permission. © University of Melbourne). In this patient with acute respiratory distress syndrome, the pleural line appears thickened and irregular

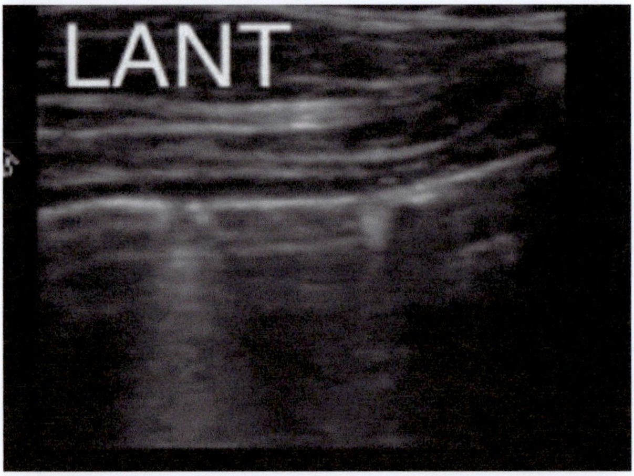

Fig. 17.31 Ultrasound of the left anterior zone (LANT) in a patient with pleural Kaposi's sarcoma. Kaposi's sarcoma (Reproduced with permission. © University of Melbourne)

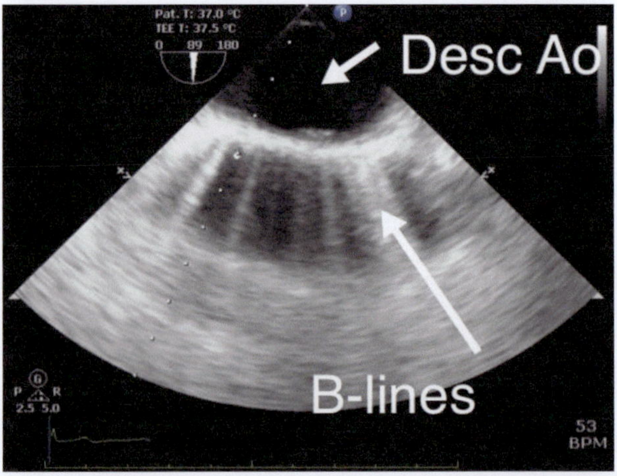

Fig. 17.32 Short-axis view of the descending aorta using transoesophageal echocardiography. B lines are seen in the dependent lung regions adjacent to the descending aorta (Reproduced with permission. © University of Melbourne)

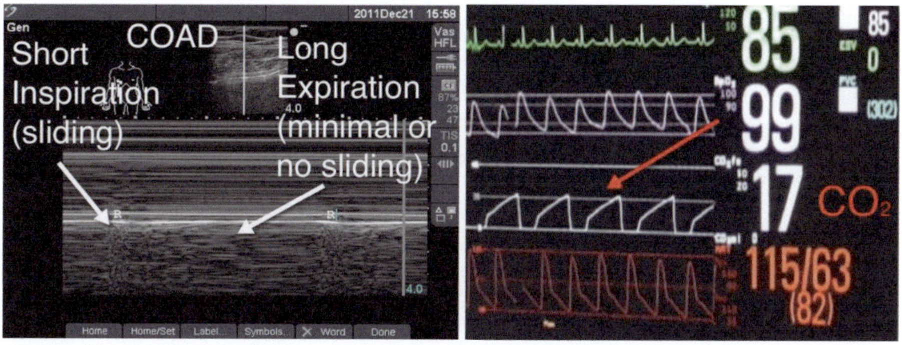

Fig. 17.33 Chronic obstructive pulmonary disease (Reproduced with permission. © University of Melbourne). Prolonged expiration appears like stratosphere or barcode sign with M-mode lung ultrasound. Note the slow upstroke in the end-tidal carbon dioxide trace on the right, which corresponds with the prolonged expiratory phase seen in chronic obstructive airways disease. Chest hyperinflation is also evident here with pulsus paradoxus of the arterial blood pressure trace reflecting raised intrathoracic pressure that impedes venous return

Video 17.15. Acute respiratory distress syndrome (Reproduced with permission. © University of Melbourne. https://s3.amazonaws.com/iTU/LU+chapter/ARDS2.mov).

Video 17.16. Wegener's granulomatosis (Reproduced with permission. © University of Melbourne. https://s3.amazonaws.com/iTU/LU+chapter/Wegeners.mp4).

Video 17.17. Kaposi's sarcoma (Reproduced with permission. © University of Melbourne. https://s3.amazonaws.com/iTU/LU+chapter/Kaposi.mp4).

17.3.8 Chronic Obstructive Pulmonary Disease

Lung ultrasound findings are usually normal in patients with chronic obstructive pulmonary disease. However an associated prolonged expiratory time and shortened inspiratory time is seen with M-mode lung surface ultrasound (Fig. 17.33). In some patient with hyperinflation, reduced diaphragmatic excursion can be observed, which is covered in the Sect. 17.3.10 below.

Other associated reversible causes of respiratory distress should be routinely excluded with ultrasound and include pneumothorax, consolidation, atelectasis from mucous plugging and AIS. Echocardiography is important to look for secondary right and left ventricular failure.

17.3.9 Internal Mammary Artery

The internal mammary artery (IMA) is seen placing the ultrasound probe lateral to the parasternal edge in the sagittal plane with a high frequency linear transducer (Fig. 17.34). Generally it is best seen in the upper chest close to the origin of the artery. The IMA can be imaged along the length of the sternum except when used for coronary

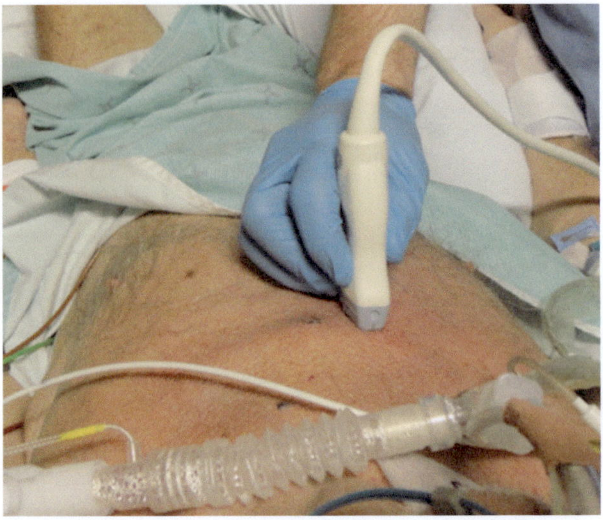

Fig. 17.34 Ultrasound of the internal mammary artery (Reproduced with permission. © University of Melbourne). Ultrasound of the internal mammary artery may be performed after coronary artery bypass surgery to confirm or exclude flow

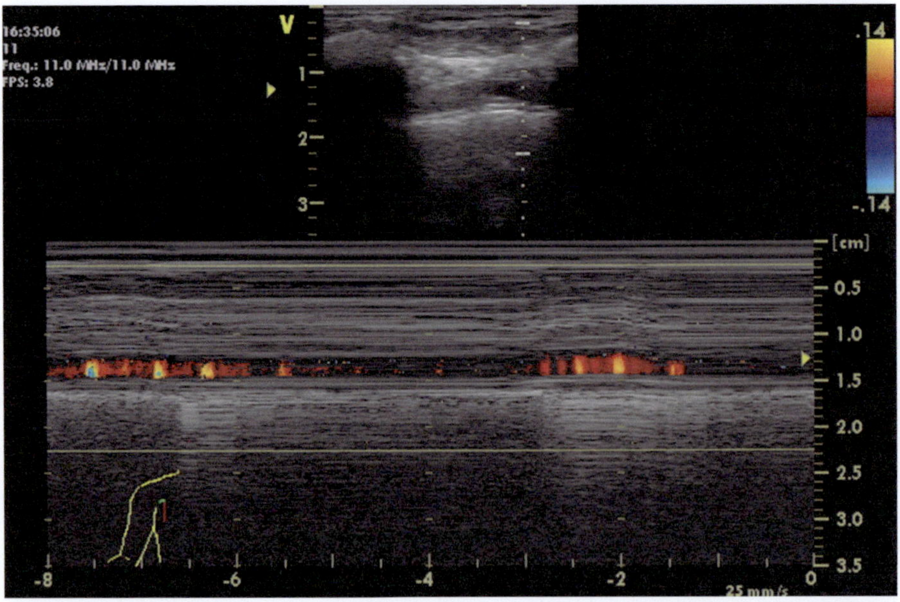

Fig. 17.35 M-mode and colour flow Doppler of the internal mammary artery (Reproduced with permission. © University of Melbourne). M-mode may be used to demonstrate flow in the internal mammary artery

artery bypass surgery, where the lung often obscures it. In this case it may be visualised closer to the subclavian artery to determine whether the artery is patent. Flow can be confirmed with colour flow Doppler (Fig. 17.35 and Video 17.18). Pulsed wave Doppler interrogation usually shows diastolic flow predominance as most coronary flow occurs during the diastolic phase when the tension in the ventricular wall is low.

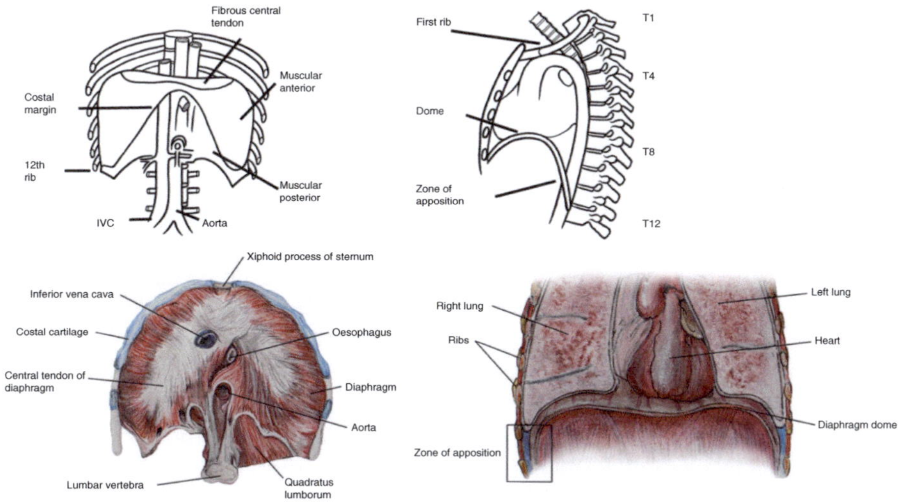

Fig. 17.36 Anatomy of the diaphragm Thoracic vertebrae levels are labelled T1 (1st thoracic vertebra) to T12 (12th thoracic vertebra). (Reproduced with permission. © University of Melbourne)

Video 17.18. Ultrasound of the internal mammary artery (Reproduced with permission. © University of Melbourne. https://s3.amazonaws.com/iTU/LU+chapter/IMA.mp4).

17.3.10 Diaphragm and Ventilation

The diaphragm performs most of the work of ventilation. Ultrasound of diaphragmatic motion is a new area of clinical use; and may be used to assess adequacy of spontaneous ventilation and may facilitate ventilatory management of patients with respiratory failure. Ultrasound assessment of the diaphragm may used be to predict successful extubation in patients who have been mechanically ventilated for a prolonged period of time or who may have become severely deconditioned. The diaphragm is a large structure and imaging the entire diaphragm is generally not possible, but visualisation is significantly improved in the presence of pleural or peritoneal effusions.

17.3.10.1 Anatomy of the Diaphragm

The diaphragm consists of a central fibrous tendon and a peripheral muscular part (Fig. 17.36). The muscles are arranged in series and parallel along with the intercostal muscles. At the end of expiration both muscular elements of the diaphragm are oriented in a parallel direction abutting the inner surface of the rib cage. Motor innervation of diaphragm is mainly from phrenic nerve with some intercostal nerve innervation of the peripheral margins.

17.3.10.2 Diaphragm Function in Breathing

During inspiration, the diaphragm contracts and moves inferiorly resulting in an increased vertical diameter of the chest cavity. Contraction of the abdominal muscles assists inspiration by increasing intra-abdominal pressure, opposing the diaphragm, resulting in a horizontal expansion of the chest cavity. During expiration, the diaphragm relaxes and elastic recoil of the lungs and chest wall occurs. In respiratory distress, there is increased assistance provided by the abdominal muscles.

Functionally there are two aspects in the diaphragm, the dome that corresponds to the central tendon and the cylindrical portion that corresponds directly to the portion opposed to inner aspect of rib cage. This is called the zone of apposition (ZOA), which extends from the diaphragm caudal insertion near the costal margin upwards to the costophrenic angle where the muscle fibres break away from the rib cage to form the dome of the diaphragm. The diaphragm decreases in surface area by 41 % with maximal inspiration. In spontaneously breathing subjects, diaphragm contraction accounts for up to 68 % of tidal volume. Increased motion of the diaphragm is associated with increased age, male gender and body mass index, which are attributed to increased thoracic volume and weight of the diaphragm. Supine posture impairs diaphragm function in the dependent regions and is attributed to gravitation effect of abdominal contents.

17.3.10.3 Performing Diaphragm Ultrasound

A low frequency probe such as a curvilinear or transthoracic probe is placed in the supine patient in a longitudinal plane with the orientation marker directed cranially on the lateral chest wall between mid and posterior axillary line (Fig. 17.37 and Video 17.19). The probe is moved inferiorly until the liver or spleen and some of the kidney is visualised. This identifies the posterior part of the hemidiaphragm, which has the most movement. The probe is aimed at the dome or the posterior part of the diaphragm, and adjusted so the beam is perpendicular to the cranio-caudal hemidiaphragmatic movement, maximising the measurement of the diaphragm movement. The measurement of the excursion can be taken using 2D or M- Mode, if the maximum movement of the diaphragm is in line with the M-mode line (Fig. 17.38). Anatomical (vertical axis) M-mode or manual entry of the Doppler angle, available on some machines, may be used to account for poor alignment. Normal values for diaphragm movement are 0.9–2.0 cm [22].

Video 17.19. Performing ultrasound assessment of the diaphragm (Reproduced with permission. © University of Melbourne. https://s3.amazonaws.com/iTU/LU+chapter/Diaphragm.mp4).

The 2-D ultrasound image of the diaphragm using the anterior approach may be divided into the lateral, middle and medial regions. The mean distance of diaphragm movement for the right and left middle and medial regions are in Fig. 17.38 [22].

Challenges to diaphragm ultrasound include lung interposition and underlying rib movement, especially with large tidal volumes. Subcostal imaging is inferior due to interference from gastric or colonic gas, reduced movement of this part of the diaphragm (central dome relatively fixed) and poor alignment with the ultrasound

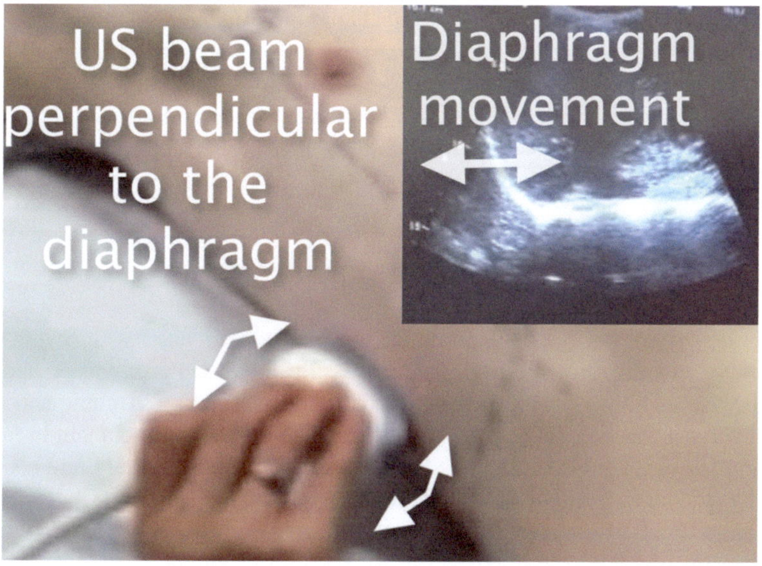

Fig. 17.37 Performing ultrasound assessment of the diaphragm (Reproduced with permission. © University of Melbourne). In ultrasound assessment of the diaphragm a curvilinear probe is placed longitudinally in the mid axillary line

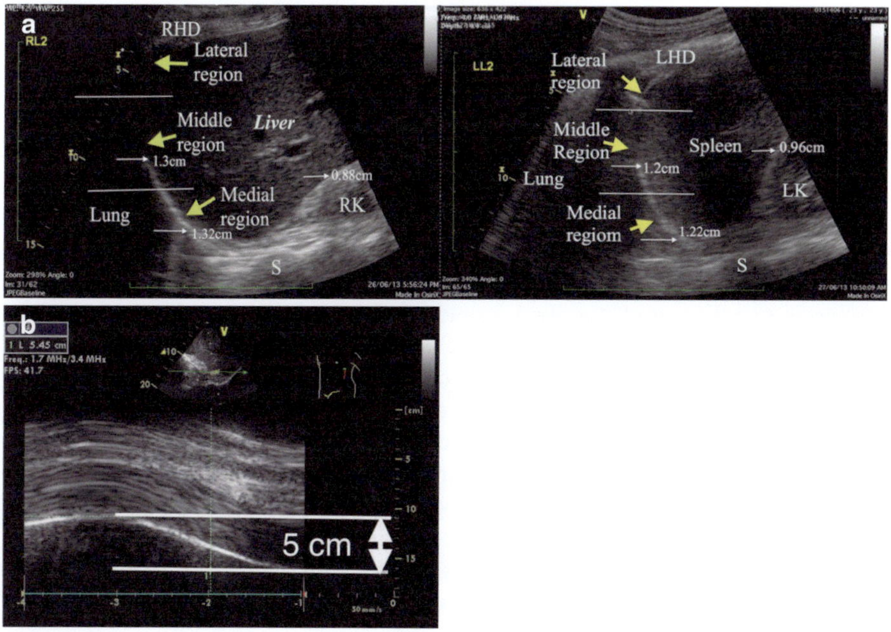

Fig. 17.38 (**a**) Two-dimensional mode ultrasound of the right and left hemidiaphragm (Reproduced with permission Elsevier (Haji et al. [22])). (**b**) M-mode ultrasound measurement of maximum diaphragm excursion (Reproduced with permission. © University of Melbourne). M-mode may be used to conveniently measure maximum diaphragm excursion, however angle correction may need to be applied if the transducer is not longitudinal to the direction of diaphragm movement. *RHD* indicates right hemidiaphragm; *RK* right kidney; *LHD* left hemidiaphragm; *LK* left kidney; *S* spines

beam. There is often no specific anatomical landmark on the diaphragm with which to measure excursion. In addition, moving the probe during the study will lead to unreliable measurements. An elevated hemidiaphragm may also arise from lung volume loss or from increased pressure from within the abdominal compartment (e.g. obesity, pregnancy, raised intra-abdominal pressure of any cause).

17.4 Lung Ultrasound Guided Procedures

Procedures within the chest are common and can be associated with severe or life threatening complications. These procedures are made simpler, easier and safer with the use of ultrasound compared to the traditional "blind" techniques. As for central venous vascular access, there is an increasingly cogent argument in favor of routine ultrasound guided or assisted procedures to improve quality and standards as well as to enhance safety. The procedures covered include thoracocentesis, intercostal catheter insertion, assessment of endotracheal intubation, use in percutaneous tracheostomy and the assessment of the vocal cord movement.

17.4.1 Pleural Effusion Drainage

Ultrasound guidance of pleural fluid aspiration has been demonstrated to increase success and reduce the incidence of complications of organ puncture [1] and there is a strong case to use ultrasound routinely when available [1, 23]. Ultrasound is especially useful in guiding small and loculated effusions. Every year there are numerous avoidable solid and hollow organ injuries from mal-directed pleural catheters [24]. In addition, use of ultrasound allows confirmation of the position of the drainage catheter and exclusion of complications such as pneumothorax or pulmonary oedema. Re-expansion pulmonary oedema may occur after drainage of a large pleural effusion or pneumothorax [25].

Needle sampling of pleural fluid, thoracocentesis, is frequently performed to provide diagnostic information of the fluid, such as to distinguish between infection, blood and tumour, and to guide antimicrobial therapy. Ultrasound may enable more frequent diagnostic aspirations due to its safety and ability to identify of radio-occult effusions.

The basic principles of lung ultrasound are similar whether it is for thoracocentesis or insertion of a pleural catheter. For patients who are able to sit up, it is preferred that a posterior site for insertion be adopted, since the effusion will collect inferiorly and thus present a larger depth which increases the safety margin when inserting needles. However, some patients must remain supine such as those in ICU or who may otherwise feel faint during a procedure.

17.4.1.1 Ultrasound

Estimation of the volume of pleural effusion size should be performed routinely before drainage (described in section 17.3.3.1). Drainage of an effusion should not be attempted if the maximum effusion dimension is less than 15 mm (300 mL).

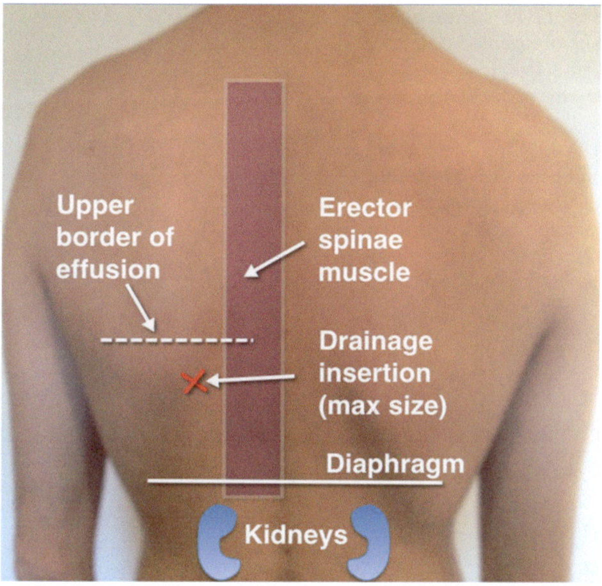

Fig. 17.39 Posterior insertion of a pleural drain in an erect patient (Reproduced with permission. © University of Melbourne). Insertion site is posterior lower chest in the region of the maximum effusion

The upper and lower limits of the effusion and the diaphragm should be located with ultrasound. The skin is then marked at the point of maximum pleural effusion size, ensuring this site is well away from intra-abdominal organs (cardiac apex, aorta and inferior vena cava) and avoiding skin folds, pleural adhesions and loculations.

17.4.1.2 Determining the Insertion Site

The insertion site depends on whether thoracocentesis or catheter insertion is to be performed. The insertion site is also determined by whether or not the patient is able to adopt the erect position, which is preferable. The traditional 'safe' triangle when using the "blind insertion" recommended by the British Thoracic Society [23] (the anterior border of latissimus dorsi, the lateral border of the pectorals major, a horizontal line from the nipple to the apex below the axilla) should be kept in mind when selecting the puncture site but with ultrasound guidance this may be of lesser importance.

Erect Position

In the erect patient, the best insertion site is lateral to the erector spinae muscle (Fig. 17.39). This is not only to reduce the amount of tissue that is needed to enter the chest wall, but also on the left it is to avoid injury to the descending aorta. It is important that a posterior, rather than a lateral approach, is selected wherever possible; as the diaphragm attachment is much lower the lateral attachment. Also the risk of a needle striking the dome of the diaphragm is greater in the lateral compared to the posterior approach.

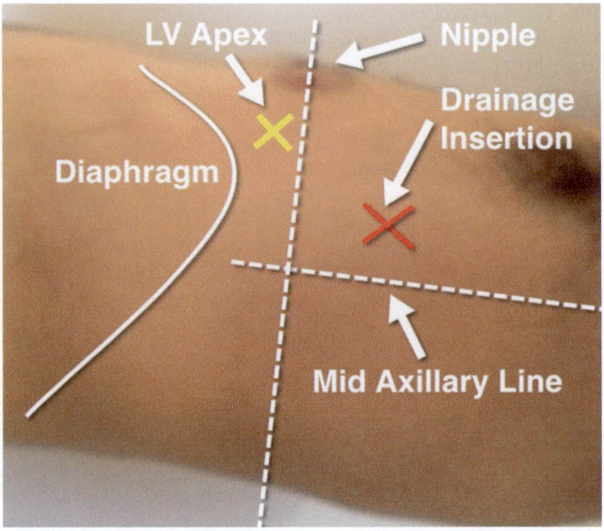

Fig. 17.40 Identifying the site of insertion of a left-sided pleural drain in a supine patient (Reproduced with permission. © University of Melbourne). Insertion site is where the effusion is detected, but superior to left ventricular (*LV*) apex and anterior to the mid axillary line

Supine Position

The lateral aspect of the chest should be selected. The entry point should be checked to ensure it is well above the diaphragm, and anterior to the mid axillary line, so as to avoid potential injury to the descending aorta from a posteriorly directed puncture (Fig. 17.40). Finally, it is mandatory on the left to ensure that the apex of the heart has been visualised using TTE, and to ensure that the puncture site is well away from it.

17.4.1.3 Performing the Procedure

Generally it is easier to adhere to asepsis by performing the insertion of the needle or catheter using the "blind" technique after ultrasound "mapping", rather than in real time to guide the needle or catheter. However it is strongly advised to perform the procedure as soon after the ultrasound marking as possible and *without any* changes in patient position (Video 17.20). If these principles are not followed then the risk of a dry tap and complication increases. However using ultrasound during the procedure allows a more precise identification of potentially dangerous structure such as the intercostal arteries when using a needle-guided approach.

Video 17.20. Performing ultrasound guided pleural fluid drainage. Reproduced with permission. © University of Melbourne. https://s3.amazonaws.com/iTU/LU+chapter/USguidedICC.mp4.

Thoracocentesis

Use of an intravenous cannula for thoracocentesis allows the withdrawal of the metal needle, minimising the duration the patient is exposed to the needle tip. Use of a "3 way tap" attached to intravenous tubing facilitates drainage of a large volume of fluid.

Intercostal Catheter Insertion

For simple pleural effusion (i.e. when it is not expected that there is much blood or the expectation of infection), there is no advantage to using a large catheter over a fine bore catheter. Therefore it is recommended that a small diameter catheter is used, and this may involve direct puncture methods (Pneumocath®) or a guide wire and dilators as per the Seldinger method (Cook® catheter). For complex pleural effusion (i.e. where there is a high probability of blood or clot, infection or loculation of the fluid), a small catheter will often occlude and it is better to use a conventional intercostal catheter as is seen in cardiac or thoracic surgery practice. The sharp trocar should never be used for insertion as this is associated with serious injuries. Instead, blunt dissection should be used to create the passage through the chest wall prior to introduction of the catheter by gentle insertion.

17.4.1.4 Pitfalls

The most important pitfall is to image the effusion and then alter the position of the patient between the scan and the insertion of the needle or catheter. The non-loculated pleural space will allow the fluid to move freely with changes of patient position and so the chest wall markings are only valid in the position of the patient at the time of the scan.

The second most common error, even with the use of the ultrasound probe, is to incorrectly identify the anatomy. For example, the left ventricle may be dilated and abut the lateral chest wall and the ventricular wall may be thin due to a large prior infarct. So it is important that the general surface markings that are "rules" with the blind technique should be kept in mind as an alert to incorrect positioning. The site inferior to the nipple should always be considered as a danger for injury to the heart on the left and for the diaphragm and abdominal organs on either side. The insertion site should not generally be performed posterior to the mid axillary line. This point is related mostly to a posteriorly directed passage of a needle or catheter, than it is related to the entry point through the chest. This is to avoid injury to the descending aorta on the left. Whilst there are not absolute rules, the lateral position for entry into the chest should typically be superior to the nipple line in males or the equivalent site in females; and anterior to the mid axillary line. Examples of an intercostal catheter inadvertently inserted into the ascending aorta are shown in Fig. 17.41.

17.4.2 Airway Management

Tracheal intubation is a commonly performed invasive procedure that may be life saving but serious complications including failed intubation continue to be a contributor to death and neurological injury [26]. Current techniques used to improve safety of intubation include fibreoptic bronchoscopy, video laryngoscopy and tracheostomy. Ultrasound can be useful in both confirming the position of the tracheal tube

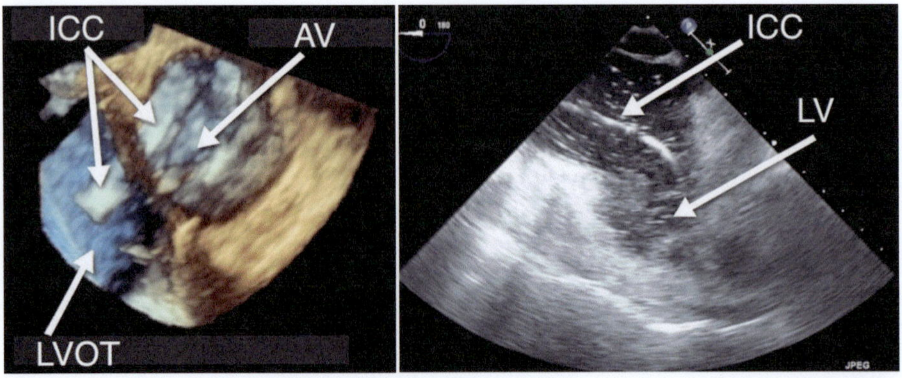

Fig. 17.41 3-D Transoesophageal echocardiography of a chest drain complication (Reproduced with permission. © University of Melbourne). The 3-D transoesophageal echocardiography image on the left demonstrates an intercostal catheter passing through the left ventricular outflow tract (*LVOT*) and aortic valve (*AV*) into the ascending aorta. The 2-D image on the right shows the passage of the catheter passing through the left ventricle (*LV*)

in the trachea and in guiding percutaneous oxygenation as with cricothyroid puncture, and percutaneous ventilation with cricothyroidotomy and tracheostomy.

The upper airway from mouth to the trachea, is easily accessible to surface ultrasound and may provide useful information used to guide intubation of the trachea either via either the supraglottic or subglottic approach. The role of ultrasound in airway management is relatively new and its role is not yet clear but has promise.

Ultrasound may be used to accurately identify the cricothyroid membrane and trachea prior to percutaneous cricothyroid puncture and tracheostomy. Blood vessels and the thyroid isthmus can easily be identified and thus avoided. Real-time ultrasound of the larynx and esophagus during oral endotracheal intubation may be used to accurately detect inadvertent esophageal intubation before insufflation of the stomach with gas occurs. Detection of endobronchial intubation and failure of ventilation of a lung or lobe can be detected by loss of lung sliding sign and reduced diaphragm movement. Assessing the degree of difficulty in oral endotracheal intubation using ultrasound has not yet been demonstrated, however shows promise. Vocal cord function can easily be seen with surface ultrasound, which may be a more convenient alternative to indirect naso-endoscopy.

17.4.2.1 Prediction of Difficult Endotracheal Intubation

Prediction of patients who are difficult to intubate is often difficult with clinical examination. Ultrasound may improve airway assessment by allowing visualisation of the airway directly from the base of the tongue, at the level of the vocal cords and down to the superior extremity of the sternum. Abnormal airway anatomy, which could make intubation difficult or impossible that are readily detected with surface ultrasound include masses and narrowing of the oropharyngeal tract, larynx and trachea. Obstructive sleep apnoea is associated with difficult intubation and may be predicted by ultrasound by measuring the thickness of the lateral pharyngeal wall [27]. The

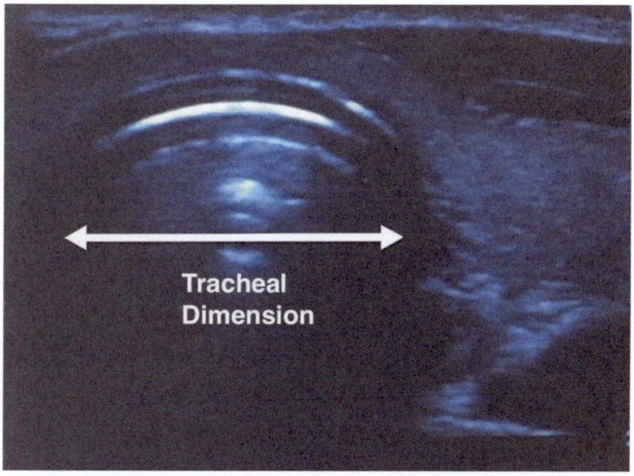

Fig. 17.42 Transverse of the subglottic trachea (Reproduced with permission. © University of Melbourne). The size of tracheostomy tube required can be predicted with the simple measurement of tracheal diameter with ultrasound

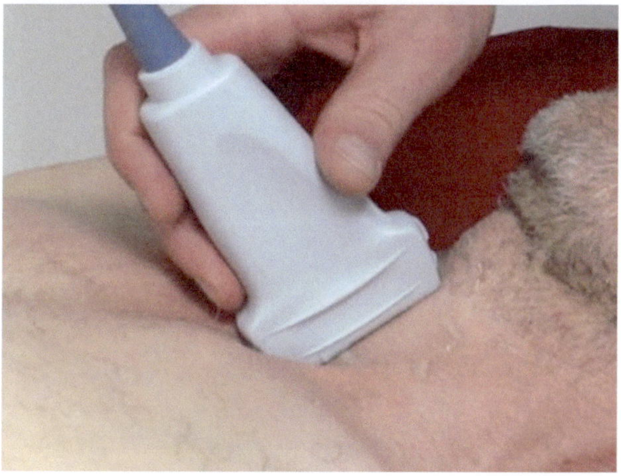

Fig. 17.43 Confirmation of tracheal intubation – probe position in long axis (Reproduced with permission. © University of Melbourne). Confirmation of tracheal placement of the tracheal tube using the long axis probe placement

appropriate diameter of tracheostomy and double lumen endotracheal tube is accurately predicted with ultrasound measurement compared with CT and magnetic resonance imaging (MRI) respectively [28, 29] (Fig. 17.42).

17.4.2.2 Confirmation of Endotracheal Intubation

Detection of tube within the trachea is difficult after it has been inserted and real-time assessment is more reliable but requires having the ultrasound probe in place prior to tube placement. The probe is placed at the level of the larynx either in long axis (longitudinal) or short axis (transverse) pointing in the cephalad direction at approximately 30° (Figs. 17.43, 17.44, and 17.45). Correct tracheal placement may be seen as a brief "fluttering" within the trachea (Videos 17.22 and 17.23). Esophageal intubation may be detected by the lack of fluttering within the trachea but more reliably demonstrated using a transverse view as a clearly visible hyperechoic curved line with distal dark shadowing (Video 17.23). The esophagus is

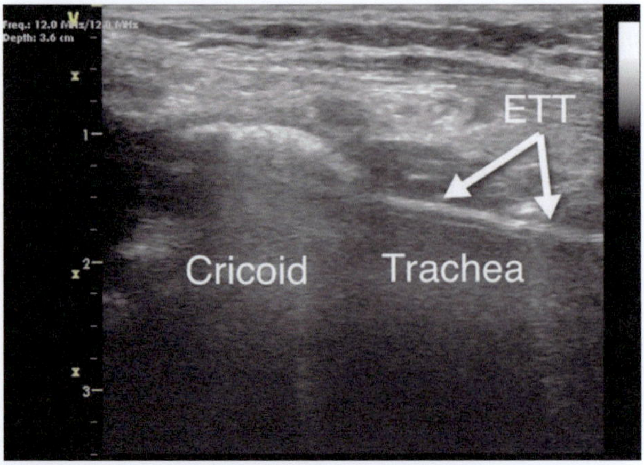

Fig. 17.44 Confirmation of tracheal intubation– ultrasound in long axis (Reproduced with permission. © University of Melbourne). Confirmation of tracheal placement of the tracheal tube using the long axis probe placement, which is demonstrated by a long thin *white line* below the tracheal beads. *ETT* endotracheal tube

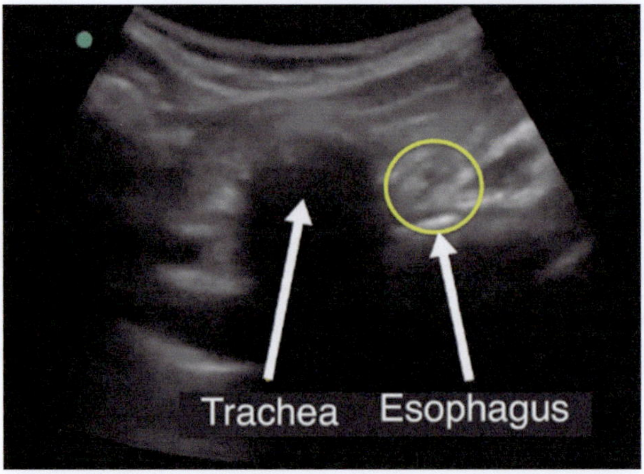

Fig. 17.45 Confirmation of tracheal intubation– ultrasound in short axis (Reproduced with permission. © University of Melbourne). Confirmation of tracheal placement of the tracheal tube using the short axis probe placement. Tracheal placement is shown here with an obvious large shadow behind the trachea (*centre*) next to the esophagus (*right*)

usually posterolateral to the trachea on the patient's left. The concentric layers of the esophagus form a characteristic "bulls eye" appearance.

Video 17.21. Long axis ultrasound confirmation of successful endotracheal intubation https://s3.amazonaws.com/iTU/LU+chapter/Intubation+LAX.mp4.

Video 17.22. Short axis ultrasound confirmation of successful endotracheal intubation (Reproduced with permission. © University of Melbourne. https://s3.amazonaws.com/iTU/LU+chapter/Intubation+SAX.mp4).

Video 17.23. Oesophageal intubation – ultrasound in short axis (Reproduced with permission. © University of Melbourne. https://s3.amazonaws.com/iTU/LU+chapter/Intubation+Esophageal.mp4).

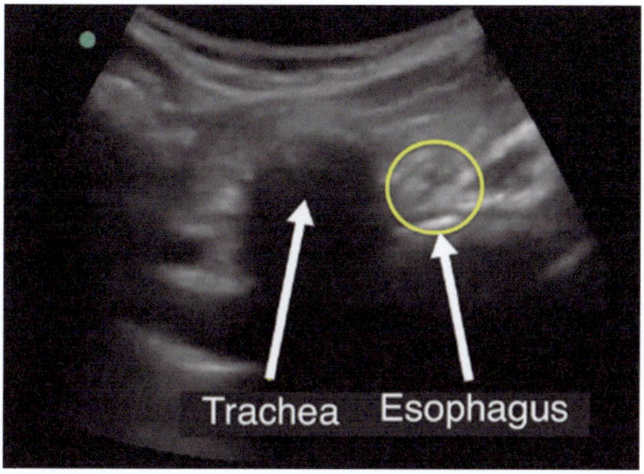

Fig. 17.46 Long axis view of the subglottic airway (Reproduced with permission. © University of Melbourne)

17.4.3 Percutaneous Cricothyroid Puncture and Tracheostomy

Surface ultrasound may be used to identify the correct site and to guide in real time percutaneous subglottic airway access. In principle, the technique is similar to ultrasound guided vascular access.

Cricothyroid subglottic tracheal access is usually performed in an emergency, with failed orotracheal intubation or when it is deemed that orotracheal intubation will fail. Although the standard method for identification of the cricothyroid membrane and trachea for planned and emergency percutaneous subglottic intubation is by palpation, correct identification is generally poor. In one study the success rate by palpation alone was only 30 % [30]. This is particularly so if there is an inflammatory process involving the neck. However the mean time to accurately identify the membrane with ultrasound by 100 % of emergency physicians was only 25 s. Ultrasound location of the subglottic airway may be particularly useful in patients with obesity or abnormal anatomy such as pre-tracheal infection or scarring. It is advocated that there be a role for routine identification and marking the site and depth of the cricothyroid membrane before attempts at intubation are made in patients with a predicted difficult intubation. This would facilitate emergency cricothyroid access if required, such as failed intubation or inability to oxygenate. Alternatively a cricothryoid cannula may be safely placed under local anaesthesia prior to oral intubation attempts.

17.4.3.1 Cricothyroid Puncture

Again either longitudinal or transverse ultrasound approach may be used. The cricothyroid membrane appears as a thin but brightly echoic line between the cricoid and thyroid cartilages and their accompanying shadowing (Figs. 17.46 and 17.47). The tracheal rings below the hyoid bone have characteristic 'line of beads' appearance.

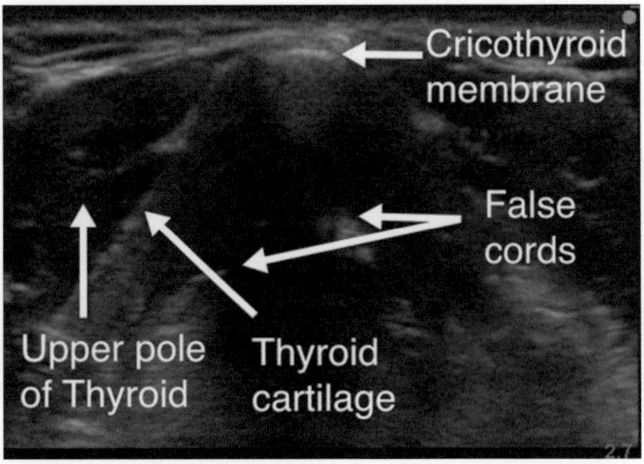

Fig. 17.47 2D transverse view – Cricothyroid membrane (Reproduced with permission. © University of Melbourne)

17.4.3.2 Procedure

The procedure is demonstrated in Video 17.4

1. Use a linear transducer for higher frequency improved resolution.
2. Stand at the head of bed with the ultrasound machine facing you; and as close to patient as possible.
3. Orient the probe such that movement of the probe laterally in one direction gives the corresponding change in image (left and right concordance).
4. Apply the probe transversely showing a short axis view of the larynx. The thyroid and cricoid cartilages are very easily palpated.
5. Look for the cricoid and thyroid cartilages with an intervening horizontal bright cricothyroid membrane.
6. Mark the skin for subsequent cricothyroid puncture in the conventional manner.
7. Alternatively:

   (a) Holding the probe in one hand, insert the needle with the other hand at the midpoint of the transducer onto the skin. Identify the needle tip on ultrasound and ensure that it is located directly above the target.
   (b) Advance the needle under direct ultrasound guidance, ensuring that the needle tip is visible at all times. Failure to do so risks needle advancement into an undesired direction.
   (c) When the needle tip reaches the airway the ultrasound equipment may be dispensed with and proceed with the procedure with your normal technique.

17.4.4 Percutaneous Tracheostomy

Percutaneous tracheostomy is commonly performed in patients who are already intubated and mechanically ventilated in ICU who are likely to require prolonged ventilatory support or are unlikely to regain a patent upper airway.

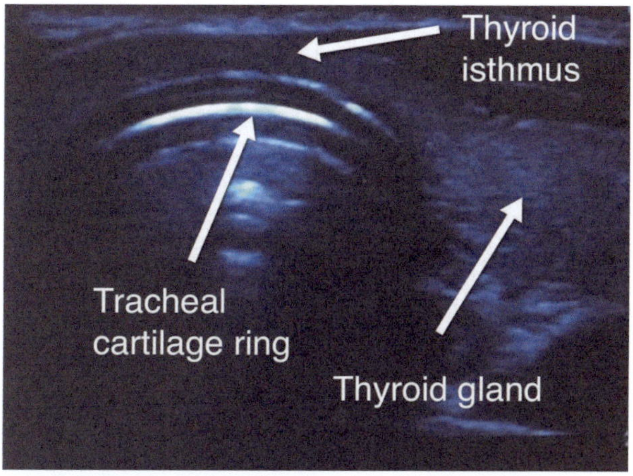

Fig. 17.48 Transverse view – Trachea (Reproduced with permission. © University of Melbourne)

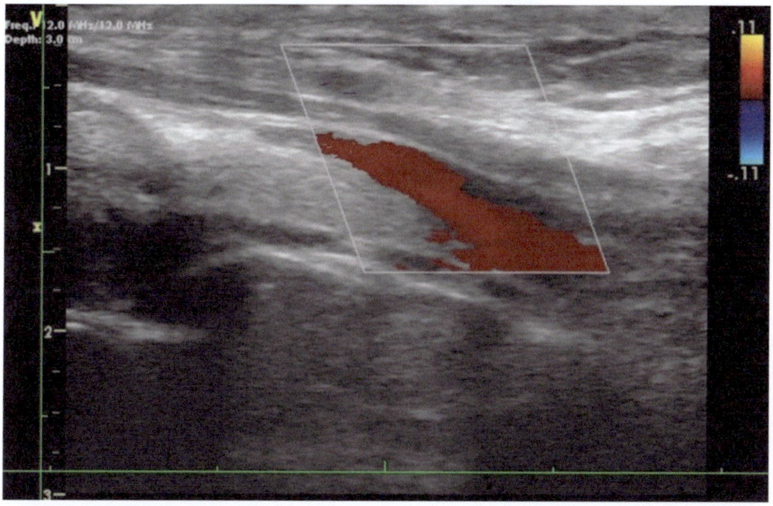

Fig. 17.49 Long axis with colour Doppler – Anterior thyroid veins (Reproduced with permission. © University of Melbourne)

Ultrasound greatly facilitates accurate location of the trachea, especially in patients with obese necks or tracheal deviation. The most important structure to identify is the size and location of the thyroid isthmus (Fig. 17.48). This is because it is both vascular and tough, making dilation of a tract through it both difficult and prone to bleeding (Figs. 17.49 and 17.50). In addition, there are often a number of moderately large anterior thyroid veins, which if torn may bleed significantly. Sometimes, these veins are visible on the ultrasound scan. The site of tracheostomy was changed in 25 % of patients by use of routine ultrasound. Demonstration of the procedure is shown in the Video 17.24.

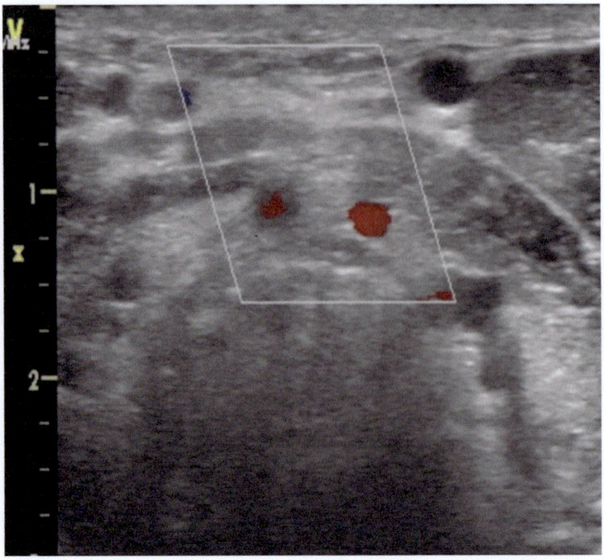

Fig. 17.50 Short axis with colour Doppler – Anterior thyroid veins (Reproduced with permission. © University of Melbourne)

17.4.4.1 Procedure

1. Suggested positioning for the procedure is shown in Fig. 17.51.
2. Scan the neck as above.
3. Identify the thyroid isthmus and note the size and locality. In some cases it may be so large as to preclude a percutaneous approach.
4. Select a site superior or inferior to the isthmus.
5. Identify and avoid significant blood vessels (generally veins).
6. Align the trachea in the exact centre of the screen.
7. Mark the skin and proceed to routine technique for insertion.
8. Alternatively (recommended):

   (a) Holding the probe in one hand, insert the needle with the other hand at the midpoint of the transducer onto the skin. Identify the needle tip on ultrasound and ensure that it is located directly above the target.
   (b) Advance the needle under direct ultrasound guidance, ensuring that the needle tip is visible at all times. Failure to do so risks needle advancement into an undesired direction.
   (c) When the needle tip reaches the airway the ultrasound equipment may be dispensed with and proceed with the procedure with your normal technique.

The procedure is demonstrated in Video 17.24.

Video 17.24. Procedure of ultrasound guided percutaneous tracheostomy (Reproduced with permission. © University of Melbourne. https://s3.amazonaws.com/iTU/LU+chapter/Perc+tracheostomy.mp4).

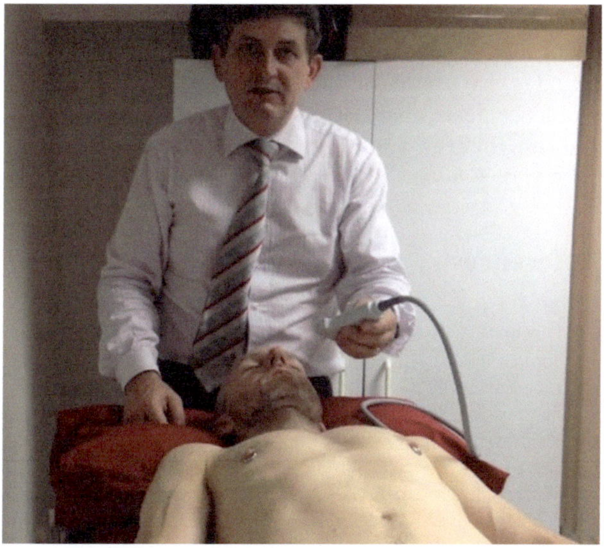

Fig. 17.51 Positioning yourself for ultrasound-guided percutaneous tracheostomy (Reproduced with permission. © University of Melbourne)

17.4.5 Assessment of the Vocal Cords

The vocal cords and function may be easily and non-invasively assessed by ultrasound, which may be useful after surgical procedures that may damage the recurrent laryngeal nerve.

To image the vocal cords, place the probe transversely at the level of the crico-thyroid membrane at a 20–30° cephalad tilt. The base of the vocal cords containing echogenic muscle and adipose tissue (the "false cords"), will be seen rather than the thin free edges (Video 17.25). The free edges of the vocal cords are surrounded by air, and hence not visible on ultrasound. Normally during phonation, both sides will be seen to move. Importantly, unlike indirect laryngoscopy, the imaging is not identifying direct vocal cord apposition; but rather is detecting movement of the cords. This is useful after surgery, such as thyroidectomy, where absence of movement on one side suggests a recurrent laryngeal nerve palsy.

Video 17.25. 2D ultrasound – Assessment of the vocal cords with ultrasound (https://s3.amazonaws.com/iTU/LU+chapter/VocalCords2.mp4).

17.5 Conclusion

Lung ultrasound is rapidly emerging as a valuable non-invasive tool to aid anaesthetists and critical care physicians to improve clinical diagnosis of important respiratory pathology. It is likely that this will fundamentally change clinical assessment of patients who are at risk of respiratory pathology and complementing transthoracic echocardiography enabling a cardiorespiratory ultrasound examination. Lung ultrasound is also useful for improving the safety of invasive thoracic procedures such as thoracocentesis, intercostal catheter insertion and shows promise for supraglottic and subglottic airway management.

References

1. Diacon AH, Brutsche MH, Soler M (2003) Accuracy of pleural puncture sites: a prospective comparison of clinical examination with ultrasound. Chest 123(2):436–441
2. Lichtenstein D, Goldstein I, Mourgeon E, Cluzel P, Grenier P, Rouby JJ (2004) Comparative diagnostic performances of auscultation, chest radiography, and lung ultrasonography in acute respiratory distress syndrome. Anesthesiology Clin Trial Comp Study Res Support Non US Govt 100(1):9–15
3. Moore CL, Copel JA (2011) Point-of-care ultrasonography. N Engl J Med 364(8):749–757
4. Lichtenstein DA, Meziere GA (2008) Relevance of lung ultrasound in the diagnosis of acute respiratory failure: the BLUE protocol. Chest 134(1):117–125
5. Volpicelli G, Elbarbary M, Blaivas M, Lichtenstein DA, Mathis G, Kirkpatrick AW et al (2012) International evidence-based recommendations for point-of-care lung ultrasound. Intensive Care Med 38(4):577–591
6. Peris A, Tutino L, Zagli G, Batacchi S, Cianchi G, Spina R et al (2010) The use of point-of-care bedside lung ultrasound significantly reduces the number of radiographs and computed tomography scans in critically ill patients. Anesth Analg 111(3):687–692
7. Royse CF, Canty DJ, Faris J, Haji DL, Veltman M, Royse A (2012) Core review: physician-performed ultrasound: the time has come for routine use in acute care medicine. Anesth Analg 115(5):1007–1028
8. Canty DJ, Royse CF, Kilpatrick D, Bowman L, Royse AG (2012) The impact of focused transthoracic echocardiography in the pre-operative clinic. Anaesthesia 67(6):618–625
9. Lichtenstein DA (2007) Ultrasound in the management of thoracic disease. Crit Care Med 35(5 Suppl):S250–S261
10. Rocco M, Carbone I, Morelli A, Bertoletti L, Rossi S, Vitale M et al (2008) Diagnostic accuracy of bedside ultrasonography in the ICU: feasibility of detecting pulmonary effusion and lung contusion in patients on respiratory support after severe blunt thoracic trauma. Acta Anaesthesiol Scand Comp Study Eval Stud 52(6):776–784
11. Sajadieh H, Afzali F, Sajadieh V, Sajadieh A (2004) Ultrasound as an alternative to aspiration for determining the nature of pleural effusion, especially in older people. Ann N Y Acad Sci 1019:585–592
12. Balik M, Plasil P, Waldauf P, Pazout J, Fric M, Otahal M et al (2006) Ultrasound estimation of volume of pleural fluid in mechanically ventilated patients. Intensive Care Med 32(2):318–321
13. Lichtenstein D, Meziere G, Seitz J (2009) The dynamic air bronchogram. A lung ultrasound sign of alveolar consolidation ruling out atelectasis. Chest 135(6):1421–1425
14. Reissig A, Kroegel C (2003) Transthoracic ultrasound of lung and pleura in the diagnosis of pulmonary embolism: a novel non-invasive bedside approach. Respiration Int Rev Thorac Dis 70(5):441–452 [Review]
15. Volpicelli G, Caramello V, Cardinale L, Cravino M (2008) Diagnosis of radio-occult pulmonary conditions by real-time chest ultrasonography in patients with pleuritic pain. Ultrasound Med Biol 34(11):1717–1723
16. Reissig A, Kroegel C (2007) Sonographic diagnosis and follow-up of pneumonia: a prospective study. Respiration Int Rev Thorac Dis 74(5):537–547

17. Parlamento S, Copetti R, Di Bartolomeo S (2009) Evaluation of lung ultrasound for the diagnosis of pneumonia in the ED. Am J Emerg Med Clin Trial Comp Study 27(4):379–384
18. Mathis G, Blank W, Reissig A, Lechleitner P, Reuss J, Schuler A et al (2005) Thoracic ultrasound for diagnosing pulmonary embolism: a prospective multicenter study of 352 patients. Chest Clin Trial Multicenter Study 128(3):1531–1538
19. Squizzato A, Rancan E, Dentali F, Bonzini M, Guasti L, Steidl L et al (2013) Diagnostic accuracy of lung ultrasound for pulmonary embolism: a systematic review and meta-analysis. J Thromb Haemost JTH 11(7):1269–1278
20. Volpicelli G, Caramello V, Cardinale L, Mussa A, Bar F, Frascisco MF (2008) Bedside ultrasound of the lung for the monitoring of acute decompensated heart failure. Am J Emerg Med 26(5):585–591
21. Picano E, Frassi F, Agricola E, Gligorova S, Gargani L, Mottola G (2006) Ultrasound lung comets: a clinically useful sign of extravascular lung water. J Am Soc Echocardiogr Off Publ Am Soc Echocardiogr 19(3):356–363 [Review]
22. Haji K, Royse A, Tharmaraj D, Haji D, Botha J, Royse C (2015) Diaphragmatic regional displacement assessed by ultrasound and correlated to subphrenic organ movement in the critically ill patients – an observational study. J Crit Care 30(2):439.e7–439.e13
23. Havelock T, Teoh R, Laws D, Gleeson F (2010) Pleural procedures and thoracic ultrasound: British Thoracic Society pleural disease guideline 2010. Thorax 65 Suppl 2:ii61–ii76
24. Harris A, O'Driscoll BR, Turkington PM (2010) Survey of major complications of intercostal chest drain insertion in the UK. Postgrad Med J 86(1012):68–72
25. Tan HC, Mak KH, Johan A, Wang YT, Poh SC (2002) Cardiac output increases prior to development of pulmonary edema after re-expansion of spontaneous pneumothorax. Respir Med 96(6):461–465
26. Cook TM, Woodall N, Harper J, Benger J, Fourth National Audit Project (2011) Major complications of airway management in the UK: results of the Fourth National Audit Project of the Royal College of Anaesthetists and the Difficult Airway Society. Part 2: intensive care and emergency departments. Br J Anaesth Res Support Non US Govt 106(5):632–642
27. Liu KH, Chu WC, To KW, Ko FW, Tong MW, Chan JW et al (2007) Sonographic measurement of lateral parapharyngeal wall thickness in patients with obstructive sleep apnea. Sleep 30(11):1503–1508
28. Hardee PS, Ng SY, Cashman M (2003) Ultrasound imaging in the preoperative estimation of the size of tracheostomy tube required in specialised operations in children. Br J Oral Maxillofac Surg Case Rep 41(5):312–316
29. Sustic A, Miletic D, Protic A, Ivancic A, Cicvaric T (2008) Can ultrasound be useful for predicting the size of a left double-lumen bronchial tube? Tracheal width as measured by ultrasonography versus computed tomography. J Clin Anesth Comp Study Eval Stud 20(4):247–252
30. Elliott DS, Baker PA, Scott MR, Birch CW, Thompson JM (2010) Accuracy of surface landmark identification for cannula cricothyroidotomy. Anaesthesia Res Support Non US Govt 65(9):889–894

Chapter 18
The Acute Pain Team

David A. Edwards, Michael Kent, Linda Le-Wendling, and Patrick J. Tighe

Abstract Aside from the moral imperative to avert needless suffering from pain, ineffective postoperative pain management may contribute to multiple adverse perioperative outcomes. The primary objective of the modern acute pain service (APS) is to diagnose and treat acute pain as a pathophysiologic process. The APS includes physicians, nurses in the procedural care area and in the units/wards, and a variety of consultant staff critical for the delivery of safe and effective care of patients suffering from acute pain. The modern APS functions across not only multiple inpatient hospital environments, but also at multiple scales of healthcare coordination and delivery. While nominally charged with improving a patient's pain, this objective is entirely predicated upon the context of optimizing patient functioning while guaranteeing patient safety. This contextualized objective has driven a reexamination of not only the evaluation of a patient's acute pain, but also measures of APS efficacy. Future work is necessary to continually optimize the effective and efficient delivery of acute pain management services in an increasingly complex healthcare system.

Keywords Acute pain • Regional anesthesia • Systems-based practice • Chronification of acute pain • Perioperative pain • Acute pain medicine

---

D.A. Edwards, MD PhD
Department of Anesthesiology, Harvard University, Cambridge, MA 02138, USA
e-mail: daedwards@mgh.harvard.edu

M. Kent, MD CDR, MC, USN
Department of Anesthesiology, Walter Reed National Military Medical Center, Bethesda, MD 20889, USA
e-mail: kentml@gmail.com

L. Le-Wendling, MD
Department of Anesthesiology, University of Florida, Gainesville, FL 32611, USA
e-mail: lle@anest.ufl.edu

P.J. Tighe, MD MS (✉)
Department of Anesthesiology, University of Florida, Gainesville, FL 32610, USA
e-mail: ptighe@anest.ufl.edu

K. Stuart-Smith (ed.), *Perioperative Medicine – Current Controversies*,
DOI 10.1007/978-3-319-28821-5_18

18.1 Team Objectives

18.1.1 Introduction

Acute postoperative pain affects 80 % of patients recovering from surgery, with 86 % of this subset reporting moderate, severe, or extreme pain [1]. Aside from the moral imperative to avert needless suffering from pain, ineffective postoperative pain management may contribute to several categories of adverse perioperative outcomes, including deep vein thrombosis, pulmonary embolism, coronary ischemia, pneumonia, poor wound healing, insomnia and demoralization [2–4], and perhaps even immunosuppression and cancer recurrence [5–7]. Indeed, commensurate with their improved analgesic effects, targeted perioperative pain management techniques have been repeatedly shown to decrease the risk of major cardiovascular and pulmonary complications after surgery [8–12].

The primary objective of the modern acute pain medicine service is to diagnose and treat acute pain as a pathophysiologic process [13]. This differs from approaches that consider acute pain as a simple symptom that might otherwise mask an underlying source of pain. For example, simply declaring a pain score provides no information as to the etiology. A patient experiencing acute coronary syndrome can easily be misdiagnosed and mistreated if the pain is simply assumed to be normal postoperative thoracotomy pain [14, 15].

In 1976, an anonymous editorial published by *Anaesthesia and Intensive Care* advocated the creation of analgesia-providing teams [16]. This was followed in 1988 by an article by Ready et al. who conceptualized a formal acute pain service (APS), defined as a postoperative pain management service staffed by anesthesiologists [17]. That year also witnessed the publication of guidelines for the management of postoperative pain in Australia, with subsequent guidelines following in the United Kingdom in 1990 and the United States in 1992 [18]. In 2002, Rawal delineated six core tenets of an APS following a series of surveys on APS implementation rates and practice patterns in healthcare systems throughout North America, Europe, and the UK [19]. These tenets are summarized as follows:

1. Provide designated specialists, available around-the-clock, for consultation or intervention in patients suffering from severe acute pain.
2. Proscribe regular pain assessments appropriate for age and comorbidity. Separate measurements should be provided for pain intensity both at rest and with movement.
3. Collaborate with surgeons, "ward nurses," and other healthcare team members to develop pathways for achieving predetermined goals for "mobilization and rehabilitation."
4. Provide continuous education for the healthcare team on safe and effective analgesia.
5. Offer patients education on pain monitoring, treatment options, benefits, adverse events, and goals of analgesia.
6. Audit the performance of the APS on cost-effectiveness and patient satisfaction with analgesic techniques.

Historically, the APS was charged with helping hospitalized patients suffering from acute pain. Given that this occurred within a hospital setting, all actions were contained within the inpatient encounter. Depending upon available resources and organizational capabilities, the APS typically ranged from a resource nurse, who could offer recommendations for therapies, to an anesthesiology-based team who could offer advanced interventions for narrowly targeted patient populations such as epidural catheters for post-thoracotomy pain management [20, 21]. Much of traditional APS work has revolved around postoperative patients given that surgery often leads to considerable pain across a wide range of diagnoses and indications [22].

The past few years witnessed a transition to formal APSs with access to a rich set of resources. The advent of expanded training, tools, and techniques in regional anesthesia lead to a progression of foci from performing blocks successfully to optimizing their role in postoperative recovery beyond minimization of pain to addressing those patients whose pain is not amenable to a single needle-based intervention [23–25]. This transition also pushed the APS out of the operating room and into a number of additional environs ranging from the presurgical clinic to the postoperative chronic pain clinic, enhancing the transition of care each step of the way.

The APS must manage its practice at different levels of healthcare delivery; the most pertinent being bedside patient care. Most APS teams recognize early on the need to enlist additional local resources such as ward and unit nurses, physical therapists, and pharmacists to coordinate safe and effective delivery of appropriate analgesics. This progression often evolves from informal issue-specific coordination to regular coordination meetings, and indeed such an organizational focus is recommended by the Fellows of the Australian and New Zealand College of Anaesthestists, the American Society of Anesthesiologists, the American Pain Society, The Joint Commission, and others [18, 26–28]. Taking this one step further, the APS now often finds itself involved in assisting with the optimization of patient care pathways across entire healthcare systems (Table 18.1).

Ideally, the APS's initial encounters with a particular patient begin well before the onset of injury. This rubric is idealized in the model of scheduled, elective surgical procedures that permit considerable lead time to optimize a patient prior to surgery. Such optimization commonly begins with offering education on available pain management options and managing subsequent patient expectations of such options [29–32]. Such discussions should involve trained individuals so that patients are not given unrealistic expectations pertaining to analgesic options [33]. Many an acute pain physician has encountered patients who have come to believe that their services would lead to a completely pain-free recovery from even major surgical procedures. Such belief patterns, once erroneously instilled, may inevitably lead to patient dissatisfaction, regardless of the physician's assessment of the success of an analgesic intervention [34, 35]. Preoperative optimization may often extend to prescribing specific multimodal analgesics and coordinating the availability of certain perioperative interventions, consultants, and expertise [36, 37]. One exciting possibility of such a screening process involves identifying patients with high degrees of pain catastrophizing, which is strongly associated with

Table 18.1 Overview of roles and responsibilities of the acute pain service

Preparing patients for surgery
Chronic pain patients
Patients on maintenance with methadone/buprenorphine
Patients with history of substance abuse
Patients with established biopsychosocial risk factors
Genetic risk factors for postoperative pain and treatment selection
Patients at risk for chronification of acute pain
Patients at risk for severe morbidity/mortality from pain treatment
Patients undergoing surgeries associated with adverse acute pain outcomes
Optimizing intraoperative management
Perioperative regional anesthetics with appropriate temporal profiles
(Procedure-specific) multimodal analgesics
Optimal intraoperative maintenance of anesthesia
Immediate PACU care
Postoperative care
Daily rounds
Medication and intervention management
Coordination with functional recovery (e.g. PT, OT)
Transitional care
Transition to home-based analgesic regimen
Provision of ambulatory regional anesthetics
Need for f/u with chronic pain team

adverse acute pain outcomes, who may be amenable to brief cognitive-behavioral therapy interventions [38–40]. This coordination itself may require further consultation with the patient's primary care team in the case of patients suffering from chronic pain and/or substance abuse. This is especially true for those patients who have been prescribed methadone or buprenorphine within the outpatient environment, especially given that the optimal method of managing these patients is currently under debate [41–45]. While discussions in these circumstances traditionally revolve around transitioning to the acute pain setting, the preoperative time frame is optimal for also discussing a transition from the acute inpatient setting back into the outpatient primary care environment.

The preoperative setting is also a common point of consultation where astute primary surgical services can identify patients for whom surgical concern of the pathophysiologic sequelae of pain, rather than pain itself, is the primary concern. Examples include vascular surgery patients suffering from limb ischemia, facilitation of limb reattachment surgery, and patients at increased risk for postoperative cardiac or pulmonary complications [46, 47]. Coming full circle, there has also been a resurgence of interest in optimizing postoperative pain management in patients undergoing minimally invasive surgeries whereby the minimal invasiveness was initially championed as a way to decrease postoperative pain. For instance, despite laparoscopic colorectal surgeries offering substantially less postoperative nociception, recent data still suggests a role for the APS to further reduce the time to return of bowel function through epidural analgesia, opioid minimization, and aggressive multimodal analgesia [48, 49]. Perhaps somewhat counterintuitively, APS's have also been regularly consulted for thoracic endovascular aortic repair for patients with an elevated risk for postoperative spinal cord ischemia [50]. Here, the APS may be asked to facilitate intrathecal drain placement and management of naloxone infusions while still optimizing patient analgesia.

Given the propensity of surgical patients to comprise the majority of the patients on an APS census, it is our firm belief that an APS must maintain a very significant presence within the intraoperative environment. Aside from offering concrete recommendations and even direct patient care, a regular intraoperative presence offers a plethora of opportunities for informal conversations, education, and feedback on matters pertaining to the APS with the surgical team. Given equipoise in other factors, there are many circumstances whereby a certain approach to anesthesia may offer an improved postoperative pain outcome [51]. Indeed, even within the realm of general anesthetics, there is mounting evidence that certain maintenance agents and adjuncts may lead to substantial improvements in postoperative pain management [24]. For instance, data from retrospective cohort studies suggest that a propofol-based maintenance strategy may be associated with less postoperative pain and opioid consumption compared with maintenance conducted with inhalational agents [52, 53]. An intraoperative presence also helps with the process of translating clinical science into clinical practice by virtue of understanding the practical concerns, limitations, and opportunities that exist on the front line of this incredibly dynamic time period.

Intraoperative involvement offers a number of opportunities to improve postoperative pain management. For instance, the APS may help select a more optimal regional anesthetic strategy (catheter vs. single shot) to better account for the anticipated temporal trajectory of postoperative pain resolution, or directly administer multimodal analgesics in the perioperative setting [54, 55]. For patients in whom the anticipated surgical plan has morphed into one with an anticipated worse postoperative pain outcome, intraoperative involvement may permit intraoperative placement or early postoperative placement of regional anesthetics and/or multimodal analgesics, although data suggest that pre-incisional timing remains optimal [56]. Intraoperative involvement should also extend to a robust presence within the recovery room, as this highly dynamic time period offers a unique opportunity to rapidly select and titrate optimal analgesic therapies, detect potential safety issues, and set the tone for successful and safe analgesia over the next several postoperative days [57–59].

Perhaps the most historically common practice arena for the APS has been in the postoperative care of surgical patients. Many APSs perform daily rounds on patients, assessing a patient's pain and pain impact on functioning, and offering recommendations for improving and/or transitioning analgesics to the next phase of recovery. Generally, APS practices fall into one of two camps. In the first, the APS assumes full responsibility and control over all analgesics, and will be the only team of physicians who modify the analgesic regimen of a patient. Such patterns may be limited to those patients receiving neuraxial opioids, or may extend to all patients in whom the APS has been consulted [60, 61]. The second camp is a consultative practice whereby recommendations are made, but the final responsibility and coordination are maintained by the primary service. Even in this camp, it is often the standard that modifications to regional anesthetics still fall under the purview of the APS.

Regardless of the practice model applied in the postoperative setting, certain facets remain common. Close coordination is necessary to avoid duplicative, or countermanding, strategies and orders that could rapidly lead to patient harm. To this end, social network analyses of even small APSs suggest a very large network of communications stemming from this need for close coordination [62]. Next, communications among physicians, although necessary, is not sufficient to ensure optimal application of chosen analgesic strategies. Rather, these efforts must include coordination with nurses, therapists, pharmacists, and other consulting teams to ensure that the analgesic strategy is not interfering with other facets of the patient's recovery.

One of the more overlooked aspects to modern APS practice involves the transition of patients to their post-admission destination and recovery. The import of this transition has received renewed focus given evidence reported by Chapman, Althus, and others that demonstrates that a significant portion of surgical patients will have a neutral, or even positive, pain trajectory slope over the first several days after surgery [54, 63, 64]. Many consultations to the APS for patients who have already received surgery focus on transitioning patients from an effective inpatient, IV-based analgesic regimen to a PO-based regimen that a patient may use at home. The APS encounters similar issues with patients whose pain has been well controlled with a regional anesthetic, but must transition to an environment where such a technique is not feasible. It is not unsurprising that many APSs thus conduct regular outreach to local rehabilitation centers, hospices, and even assisted living facilities to improve the likelihood that a patient's perineural catheter infusions may follow them beyond the hospital setting [65, 66]. One intriguing option is the use of an APS to triage and refer select patient populations for early, aggressive interventional procedures in the chronic pain clinic for pain in the midst of transitioning from acute to chronic.

18.2 Team Composition

Every individual in a team should play a contributory role toward optimizing patient comfort during his/her progress through the healthcare system [67]. The group of individuals on the APS includes physicians, nurses both in the procedural care area and in the units/wards, and consultant staff (e.g., surgical team, physical therapy/occupational therapy, psychiatry, palliative care medicine, pharmacy, social work).

The large number of social connections results in complexity in communication [62]. Yet, solid communication is key for the team to function optimally. This communication can take place in the form of written frameworks or protocols set in place to minimize variability of patient care, in addition to verbal communication (both through group discussions or direct communication with the team members).

18.2.1 Physicians

A physician comprehensively trained in Acute Pain Medicine should be the cornerstone in a functional APS and needs to be involved during the entire perioperative course, not just the intraoperative period. Preoperative education and optimization of patients is vital preparatory work to reduce delays and cancellations on the day of surgery. Patient optimization includes addressing issues such as anticoagulation, infection, and cardiovascular stability, allowing the physician to appropriately decide anesthetic and analgesic modalities that optimize benefit:risk ratios [13]. Preoperative planning also includes discussion with surgical colleagues on potential factors that may complicate care of the patient (planned postoperative anticoagulation, anticipated duration of hospital stay, extent of surgery) as well as resource management (nursing staff and equipment availability).

The physician's role in the postoperative period cannot be emphasized enough. The placement of a regional anesthetic is but one aspect of postoperative multimodal pain management. The physician must continue his/her involvement in the postoperative period with the following goals:

- To continually assess the efficacy of the regional analgesic catheter and determine and manage etiologies for breakthrough pain (e.g., compartment syndrome, inadequate coverage by the regional anesthetic catheter, pump malfunctions, catheter migration, unanticipated extensive surgery, failure to reinitiate home opioid regimen)
- To optimize functional recovery of the patient (improve physical function, assist with sleep hygiene, minimize undesired side effects of medications)
- To address postoperative issues with the surgical team (e.g., optimal timing of postoperative anticoagulation, discharge planning)
- To detect and manage post-procedural complications (infection, nerve injury, bleeding, local anesthetic toxicity)
- To transition patients to oral analgesics and plan for regional analgesic catheter discontinuation

18.2.2 Nursing Staff

Different subsets of nurses play crucial roles in the pain management team. These subsets include nurses assisting in the procedural area, dedicated acute pain nurses who assist physicians with post-procedural rounds, and the floor and unit nurses who receive and care for the patients in the post-procedural period. These dedicated nurses require specific skill sets in order to assist the physician in safe and effective delivery of analgesic modalities while minimizing variability in practice.

Dedicated nurses in the procedural area may allow for improved patient safety and efficiency when performing regional anesthetics, especially in the context of a dedicated block area to perform procedures in the preoperative setting. Dedicated nurses in these areas can be involved in formal time-out processes to ensure placement of a regional anesthetic in the correct location of surgery and to confirm indications and contraindications to procedures (e.g., potent anticoagulation). They need to have a basic understanding of hemodynamic monitoring, sedation, and airway management in order to aid the physician during placement of a nerve block. They must understand nerve stimulation and be able to recognize block-related events (e.g. local anesthetic systemic toxicity) and be familiar with equipment such as ultrasound machines.

In one practice model outside the procedural area the role of the APS nurse is multidimensional, and a complement to the pain physician on rounds. The nurse assists with consult triage to determine the needs of the primary service and as an educational liaison for patients and their families as well as other support staff caring for the patient to communicate with them about topics relevant to the patient's pain management (such as avoiding potent anticoagulation in patients with neuraxial or deep blocks, optimizing multimodal analgesia, and anticipating motor and sensory changes with regional anesthesia). The nurse gathers focused patient information, including pertinent medical history, active medications, and a history of chronic opioid requirements. The nurse communicates with other medical staff (ward nurses, surgical teams) to fully appreciate the patient's course and conditions, round-the-clock efficacy of regional analgesics, and the anticipated discharge time, enabling the APS team to make more informed decisions regarding pain management. The APS nurse also is responsible for data collection for quality improvement to substantiate future management decisions [13, 67, 68].

Ward and unit nurses should be considered part of the acute pain team and be encouraged to participate on rounds in order to communicate with the physician issues that need to be addressed (e.g., adequacy of pain management, poorly tolerated side effects of systemic drugs, concerns for infection, equipment issues). These nurses should have a fair understanding of pain assessment, multimodal analgesia, basic pharmacology, and concepts related to regional anesthesia (e.g., motor weakness in the distribution of the nerve blocked, avoidance of potent anticoagulants with neuraxial catheters, pump mechanics, recognition of local anesthetic toxicity, and dressing maintenance).

18.2.3 Consult Services

The role of the pain team is to optimize the patient's functional status and not just to reduce pain scores to a minimum. The pain team operates in a multidisciplinary milieu where consult services should be utilized to improve the patient's overall perioperative outcome. This includes working in concert with physical and occupational therapy to enable successful rehabilitation for the patient (maximize regional anesthetic efficacy while minimizing motor weakness, optimize fluid status and local anesthetic infusions in patients receiving epidural catheters to reduce the incidence of orthostatic hypotension, etc.). Psychiatry should be consulted in patients with conditions such as depression or post-traumatic stress disorder that may hinder functional recovery. Pharmacy should be involved in processes to minimize medication errors (safe labeling of medications such as local anesthetics), to supply appropriate types and concentrations of medications, and to design underlying systems to minimize clinical errors (clarify necessity of potent anticoagulation prescriptions in patients with neuraxial or deep blocks) [13]. Services such as palliative care and social work can be consulted for further specialized care and to aid in a smooth transition from the inpatient to outpatient setting.

18.3 Daily Workflow

18.3.1 Environmental Assumptions

Here, we describe a typical day on a modern APS. Note that this "typical day" carries many assumptions. The day is modeled on an academic medical center in the United States. The APS is administered by the Department of Anesthesiology with additional resources from the hospital itself. It is staffed by attending anesthesiologists who supervise fellows, residents, and undergraduate trainees of all stripes. The APS here also offers side duties, including regional anesthesia, intraoperative, and recovery room anesthetic care, and pre-anesthetic evaluation. Attending anesthesiologists are also responsible for administrative, education, quality improvement, and research duties. While the APS may offer care across separate facilities, each facility is geographically co-located to be within walking distance. The typical patient census for this APS ranges from 20 to 50 patients and consists of a mix surgical patients, medical patients, and patients from special populations such as pediatrics, obstetrics, and palliative care.

18.3.1.1 Pre-rounds

We begin this walk through at 0600, whereby the in-house overnight resident for the APS provides sign out to the oncoming team. The oncoming team is split between those placing nerve blocks and consults for new patients (the block team) and those providing care for existing patients (the rounding team). The rounding team initially assists with care of patients presenting for nerve block until 0800 and then collects data on overnight pain evaluations, analgesic requirements, hemodynamic hepatorenal and coagulation status, progress with recovery, and disposition planning. There is typically then an hour of formal education for fellows, residents, and trainees.

18.3.1.2 Blocks

Nerve blocks are a key tool for a modern, effective APS. Here, we focus on organizational issues surrounding block placement. Patient candidacy for a nerve block is initially assessed the day before surgery. Note than with over 200 patients presenting for surgical procedures each day, an in-depth chart review and interview for all patients by the APS is not feasible, therefore a number of heuristics are involved. For high-volume, low-variance procedures such as total joint arthroplasty, blocks are placed according to established frameworks. A similar approach is followed for low-volume, high-variance services such as orthopedic oncology, where the majority of procedures are likely to benefit from a nerve block. However, in these populations, following a review of the medical record and imaging, there is a discussion with the surgical attending on the planned approach and anticipated recovery profile to better match block selection to each patient. For relatively rate, one-off combinations of patients/procedures that are highly likely to benefit from a nerve block, there is a lengthy discussion with the surgical team that includes an in-depth review of anticipated postoperative analgesia, anticoagulation, and functional recovery before the decision for nerve block placement.

18.3.1.3 Intraoperative Consults/Blocks

Occasionally, the surgical plan will need to be heavily modified intraoperatively due to unanticipated events. For example, this may include aborting a Whipple procedure due to advanced metastasis, or conversion from a mass excision to an amputation. Intraoperative changes may also be subtle, such as a requirement for substantially more anti-nociceptive agents than anticipated, or even the possibility of a failed preoperative nerve block. Regardless, it is not uncommon for the APS to be summoned to the operating room prior to emergence to either assist with planning of intraoperative multimodal analgesia, planning for or placing of postoperative regional anesthetics, or troubleshooting of existing nerve blocks. Even in the absence of changes to the surgical plan, intraoperative assessment of block performance by the APS block team serves as an internal quality control. Additionally, these encounters offer opportunities for informal consults, queries, and education. They can also demonstrate a commitment to follow-through and excellence, alleviating surgical fears of “fire and forget” nerve blocks that leave any adverse outcomes or block failures to the already busy surgical team to manage. Such intraoperative consultations point to the need for the APS to be both familiar with and accessible to the intraoperative environment to support the continuity of analgesic care.

18.3.1.4 Rounds

The APS rounds on each patient occur at least twice a day. The first set of rounds is teaching rounds, where a resident or fellow to the APS attending presents each patient. The APS attending sometimes directs that the fellow perfunctorily lead rounds, a transition generally occurring later in the academic year. The presentation will generally focus on the patient's documented pain evaluations from the prior evening, including any salient details passed on from the night resident. The presentation will also cover the prescribed analgesics, as well as any hepatic, renal, hemodynamic, or coagulation issues that may impact any analgesic decisions. Updates to the surgical recovery plan and disposition goal are addressed during this pre-interview discussion as well. Once the established data have been reviewed, the APS then enters the patient room and interviews the patient. Aside from inquiries on the patient's pain, recovery, and response to analgesics, the patient is also given a thorough neurovascular examination depending upon the type of regional anesthetic. This interview should move well beyond simple numeric rating scale (NRS)-based inquiries of the patient's pain intensity at rest and with movement. Rather, any inquiries regarding the patient's pain intensity are placed within the context of recovery goals. Has the patient been able to breathe deeply and cough? How was it getting out of bed? Did the patient have any difficulty ambulating, voiding, or moving the afflicted surgical areas? Were there any adverse events to the prescribed analgesics; if any analgesics were modified, how did the updated plan work? How well did the patient sleep, and how is pain influencing the patient's mood?

Note that such a shift to assessment across patient-reported outcomes could theoretically consume an inordinate volume of time for each patient. Indeed, inefficient survey-based methodologies of established patient-reported outcome (PRO) reporting tools can require per-subject time allocation even when subjects are not recovering from surgery [69]. Thus, the patient interview necessitates mediation of tension between rapid, highly efficient assessments of patient well-being versus those questions likely to be of actual value from the patient's perspective. Recent work on the use of computer-adaptive testing to rapidly sort through the myriad possible PRO items may offer an exciting opportunity to rapidly yet effectively assess PROs [70–72].

To add further context, any patient-controlled analgesic (PCA) devices are also queried to determine utilization rate. Note that the PCA in this instance serves as an important behavioral indicator of the patient's response to pain. Given that the PCA is often applied as a means of achieving, rather than maintaining, a minimum effective analgesic concentration, the patient interview is an important opportunity to offer ongoing education on the role and purpose of the PCA device, although such education must also be directed toward members of the healthcare team charged with choosing among multiple available analgesic options in the face of an uncomfortable patient [73].

18.3.1.5 Post-anaesthesia Care Unit (PACU) Evaluations

While the rounding team is underway, the block team continues its work with patients presenting for surgery and offers intraoperative consultation. As patients present to the PACU for recovery from surgery, the APS block team will frequently walk through the PACU to preemptively identify patients with aberrant analgesic requirements and/or offer troubleshooting of nerve blocks. At our institution, all patients with nerve blocks must be seen by the APS prior to discharge from the PACU, whether for ambulatory surgery or inpatient recovery. Even for patients with adequate pain control, the APS may be consulted for patients experiencing difficulties with oxygenation or ventilation to help rearrange and optimize analgesics.

18.3.1.6 Post-block Rounds

Following formal rounds, a "to do" list of tangible activities is reviewed. Responsibilities are divvied among the team according to suitability, availability, and priority. A "list" of all patients on the service is maintained and updated to reflect the observations and treatment plans reviewed during rounds. Blocks placed by the block team during rounds are reviewed with the rounding team. The rounding team sees new consults. By evening, most of the task list is complete, and progress notes are completed. Each evening, there is a designated physician on the APS who provides in-house coverage for the APS. This individual rounds on all new and established patients in the late afternoon, and often again during the evening. This ensures that the APS is able to achieve multiple diagnose-treat-evaluate cycles throughout the day, "pressing" progress forward and ensuring that appropriate monitoring exists on patients beyond the presence of a physician's order. Even patients for whom the team has difficulty achieving safe and effective analgesia, we firmly believe this multi-round approach sends the informal message that the APS has not forgotten about nor ignored them, but is continuing to press forward with new therapies. This multi-round approach is also critical for certain populations such as orthopedic trauma to prevent masking of complications and to offer increased numbers of neurovascular exams for at-risk patients.

18.3.1.7 Ambulatory Catheter Care

Many patients receiving ambulatory surgery are candidates for ambulatory perineural catheters. This topic has been exceedingly well covered in reviews by Enneking and Ilfeld [66, 74]. Organizationally, the assessment for candidacy for ambulatory perineural catheters begins with the preoperative assessment. Postoperatively, patients are transitioned to an ambulatory perineural catheter infusion device on the day of discharge. Follow-up is conducted at least once via telephone, including a telephone discussion during the catheter removal process. The APS will contact the patient at least once per day via telephone, and the patient is instructed to contact the APS at least once per day via the telephone number provided, even if there are no appreciated issues, if they have not heard from the APS. Patients are also provided a very detailed set of instructions on maintenance and troubleshooting of perineural catheters, with explicit instructions to call the APS with any questions they may have.

18.4 Pain Evaluation

A cornerstone of APS rounds is the evaluation of pain, which is not an easy thing. In the late nineteenth century, psycho-physical analysis was an attempt to correlate a patient's subjective expression of pain intensity with actual physical level of pain [70]. In a controlled laboratory setting, this may have been feasible, but it became less useful when used in the clinic. There were just too many confounding factors that influenced the subjective pain experience: the surgical experience, comorbid conditions, medications, and psychological stressors to name a few. Pain rating scales were developed out of the need for a simple and efficient method to evaluate pain intensity over time, as a way to gauge the efficacy of treatment [71]. In the setting of the increased recognition of pain in the backdrop of the congressional designation of the Decade of Pain or the Joint Commission's mandate to consider pain as the "fifth vital sign," accurate assessment of pain has therefore arisen as a key issue in acute pain medicine [75].

18.4.1 Pain Intensity Measurement Tools

Often, the assessment of pain will default to the question, "What is the best scale?" However, a comprehensive pain assessment should include a precise pain history, a measurement of intensity, measurement(s) of pain impact, and follow-up assessments for treatment response. Although the frequency of such assessments should be regular, assessments must be tailored to the patient's pain intensity (more intense = more frequent) and treatment (follow-up assessments after intervention).

The initial evaluation should consist of a review of salient medical history along with medications currently being administered for pain and non-pain-related conditions. Investigation into pain characteristics should subsequently involve defining the location of the pain, exacerbation, alleviating factors, quality, etc., as is standard. Such an initial evaluation allows for the generation of differential mechanisms for a given patient's pain such as nociceptive, neuropathic, or mixed etiology.

Numerous intensity scales have a long history of use and validation, including the Visual Analog Scale (0–100 mm), the NRS (0–10), and the Verbal Rating Scale (VRS; No Pain, Mild, Moderate, Severe, and Excruciating) [76]. Hewer and Keele developed one of the earliest versions of the VRS, a categorical scale of pain intensity where patients assign their pain to one of three categories: "0" being no pain, "1" is mild, "2" is moderate, and "3" is severe pain [74]. In the 1960s, Bond, Pilowsky, and Spear created the Visual Analog Scale (VAS), a 10-cm line (100-mm) on paper with "0 cm" being defined as "no pain" at one end, and "10 cm" being defined as "the pain is as much as I can bear." The patient draws through the line to express the level of pain they are experiencing. Given that they can draw at any point along the line, linear data on the patient's pain are obtained. As a result, in comparison to the categorical scales, the VAS is thought to be more sensitive at detecting a change in pain intensity with treatment (in persons capable of performing the task), and may reflect ratios of pain intensity [77]. Still, the major drawback is that written scales are an inefficient way to regularly assess pain.

Verbal assessment is more practical on APS rounds and with repeated, regular assessment by the bedside nurse. Although the VRS, VAS, and NRS are traditionally written scales, the NRS is more commonly applied verbally. To use the NRS, a patient is asked to rate the intensity of his/her pain on a scale from 0 to 10. At many institutions, including ours, the NRS pain scale is documented after recovery from surgery, on rounds, and before and after pain medications are given, to assess efficacy of treatments over time. While practical and efficient, the NRS may not be statistically sensitive enough for clinical research to detect treatment effects [78].

The NRS has a long history, thereby establishing reliability and validity. Specifically, the NRS has acceptable discriminative capability, but also enough options to have the ability to detect change over time. However, while the NRS is an important tool, its unidimensional nature significantly limits the characterization of the multidimensional nature of pain. As an expansion of such unidimensional scales, numerous functional and multidimensional scales exist such as the Brief Pain Inventory or the McGill Pain Questionnaire [79, 80]. Such scales capture additional insights into pain interference, impact on sleep, and anxiety. However, in a clinical setting, such scales have a limited role secondary to lengthy time of administration, which is impractical on a daily basis. Thus, in an inpatient setting, the NRS remains the most consistently used measure for daily measures of pain.

18.4.2 Development of the Defense and Veterans Pain Rating Scale

Current pain assessment practices within inpatient settings, while commonly employed, have not been shown to improve pain outcomes or the safety of care [81]. Furthermore, in a survey of 77 Veterans Administration providers, little confidence was expressed regarding the accuracy of nurse-acquired pain ratings [82]. Such insights reinforce the concept that current pain assessment practices lack the ability to capture the multifaceted patient experience of pain.

Stemming from the limitations of unidimensional scales, a 22-member working group chartered by the US Army Surgeon General in 2008 developed a novel scale focused on capturing the biopsychosocial impact of pain [81]. This scale, the Defense and Veterans Pain Rating Scale (DVPRS), focuses on expanding the NRS by integrating visual variables (faces/colors) along with descriptive functional anchors assigned to pain intensity designations (Fig. 18.1a). In addition, four supplemental questions focused on sleep, anxiety, activity, and mood offer further multidimensional information regarding the impact of pain (Fig. 18.1b). Such a tool can

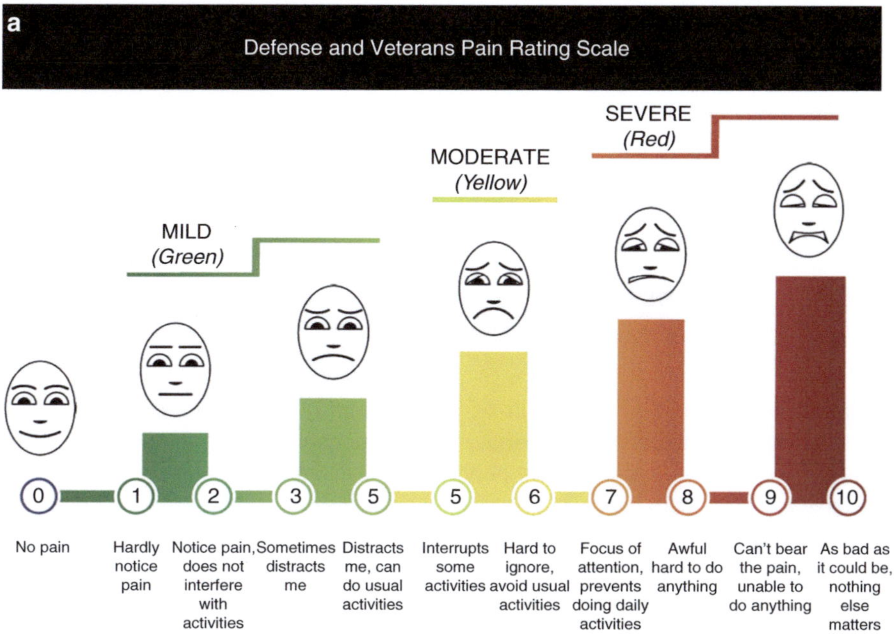

b DoD/VA PAIN SUPPLEMENTAL QUESTIONS

For clinicians to evaluate the biopsychosocial impact of pain

1. Circle the one number that describes how, during the past 24 hours, pain has interfered with your usual ACTIVITY:

0 — 1 — 2 — 3 — 4 — 5 — 6 — 7 — 8 — 9 — 10

Does not interfere   Completely interferes

2. Circle the one number that describes how, during the past 24 hours, pain has interfered with your SLEEP:

0 — 1 — 2 — 3 — 4 — 5 — 6 — 7 — 8 — 9 — 10

Does not interfere   Completely interferes

3. Circle the one number that describes how, during the past 24 hours, pain has affected your MOOD:

0 — 1 — 2 — 3 — 4 — 5 — 6 — 7 — 8 — 9 — 10

Does not affect   Completely affects

4. Circle the one number that describes how, during the past 24 hours, pain has contributed to your STRESS:

0 — 1 — 2 — 3 — 4 — 5 — 6 — 7 — 8 — 9 — 10

Does not contribute   Contributes a graet deal

Fig. 18.1 Defense and Veterans Pain Rating Scale

be quickly administered at the bedside and on a daily basis. The goal of such a scale is to further standardize pain assessments, improve the ability to identify "at risk" patients, and provide a tool that is applicable in numerous environments. Although new within the landscape of pain assessments, the DVPRS scale has been found to have acceptable reliability and validity in a sample of 350 inpatient/outpatient active duty or retired military personnel [83].

18.4.3 Pain Assessment in Special Populations (Neuropathic Pain, Elderly, Pediatrics, Non-verbal)

Pediatrics An important limitation of the commonly used pain intensity scales designed for use in healthy adults is that they require patient insight and participation. In most cases, verbal and preverbal children are incapable of using these scales. In preverbal children, pain has to be expected in conditions that are assumed to be painful, prevented, and treated [83]. Multidimensional pain scales such as the Face, Legs, Activity, Cry, Consolability (FLACC) Pain Assessment Tool provide a standardized method for use by nurses and clinicians to assess pain in preverbal children in the clinic. FLACC correlates well with other pain assessment tools such as the Coloured Analogue Scale, the COMFORT scale, and the NRS in verbal children as well [84].

Elderly There is wide recognition that pain is under-recognized and undertreated in the elderly population [85]. Various factors such as cognitive impairment, altered sensation, or atypical manifestations all contribute to the variability in accurate pain assessments within this population. For patients with cognitive impairment that is mild or even moderate, scales such as the NRS or the Faces Pain Scale have shown good reliability and can be utilized in the inpatient setting [86]. However, for nonverbal or severely cognitively impaired individuals, observational scales such as the Checklist of Nonverbal Pain Indicators or the Pain Assessment in Advanced Dementia scale can be used [87]. Such observational tools can serve as a platform for empiric analgesic interventions followed by reassessment of such behavioral manifestations of pain. Of note, these observational tools are not applicable to unresponsive or sedated patients.

Critically Ill Pain in a critical care setting is often complicated by coexisting physiologic derangements, agitation, sedation, delirium, and a variety of comorbidities that cloud both the assessment of pain and response to treatment. Numerous series have shown that large percentages of ICU patients experience significant levels of pain, which highlights the need for accurate assessments of pain [88]. Similar to elderly patients, patients with minimal cognitive deficits are able to accurately report pain intensity with the NRS scale or other unidimensional tools [89]. However, for patients that are either sedated or nonverbal, the Behavioral Pain Scale or Critical Care Pain Observation Tool are two validated tools used within an intensive care environment [90]. Both tools rely on observational data from behaviors such facial expression, body tone, or ventilator compliance.

18.4.4 Future Considerations: Pain Trajectories

Acute postoperative pain has remained a significant issue despite decades of advancing analgesic practices. Furthermore, suboptimal perioperative pain management has been suggested to strongly correlate with persistent post-surgical pain. While scales such as the DVPRS have embraced the multidimensional nature of pain, the most commonly used scales are unidimensional pain intensity scales such as the NRS. As noted above, the NRS clearly lacks the ability to convey the complex nature of acute pain and is often unable to lead to meaningful treatment decisions linked to the assessment of pain.

To counteract this, Chapman et al. measured the pain trajectories of 502 surgical patients over a 6-day period following a variety of surgeries. While the majority of the sample (63 %) displayed negative slopes (improving pain), 37 % of patient had either a flat (no improvement) or positive (worsening pain) slope [54]. Noting a significantly small standard error of measurement, measurement of such pain trajectories improved precision when compared to NRS scores. Such a study highlights the notion that pain in the perioperative setting is a "self-limiting" phenomenon and that the measurement of its trajectory improves the precision of pain assessments, can be applied on an individualized basis, and assists treatment decisions based on direction and magnitude of the slope. In addition, as an extension of this study, Althaus et al. measured similar trajectories in 199 surgical patients and noted that the rate of pain resolution clearly predicted persistent pain at 6 months [63]. Such a finding has profound implications in regard to early recognition of at-risk patients that may need to be stratified to more aggressive treatment regimens and follow-up post-discharge.

Accurate and timely pain assessments that focus on the multidimensional nature of acute pain are essential to guide effective treatment regimens. Currently, the vast majority of pain assessments consist of unidimensional pain intensity scales that have not significantly improved pain outcomes or safety measures. With the wide prevalence of acute pain and the significant percentage of surgical patients that proceed to experience persistent pain, further research in novel functional measures and how these measures are analyzed over time is needed. Such a course will hopefully allow for increased precision in acute pain assessments and targeted therapies not only addressing the immediate setting, but also stratifying patients who are at risk for persistent postsurgical pain.

18.5 Environmental Factors of Acute Pain Care

18.5.1 Introduction

In principle, minimizing postoperative pain should perhaps be one of the simplest tasks in all of modern medicine: simply continue to infuse an opioid of choice at ever-increasing doses until a patient is rendered somnolent. Indeed, given the traditional focus of APSs on pain intensity ratings, and the oft-cited goal of aggressively treating acute pain until a patient reports 0/10 on a numerical rating scale, one wonders about the need for an entire subspecialty of physicians whose sole skill would then be cautiously opposing their thumbs and forefingers around a syringe of fentanyl, stopping only when a patient ceases their screaming.

Obviously, the above observation is hyperbolic ad absurdity. But it does highlight a common misconception surrounding the modern APS: while nominally charged with improving a patient's pain, this charge is entirely predicated upon the context of optimizing patient functioning while guaranteeing patient safety.

18.5.2 Patient Safety

All analgesic interventions represent a balance between risk and benefit. In a retrospective cohort study of administrative data from 37,031 patients from 26 hospitals, 98.6 % of patients received opioids, with 13.6 % of opioid recipients suffering at least one opioid-related adverse drug event (ORADES) [91]. Overall, ORADES were associated with a 55 % longer length of stay, a 47 % increase in cost of care, a 36 % increase in risk of 30-day readmission, and a 3.4-fold increased risk of inpatient mortality compared with those patients who did not experience an ORADE during postoperative hospitalization.

One of the most important and pervasive patient safety concerns across all of healthcare involves opioid-associated respiratory depression, although identifying a consistent and clinically useful indicator for respiratory depression remains elusive. A 2010 estimate in U.S. hospitals suggested that postoperative respiratory failure was the third most common patient safety-related adverse event affecting the Medicare population, and that over half of these events occurred in patients receiving postoperative opioids [92]. Work by Cashman and Dolin, reviewing data from approximately 20,000 patients pooled for 165 studies in patients receiving intramuscular analgesia, PCA, and/or epidural analgesia, identified an overall risk of respiratory depression of 0.3 % (95 % CI 0.1–1.3 %) when depression was defined by the requirement of opioid reversal with naloxone [93]. This incidence increased to 1.7 % (95 % CI 0.7–1.7 %) when hypoventilation was used as an indicator of respiratory depression, 3.3 % (95 % CI, 1.4–7.6 %) when indicated by hypercarbia, and 17 % (95 % CI 10.2–26.9 %) using oxygen desaturation as an indicator. A review of the American Society of Anesthesiologist's closed claims database suggested that the majority of opioid-associated postoperative respiratory depression (PORD) events occurred within 24 h of surgery, a finding in agreement with the work of Weingarten demonstrating 58 % percent of naloxone administrations occurred within 12 h following discharge from anesthesia care [94].

Given the prevalence and severity of opioid-associated PORD and other ORADES, efforts to improve primary and secondary prevention strategies have appropriately received considerable attention by the APS community. Kessler et al. identified the following risk factors for ORADES: male sex, age ≥65, obesity, preoperative opioid use, and Charlson Comorbidity Index [91]. Although obesity and sleep apnea have been repeatedly associated with the risk for PORD, prior data suggests that over 75 % of individuals with moderate to severe obstructive sleep apnea remain undiagnosed [95]. Further data from The Joint Commission, in reviewing the sentinel events from 2004 to 2011 of patients suffering from opioid-associated PORD, suggest that over 40 % of these patients received incorrect doses of an opioid analgesic that was temporally linked to their episode of respiratory depression (http://www.jointcommission.org/assets/1/18/SEA_49_opioids_8_2_12_final.pdf). It is worth noting that from a decision-theoretic standpoint, opioid-associated PORD remains a "rare event" with an outcome that entails a certain difficulty in predicting with a high degree of certainty.

Indeed, these findings called into question the feasibility of reliably predicting which patients were likely to suffer from opioid-associated PORD using currently available patient data in electronic medical record systems. Instead, Taenzer et al. embarked upon a remarkable quality improvement project that tested the efficacy of first optimizing the use of SpO_2 monitoring through establishment of intelligent alarm thresholds, culture readjustment consummate with the updated alarm utilization, and universal application of continuous SpO_2 regardless of perceived risk of opioid-associated PORD [96]. The results of this process improvement project were shocking. A before-and-after concurrence study demonstrated a reduction in the number of rescue events from 3.4 to 1.2 per 1000 patient discharges and a decrease in ICU transfers from 5.6 to 2.9 per 1000 patient days.

Analgesic risks are not limited to opioid analgesics. Non-selective COX inhibitors have been associated with increased risks of bleeding, and at least one prospective study in healthy volunteers uncovered a 20–30 % incidence of gastric ulcers on gastroscopy after only 5 days of nonselective NSAIDS [97]. While there appears to be a reduced incidence of bleeding with COX-2-selective inhibitors compared with COX-1, the risk of renal dysregulation appears to be similar between nonselective and COX-2-selective NSAIDS [98]. Regarding acetaminophen, there has been increasing attention to the prevalence of APAP-associated hepatic toxicity, which has led to the uncoupling of APAP and opioids by many APS teams [99, 100]. Contemporary estimates of the risk of regional anesthetics resulting in severe neurologic injury is less than 4:10,000 for neuraxial anesthetics and less than 3:100 for neuropathy following peripheral nerve blocks, although the likelihood of neuropathy following peripheral nerve block persisting beyond 6 months appears quite rare [101].

These findings carry important considerations for the APS. First, while a properly organized APS can achieve better pain relief while also demonstrating a lower incidence of side effects and even lower postoperative morbidity and mortality [102–104], an approach to simply offering additional analgesic interventions without considering systems-level effects may be fraught with danger. Witness the results of Vila et al.: following the institution of a numerical pain treatment algorithm in a large academic medical center that was heavily predicated upon aggressive opioid titration based upon a single pain intensity indicator, patient satisfaction increased from 4.13 to 4.38 on a 5-point scale ($P<0.001$), yet patients also suffered over twice as many ORADES despite a stable patient demographic over the same time interval [105].

Such systems-level deviations from the expected benefit of an APS reinforce the need for the APS to be involved in several different scales of healthcare delivery. It is worth noting that in some sectors, the APS was born out of a combined effort with critical care service provisions in recognition of the overlapping nature of these specialties [104]. The Anesthesia Patient Safety Foundation has offered numerous overviews of the risks of opioid-induced respiratory depression, emphasizing the need for continuous monitoring of oxygenation rather than intermittent spot-checks of patient status [106]. Furthermore, when supplemental oxygenation is indicated, the Anesthesia Patient Safety Foundation has suggested the need for continuous monitoring of ventilation as well as oxygenation. Notably, the impact of such recommendations may be impaired if only initiated on the subset of hospitalized patients seen by the APS. To improve the safety of all hospitalized patients with postoperative pain, such recommendations need to be implemented and coordinated at higher organizational levels of a healthcare delivery system. Altogether, these considerations point to the need to involve patients in a careful discussion on the relative risks, benefits, and feasible alternatives regarding postoperative analgesia in an effort to match the risk of the therapy to the risk of non-therapy. Otherwise, aside from endangering patient safety, such aberrant approaches can lead to a significant pushback such as overt concerns over the need for optimized postoperative analgesia in the surgical literature [107, 108].

18.5.3 Evolution of the APS

The capabilities of the APS are often a function of the environment in which the APS practices. This does not necessarily obviate the responsibility of an APS to promote advanced capabilities, but rather points to the need for the APS to be involved at multiple levels of healthcare delivery systems to enable an economically feasible approach to the safe, effective delivery of advanced analgesic therapies. Many APSs originated from anesthesiology practices, often due to facility with epidurals and high doses of opioids and comfort of rescuing from respiratory depression.

In many centers, a progression that begins with occasional applications of regional anesthesia [13] progresses to the predictable use of nerve blocks and epidurals, and segues into a protocol-based approach such that the use of nerve blocks and epidurals is not dependent upon a single individual's practice preference but instead becomes team-based (Fig. 18.2). Further progression with structured follow-up integrates a more classical framework of ward-based APS practice with intraoperative origins of some APS systems into a perioperative pain medicine service. This stage also recognizes the impact of continuous regional anesthesia techniques on nursing, pharmacy, and physical therapy practices and involves these specialists in a coordinated strategy to enhance recovery from surgery. This stage also permits the use of ambulatory continuous nerve blocks, as well as a resource serving preoperative and postoperative consultations. The epitome of an idealized

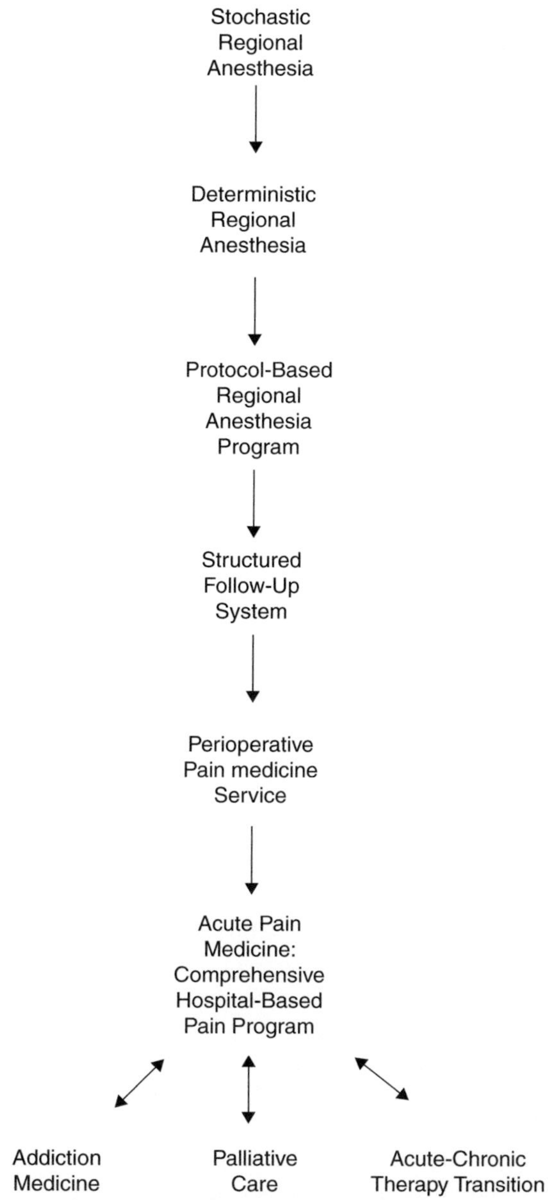

Fig. 18.2 Evolution of Acute Pain Medicine Programs. While many patterns of progression are observed, the evolution of stochastic nerve-block placement to comprehensive, high-quality, evidence-based acute pain medicine programs often requires many iterations through several common stages

APS occurs with further integration with consultants in addiction medicine, chronic pain, palliative care, and psychiatry to enhance care delivery to special populations.

Continued integration also confers additional, if shared, resources that accelerate the number of diagnose-treat-evaluate cycles that can be accomplished during a given time period. For refractory and/or extreme pain, the APS can also access

Fig. 18.3 The Acute Pain Medicine Team Contributes to Acute Pain Management Systems at Multiple Scales of Healthcare Delivery

resources in the PACU, block room, and even the ICU to further accelerate the diagnose-treat-evaluate cycle and include certain agents such as ketamine that might otherwise encounter regulatory restrictions in ward environments. Note that these systems can even be extended to serve as a resource for chronic pain patients suffering acute pain exacerbations due to injury, whereby non-opioid adjuncts can be used to rapidly stabilize certain carefully selected patient populations in concert with a chronic pain physician. In some centers, this has even served as a nidus for accelerating entry into rehabilitation.

Evolution of the APS occurs not just in the complexity of direct patient care capabilities, but also with regard to optimizing the delivery of care across an ever broader and more diverse set of patient populations. This necessitates a co-evolution in the organizational structure of the APS moving beyond silos of care delivery (Fig. 18.3). For instance, with a very small number of patients, an APS can afford to devote a large amount of resources toward tailoring the analgesic plan to each individual patient, and can even theoretically do so without any prior assumptions on the healthcare delivery system. However, with excessively siloed patient-level coordination, the APS may find itself reinventing the wheel with each and every patient. Moving up one level in the scale of healthcare delivery is the coordination at the ward or unit scale. This level of coordination offers many opportunities such as standardizing frameworks across common patient populations, providing unified education for all healthcare team members involved in direct patient care, and permitting the tracking and updating of analgesic frameworks based on both international as well as local evidence. Coordination across the hospital permits access to additional expertise from those areas not directly particular to any given unit or ward, such as pharmacy, equipment, and monitoring. Hospital-level coordination also permits consideration of frameworks within the context of optimizing short- and long-term goals for the patient and healthcare delivery system, rather than silo-based optimization of any given unit at the cost of other aspects of the hospital.

18.6 Evaluating Performance of an APS

18.6.1 Introduction

Rawal's sixth tenet of responsibility for an APS is to "audit the performance of the APS on cost-effectiveness and patient satisfaction with analgesia techniques" [19]. In other words, the APS should be evaluated in terms of the benefit it brings both to the institution and to the patient. Benefit to the institution requires gathering data relative to the cost of providing the service in manpower and materials, and comparing it to the efficiency obtained by providing specialty pain care. Benefit to the patient requires gathering data relative to patient outcomes such as pain control, quality of recovery, and satisfaction. Cost-utility, the value of the service in terms of patient health improvement, and cost-effectiveness, the monetary benefit or cost of a new intervention, are health economics measures useful for value comparison across institutions that practice in different models [109, 110]. Although the cost of an APS is important, as Stadler puts it, "in our society, there exists an ethical imperative to control pain and suffering" [109], so cost alone should not be the primary consideration. In fact, patients feel that pain control is a priority and expect there to be a reasonable associated cost [111]. From the Joint Commission of Accreditation of Healthcare Organizations: "Excuses for inadequate pain control will no longer be accepted because poor pain control is unethical, clinically unsound, and economically wasteful" [112].

18.6.2 Economics Under Different Models

Fiscal viability of the APS should be monitored over time to inform the department and institution to guide cost-effective changes [113]. In fee-for-service models, a minimum goal would be for the APS to generate sufficient service fees to cover its own costs. As an example, at one of the author's institutions, the APS is a standalone, 24-h service. Staffing consists of an attending anesthesiologist, a fellow, two residents, and shared administrative assistants and billing staff. The main costs to run the service include staff salaries and benefits, supplies (epidural and block supplies, etc.), and teaching and training materials. The only source of revenue is what the attending anesthesiologist bills for, and includes consultation fees, daily follow-up fees, epidural management fees, and procedure fees. Reimbursement for fees depends on insurance contracts. The average monthly costs are compared to the average monthly revenue. The fiscal viability of this model directly depends on surgical volume, reimbursement rate, and cost containment. Under regionalized or nationalized health models, APS costs are still likely to include employment and material costs. The main difference is in the source of revenue, be it institution- or department-allocated funds out of a larger operating budget, or fees paid out based upon services rendered. The APS director must understand the service structure and budget for cost-efficient use of resources.

18.6.3 Outcomes

Value of the APS is in the services it renders. Under any practice model, it is especially important to communicate value to the institution (Table 18.1), especially if the service runs a monetary deficit. The APS may impact hospital length of stay [114], readmission rates for high resource utilizers [99], patient satisfaction, and hospital rating [115]. The APS follows patients longitudinally and is well positioned to gather data on the success or failure of acute pain-related interventions such as epidural or block placement. Feedback to individual providers would be a valuable quality metric to guide individualized education and improvement.

The main purpose of the APS is to reduce pain in the perioperative period. Many factors that we know of, and many that we have not yet discovered, may impact a patient's experience both in the acute and chronic timeframe. To measure the effectiveness of the services provided, specific data about each patient, the treatments, and the outcomes should be collected and measured. Patient-related outcomes can be divided into objective and subjective (patient-reported) measures (Table 18.2) [101]. Some outcomes of interest include daily pain intensity [110], presence of medication-related side effects or adverse events [103], success, failure, or complications related to performance of regional anesthesia, quality of recovery, and patient satisfaction.

With the increase in the number of APS services globally, organizations like the American Academy of Pain Medicine support the development of uniform acute pain taxonomy and of patient data registries to facilitate the collection of similar outcome parameters. This will be important for the strength of future studies and meta-analyses that inform the evolution of our practice [110, 113].

Table 18.2 General categories of acute pain outcomes

Institutional
Pre-hospital (plans for pain management)
Hospital (LOS, APS consult volume/type, APS procedures, individual staff procedure performance)
Post-hospital (readmissions, satisfaction)
Patient objective
Pre-hospital (demographics, co-morbidities, medical and social history, etc.)
Hospital (surgery/procedures, treatments (PCA, catheters, medications, procedures), rehabilitation, vital signs, etc.)
Post-hospital (chronification of pain, rehabilitation, medications, etc.)
Patient subjective
Pre-hospital (pain scores, PRO's, etc.)
Hospital (pain scores, quality of recovery, PRO's, etc.)
Post-hospital (chronification of pain, pain scores, quality of recovery, PRO's, etc.)
Quality
Adverse events (opioid-related, procedure related, infection, complications, etc.)
Patient satisfaction

References

1. Apfelbaum JL, Chen C, Mehta SS, Gan AT (2003) Postoperative pain experience: results from a national survey suggest postoperative pain continues to be undermanaged. Anesth Analg 97(2):534–540
2. Lasagna L (1991) Clinical analgesic research: a historical perspective. In: Max M, Portenoy R, Laska E (eds) The design of analgesic clinical trials. Raven, New York, pp 1–7
3. Carr DB, Goudas LC (1999) Acute pain. Lancet 353(9169):2051–2058
4. Breivik H (1998) Postoperative pain management: why is it difficult to show that it improves outcome? Eur J Anaesthesiol 15(6):748–751
5. Ahlers O, Nachtigall I, Lenze J, Goldmann A, Schulte E, Hohne C et al (2008) Intraoperative thoracic epidural anaesthesia attenuates stress-induced immunosuppression in patients undergoing major abdominal surgery. Br L Anesth 101(6):781–787
6. Biki B, Mascha E, Moriarty DC, Fitzpatrick JM, Sessler DI, Buggy DJ (2008) Anesthetic technique for radical prostatectomy surgery affects cancer recurrence: a retrospective analysis. Anesthesiology 109(2):180–187
7. Gottschalk A, Sharma S, Ford J, Durieux ME, Tiouririne M (2010) The role of the perioperative period in recurrence after cancer surgery. Anesth Analg 110(6):1636–1643
8. Beattie WS, Badner NH, Choi P (2001) Epidural analgesia reduces postoperative myocardial infarction: a meta-analysis. Anesth Analg 93(4):853–858
9. Beattie WS, Buckley DN, Forrest JB (1993) Epidural morphine reduces the risk of postoperative myocardial ischaemia in patients with cardiac risk factors. Can J Anaesth 40(6):532–541
10. Rodgers A1, Walker N, Schug S, McKee A, Kehlet H, van Zundert A et al (2000) Reduction of postoperative mortality and morbidity with epidural or spinal anaesthesia: results from overview of randomised trials. BMJ 16;321(7275):1493
11. Capdevila X, Barthelet Y, Biboulet P, Ryckwaert Y, Rubenovitch J, d'Athis F (1999) Effects of perioperative analgesic technique on the surgical outcome and duration of rehabilitation after major knee surgery. Anesthesiology 91(1):8–15
12. Ballantyne JC, Carr DB, de Ferranti S, Suarez T, Lau J, Chalmers TC et al (1998) The comparative effects of postoperative analgesic therapies on pulmonary outcome: cumulative meta-analyses of randomized, controlled trials. Anesth Analg 86(3):598–612
13. Upp J, Kent M, Tighe PJ (2013) The evolution and practice of acute pain medicine. Pain Med 14(1):124–144
14. Kucera TJ, Boezaart AP (2014) Regional anesthesia does not consistently block ischemic pain: two further cases and a review of the literature. Pain Med 15(2):316–319
15. Kehlet H, Jensen TS, Woolf CJ (2006) Persistent postsurgical pain: risk factors and prevention. Lancet 367(9522):1618–1625
16. Anonymous (1976) Postoperative pain (Editorial) 4(95)
17. Ready LB, Oden R, Chadwick HS, Benedetti C, Rooke GA, Caplan R et al (1988) Development of an anesthesiology-based postoperative pain management service. Anesthesiology 68(1):100–106
18. American Society of Anesthesiologists Task Force on Acute Pain Management (2004) Practice guidelines for acute pain management in the perioperative setting: an updated report by the American Society of Anesthesiologists Task Force on Acute Pain Management. Anesthesiology 100(6):1573–1581
19. Rawal N (2002) Acute pain services revisited–good from far, far from good? Reg Anesth Pain Med 27(2):117–121
20. Rawal N (1997) Organization of acute pain services–a low-cost model. Acta Anesthesiol Scand Suppl 111:188–190
21. Sun E, Dexter F, Macario A (2010) Can an acute pain service be cost-effective? Anesth Analg 111(4):841–844
22. Gerbershagen HJ, Pogatzki-Zahn E, Aduckathil S, Peelen LM, Kappen TH, van Wijck AJ et al (2014) Procedure-specific risk factor analysis for the development of severe postoperative pain. Anesthesiology 120(5):1237–1245

23. Kehlet H (1997) Multimodal approach to control postoperative pathophysiology and rehabilitation. Br J Anaesth 78(5):606–617
24. Kehlet H, Dahl JB (2003) Anaesthesia, surgery, and challenges in postoperative recovery. Lancet 362(9399):1921–1928
25. Kehlet H, Wilmore DW (2008) Evidence-based surgical care and the evolution of fast-track surgery. Ann Surg 248(2):189–198
26. Max M, Donovan M, Miaskowski C, Ward SE, Gordon D, Bookbinder M et al (1995) Quality improvement guidelines for the treatment of acute pain and cancer pain. JAMA 274(23):1874–1880
27. Gordon DB, Dahl JL, Miaskowski C, McCarberg B, Todd KH, Paice JA et al (2005) American pain society recommendations for improving the quality of acute and cancer pain management: American Pain Society Quality of Care Task Force. Arch Intern Med Am Med Assoc 165(14):1574–1580
28. Macintyre PE, Scott DA, Schug SA, Visser EJ, Walker SM, eds. (2010) Acute Pain Management: Scientific Evidence, Melbourne, Australia: Australian and New Zealand College of Anaesthetists, Australian and New Zealand College of Anaesthetists and Faculty of Pain Medicine.
29. Hamilton DF, Lane JV, Gaston P, Patton JT, Macdonald D, Simpson AH et al (2013) What determines patient satisfaction with surgery? A prospective cohort study of 4709 patients following total joint replacement. BMJ Open 3(4):e002525, British Medical Journal Publishing Group
30. Waljee J, McGlinn EP, Sears ED, Chung KC (2014) Patient expectations and patient-reported outcomes in surgery: a systematic review. Surgery 155(5):799–808
31. Danforth RM, Pitt HA, Flanagan ME, Brewster BD, Brand EW, Frankel RM (2014) Surgical inpatient satisfaction: what are the real drivers? Surgery 156(2):328–335
32. Farr J, Jaggers R, Lewis H, Plackis A, Sim SB, Sherman SL (2014) Evidence-based approach of treatment options for postoperative knee pain. Phys Sportsmed 42(2):58–70, Informa UK Ltd UK
33. Ahsan ZS, Carvalho B, Yao J (2014) Incidence of failure of continuous peripheral nerve catheters for postoperative analgesia in upper extremity surgery. J Hand Surg Am 39(2):324–329
34. Lebovits AH, Zenetos P, O'Neill DK, Cox D, Dubois MY, Jansen LA et al (2001) Satisfaction with epidural and intravenous patient-controlled analgesia. Pain Med 2(4):280–286, Blackwell Science Inc
35. Sullivan M, Tanzer M, Reardon G, Amirault D, Dunbar M, Stanish W (2011) The role of presurgical expectancies in predicting pain and function one year following total knee arthroplasty. Pain 152(10):2287–2293
36. Sharma S, Balireddy RK, Vorenkamp KE, Durieux ME (2012) Beyond opioid patient-controlled analgesia: a systematic review of analgesia after major spine surgery. Reg Anesth Pain Med 37(1):79–98
37. Hebl JR, Dilger JA, Byer DE, Kopp SL, Stevens SR, Pagnano MW et al (2008) A pre-emptive multimodal pathway featuring peripheral nerve block improves perioperative outcomes after major orthopedic surgery. Reg Anesth Pain Med 33(6):510–517
38. Pavlin DJ, Sullivan MJ, Freund PR, Roesen K (2005) Catastrophizing: a risk factor for postsurgical pain. Clin J Pain 21(1):83–90
39. Sullivan MJ, Thorn B, Haythornthwaite JA, Keefe F, Martin M, Bradley LA et al (2001) Theoretical perspectives on the relation between catastrophizing and pain. Clin J Pain 17(1):52–64
40. Sullivan MJ, Bishop SR, Pivik J (1995) The pain catastrophizing scale: development and validation. Psychol Assess 7(4):524–532, American Psychological Association
41. Huxtable CA, Roberts LJ, Somogyi AA, Macintyre PE (2011) Acute pain management in opioid-tolerant patients: a growing challenge 39(5):804–823
42. Macintyre PE, Russell RA, Usher KA, Gaughwin M, Huxtable CA (2013) Pain relief and opioid requirements in the first 24 hours after surgery in patients taking buprenorphine and methadone opioid substitution therapy. Anaesth Intensive Care 41(2):222–230

43. Huxtable CA, Macintyre PE (2013) An alternative way of managing acute pain in patients who are in buprenorphine opioid substitution therapy programs. Eur J Anaesthesiol 30(11):717–718
44. Stromer W, Michaeli K, Sandner-Kiesling A (2013) Perioperative pain therapy in opioid abuse. Eur J Anaesthesiol 30(2):55–64
45. Stromer W, Michaeli K, Sandner-Kiesling A (2013) Reply to: an alternative way of managing acute pain in patients who are in buprenorphine opioid substitution therapy programmes. Eur J Anaesthesiol 30(11):718–719
46. Breschan C, Kraschl R, Jost R, Marhofer P, Likar R (2004) Axillary brachial plexus block for treatment of severe forearm ischemia after arterial cannulation in an extremely low birth-weight infant. Paediatr Anaesth 14(8):681–684, Blackwell Publishing Ltd
47. Atkinson CJ, Ramaswamy K, Stoneham MD (2013) Regional anesthesia for vascular surgery. Semin Cardiothorac Vasc Anesth 17(2):92–104
48. Miller TE, Thacker JK, White WD, Mantyh C, Migaly J, Jin J et al (2014) Reduced length of hospital stay in colorectal surgery after implementation of an enhanced recovery protocol. Anesth Analg 118(5):1052–1061
49. Larson DW, Lovely JK, Cima RR, Dozois EJ, Chua H, Wolff BG et al (2014) Outcomes after implementation of a multimodal standard care pathway for laparoscopic colorectal surgery. Br J Surg 101(8):1023–1030, John Wiley & Sons Ltd
50. Etz CD, Weigang E, Hartert M, Lonn L, Mestres CA, Di Bartolomeo R et al (2015) Contemporary spinal cord protection during thoracic and thoracoabdominal aortic surgery and endovascular aortic repair: a position paper of the vascular domain of the European Association for Cardio-Thoracic Surgery†. Eur J Cardiothorac Surg 47(6):943–957, Oxford University Press
51. Swenson BR, Gottschalk A, Wells LT, Rowlingson JC, Thompson PW, Barclay M et al (2010) Intravenous lidocaine is as effective as epidural bupivacaine in reducing ileus duration, hospital stay, and pain after open colon resection: a randomized clinical trial. Reg Anesth Pain Med 35(4):370–376
52. Cheng SS, Yeh J, Flood P (2008) Anesthesia matters: patients anesthetized with propofol have less postoperative pain than those anesthetized with isoflurane. Anesth Analg 106(1):264–269
53. Tan T, Bhinder R, Carey M, Briggs L (2009) Day-surgery patients anesthetized with propofol have less postoperative pain than those anesthetized with sevoflurane. Anesth Analg 2010;111(1):83–85
54. Chapman CR, Donaldson GW, Davis JJ, Bradshaw DH (2011) Improving individual measurement of postoperative pain: the pain trajectory. J Pain 12(2):257–262
55. Gerbershagen HJ, Aduckathil S, van Wijck AJ, Peelen LM, Kalkman CJ, Meissner W (2013) Pain intensity on the first day after surgery: a prospective cohort study comparing 179 surgical procedures. Anesthesiology 118(4):934–944
56. Katz J, Clarke H, Seltzer Z (2011) Review article: preventive analgesia: quo vadimus? Anesth Analg 113(5):1242–1253
57. Tighe PJ, Harle CA, Boezaart AP, Aytug H, Fillingim R (2014) Of rough starts and smooth finishes: correlations between post-anesthesia care unit and postoperative days 1–5 pain scores. Pain Med 15(2):306–315
58. Leykin Y, Pellis T (2006) A model of postanesthesia care unit-based acute pain service. Anesth Analg 103(1):263–264
59. Leykin Y, Pellis T, Ambrosio C, Zanette G, Malisano A, Rapotec A, Casati A (2007) A recovery room-based acute pain service. Minerva Anestesiol 73(4):201–206
60. Richman JM, Wu CL (2005) Epidural analgesia for postoperative pain. Anesthesiol Clin N Am 23(1):125–140
61. Sultan P, Gutierrez MC, Carvalho B (2011) Neuraxial morphine and respiratory depression: finding the right balance. Drugs 71(14):1807–1819, Springer International Publishing

62. Tighe PJ, Smith JC, Boezaart AP, Lucas SD (2012) Social network analysis and quantification of a prototypical acute pain medicine and regional anesthesia service. Pain Med 13(6):808–819
63. Althaus A, Arránz Becker O, Neugebauer E (2013) Distinguishing between pain intensity and pain resolution: using acute post-surgical pain trajectories to predict chronic post-surgical pain. Eur J Pain 2014;18(4):513–521
64. Lavand'homme PM, Grosu I, France M-N, Thienpont E (2014) Pain trajectories identify patients at risk of persistent pain after knee arthroplasty: an observational study. Springer US 472(5):1409–1415
65. Esch AT, Esch A, Knorr JL, Boezaart AP (2010) Long-term ambulatory continuous nerve blocks for terminally ill patients: a case series. Pain Med 11(8):1299–1302
66. Enneking FK, Ilfeld BM (2002) Major surgery in the ambulatory environment: continuous catheters and home infusions. Best Pract Res Clin Anaesthesiol 16(2):285–294
67. Boezaart AP, Munro AP, Tighe PJ (2013) Acute pain medicine in anesthesiology. F1000Prime Rep 5(54):54
68. Boezaart AP, Tighe PJ (2011) The progression of regional anesthesia into acute and perioperative pain medicine. Int Anesthesiol Clin 49(3):104–109
69. Döring A-C, Nota SPFT, Hageman MGJS, Ring DC (2014) Measurement of upper extremity disability using the patient-reported outcomes measurement information system. J Hand Surg Am 39(6):1160–1165
70. Cella D, Yount S, Rothrock N, Gershon R, Cook K, Reeve B et al (2007) The Patient-Reported Outcomes Measurement Information System (PROMIS): progress of an NIH Roadmap cooperative group during its first two years. Med Care 45(5 Suppl 1):S3–S11
71. Cook KF, Dunn W, Griffith JW, Morrison MT, Tanquary J, Sabata D, Victorson D, Carey LM, Macdermid JC, Dudgeon BJ, Gershon RC (2013) Pain assessment using the NIH Toolbox. Neurology 80(11 Suppl 3):S49–S53
72. Revicki DA, Cook KF, Amtmann D, Harnam N, Chen WH, Keefe FJ (2014) Exploratory and confirmatory factor analysis of the PROMIS pain quality item bank. Qual Life Res 23(1):245–255
73. Grass JA (2005) Patient-controlled analgesia. Anesth Analg 101(5 Suppl):S44–S61
74. Ilfeld BM, Enneking FK (2005) Continuous peripheral nerve blocks at home: a review. Anesth Analg 100(6):1822–1833
75. Boswell MV, Giordano J (2009) Reflection, analysis and change: the decade of pain control and research and its lessons for the future of pain management. Pain Phys 12(6):923–928
76. Hjermstad MJ, Fayers PM, Haugen DF, Caraceni A, Hanks GW, Loge JH et al (2011) Studies comparing Numerical Rating Scales, Verbal Rating Scales, and Visual Analogue Scales for assessment of pain intensity in adults: a systematic literature review. J Pain Symptom Manage 41(6):1073–1093
77. Price DD, Bush FM, Long S, Harkins SW (1994) A comparison of pain measurement characteristics of mechanical visual analogue and simple numerical rating scales. Pain 56(2):217–226
78. Hartrick CT, Kovan JP, Shapiro S (2003) The numeric rating scale for clinical pain measurement: a ratio measure? Pain Pract 3(4):310–316
79. Melzack R (1987) The short-form McGill Pain Questionnaire. Pain 30(2):191–197
80. Cleeland CS, Ryan KM (1994) Pain assessment: global use of the brief pain inventory. Ann Acad Med Singap 23(2):129–138
81. Helfand M, Freeman M (2011) Assessment and management of acute pain in adult medical inpatients. A Systematic Review. Washington, DC: Department of Veterans Affairs
82. Zubkoff L, Lorenz KA, Lanto AB, Sherbourne CD, Goebel JR, Glassman PA et al (2010) Does screening for pain correspond to high quality care for veterans? J Gen Intern Med 25(9):900–905, Springer-Verlag
83. Buckenmaier CC, Galloway KT, Polomano RC, McDuffie M, Kwon N, Gallagher RM (2013) Preliminary validation of the Defense and Veterans Pain Rating Scale (DVPRS) in a military population. Pain Med 14(1):110–123

84. Nilsson S, Finnström B, Kokinsky E (2008) The FLACC behavioral scale for procedural pain assessment in children aged 5–16 years. Paediatr Anaesth 18(8):767–774, Blackwell Publishing Ltd
85. Malec M, Shega JW (2015) Pain management in the elderly. Med Clin North Am 99(2):337–350
86. Pautex S, Herrmann F, Le Lous P, Fabjan M, Michel J-P, Gold G (2005) Feasibility and reliability of four pain self-assessment scales and correlation with an observational rating scale in hospitalized elderly demented patients. J Gerontol A Biol Sci Med Sci 60(4):524–529
87. Zwakhalen SM, Hamers JP, Abu-Saad HH, Berger MP (2006) Pain in elderly people with severe dementia: a systematic review of behavioural pain assessment tools. BMC Geriatr 6(1):3, BioMed Central Ltd
88. Puntillo K, Nelson JE, Weissman D, Curtis R, Weiss S, Frontera J et al (2014) Palliative care in the ICU: relief of pain, dyspnea, and thirst–a report from the IPAL-ICU Advisory Board. Inten Care Med 40:235–248, Springer Berlin Heidelberg
89. Pun BT, Dunn J (2007) The sedation of critically ill adults: Part 1: assessment. The first in a two-part series focuses on assessing sedated patients in the ICU. Am J Nurs 107(7):40–48, quiz 49
90. Lindenbaum L, Milia DJ (2012) Pain management in the ICU. Surg Clin North Am 92(6):1621–1636
91. Kessler ER, Shah M, Gruschkus SK, Raju A (2013) Cost and quality implications of opioid-based postsurgical pain control using administrative claims data from a large health system: opioid-related adverse events and their impact on clinical and economic outcomes 33(4):383–391
92. Overdyk FJ (2010) Postoperative opioids remain a serious patient safety threat. Anesthesiology Am Soc Anesthesiol 113(1):259–260, author reply 260–261
93. Cashman JN, Dolin SJ (2004) Respiratory and haemodynamic effects of acute postoperative pain management: evidence from published data. Br J Anaesth 93(2):212–223, Oxford University Press
94. Weingarten TN, Herasevich V, McGlinch MC, Beatty NC, Christensen ED, Hannifan SK et al (2015) Predictors of delayed postoperative respiratory depression assessed from naloxone administration. Anesth Analg 121:422–429
95. Ankichetty S, Chung F (2011) Considerations for patients with obstructive sleep apnea undergoing ambulatory surgery. Curr Opin Anaesthesiol 24(6):605–611
96. Taenzer AH, Pyke JB, McGrath SP, Blike GT (2010) Impact of pulse oximetry surveillance on rescue events and intensive care unit transfers: a before-and-after concurrence study. Anesthesiology 112(2):282–287
97. Geis GS, Stead H, Wallemark CB, Nicholson PA (1991) Prevalence of mucosal lesions in the stomach and duodenum due to chronic use of NSAID in patients with rheumatoid arthritis or osteoarthritis, and interim report on prevention by misoprostol of diclofenac associated lesions. J Rheumatol Suppl 28:11–14
98. Elia N, Lysakowski C, Tramèr MR (2005) Does multimodal analgesia with acetaminophen, nonsteroidal antiinflammatory drugs, or selective cyclooxygenase-2 inhibitors and patient-controlled analgesia morphine offer advantages over morphine alone? Meta-analyses of randomized trials. Anesthesiology 103(6):1296–1304
99. Blieden M, Paramore LC, Shah D, Ben-Joseph R (2014) A perspective on the epidemiology of acetaminophen exposure and toxicity in the United States. Expert Rev Clin Pharmacol 7(3):341–348, Informa UK, Ltd. London
100. Stephens C, Andrade RJ, Lucena MI (2014) Mechanisms of drug-induced liver injury. Curr Opin Allergy Clin Immunol 14(4):286–292
101. Brull R, McCartney CJL, Chan VW, El-Beheiry H (2007) Neurological complications after regional anesthesia: contemporary estimates of risk. Anesth Analg 104(4):965–974
102. Stadler M, Schlander M, Braeckman M, Nguyen T, Boogaerts JG (2004) A cost-utility and cost-effectiveness analysis of an acute pain service. J Clin Anesth 16(3):159–167

103. Werner MU, Søholm L, Rotbøll-Nielsen P, Kehlet H (2002) Does an acute pain service improve postoperative outcome? Anesth Analg 95(5):1361–1372, table of contents
104. Story DA, Shelton AC, Poustie SJ, Colin-Thome NJ, McIntyre RE, McNicol PL (2006) Effect of an anaesthesia department led critical care outreach and acute pain service on postoperative serious adverse events. Anaesthesia 61(1):24–28, Blackwell Science Ltd
105. Vila H, Smith RA, Augustyniak MJ, Nagi PA, Soto RG, Ross TW et al (2005) The efficacy and safety of pain management before and after implementation of hospital-wide pain management standards: is patient safety compromised by treatment based solely on numerical pain ratings? Anesth Analg 101(2):474–480, table of contents
106. Weinger MB, Lee LA (2011) No patient shall be harmed by opioid-induced respiratory depression. APSF Newsletter, Fall, 26(2):21–40
107. Kozol RA, Voytovich A (2007) Misinterpretation of the fifth vital sign. Arch Surg Am Med Assoc 142(5):417–419; discussion 419–420
108. Taylor S, Voytovich AE, Kozol RA (2003) Has the pendulum swung too far in postoperative pain control? Am J Surg 186(5):472–475
109. Stadler M, Schlander M, Braeckman M, Nguyen T, Boogaerts JG (2004) A cost-utility and cost-effectiveness analysis of an acute pain service. J Clin Anesth 16(3):159–167
110. Lee A, Chan SK, Ping Chen P, Gin T, Lau AS, Hung Chiu C (2010) The costs and benefits of extending the role of the acute pain service on clinical outcomes after major elective surgery. Anesth Analg 111(4):1042–1050
111. Owen H, McMillan V, Rogowski D (1990) Postoperative pain therapy: a survey of patients' expectations and their experiences. Pain 41(3):303–307
112. Phillips DM (2000) JCAHO pain management standards are unveiled. Joint Commission on Accreditation of Healthcare Organizations. JAMA 284(4):428–429
113. Lee A, Chan S, Chen PP, Gin T, Lau AS (2007) Economic evaluations of acute pain service programs: a systematic review. Clin J Pain 23(8):726–733
114. Kehlet H, Mogensen T (1999) Hospital stay of 2 days after open sigmoidectomy with a multimodal rehabilitation programme. Br J Surg 86(2):227–230
115. Gupta A, Daigle S, Mojica J, Hurley RW (2009) Patient perception of pain care in hospitals in the United States. J Pain Res 2:157–164, Dove Press

Chapter 19
The Transition of Acute Postoperative Pain to Acute Persistent Pain to Chronic Pain: Assessing and Managing the Risks

Edward A. Shipton

Abstract The transition of acute pain to pathological chronic pain is a complex and poorly understood process. Biological, psychological, and social-environmental factors and the known polymorphisms in human genes are all involved in perpetuating the pain. The prevalence at a year of adult chronic post-surgical pain (CPSP) characteristically ranges between 1.5 and 10 %. About 13 % of children report the existence of symptoms of chronic postoperative pain. Understanding the factors that contribute to the transition from acute, time-limited post-surgical pain to chronic pathological pain is essential to prevent CPSP. Preoperative, intraoperative, and postoperative risk factors determine the development of chronic postoperative pain. Much has been written about the association of genetic polymorphisms and the development of chronic pain.

The early identification of patients at high risk of developing CPSP enables early interventions to lessen the impact on these patient populations Predictors of CPSP fall into five domains, namely, demographic, pain, clinical, psychological, and surgery-related. Preventive analgesia focuses on attenuating the impact of the peripheral nociceptive barrage associated with noxious perioperative stimuli, in order to reduce pain and analgesic requirements well beyond the surgical period. It prevents central sensitisation

In multimodal analgesia, the patient is administered a combination of opioid and non-opioid analgesic drugs that act at different sites within the central and peripheral nervous systems in an effort to minimize opioid use and decrease opioid-related adverse effects. Regional anaesthesia should be used whenever possible on a procedure-specific basis as it extends analgesia into the postoperative period. There is now more focus on less invasive surgical techniques, multimodal pain management protocols, more rapid functional recovery and reduced length of stay. These reduce the adverse effects of treatment.

E.A. Shipton, FANZCA, FRCA, FFPMANZCA, DM
Department of Anaesthesia, University of Otago, Christchurch,
4345, Christchurch 8001, New Zealand
e-mail: shiptonea@xtra.co.nz; ted.shipton@otago.ac.nz

K. Stuart-Smith (ed.), *Perioperative Medicine – Current Controversies*,
DOI 10.1007/978-3-319-28821-5_19

Optimal postoperative pain management that is specific for different surgical procedures can be obtained. Enhanced postoperative recovery programmes have evolved during the past 15 years. A multidisciplinary approach integrating expertise within different health care professions is essential for optimising recovery based on evidence-based perioperative care components.

The task is to aim for pain-free and risk-free surgery. The recognition and treatment of sub-acute pain is important, allowing the secondary prevention of chronic pain.

Keywords Perioperative risk factors • Mechanisms of transition • Persistent postoperative pain • Strategies of constraint

19.1 Introduction

Postoperative pain comes from tissue damage that causes the release of inflammatory chemicals that extend to sensitise surrounding nociceptors [79]. Acute pain is generally defined as pain with a precise onset and duration limited to the first few weeks (usually within 30 days) after surgery, injury, or a painful illness. The resolution of pain is associated with tissue healing [25]. In contrast to this, chronic pain is maladaptive pain that becomes persistent or recurrent; it persists beyond the expected period of healing' (3–6 months) after surgery, injury, or a painful illness [107]. Chronic post-surgical pain (CPSP) is pain attributable to the surgical procedure that persists for at least 2 months postoperatively, with exclusion of other causes and pre-existing problems [3, 41, 80].

The transition of acute pain to pathological chronic pain is a complex and poorly understood process. Poorly controlled and severe pain that occurs after surgery and trauma increases the risk of a chronic pain state. Biological, psychological, and social-environmental factors and the known polymorphisms in human genes are all involved in perpetuating the pain [107, 126]. 'Subacute' pain is a concept poorly defined in the literature, being loosely seen as term used to describe the stage between 'acute' pain and 'chronic' pain such as acute persistent pain [79].

Acute postoperative pain is associated with autonomic, endocrine-metabolic, physiological, and behavioural responses [98]. Acute postoperative pain often continues past hospital discharge [25]. A survey of 250 adult surgical patients discovered that 82 % experienced acute pain after surgery through the first 2 weeks after hospital discharge. Eighty-six per cent of these patients experienced pain that was classified as moderate, severe, or extreme [5, 25]. In addition, more patients reported pain during the first 2 weeks after discharge than before discharge (75 % vs 58 %), and 81 % of patients continued to experience moderate to severe pain once leaving the hospital [5, 25].

A cross-sectional, population-based survey of 12,982 Norwegian patients undergoing surgery ≥3 months beforehand revealed that 40.4 % of respondents reported chronic postoperative pain that included 18.3 % reporting pain of moderate or severe intensity [5, 55]. In a prospective cross-sectional study, 911 patients at a

Table 19.1 Estimated incidence of chronic postoperative pain following various surgical procedures [107]

Type of surgery	Incidence of chronic postoperative pain (estimated) (%)
Coronary artery bypass surgery	30–50
Amputation of an extremity	30–50
Thoracotomy	30–40
Breast surgery	20–30
Bone and joint surgery	20
Iliac crest bone harvest	19
Hysterectomy	17
Inguinal herniorrhaphy	10
Surgeries on abdominal visceral structures	10–14

See Text for References

university clinic/level 1 trauma centre were followed up at a year postoperatively. Some 83 patients (14.8 %) were identified as having CPSP [113]. The prevalence at a year of adult CPSP characteristically ranges between 1.5 and 10 % [59, 106]. About 13 % of children report the existence of symptoms of chronic postoperative pain [39, 106]. However, in very young children, bones, tendons, and ligaments are more indulgent and flexible. An immature peripheral and central nervous system and an enhanced neuronal plastic capacity in the child's brain may result in a lower risk of developing chronic pain [89].

All surgical procedures can potentially result in CPSP [3]. The majority of patients heal without long-term sequelae. However, procedures such as coronary artery bypass, and mastectomy, amputation, thoracotomy, and hernia surgery, can inflict a significant burden of persistent postsurgical pain [14, 41, 72, 125]. The wide range in the incidence of CPSP in the literature is owing to varying definitions of chronic pain, different measurements of pain and diverse adjuvant treatments [106]. Previous literature reviews have found that postoperative pain progresses to chronic pain in the following: in 10 % of patients after inguinal hernia repair [95]; in 10–14 % in surgeries on abdominal visceral structures [49]; in 19 % after iliac crest bone harvest [49]; in about 20 % after bone and joint surgery [49]; in 20–30 % after breast surgery [80]; in 30–40 % after thoracotomy [129]; and in 30–50 % after amputation or coronary artery bypass surgery; and in 17 % after hysterectomy (Table 19.1) [10, 25]. Sometimes severe pain can occur in the "minor" procedures (appendectomy, cholecystectomy, haemorrhoidectomy, and tonsillectomy) [42].

19.2 Mechanisms of Transition for Acute to Chronic Postsurgical Pain

Understanding the factors that contribute to the transition from acute, time-limited post-surgical pain to chronic pathological pain is essential to prevent CPSP [23]. It is not particularly clear how the transition from acute to chronic pain occurs. The pathological mechanisms leading to chronic CPSP seem to be inflammation, tissue and nerve damage and alterations in central pain processing [124].

19.2.1 Peripheral and Central Sensitisation

Post-surgical pain arises from nociception (from sensory receptors on C fibres and A-delta fibres that respond to intense thermal or mechanical stimuli) and from the detection of inflammation by nociceptors (a consequence of trauma to peripheral tissues) [23].

Following incision, inflammatory mediators (prostaglandins, bradykinin, histamine, cytokines) are released at the periphery by damaged tissue triggering an inflammatory cascade [23]. These act directly on the primary afferent terminals and reduce their excitation threshold. This is known as peripheral sensitization [28]. Substance P (Sub P) and Calcitonin-Gene-Related Peptide (CGRP) are expressed within the superficial dorsal horn [28].

With modification in the response properties of central neurones, central sensitization can occur as well. After an injury or surgery central sensitization results in an enhanced responsiveness to activation, a decreased threshold, expanded receptive fields, and spontaneous activity that create increased pain after surgery [23]. Peripheral and central sensitization is meant to be protective but can become maladaptive, with damaging consequences.

A sign of central sensitization is allodynia where a non-painful stimulus, such as touch becomes painful. Another sign is secondary hyperalgesia where an increase in pain sensitivity occurs in non-injured areas outside the area of primary injury [71]. Peripheral and central sensitization remains critical in the development of pain chronicity [41]. When long-term potentiation of synaptic responses amalgamates with weakened nociceptive inhibitory modulation, enhanced nociceptive facilitatory modulation, and injury-induced proliferation of immunologically active microglia and astrocytes, central sensitisation is maintained and results in chronic pain [71].

Patients with poor inhibitory systems and enhanced excitatory processes experience higher postoperative pain, and are more prone to CPSP. Continuing translation of mRNA in the peripheral terminal of the primary afferent nociceptor plays a role in the transition from acute to chronic pain [34].

19.2.2 Hyperalgesic Priming

In the animal model hyperalgesic priming is used to exhibit the transition from acute to chronic pain. Hyperalgesic priming depends on the epsilon isoform of protein kinase C (PKC epsilon). Priming can be detected as a boosted and prolonged hyperalgesic response to the pro-inflammatory cytokine prostaglandin E2 (PGE2) switch that occurs in the intracellular signalling pathway mediating PGE2-induced hyperalgesia [34, 106].

Over a period of time, molecular memory is formed in primary afferent nociceptors [34]. Studies in the rat model confirm a switch in primary mechanical hyperalgesia to a G (i)-activated and PKC epsilon-dependent signalling pathway that is induced by stress or inflammation [106]. Priming in the rat model can be reversed by the translation inhibitors, rapamycin and cordycepin at the peripheral terminal of the primed nociceptor [34].

Acute insults produce hyperalgesic priming, a neuroplastic change in nociceptors that markedly prolongs inflammatory mediator-induced hyperalgesia. After an acute initiating insult, there is a 72 h delay to the onset of priming, for which the underlying mechanism remains unknown [35]. Spinally induced priming is detected not only when prostaglandin E2 (PGE2) is presented to the peripheral nociceptor terminals, but also when it is presented intrathecally to the central terminals in the spinal cord [35]. It has recently been shown that cytoplasmic polyadenylation element binding protein (CPEB) expressed by isolectin B4-positive nociceptors, a subset of nociceptors in which hyperalgesic priming occurs, are able to be co-immuno-precipitated with protein kinase C epsilon (PKCε) [34]. Both spinally and peripherally induced hyperalgesic priming can be prevented by intrathecal antisense oligodeoxynucleotide to the nuclear transcription factor CREB mRNA [34, 35].

19.2.3 Protease-activated Receptor Type 2 and Hyperalgesic Priming

The protease-activated receptor type 2 (PAR2) plays an important role in inflammatory, visceral, and cancer-evoked pain based on studies using PAR2 knockout [or PAR2(−/−)] mice [119]. Intraplantar injection of the novel highly specific PAR2 agonist, 2-aminothiazol-4-yl-LIGRL-NH2 (or 2-at) arouses a long-lasting acute mechanical hypersensitivity, facial grimacing, and causes robust hyperalgesic priming. This is shown by subsequent mechanical hypersensitivity and facial grimacing to prostaglandin E2 (PGE2). The promechanical hypersensitivity effect of 2-at is completely absent in PAR2 knockout mice, as is hyperalgesic priming as well. PAR2 activation is sufficient to induce neuronal plasticity. This leads to a chronic pain state, the maintenance of which is dependent on a BDNF/trkB/aPKC signalling axis [119].

19.2.4 Neuroplastic Changes

Neuroplastic changes within different areas of the central nervous system may help to explain the transition from acute to chronic conditions [92].

An important property of the nociceptive system is its plasticity. This is the ability to change in an experience-dependent manner; this is implicated in the transition from acute pain to chronic pathological pain. Disease-induced plasticity can occur at both structural and functional levels. It manifests as changes in individual molecules, cellular function, network activity, and synapses [70]. Synaptic plasticity may mediate pathophysiological alterations linked to chronic pain. It does this by virtue of shifting the balance between excitation and inhibition, with a particular emphasis on the spinal dorsal horn [70]. Synapse-to-nucleus communication can mediate long-term changes in nociceptive sensitivity [70].

19.2.5 Genetics

Much has been written about the association of genetic polymorphisms and the development of chronic pain [15]. Epigenetic regulation governs gene expression in response to environmental cues. Recent animal model and clinical studies indicate that epigenetic regulation plays an important role in the development or maintenance of persistent pain and possibly the transition of acute pain to chronic pain [8].

19.2.6 Stress

Social defeat stress enhanced α-amino-3-hydroxy-5-methyl-4-isoxazolepropionic acid (AMPA) receptor GluA1 phosphorylation at the Ser831 site in the spinal cord greatly prolonged plantar incision-induced pain [76]. Spinal AMPA receptor phosphorylation contributes to the mechanisms underlying stress-induced pain transition [76].

19.3 Perioperative Risk Factors

Preoperative, intraoperative, and postoperative risk factors determine the development of chronic postoperative pain (Table 19.2) but are often derived from small samples of surgical patients [123]. The early identification of patients at high risk of developing CPSP becomes important to enable early interventions to lessen the impact on these patient populations [3]. Predictors of CPSP fall into five domains, namely, demographic, pain, clinical, psychological, and surgery-related [123].

19.3.1 Preoperative Risk Factors

A higher level of preoperative pain (intensity and duration) has frequently been identified as a risk factor for early severe acute postsurgical pain across a range of surgery types in the days and weeks after surgery, and of long term

Table 19.2 Perioperative risk factors for developing chronic pain [107]

Preoperative risk factors
Preoperative pain state and demographics
Female gender
Younger age
Preoperative pain at the surgical site
Preoperative pain distant from surgical site
Genetics
Pharmacogenomics
Genetic polymorphisms
Intraoperative risk factors
Open versus minimally invasive surgery
Site (e.g., mastectomy, major limb amputation, thoracotomy, and sternotomy)
Use of electrocautery
Excess of intravenous fluid
Low (compared to high) volume surgical unit
Damage to nerves occurs by compression, stretching, ischaemia, and surgical section
Length of surgery
Pericostal (versus intracostal) stitches
Postoperative risk factors
Red flags (infection, bleeding, organ rupture, compartment syndrome) should be excluded
Unrelieved and severe pain
Repeated surgery in the same area
Radiation therapy and chemotherapy
High postoperative use of analgesics
Surgery performed in a previously injured area

post-surgical pain [23, 74, 123]. Multiple sites of pain suggest widespread sensitization in the nociceptive system and potentially indicate individuals who are predisposed to the development of chronic pain. Female gender and a younger age have frequently been identified as risk factors for persistent postoperative pain [74]. In a review of 333,000 pain scores in those with severe pain events following surgery, females experienced greater mean pain scores on postoperative day 1 for a variety of surgical procedures [118]. Younger age is one of the most consistently reported factors associated with an increased risk of development of CPSP [107, 123]. Comorbid conditions have a relatively minor influence on pain outcomes [74].

The type of pain is an important determinant of whether the patient will develop chronic pain. For example, sensory abnormalities such as hyperaesthesia and hypoaesthesia may indicate that the patient had pre-existing neuropathic pain before the procedure, or developed neuropathic pain consequent to the procedure [23, 25].

Chronic pain unrelated to the surgical site may include fibromyalgia, irritable bowel syndrome, chronic headaches, and back pain. These may be characterized by an underlying state of pain amplification or central sensitization seen in patients suffering idiopathic pain disorders [72, 106].

19.3.2 Intraoperative Risk Factors

Surgery-related factors increase the risk of CPSP as well. These include the following: [60, 106, 107]; site (e.g. major limb amputation, thoracotomy, mastectomy, sternotomy); extent of surgery [77, 108]; a low (compared to a high) volume surgical unit [108, 117]; pericostal (versus intracostal) stitches; and intraoperative nerve damage [77, 108]. Nerve damage can occur from surgical section, compression (in a suture or from clips, or following rib retraction), stretching, ischaemia, or infection [98, 108]. Other factors taken into account include the duration of surgery, and the specific surgical techniques used [1, 106].

Surgical techniques include open versus laparoscopic or arthroscopic surgery, the type of implant, and the newer developments in minimally invasive surgery. Minimally invasive surgery reduces the inflammatory response [63, 106]. Examples consist of natural orifice transluminal endoscopic surgery, single-incision laparoscopic surgery, imaging-assisted lesion localization for single-port video assisted thoracic surgery (lung resection), endoscopy-assisted surgery, minimally invasive mitral valve surgery, and robotic surgery [45, 66, 87, 123].

A recent systematic review of meta-analyses of randomised controlled trials in appendectomies showed that the laparoscopic approach shortened hospital stay from 0.16 to 1.13 days in seven out of eight meta-analyses [54]. Mini percutaneous nephrolithotomy in the treatment of renal and upper ureteral stones has been found to be related to less blood loss and shorter hospital stay [33]. In lumbar surgery, minimally invasive techniques have required smaller incisions. They have decreased approach-related morbidity by avoiding muscle crush injury by self-retaining retractors, and by preventing the disruption of tendon attachment sites of important muscles at the spinous processes [115]. A recent meta-analysis showed minimally invasive transforaminal lumbar interbody fusion to be associated with lower blood loss and infection rates [94]. In caesarean sections, the Misgav-Ladach incision was associated with a significant advantage in terms of reduction of post-surgical acute and chronic pain [44]. In open incisional hernia repair, the use of lightweight mesh seems to be associated with less chronic pain without the increase of recurrence [75]. Laparoscopic resection and transvaginal specimen extraction looks like a promising technique for some right-sided colon pathologies [61].

Mixed evidence is found for the role of the anaesthetic (e.g. regional anaesthesia) and for analgesic regimens (e.g. pre-emptive analgesia, epidural analgesia), and for the experience of the surgeon [19, 106]. High intra-operative doses of remifentanil are associated with small but significant increases in acute pain after surgery [36]. Perioperative fluid management is important to improve outcomes, especially by avoiding fluid excess [66].

19.3.3 Postoperative Risk Factors

Postoperative risk factors include the following: unrelieved and severe pain; surgery performed in a previously injured area with pain or repeated surgery in the same area; preoperative chronic pain, high postoperative use of analgesics, co-morbid stress symptoms, capacity overloads, and radiation therapy and chemotherapy [3, 106, 108]. The severity of acute postoperative pain in the days and weeks after surgery remains one of the most consistent and strongest predictors of CPSP [23, 96]. Adequate postoperative analgesia allowing early mobilization is a prerequisite [66, 106]. However, red flags should be excluded. These include infection, bleeding, organ rupture, and the onset of a compartment syndrome. Better clarification is needed on the underlying cause of the postoperative pain (i.e., infection versus mechanical versus neuropathic [23, 106].

Prolonged postoperative opioid use occurs with legitimate preoperative opioid use, self-perceived susceptibility to pain medication, addiction, and preoperative depressive symptoms [17, 25]. Patients who are unable to self-implement a prescribed pain management regimen are at high risk for transitioning to chronic pain from unresolved acute pain [25]. For example, in ophthalmic surgery a recent systematic review has shown that female sex, longer duration of surgical procedure, second eye surgery as a consecutive procedure, type of surgery, general anaesthesia, lower satisfaction with anaesthesia, and postoperative nausea may contribute to increased postoperative pain intensity [73]. The type of surgery, the type of anaesthesia, and patient satisfaction with anaesthesia were factors associated with increased analgesic consumption as well [73].

There should be an ongoing assessment of pain, both at rest and with movement. Not only pain severity but pain qualities, duration, impact on functional capabilities, and underlying cause(s) need to be considered as well [99]. A longer assessment of postoperative pain that relies on the development of pain trajectories (e.g. from day 1 to day 5) is required [72].

Exposure to intense pain and stress during medical and nursing procedures can contribute to the transition from acute to chronic pain [72]. The presence of large areas of secondary hyperalgesia in the early postoperative phase has been found to be a major predictor of CPSP [28].

19.3.4 Psychosocial Risk Factors

The most common emotions experienced by patients scheduled for major surgery are fear and anxiety. High preoperative anxiety, surgical fear and depression are predictors of severe postoperative pain (Table 19.3) [19, 72]. The fear of pain and pain anxiety play a significant role in the development of chronic pain and disability, although underlying mechanisms remain widely unknown [50]. The preoperative psychological state of a patient has been found to affect the outcome after total

Table 19.3 Psychosocial factors for developing chronic pain [107]

Preoperative risk factors
Pain catastrophizing
High preoperative surgical fear
High preoperative depression
High preoperative anxiety
Postoperative risk factors
Emotional numbing
Low expectation of return to work
Pain hypervigilance
Social and environmental factors
Stressful life events
Low self-rated health
Lower socioeconomic status
Lower education
Social support
Solicitous responding from significant others

knee arthroplasty [67]. Poorly relieved long-lasting postoperative pain not only has a negative psychological impact but contributes to maintaining an underlying central sensitization process. This interferes with rehabilitation.

More recently, pain catastrophizing has been studied as a risk factor for CPSP (after knee arthroplasty and shoulder surgery) [19]. Catastrophizing continues to exert an effect many months to years after surgery [74]. Catastrophizing has been identified as the clearest predictor of chronic pain after total knee arthroplasty. Catastrophizing is associated with the severity of acute and chronic pain, altered central nociceptive processing (e.g. decreased conditioned pain modulation) exaggerated healthcare use, postoperative disability, and reduction in function [31, 38, 46, 74]. The critical elements needed to effectively treat or modify catastrophizing need to be better understood [74].

Less frequently identified factors include somatization, hypochondriasis, avoidant coping, psychological vulnerability, neuroticism, and a low expectation of return to work [59]. Many of these factors occur in the paediatric patient-related population as well [59].

Pain hypervigilance with a strong attentional bias toward pain, and emotional numbing measured after hospital discharge, predict subsequent CPSP and pain disability as well [23, 72]. Most changes in health-related quality of life occur during the first weeks and months after surgery [72].

19.3.5 Social and Environmental Factors

Social and environmental factors have been associated with persistent postsurgical pain. These include solicitous responding from significant others, stressful life events, and social support [108]. Lower education and lower socioeconomic status have been identified as risk factors for CPSP as well [11, 123]. The evidence is mixed for other factors such as employment status, marital status, and for workers compensation [19, 123].

19.3.6 Genetic Risk Factors

A patient's gene expression profile can change rapidly in the post-injury period. Over 1000 genes are activated in the dorsal root ganglion alone after nerve injury [15]. Genetic variation can influence pharmacokinetics (drug transporters and drug-metabolizing enzymes) and pharmacodynamics (opioid receptor and catechol-O-methyltransferase enzymes). A few candidate gene polymorphisms have been linked to pain susceptibility, including catechol-O-methyltranferase (COMT). This enzyme affects nociceptive and inflammatory pain, and its variants (particularly the Val158Met variant) have been linked to temporomandibular joint pain syndromes [15].

Variants of the guanosine triphosphate cyclohydrolase 1 (GTP-cyclohydrolase 1) that is involved on the synthesis of tetrahydrobiopterin affects cancer pain severity and the development of persistent peripheral neuropathic low back pain [108]. A recent meta-analysis has shown the A118G allele variant of opioid receptor mu 1 (OPRM1) to have the most potent influence on pain management of postoperative patients [102]. Other candidate genes are the transient receptor potential cation channel, subfamily 5 member 1 gene, and the melanocortin-1 receptor gene [108].

The SCN9A gene has been looked at as a marker of pain sensitivity. Mutations in this gene that codes for the alpha-subunit of voltage gated sodium channel (Nav1.7) that results in alterations of pain perception. These are found in rare pain disorders such as erythromelalgia and in paroxysmal extreme pain disorder [15]. Five hereditary sensory and autonomic neuropathy syndromes have been described [108]. SCN9A polymorphisms have been described in individuals who are insensitive to pain as well [15].

The induction of inflammatory or neuropathic pain states is known to involve molecular activity in the spinal superficial dorsal horn and dorsal root ganglia, including intracellular signalling events which lead to changes in gene expression [40].

The incorporation of genetic knowledge is expected to lead to the development of more effective means to prevent and manage CPSP using tools of personalized pain medicine [21]. There is a large variability in CPSP amongst patients undergoing similar surgery [21]. Heritability estimates suggest that about half of the variance in CPSP levels is attributable to genetic variation [21]. Analysing patients' DNA sequences, blood and salivary pain biomarkers, as well as their analgesic responses to medications will facilitate developing insights into CPSP pathophysiology, and inform predictive algorithms to determine a patient's likelihood of developing CPSP even prior to surgery [21]. As genetic testing allow prediction as to which patients may have greater problems with pain control, more logical medication strategies would be selected as well [62].

Epigenetics is the study of heritable modifications in gene expression and phenotype that do not require changes in genetic sequence to manifest their effects [15]. Epigenetic mechanisms are essential for long-term synaptic plasticity and modulation of gene expression. This is because epigenetic modifications are known to regulate gene transcription [40]. Epigenetic modifications potentially play an important role in inflammatory cytokine metabolism, in steroid responsiveness, and in opioid sensitivity [15]. DNA methylation, histone acetylation, and RNA interference can alter epigenetic processes. Long-lasting epigenetic modifications could contribute to the transition from acute to chronic pain states by supporting maladaptive molecular changes [40]. At present there is an inability to prevent the epigenetic changes that occur following injury and surgery.

19.4 Preoperative Experimental Predictors

To test the correlation between preoperative responses to experimental pain stimuli and clinical postoperative pain, use has been made of preoperative experimental stimulation methods. Such methods include the heat immersion test, the cold-pressor test, pressure algometry, and cutaneous electrical stimulation [108]. A systematic review found the predictive strength to be much higher than previously reported for single factor analyses of demographics and psychological factors [127].

19.5 Prevention or Risk Limitation Strategies

Perioperative interactions between the nervous and immune systems are complex and can result in peripheral and central sensitization that is crucial to the development of chronic pain [43]. The extent to which perioperative peripheral nociceptive inputs contributes to central sensitization and postoperative pain remains unknown [23]. On average, one in five patients undergoing surgery will develop chronic postsurgical pain [43]. Early identification of factors that may lead to suboptimal pain control in the perioperative period may mitigate these factors [22].

Pre-emptive analgesia whereby preoperative anti-nociceptive treatments prevents surgery-induced central sensitization and increased postoperative pain intensity has not proven clinically effective [23]. Preventive analgesia, however, focuses on attenuating the impact of the peripheral nociceptive barrage associated with noxious perioperative stimuli, in order to reduce pain and analgesic requirements well beyond the surgical period [23]. This prevents central sensitisation [57]. For example, recent preliminary results show that after limb amputation, prolonged and individualized (from 4 to 83 days) perineural infusion of local anaesthetic significantly reduces the incidence of phantom pain (16 % instead of the 67 % usually reported with shorter lasting therapeutic interventions) [72].

The basic mechanism of postoperative risk is the surgical stress response and subsequent organ dysfunction [64]. Hence use is made of regional anaesthetic techniques, minimally invasive surgery, and pharmacological modification of the inflammatory responses to decrease this. Empirical evidence from clinical trials has shown that multi-modal analgesic therapy with a combination of analgesic agents, each with a different mechanism of action, is effective in blocking the various inputs/receptors related to neural and inflammatory processes [23]. In the elderly population, multimodal analgesic techniques have been shown to facilitate the recovery process despite age- and disease-related changes in physiology, and disease-drug and drug-drug interactions [47].

19.5.1 Opioid Reduction

There is an increased postoperative morbidity associated with opioid-only strategies. Adverse effects of opioids include nausea, vomiting, sedation, pruritus, constipation, urinary retention, and respiratory depression) [23]. Opioid medications tend not to be effective in relieving movement-evoked pain. Opioids enhance central sensitization. Multimodal analgesia uses more than one analgesic modality to achieve effective pain control while reducing opioid-related adverse effects [116].

Opioid adverse effects (sedation, postoperative nausea and vomiting, urinary retention, ileus, respiratory depression) can delay discharge [116]. The long-term consequences of perioperative opioids (development of chronic postsurgical pain syndromes, hyperalgesia, and immunomodulation) are increasingly being recognized [109]. There is emerging evidence that opioids may influence cancer outcomes by directly stimulating tumour growth and inhibiting the body's immune response [62]. For patients with severe postoperative pain that is non-neuropathic requiring around-the-clock analgesia, data show that cyclooxygenase-2 inhibition may reduce opioids and improve long-term outcomes [25].

Combining opioids with non-opioid analgesics limits their dose and boosts recovery [66]. Many clinical studies have reported an opioid sparing effect using a wide variety of non-opioid adjuvants (i.e. paracetamol, local anaesthetics, non-steroidal anti-inflammatory drugs [NSAIDs], cyclooxygenase-2 inhibitors, paracetamol, ketamine, clonidine, dexmedetomidine, adenosine, gabapentin, pregabalin, glucocorticoids, esmolol, neostigmine, and magnesium) [81, 108, 109]. Implementing a multi-modal regimen at the outset can diminish the unwanted adverse effects of opioid analgesics and reduce intraoperative and long-term tolerance [23].

Patient-controlled analgesia (PCA) empowers patients to play a role in their own postoperative pain management. PCA routes include PCA-intravenous or subcutaneous (morphine, fentanyl with adjuvants such as clonidine, dexmedetomidine, and ketamine), PCA-oral, patient-controlled regional analgesia, or patient controlled epidural analgesia (PCEA) [92]. A recent Cochrane review provides moderate to low quality evidence that PCA is an efficacious alternative to non-patient controlled systemic analgesia [83]. A recent meta-analysis has shown that evidence of a clinically relevant analgesic effect of peripherally applied opioids for acute postoperative pain is lacking; this may depend on the presence of preoperative inflammation [88].

19.5.2 Postoperative Nausea and Vomiting

Postoperative nausea and vomiting (PONV) can delay the time to discharge and resumption of activities of daily living [91]. Multi-modal opioid sparing analgesia should be provided and the Apfel risk score and a risk based multi-modal PONV prophylaxis followed [91]. Pharmacogenetics in PONV-related studies is attracting attention [86].

19.5.3 Nerve Protection

Surgery and its associated requirements such as positioning and tourniquet have specific risks. Patients with pre-existing neuropathy may be at an increased risk of postoperative neurological dysfunction [13]. Animal studies of posttraumatic nerve injury pain demonstrate that there is a critical time frame before and immediately after nerve injury in which specific interventions can reduce the incidence and intensity of chronic neuropathic pain behaviours-so called "preventative analgesia" [18].

Several studies using quantitative sensory testing and specific questionnaires have reported a major neuropathic component in CPSP [72]. Following surgical nerve repair, cofactors like poor capacity of nerve regeneration (a suspected genetic origin), decreased pain tolerance (due to altered endogenous pain processing) and a particular psychological profile played a major role as shown in the 33 % of patients who developed CPSP [72].

In animal models, perineural local anaesthetic, systemic intravenous local anaesthetic, perineural clonidine, systemic gabapentin, systemic tricyclic antidepressants, and minocycline have each been shown to reduce pain behaviours days to weeks after treatment [18]. The translation of this work to humans also suggests that brief perioperative interventions may protect patients from developing new chronic postsurgical pain [18]. Patients with a neuropathic component, first-line treatments such as topical lignocaine, tricyclic antidepressants, and alpha-2-delta calcium channel modulators are used before opioids [25]. A recent Cochrane review found no evidence from randomised controlled studies supporting the use of topical lignocaine for neuropathic pain, although individual studies indicated that it was effective for relief of pain [27]. More intraoperative data on handling of tissue and nerves needs to be collected.

19.5.4 Psychological Strategies

Some of the psychological strategies that are in use include cognitive-behavioural therapy, acceptance commitment therapy, hypnosis, electro-myographic biofeedback, and coping strategies (Table 19.4). Modern treatment approaches focus on the extinction of aversive memories, and the restoration of the body image and normal brain function. Such approaches include brain stimulation, mirror training, virtual reality applications, or behavioural extinction training [37]. A recent

Table 19.4 Prevention and management of risk factors in acute perioperative pain for developing chronic pain [108]

Preoperative prevention and management
Educate patients and staff (regarding procedure and intended pain management)
Applicable patient information should be supplied
Attitudes and concerns should be addressed
Surgical procedures resulting in severe pain should be identified
Optimise analgesia before surgery and for several days after surgery
Pain evaluation and treatment protocols should be standardised
Patients with modifiable risk factors giving rise to acute persistent and later chronic pain should be identified. They should be followed up and managed after discharge
Psychological pain-management interventions should be provided around surgery as necessary
Intraoperative prevention and management
Known surgical risk factors should be minimised
Obtain acceptable exposure via least painful surgical approach
Minimise neural and tissue damage
Multimodal opioid-sparing analgesic pharmacotherapy should be afforded
Afferent neural blockade should be used where appropriate (for example, paravertebral blockade for breast surgery, and epidural analgesia for thoracotomy and major laparotomy)
Local anaesthesia should be given at incision sites
Use a procedure-specific analgesic regimen (from the PROSPECT group)
Postoperative prevention and management
Pain levels should be measured at rest and with motion
Use protective multimodal opioid-sparing analgesic pharmacotherapy (such as gabapentin) to optimise analgesia; pain levels should be less than 5/10 on days 1–5 after surgery
Make use of a multidisciplinary enhanced postoperative recovery programme
There should be evidence-based adjustments in the usage of nasogastric tubes, drains, and urinary catheters, and in preoperative bowel preparation, as well as the early start of oral feeding and mobilisation
If neuropathic pain is suspected, a bedside neurological examination should be undertaken
Analgesia should be administered well into the postoperative period
Discharge plan
At discharge both analgesic packages and home follow-up should be individualised

systematic review has found that the effectiveness of therapeutic play intervention in children's perioperative anxiety, negative behaviours, and postoperative pain is inconclusive [51].

19.5.5 Multimodal Pharmacological Strategies

In multimodal (or "balanced") analgesia, the patient is administered a combination of opioid and non-opioid analgesic drugs that act at different sites within the central and peripheral nervous systems in an effort to minimize opioid use and decrease opioid-related adverse effects (Table 19.1) [109]. Individualized pharmacologic treatments require a complete characterization of a patient's pain profile, in terms of frequency of pain over the course of a 24-h day and over time thereafter, frequency and duration of pain flares, and presence of neuropathic pain [25]. In the elderly, postoperative pain may be complicated by a number of factors, including a higher risk of age-related and disease-related changes in physiology and disease-drug and drug-drug interactions. Early studies evaluating approaches to facilitating the recovery process have demonstrated the benefits of multimodal analgesic techniques [47]. Recent advances have been due to new ways to use old drugs in a variety of settings, often as components of a multimodal approach to pain relief [104].

Multimodal analgesia uses a combination of delivery routes (i.e. enteral, parenteral, epidural, intrathecal). These are administered at variable time points (i.e. preoperative, intraoperative, and postoperative) to optimize outcomes in the treatment of acute pain and in the prevention of chronic pain [82].

19.5.5.1 Paracetamol (Acetaminophen)

Paracetamol (acetaminophen) modestly inhibits peripheral prostaglandin synthesis; it centrally blocks the formation and release of prostaglandins inhibiting action of endogenous pyrogens on the heat-regulating centres in the brain [23]. Recent research suggests that a COX-3 pathway inhibits a central prostaglandin synthesis in the hypothalamus and decreases prostaglandin E in cerebrospinal fluid, to produce analgesic and antipyretic effects [23]. The optimal use of paracetamol (acetaminophen) by mouth, rectally, or intravenously (as the prodrug proparacetamol) can improve pain control [109].

19.5.5.2 Non-steroidal Anti-inflammatory Drugs (NSAIDs) and Coxibs (COX-2 Inhibitors)

Adverse effects of NSAIDs include gastric ulceration, haemorrhage, renal dysfunction, and platelet inhibition [23]; NSAIDs should be used cautiously in patients with renal dysfunction, sepsis, end stage liver disease, or cardiac disease [23]. Selective

COX-2 inhibitors have been introduced into perioperative clinical practice; they have similar analgesic efficacy to that of NSAIDs, but a lower bleeding risk (no platelet inhibition), and are presumed to have less gastrointestinal toxicity (gastropathy is approximately half of the NSAIDS) [109]. Of note is that the COX-2 selective inhibitors (celecoxib, etoricoxib) do not inhibit bone healing [23]. There is fair (Grade B) evidence that short-term use of NSAIDs result in no long-term reduction in bone healing or fusion rates [29]. Currently celecoxib, etoricoxib, and lumiracoxib (in low doses) are used. Parecoxib is the only parenterally administered coxib available to date [109]. Inflammatory responses contribute to pain, fatigue and organ dysfunction. They can be modified pharmacologically by preoperative glucocorticoids [39]. Whilst multiple studies investigating the role of NSAIDs and steroids for acute postoperative pain have consistently shown benefit, many of these studies have not specifically investigated their longer term effects on the development of chronic pain [82].

19.5.5.3 N-Methyl-D-aspartate (NMDA) Antagonists

Low-dose ketamine (N-Methyl-D-aspartate antagonist) in the range of 0.25–0.5 mg/kg as an initial bolus followed by 50–500 microgram/kg/h is used as an adjuvant for postoperative analgesia and for the reduction of exogenous opioid-induced hyperalgesia [23]. Ketamine in sub-anaesthetic doses is effective in reducing morphine requirements in the first 24 h after surgery [109]. The mean reduction of opioid consumption when using low-dose intravenous infusion ketamine (infusion rate less than 1.2 mg/kg/h) is 40 % [58]. More research into the effectiveness of dextromethorphan, another N-Methyl-D-aspartate antagonist is needed. Ketamine may attenuate tolerance to opioids. It has a definite role in preventing opioid-induced hyperalgesia in patients receiving high doses of opioid for their postoperative pain relief [109].

19.5.5.4 Pregabalin and Gabapentin

Adjuvant analgesics are drugs with non-pain primary indications but which are shown to have analgesic efficacy in painful conditions [109]. An example of adjuvant analgesics are gabapentin and pregabalin. In a recent meta-analysis, pregabalin was found to improve postoperative analgesia compared with placebo [85]. Pregabalin and gabapentin have been found to be useful components of multimodal analgesia for lumbar spine surgery [57, 131]. The mechanisms of action of gabapentin or pregabalin involve binding to presynaptic alpha-2-delta subunit of N-type voltage-gated calcium channels (VGCC) in the dorsal root ganglia, which ultimately reduces calcium-dependent release of excitatory neurotransmitters such as glutamate, noradrenaline, calcitonin gene-related-peptide (CGRP), and substance P (Sub P). In addition, gabapentin has been shown to exert modulatory effects at the level of the periaqueductal grey nucleus [82].

Several meta-analyses have confirmed that gabapentin and pregabalin reduce both postoperative opioid use and postoperative pain scores [23]. Pregabalin, when administered before surgery, and for 14 days after surgery, resulted in a reduction in both postoperative and chronic neuropathic pain following total knee arthroplasties [82]. At both 3 and 6 months postoperatively, there was a statistically significant reduction in the incidence of neuropathic pain in the pregabalin group (0 %) compared with the placebo group (8.7 % and 5.2 % at 3 and 6 months, respectively), as measured by Leeds Assessment of Neuropathic Symptoms and Signs scale [82]. Patients receiving pregabalin also had statistically significant reductions in epidural opioids, oral opioid pain medication while hospitalized, and obtained greater active knee flexion over the first 30 postoperative days [82].

19.5.5.5 Glucocorticoids

The preoperative use of a high-dose glucocorticoid has potential as an analgesic technique that leads to enhanced recovery and less fatigue due to reduced inflammatory responses [64]. Dexamethasone and methylprednisolone offer analgesia with anti-emesis. Their doses are limited by steroid-related side effects. Practically, a dose of dexamethasone 0.1 mg/kg is given [109]. A recent meta-analysis showed that perineural dexamethasone improves postoperative pain outcomes when given as an adjunct to brachial plexus blocks [26].

19.5.5.6 Others

The impact on the use of the tricyclic antidepressants on descending modulation of pain provides an ideal target for the prevention of chronic pain [82]. After limb amputation, amitriptyline showed efficacy in patients with minimal analgesic treatments or who had obtained no relief with tramadol [82]. Tapentadol is a new centrally acting analgesic with a unique dual mode of action as a noradrenaline (norepinephrine) reuptake inhibitor, and as an agonist at the mu-opioid receptor [109].

Magnesium is beneficial especially in abdominal surgery. Toxicity is rare when administered as pre-incision loading followed by infusion under vigilant monitoring [110]. Adenosine improves postoperative recovery, as indicated by lower pain scores and less opioid consumption [109].

19.5.6 Regional Analgesia/Anaesthesia Adjuncts

Meta-analyses of the preventive analgesic effects of regional anaesthetic techniques have been show extended post-operative analgesia as well as reduced incidence of persistent post-surgical pain [4, 57]. Regional anaesthesia should be used whenever possible on a procedure-specific basis and it extends analgesia into the postoperative period [39]. This is an important technique in patients with pre-existing pain [121].

Techniques range from simple subcutaneous wound infiltration (by the surgeon) to complex peripheral (plexus), paravertebral and central (spinal and epidural) blocks (by the anaesthetist). Block duration is determined by the properties of the local anaesthetic used, the concentration of the local anaesthetic, and the volume injected [66]. Local anaesthesia should be used at incision sites but not in joints.

Regional anaesthetic techniques have a powerful effect on the catabolic response. They decrease cardiac demands, pain, ileus, and pulmonary dysfunction. Regional anaesthesia attenuates the endocrine-metabolic effects (the rise in glucagon, cortisol, catecholamines, hyperglycaemia, insulin resistance, and negative nitrogen balance) on stress-induced organ dysfunction [109].

Local infiltration analgesia can be used for patients undergoing hip arthroplasty because of its ability to reduce pain scores and analgesic consumption without any additional adverse effects [130]. Femoral and sciatic nerve blocks remain very popular after total knee arthroplasty as well [23]. The analgesic efficacy and safety of periarticular multimodal drug injection compared with femoral nerve block for postoperative pain management in total knee arthroplasty remains controversial [127]. Pre-emptive nebulization of local anaesthetic in the abdominal cavity has been used to control postoperative pain after laparoscopy [2]. A meta-analysis showed that transversus abdominis plane (TAP) block for postoperative analgesia after laparoscopic surgery resulted in less opioid consumption up to 24 h after laparoscopic surgery in comparison with usual care alone [132]. It affords technical simplicity, and long duration of action [62]. In a meta-analysis of a TAP block versus local anaesthetic wound infiltration in lower abdominal surgery, the TAP block had better long-lasting effect [130].

Continuous peripheral nerve or plexus catheters placed in the perineural space result in better pain relief at rest and with movement, and earlier discharge [23]. Although single-dose or continuous peripheral nerve blocks can optimise analgesia, the risk of muscle weakness and falls should not be forgotten. Perineural patient-controlled regional analgesia (PCRA) allows patients to self-titrate local anaesthetic peripheral nerve blocks in order to achieve comfort. Intravenous lignocaine infusion can be used when regional anaesthesia or techniques are not practical.

An advance is the liposome or polymer (microsphere) encapsulation of local anaesthetics [109]. Several drug delivery systems such as liposomes, micro particles and nanoparticles have been investigated as local anaesthetic delivery vehicles to achieve prolonged anaesthesia. Recently, injectable responsive hydrogels have been used for the localised delivery of anaesthetic molecules [7]. Extended-release epidural morphine has been used as well [6]. Clinical trials with liposomal bupivacaine and SABER-Bupivacaine show that both safely prolong analgesia, while decreasing opioid requirements when compared with placebo [114]. Preliminary clinical trials suggest that liposomal bupivacaine has predictable pharmacokinetics and a similar adverse effect profile when compared to bupivacaine hydrochloride [120].

19.5.6.1 Epidural

Continuous epidural infusions of local anaesthetics reduce the stress response to surgery and allow rapid mobilization after major surgical procedures [110]. The efficacy of pain relief with epidural infusions can be outstanding [101, 107, 110]. There are some benefits in decreased cardiovascular and pulmonary morbidity in high-risk patients undergoing major open vascular or cardiac surgery [101, 110]. After major thoracic or abdominal surgery, patients at risk of postoperative respiratory and cardiac complications should be offered optimal analgesia with thoracic (or thoracolumbar) epidural infusions. Abdominal surgery induces postoperative respiratory dysfunction related to a combination of reflex diaphragmatic inhibition, respiratory muscle injury and pain. Thoracic epidural analgesia using local anaesthetics is able to partially reverse the diaphragmatic dysfunction. Postoperative epidural analgesia has been associated with a reduced incidence of CPSP after abdominal surgery [9]. However, paravertebral blockade have been found to provide comparable analgesia with a better preservation of pulmonary function and a lower incidence of hypotension than thoracic epidurals [69].

Low doses of local anaesthetic (e.g. levobupivacaine, ropivacaine), a lipophilic opioid (e.g., fentanyl, sufentanil), and adrenaline (epinephrine) are used [12, 107, 110]. The addition of opioids (acting on the pre and postsynaptic opioid receptors) improves the quality of postoperative epidural analgesia, and. Adrenaline acts on the pre- and postsynaptic alpha-2 receptors and has a dose-related analgesic effect of its own [107, 110]. The alpha (2)-adrenergic agonists, clonidine and dexmedetomidine are added in thoracic epidurals to provide neuraxial analgesia via alpha-adrenergic receptors. They produce sedation, anxiolysis, sympatholysis, and analgesia but no respiratory depression, and are increasingly being used as adjuvants to local anaesthetics and opioids [107, 110].

Adverse effects of local anaesthetics include hypotension, motor blockade, and urinary retention (if the catheter is in the low thoracic or thoracolumbar area) [12, 107, 110]. Epidural opioids are associated with distressing adverse effects. These include nausea, vomiting, pruritus, urinary retention, gastrointestinal immobility, and respiratory depression [48, 110].

19.5.6.2 Intrathecal

The intrathecal route is simple, reliable, and requires low doses of drugs. It is distinguished by rapid onset and offset, easy administration, minimal expense, and minimal adverse effects or complications [111]. Inserted into the intrathecal space are multimodal combinations of local anaesthetics (bupivacaine, Ropivacaine) with opioids (fentanyl, sufentanil and morphine) [111].

Adding intrathecal opioids result in an improved analgesia, a reduction in local anaesthetic requirements, and a shorter duration of motor blockade [111]. The speed of onset and duration of intrathecal analgesia is regulated by the lipid solubility of opioids. Hydrophilic drugs (morphine) have a slower onset of action and longer half-lives in cerebrospinal fluid with greater dorsal horn bioavailability and greater cephalad migration compared with lipophilic opioids (fentanyl, sufentanil) [111]. Preservative-free morphine in doses from 100 to 200 mg provides excellent postoperative analgesia [111]. The combined spinal-epidural (CSE) technique has been developed to counteract the short duration of effect (90–180 min) of intrathecal opioids [111].

Intrathecal and epidural bleeding producing spinal cord or cauda equina compression and paresis (usually after coagulopathies or anticoagulant therapies) remain a rare but major complication [112]. Randomized, placebo controlled studies have shown a reduction in the incidence of chronic pain at 6 months in patients receiving intrathecal clonidine 300 mg prior to surgery [24, 32, 82].

The use of non-opioid analgesic techniques together with an improved understanding of early loss of muscle function, orthostatic intolerance, perioperative blood management, mechanisms of postoperative delirium and cognitive dysfunction, and the need for thromboembolic prophylaxis can all result in a reduction of postoperative hospital stay to 1–2 days with discharge to home [63]. This is seen with fast-track total hip and knee arthroplasties [63]. Thromboembolic complications were less likely to occur in enhanced recovery after surgery with its rapid mobilisation, and do not require the conventional extended prophylaxis [64].

19.5.7 New Techniques

New, targeted, analgesic medications and techniques may provide a safer and more effective approach to the management of acute postoperative pain. These include preoperative nonsteroidal anti-inflammatory drugs, anxiolytics, and anticonvulsants; intraoperative approaches such as neuraxial analgesia, continuous local anaesthetic wound infusion, transversus abdominis plane block, extended-release epidural morphine, intravenous acetaminophen (paracetamol), and intravenous ketamine; and postoperative use of intravenous ibuprofen, new opioids (e.g. tapentadol) or opioid formulations (morphine-oxycodone), and patient-controlled analgesia [6]. There is moderate level evidence supporting the use of a 2 g preoperative dose of vitamin C as an adjunct for reducing postoperative morphine consumption, and high level evidence supporting perioperative vitamin C supplementation of 1 g per day for 50 days for prevention of CRPS type 1 after extremity surgery [20].

19.5.8 Wound Infusions

A meta-analysis has shown that ropivacaine for continuous wound infusion for postoperative pain management provides clinically meaningful reductions in opioid use and pain outcomes [100].

19.5.9 Procedure Specific Postoperative Pain Management

Optimal postoperative pain management that is specific for different surgical procedures can be obtained from the PROSPECT group (PROcedure-SPECific Postoperative Pain ManagemenT), a collaboration of anaesthesiologists, surgeons, and surgical scientists (Table 19.1) [56, 107]. Their recommendations are based on procedure specific evidence from systematic reviews (using the protocol of the Cochrane Collaboration to evaluate randomized controlled trials of analgesic, anaesthetic and surgical interventions affecting post-operative pain) performed using a rigorously defined methodological process, and supplemented with evidence from other similar surgical procedures plus clinical practice information [56, 57]. It provides evidence-based, procedure-specific information for the use of analgesic interventions.

The PROSPECT group has systematically reviewed pain and other outcomes for several surgical procedures. These include total knee arthroplasty thoracotomy, colonic resection, laparoscopic cholecystectomy, total hip arthroplasty, abdominal hysterectomy, haemorrhoid surgery, non-cosmetic breast surgery, herniorrhaphy, and radical prostatectomy [57]. For example a PRISMA-compliant and PROSPERO-registered systematic review was conducted of randomized placebo-controlled trials (RCTs) of medication-based analgesic interventions after total hip arthroplasty. Non-steroidal anti-inflammatory drugs and lumbar plexus block were demonstrated to provide reductions in postoperative pain scores [52].

Such an optimal procedure-specific analgesic regimen (see PROSPECT Web site: www.postoppain.org) should be integrated into an enhanced recovery care programme [107, 108]. Gabapentin and COX-2 inhibitors have been found to be the most effective pre-emptive analgesics [93]. Future improvements may include the use of preoperative high dose glucocorticoids, gabapentinoids, duloxetine, and ketamine [63]. Prospective Randomised Controlled Trials evaluating such strategies are warranted [82].

19.5.10 Enhanced Recovery After Surgery (ERAS) Programmes

The relief from suffering leads to early mobilization, less hospital stay, reduced hospital costs, and increased patient satisfaction [68]. Fast-track surgery or enhanced postoperative recovery programmes have evolved during the past 15 years (Table 19.4). A multidisciplinary approach integrating expertise within different health care professions (anaesthesiologists, surgeons, surgical nurses, and physiotherapists) is essential for optimising recovery based on evidence-based perioperative care components [57, 64, 65].

Essential components of ERAS include preoperative information, thoracic epidural analgesia, avoidance of fluid overload, early mobilization, oral feeding, and elimination of nasogastric tubes, and drains [64]. Preoperative carbohydrate loading and bowel preparation may be appropriate as well [64]. Optimising perioperative fluid therapy avoids both fluid overload and hypovolemia [64]. Patients' quality of sleep is important as it can determine an individual's risk of developing persistent pain [104]. Other factors such as postoperative delirium, postoperative orthostatic intolerance (from hypovolemia or change in sympatho-vagal balance), anaemia, and sleep disturbances should be addressed as well [64].

These components have proved effective across a range of procedures in terms of minimizing postoperative morbidity, decreasing variability in surgical care, reducing length of hospital stay and containing hospital costs, and improving convalescence [63]. For example, a retrospective cohort study showed that increased postoperative day one discharges occurred after implementation of a hysterectomy enhanced recovery pathway [84].

Hormonal release and various inflammatory responses inherent in the stress response to surgery can contribute to a state of insulin resistance in the postoperative period. This forms one of the main pathogenic factors modulating perioperative outcome [16]. Insulin resistance increases endogenous hepatic glucose, decreases uptake of peripheral glucose, and increases protein breakdown leading to the overall loss of lean muscle tissue [16]. A targeted ERAS program that modulates perioperative insulin sensitivity will improve postoperative outcomes [16].

Faster recovery, reduced hospital stay, and decreased length of convalescence can occur if multimodal analgesia is combined with a rehabilitation program that is both multidisciplinary and multimodal [109]. Many studies, however, have not implemented the most important aspects in surgical care programmes that were originally described. These include patient information, effective thoracic epidural analgesia with sufficient local anaesthetic to facilitate early mobilization, reduced ileus and rapid reintroduction of oral feeding, and the use of non-opioid analgesia [66]. Fast-track techniques use evidence-based adjustments to the use of nasogastric tubes, drains, and urinary catheters, preoperative bowel preparation, and the early initiation of oral feeding and mobilization [109].

19.6 Prevention and Management

It is essential to ensure patients are adequately assessed and adequately educated and prepared (both clinically and psychologically) for surgery [112]. Psychosocial screening questionnaires should be used pre-operatively to pick up acute psychological distress and psychosocial factors that make patients at high risk for developing acute persistent pain [108]. Patients should be screened for the presence of major depression and other psychiatric conditions with aggressive treatment strategies concomitant to analgesic strategies. Preoperative interventions should be undertaken to reduce pain catastrophizing [74].

Individually tailored preoperative education should be provided to both patient and clinical staff about the procedure and intended pain management. A thorough explanation of the surgery and the expected outcome should be given [108]. Patient information should include procedural information (current and reflect best practice), sensory information (what the patient can expect), and physiological coping information [112]. Analgesia in the acute injury and preoperative phase should be aggressively optimised. This should be extended into the postoperative period where healing takes place [82]. In summary, patients at risk for uncontrolled postoperative pain should be identified; effective preventative multimodal analgesia should be implemented; intervention with rescue regional analgesia, additional opioids, or non-opioid agents should take place if needed [30].

There is now more focus on less invasive surgical techniques, multimodal pain management protocols, more rapid functional recovery and reduced length of stay. These reduce the adverse effects of treatment [122]. Known surgical risk factors should be modified. The use of a procedure-specific, multimodal perioperative pain management offers a rational basis for enhanced postoperative pain control, optimal analgesia, decrease in adverse effects, and improved patient satisfaction [78]. The prevention of nerve and tissue damage (such as nerve-sparing techniques) during surgery is probably the most important preventive factor [108]. The least painful surgical approach (e.g. keyhole) with acceptable exposure should be used [108]. Protective pharmacotherapy and the provision of perioperative psychological pain-management interventions aim at preventing and treating CPSP whilst optimizing functional recovery through safe, early mobilization [23]. Early identification of a neuropathic aetiology of pain is necessary to guide selection of treatment. Given the substantial proportion of patients who develop neuropathic pain after surgery, the presence of pain 2 weeks after surgery should prompt a thorough investigation for a neuropathic pain component [25].

The major goal in the management of postoperative pain remains minimising medication dose to lessen adverse effects (especially of opioids) while still providing adequate analgesia. Comprehensive pain management with cohesive interdisciplinary teams (anaesthesiologists, surgeons, psychologists, physiotherapists) that address the needs of each individual should be initiated [82]. The pain relief plan should be formulated by this multidisciplinary team [68].

For patients there remains a need to develop and evaluate truly multidimensional preventive pain-management strategies and their effects on surgical and rehabilitation outcomes, including return to activities of daily living and incidence of chronic pain (Table 19.1) [23, 109].

Standardized, institutional policies and procedures should be developed and an integrated approach used for pain management [30]. Clinical investigation should always be thorough to exclude other physical causes of a chronic pain state [3]. Individualized discharge analgesic packages, and expectations at 3 months after surgery should be discussed [113]. At discharge the patient and family should be provided with a written workable, effective, and safe pain management programme for use at home.

19.7 What of the Future?

Strategies to prevent the development of chronic pain involves better understanding of the mechanisms involved in the transition from acute to chronic pain [43]. Patient responses are variable [56]. A direct correlation between noxious stimuli and perceived pain does not occur [56]. Not all pain can be eliminated, and a goal-related therapy may be appropriate instead [56].

Developing more comprehensive tools for patient assessment remains the first step in achieving the ultimate goal of effective acute pain management [99]. Future studies should include patients at high risk for prolonged postoperative opioid use [25]. Large, well-designed trials need to establish the relative benefits and risks of the various approaches used in improving postoperative pain management [57]. New applications of local anaesthetics and other drugs, new combinations of existing drugs and new developments in predicting an individual's response to pain provide tools to minimise opioid use for postoperative pain relief [105]. Optimal analgesic combinations and routes of administration to optimise current concepts of multimodal analgesia need to be learned.

The impact of preventive analgesia on long-term outcomes including return to activities of daily living and incidence of chronic pain needs further evaluation. The relationship between a patient's experience of pain and function needs to be better understood to reduce the development of CPSP [23]. The role of protective resilience factors needs to be established [123].

Recent studies have underlined the importance of focusing on the acute and sub-acute post injury periods. Neuroimaging of pain and on learning processes shows that learning-induced functional and structural brain changes involving sensorimotor, as well as limbic and frontal areas are important in the transition from acute to chronic pain [37]. Understanding the part played by the individual's endogenous pain modulatory processes and psychological functioning should improve the understanding of the progression from acute to chronic pain [79].

There is a need to improve the methodology within studies of surgical pain [123]. To identify the contribution of relevant determinants and modulators of postoperative pain, large prospective-controlled clinical studies are urgently required that include measurement and documentation of all risk factors and procedure-specific aspects [108]. Establishing a core minimum dataset for future epidemiologic studies will facilitate aggregation of CPSP risk and outcome data from different centres. Then suitable comparisons across surgical populations from different geographical and health care settings can take place [123].

19.8 Conclusion

The task is to aim for pain-free and risk-free surgery. The risk of progressing from acute postoperative pain to more chronic pain is dependent on multiple factors. These include, patient-specific factors, surgical factors, types of pain (nociceptive versus neuropathic), and the extent of postoperative pain resolution. The single best approach to persistent acute postoperative pain is to prevent it. This should include a procedure-specific, evidence-based, updated and multidisciplinary approach within the fast-track methodology [66]. Standardized pain evaluation and treatment protocols are strongly recommended.

The recognition of sub-acute pain as a new entity is important as many surgical patients will need therapy beyond the first 8 days after surgery. In this group of patients the early diagnosis of a neuropathic pain component is vital so that appropriate multimodal therapy may be implemented [25]. The introduction of a transitional pain service, whose aim is to optimise pain control, monitor and appropriately wean patients off opioid medications, prevent unnecessary readmissions post-discharge, and reduce disability (associated with the development of chronic post-surgical pain), should benefit to patients and the healthcare system in general [53].

References

1. Ahn ST, Kim JH, Park JY, du Moon G, Bae JH (2012) Acute postoperative pain after ureteroscopic removal of stone: incidence and risk factors. Korean J Urol 53:34–39
2. Allegri M, Clark MR, De Andres J, Jensen TS (2012) Acute and chronic pain: where we are and where we have to go. Minerva Anestesiol 78:222–235
3. Althaus A, Hinrichs-Rocker A, Chapman R et al (2012) Development of a risk index for the prediction of chronic post-surgical pain. Eur J Pain 16:901–910
4. Andreae MH, Andreae DA (2012) Local anaesthetics and regional anaesthesia for preventing chronic pain after surgery. Cochrane Database Syst Rev (10):CD007105
5. Apfelbaum JL, Chen C, Mehta SS, Gan TJ (2003) Postoperative pain experience: results from a national survey suggest postoperative pain continues to be undermanaged. Anesth Analg 97(2):534–540
6. Argoff CE (2014) Recent management advances in acute postoperative pain. Recent management advances in acute postoperative pain. Pain Pract 14(5):477–487
7. Bagshaw KR, Hanenbaum CL, Carbone EJ, Lo KW, Laurencin CT, Walker J, Nair LS (2015) Pain management via local anesthetics and responsive hydrogels. Ther Deliv 6(2):165–176
8. Bai G, Ren K, Dubner R (2015) Epigenetic regulation of persistent pain. Transl Res 165(1):177–199
9. Bouman EA, Theunissen M, Bons SA, van Mook WN, Gramke HF, van Kleef M, Marcus MA (2014) Reduced incidence of chronic postsurgical pain after epidural analgesia for abdominal surgery. Pain Pract 14(2):E76–E84
10. Brandsborg B, Dueholm M, Nikolajsen L, Kehlet H, Jensen TS (2009) A prospective study of risk factors for pain persisting 4 months after hysterectomy. Clin J Pain 25:263–268
11. Brandsborg B, Nikolajsen L, Kehlet H, Jensen TS (2008) Chronic pain after hysterectomy. Acta Anaesthesiol Scand 52:327–331
12. Breivik H (2002) Postoperative pain: toward optimal pharmacological and epidural analgesia. In: Giamberardino MA (ed) Pain 2002 – an updated review. IASP Press, Seattle, pp 337–349
13. Brull R, Hadzic A, Reina MA, Barrington MJ (2015) Pathophysiology and etiology of nerve injury following peripheral nerve blockade. Reg Anesth Pain Med 40(5):479–490
14. Buchheit T, Pyati S (2012) Prevention of chronic pain after surgical nerve injury: amputation and thoracotomy. Surg Clin North Am 92(2):393–407
15. Buchheit T, Van de Ven T, Shaw A (2012) Epigenetics and the transition from acute to chronic pain. Pain Med 13:1474–1490
16. Carli F (2015) Physiologic considerations of Enhanced Recovery After Surgery (ERAS) programs: implications of the stress response. Can J Anaesth 62(2):110–119
17. Carroll I, Barelka P, Wang CK et al (2012) A pilot cohort study of the determinants of longitudinal opioid use after surgery. Anesth Analg 115(3):694–702

18. Carroll I, Hah J, Mackey S, Ottestad E, Kong JT, Lahidji S, Tawfik V, Younger J, Curtin C (2013) Perioperative interventions to reduce chronic postsurgical pain. J Reconstr Microsurg 29(4):213–222
19. Celestin J, Edwards RR, Jamison RN (2009) Pretreatment psychosocial variables as predictors of outcomes following lumbar surgery and spinal cord stimulation: a systematic review and literature synthesis. Pain Med 10:639–653
20. Chen S, Roffey DM, Dion CA, Arab A, Wai EK (2016) Effect of perioperative vitamin C supplementation on postoperative pain and the incidence of chronic regional pain syndrome: a systematic review and meta-analysis. Clin J Pain 32(2):179–185
21. Clarke H, Katz J, Flor H, Rietschel M, Diehl SR, Seltzer Z (2015) Genetics of chronic post-surgical pain: a crucial step toward personal pain medicine. Can J Anaesth 62(3):294–303
22. Clark AJ, Spanswick CC (2014) Why anesthesiologists need to care about the way chronic pain is managed. Can J Anaesth 61:95–100
23. Clarke H, Woodhouse LJ, Kennedy D, Stratford P, Katz J (2011) Strategies aimed at preventing chronic post-surgical pain: comprehensive perioperative pain management after total joint replacement surgery. Physiother Can 63:289–304
24. De Kock M, Lavand'homme P, Waterloos H (2005) The short-lasting analgesia and long-term antihyperalgesic effect of intrathecal clonidine in patients undergoing colonic surgery. Anesth Analg 101:566–572
25. de Leon-Casasola O (2014) A review of the literature on multiple factors involved in postoperative pain course and duration. Postgrad Med 126(4):42–52
26. De Oliveira GS Jr, Castro Alves LJ, Nader A, Kendall MC, Rahangdale R, McCarthy RJ (2014) Perineural dexamethasone to improve postoperative analgesia with peripheral nerve blocks: a meta-analysis of randomized controlled trials. Pain Res Treat 2014:179029
27. Derry S, Wiffen PJ, Moore RA, Quinlan J (2014) Topical lidocaine for neuropathic pain in adults. Cochrane Database Syst Rev (7):CD010958
28. Deumens R, Steyaert A, Forget P, Schubert M, Lavand'homme P, Hermans E et al (2013) Prevention of chronic postoperative pain: cellular, molecular, and clinical insights for mechanism-based treatment approaches. Prog Neurobiol 104:1–37
29. Devin CJ, McGirt MJ (2015) Best evidence in multimodal pain management in spine surgery and means of assessing postoperative pain and functional outcomes. J Clin Neurosci 22(6):930–938
30. Dickerson DM (2014) Acute pain management. Anesthesiol Clin 32(2):495–504
31. Edwards RR, Smith MT, Stonerock G, Haythornthwaite JA (2006) Painrelated catastrophizing in healthy women is associated with greater temporal summation of and reduced habituation to thermal pain. Clin J Pain 22:730–737
32. Eisenach JC, Pan PH, Smiley R, Lavand'homme P, Landau R, Houle TT (2008) Severity of acute pain after childbirth, but not type of delivery, predicts persistent pain and postpartum depression. Pain 140:87–94
33. Ferakis N, Stavropoulos M (2015) Mini percutaneous nephrolithotomy in the treatment of renal and upper ureteral stones: Lessons learned from a review of the literature. Urol Ann 7(2):141–148
34. Ferrari LF, Bogen O, Chu C, Levine JD (2013) Peripheral administration of translation inhibitors reverses increased hyperalgesia in a model of chronic pain in the rat. J Pain 14:731–738
35. Ferrari LF, Bogen O, Reichling DB, Levine JD (2015) Accounting for the delay in the transition from acute to chronic pain: axonal and nuclear mechanisms. J Neurosci 35(2):495–507
36. Fletcher D, Martinez V (2014) Opioid-induced hyperalgesia in patients after surgery: a systematic review and a meta-analysis. Br J Anaesth 112(6):991–1004
37. Flor H (2012) New developments in the understanding and management of persistent pain. Curr Opin Psychiatry 25:109–113
38. Forsythe ME, Dunbar MJ, Hennigar AW, Sullivan MJL, Gross M (2008) Prospective relation between catastrophizing and residual pain following knee arthroplasty: two-year follow-up. Pain Res Manag 13:335–341

39. Fortier MA, Chou J, Maurer EL, Kain ZN (2011) Acute to chronic postoperative pain in children: preliminary findings. J Pediatr Surg 46:1700–1705
40. Géranton SM, Tochiki KK (2015) Regulation of gene expression and pain states by epigenetic mechanisms. Prog Mol Biol Transl Sci 131:147–183
41. Gerbershagen HJ (2013) Transition from acute to chronic postsurgical pain. Physiology, risk factors and prevention. Schmerz 27:81–93
42. Gerbershagen HJ, Aduckathil S, van Wijck AJ, Peelen LM, Kalkman CJ, Meissner W (2013) Pain intensity on the first day after surgery: a prospective cohort study comparing 179 surgical procedures. Anesthesiology 118:934–944
43. Gilron I, Kehlet H (2014) Prevention of chronic pain after surgery: new insights for future research and patient care. Can J Anaesth 61:101–111
44. Gizzo S, Andrisani A, Noventa M, Di Gangi S, Quaranta M, Cosmi E, D'Antona D, Nardelli GB, Ambrosini G (2015) Caesarean section: could different transverse abdominal incision techniques influence postpartum pain and subsequent quality of life? A systematic review. PLoS One 10(2):e0114190
45. Goldstone AB, Joseph WY (2014) Minimally invasive surgical treatment of valvular heart disease. Semin Thorac Cardiovasc Surg 26(1):36–43
46. Goodin BR, McGuire L, Allshouse M et al (2009) Associations between catastrophizing and endogenous pain-inhibitory processes: sex differences. J Pain 10:180–190
47. Grifasi C, Calogero A, Esposito A, Dodaro C (2015) Perioperative care of elderly outpatient. Ann Ital Chir 86(2):100–105
48. Guay J (2006) The benefits of adding epidural analgesia to general anesthesia: a meta analysis. J Anesth 20:335–340
49. Haroutiunian S, Nikolajsen L, Finnerup NB, Jensen TS (2013) The neuropathic component in persistent postsurgical pain: a systematic literature review. Pain 154:95–102
50. Hasenbring MI, Chehadi O, Titze C, Kreddig N (2014) Fear and anxiety in the transition from acute to chronic pain: there is evidence for endurance besides avoidance. Pain Manag 4(5):363–374
51. He HG, Zhu L, Chan SW, Klainin-Yobas P, Wang W (2015) The effectiveness of therapeutic play intervention in reducing perioperative anxiety, negative behaviors, and postoperative pain in children undergoing elective surgery: a systematic review. Pain Manag Nurs 16(3):425–439
52. Højer Karlsen AP, Geisler A, Petersen PL, Mathiesen O, Dahl JB (2015) Postoperative pain treatment after total hip arthroplasty: a systematic review. Pain 156(1):8–30
53. Huang A, Katz J, Clarke H (2015) Ensuring safe prescribing of controlled substances for pain following surgery by developing a transitional pain service. Pain Manag 5(2):97–105
54. Jaschinski T, Mosch C, Eikermann M, Neugebauer EA (2015) Laparoscopic versus open appendectomy in patients with suspected appendicitis: a systematic review of meta-analyses of randomised controlled trials. BMC Gastroenterol 15:48
55. Johansen A, Romundstad L, Nielsen CS, Schirmer H, Stubhaug A (2012) Persistent postsurgical pain in a general population: prevalence and predictors in the Tromsø study. Pain 153(7):1390–1396
56. Joshi GP, Kehlet H (2013) Procedure-specific pain management: the road to improve postsurgical pain management? Anesthesiology 118:780–782
57. Joshi GP, Schug SA, Kehlet H (2014) Procedure-specific pain management and outcome strategies. Best Pract Res Clin Anaesthesiol 28(2):191–201
58. Jouguelet-Lacoste J, La Colla L, Schilling D, Chelly JE (2015) The use of intravenous infusion or single dose of low-dose ketamine for postoperative analgesia: a review of the current literature. Pain Med 16(2):383–403
59. Katz J, Pagé MG (2010) Identification of risk and protective factors in the transition from acute to chronic post surgical pain. In: Lynch ME, Craig KD, Peng PWH (eds) Clinical pain management: a practical guide. Wiley-Blackwell, Oxford, pp 32–41
60. Katz J, Seltzer Z (2009) Transition from acute to chronic postsurgical pain: risk factors and protective factors. Expert Rev Neurother 9:723–744

61. Kayaalp C, Yagci MA (2015) Laparoscopic right colon resection with transvaginal extraction: a systematic review of 90 Cases. Surg Laparosc Endosc Percutan Tech 25(5):384–391.
62. Kaye AD, Urman RD (2014) Perioperative analgesia and future directions. Best Pract Res Clin Anaesthesiol 28(2):103–104
63. Kehlet H (2013) Fast-track hip and knee arthroplasty. Lancet 381:1600–1602
64. Kehlet H (2015) Enhanced Recovery After Surgery (ERAS): good for now, but what about the future? Can J Anaesth 62(2):99–104
65. Kehlet H, Delaney CP, Hill AG (2015) Perioperative medicine- the second round will need a change of tactics. Br J Anaesth 115(1):13–14
66. Kehlet H, Slim K (2012) The future of fast-track surgery. Br J Surg 99:1025–1026
67. Khatib Y, Madan A, Naylor JM, Harris IA (2015) Do psychological factors predict poor outcome in patients undergoing TKA? A systematic review. Clin Orthop Relat Res 473(8):2630–2638
68. Kolettas A, Lazaridis G, Baka S et al (2015) Postoperative pain management. J Thorac Dis 7(Suppl 1):S62–S72
69. Koyyalamudi VB, Arulkumar S, Yost BR, Fox CJ, Urman RD, Kaye AD (2014) Supraclavicular and paravertebral blocks: are we underutilizing these regional techniques in perioperative analgesia? Best Pract Res Clin Anaesthesiol 28(2):127–138
70. Kuner R (2015) Spinal excitatory mechanisms of pathological pain. Pain 156(Suppl 1): S11–S17
71. Kyranou M, Puntillo K (2012) The transition from acute to chronic pain: might intensive care unit patients be at risk? Ann Intensive Care 2:36
72. Lavand'homme P (2011) The progression from acute to chronic pain. Curr Opin Anaesthesiol 24:545–550
73. Lesin M, Domazet Bugarin J, Puljak L (2015) Factors associated with postoperative pain and analgesic consumption in ophthalmic surgery: a systematic review. Surv Ophthalmol 60(3): 196–203
74. Lewis GN, Rice DA, McNair PJ, Kluger M (2015) Predictors of persistent pain after total knee arthroplasty: a systematic review and meta-analysis. Br J Anaesth 114(4):551–561
75. Li J, Ji Z, Zhang W, Li L (2015) The comparison of lightweight mesh and standard mesh in incisional hernia repair with the open sublay technique: the results of a meta-analysis. Surg Laparosc Endosc Percutan Tech 25(3):238–244
76. Li C, Yang Y, Liu S, Fang H, Zhang Y, Furmanski O, Skinner J, Xing Y, Johns RA, Huganir RL, Tao F (2014) Stress induces pain transition by potentiation of AMPA receptor phosphorylation. J Neurosci 34(41):13737–13746
77. Liem MS, van Duyn EB, van der Graaf Y, van Vroonhoven TJ, Coala Trial Group (2003) Recurrences after conventional anterior and laparoscopic inguinal hernia repair: a randomized comparison. Ann Surg 237:136–141
78. Lovich-Sapola J, Smith CE, Brandt CP (2015) Postoperative pain control. Surg Clin North Am 95(2):301–318
79. MacLauchlan C, Shipton EA, Wells JE (2015) Perioperative pain correlates and prolonged post-operative pain predictors: demographics and psychometric questionnaires. Pain Ther 4(1):119–133
80. Macrae WA (2001) Chronic pain after surgery. Br J Anaesth 87:88–98
81. Mariano ER, Marshall ZJ, Urman RD, Kaye AD (2014) Multimodal therapy in perioperative analgesia. Best Pract Res Clin Anaesthesiol 28(1):29–39
82. McGreevy K, Bottros MM, Raja SN (2011) Preventing chronic pain following acute pain: risk factors, preventive strategies, and their efficacy. Eur J Pain Suppl 5:365–372
83. McNicol ED, Ferguson MC, Hudcova J (2015) Patient controlled opioid analgesia versus non-patient controlled opioid analgesia for postoperative pain. Cochrane Database Syst Rev (6):CD003348
84. Miller EC, McIsaac DI, Chaput A, Antrobus J, Shenassa H, Lui A (2015) Increased postoperative day one discharges after implementation of a hysterectomy enhanced recovery pathway: a retrospective cohort study. Can J Anaesth 62(5):451–460

85. Mishriky BM, Waldron NH, Habib AS (2015) Impact of pregabalin on acute and persistent postoperative pain: a systematic review and meta-analysis. Br J Anaesth 114(1):10–31
86. Moon YE (2014) Postoperative nausea and vomiting. Korean J Anesthesiol 67(3):164–170
87. Ng CS (2014) Single-port thoracic surgery: a new direction. Korean J Thorac Cardiovasc Surg 47(4):327–332
88. Nielsen BN, Henneberg SW, Schmiegelow K, Friis SM, Rømsing J (2015) Peripherally applied opioids for postoperative pain: evidence of an analgesic effect? A systematic review and meta-analysis. Acta Anaesthesiol Scand 59(7):830–845
89. Nikolajsen L, Brix LD (2014) Chronic pain after surgery in children. Curr Opin Anaesthesiol 27(5):507–512
90. Öbrink E, Jildenstål P, Oddby E, Jakobsson JG (2015) Post-operative nausea and vomiting: update on predicting the probability and ways to minimize its occurrence, with focus on ambulatory surgery. Int J Surg 15:100–106
91. Pelletier R, Higgins J, Bourbonnais D (2015) Is neuroplasticity in the central nervous system the missing link to our understanding of chronic musculoskeletal disorders? BMC Musculoskelet Disord 16:25
92. Peng K, Liu HY, Wu SR, Cheng H, Ji FH (2015) Effects of combining dexmedetomidine and opioids for postoperative intravenous Patient-controlled analgesia: a systematic review and meta-analysis. Clin J Pain 31(12):1097–1104
93. Penprase B, Brunetto E, Dahmani E, Forthoffer JJ, Kapoor S (2015) The efficacy of preemptive analgesia for postoperative pain control: a systematic review of the literature. AORN J 101(1):94–105
94. Phan K, Rao PJ, Kam AC, Mobbs RJ (2015) Minimally invasive versus open transforaminal lumbar interbody fusion for treatment of degenerative lumbar disease: systematic review and meta-analysis. Eur Spine J 24(5):1017–1030
95. Poobalan AS, Bruce J, Smith WC, King PM, Krukowski ZH, Chambers WA (2003) A review of chronic pain after inguinal herniorrhaphy. Clin J Pain 19:48–54
96. Powell R, Johnston M, Smith WC et al (2012) Psychological risk factors for chronic postsurgical pain after inguinal hernia repair surgery: a prospective cohort study. Eur J Pain 16:600–610
97. Power I, McCormack JG, Myles PS (2010) Regional anaesthesia and pain management. Anaesthesia 65:38–47
98. Prabhakar A, Mancuso KF, Owen CP, Lissauer J, Merritt CK, Urman RD, Kaye AD (2014) Perioperative analgesia outcomes and strategies. Best Pract Res Clin Anaesthesiol 28(2): 105–115
99. Radnovich R, Chapman CR, Gudin JA, Panchal SJ, Webster LR, Pergolizzi JV Jr (2014) Acute pain: effective management requires comprehensive assessment. Postgrad Med 126(4): 59–72
100. Raines S, Hedlund C, Franzon M, Lillieborg S, Kelleher G, Ahlén K (2014) Ropivacaine for continuous wound infusion for postoperative pain management: a systematic review and meta-analysis of randomized controlled trials. Eur Surg Res 53(1–4):43–60
101. Rawal N (2012) Epidural technique for postoperative pain: gold standard no more? Reg Anesth Pain Med 37:310–317
102. Ren ZY, Xu XQ, Bao YP, He J, Shi L, Deng JH, Gao XJ, Tang HL, Wang YM, Lu L (2015) The impact of genetic variation on sensitivity to opioid analgesics in patients with postoperative pain: a systematic review and meta-analysis. Pain Physician 18(2):131–152
103. Schreiber KL, Kehlet H, Belfer I, Edwards RR (2014) Predicting, preventing and managing persistent pain after breast cancer surgery: the importance of psychosocial factors. Pain Manag 4(6):445–459
104. Schug SA, Goddard C (2014) Recent advances in the pharmacological management of acute and chronic pain. Ann Palliat Med 3(4):263–275
105. Schyns-van den Berg AM, Huisjes A, Stolker RJ (2015) Postcaesarean section analgesia: are opioids still required? Curr Opin Anaesthesiol 28(3):267–274

106. Shipton EA (2014) The transition of acute postoperative pain to chronic pain: part 1 – risk factors for the development of postoperative acute persistent pain. Trends Anaesth Crit Care 4:67–70
107. Shipton EA (2014) The transition of acute postoperative pain to chronic pain: part 2 – limiting the transition. Trends Anaesth Crit Care 4:71–75
108. Shipton EA (2013) Postoperative pain, persistent acute pain. In: Schmidt RF, Willis WD (eds) Encyclopedic reference of pain. Springer, Berlin/Heidelberg, pp 3079–3083
109. Shipton EA (2013) Postoperative pain and multimodal analgesia. In: Schmidt RF, Willis WD (eds) Encyclopedic reference of pain. Springer, Berlin/Heidelberg, pp 1932–1938
110. Shipton EA (2013) Postoperative pain, epidural infusions. In: Schmidt RF, Willis WD (eds) Encyclopedic reference of pain. Springer, Berlin/Heidelberg, pp 3004–3009
111. Shipton EA (2013) Postoperative pain, intrathecal drug administration. In: Schmidt RF, Willis WD (eds) Encyclopedic reference of pain. Springer, Berlin/Heidelberg, pp 3019–3028
112. Shipton EA (2013) Postoperative pain and preoperative education. In: Schmidt RF, Willis WD (eds) Encyclopedic reference of pain. Springer, Berlin/Heidelberg, pp 3093–3099
113. Simanski CJ, Althaus A, Hoederath S, Kreutz KW, Hoederath P, Lefering R, Pape-Köhler C, Neugebauer EA (2014) Incidence of chronic postsurgical pain (CPSP) after general surgery. Pain Med 15(7):1222–1229
114. Skolnik A, Gan TJ (2014) New formulations of bupivacaine for the treatment of postoperative pain: liposomal bupivacaine and SABER-Bupivacaine. Expert Opin Pharmacother 15(11):1535–1542
115. Skovrlj B, Gilligan J, Cutler HS, Qureshi SA (2015) Minimally invasive procedures on the lumbar spine. World J Clin Cases 3(1):1–9
116. Tan M, Law LS, Gan TJ (2015) Optimizing pain management to facilitate Enhanced Recovery After Surgery pathways. Can J Anaesth 62(2):203–218
117. Tasmuth T, Blomqvist C, Kalso E (1999) Chronic post-treatment symptoms in patients with breast cancer operated in different surgical units. Eur J Surg Oncol 25:38–43
118. Tighe PJ, Riley JL 3rd, Fillingim RB (2014) Sex differences in the incidence of severe pain events following surgery: a review of 333,000 pain scores. Pain Med 15(8):1390–1404
119. Tillu DV, Hassler SN, Burgos-Vega CC, Quinn TL, Sorge RE, Dussor G, Boitano S, Vagner J, Price TJ (2015) Protease-activated receptor 2 activation is sufficient to induce the transition to a chronic pain state. Pain 156(5):859–867
120. Tong YC, Kaye AD, Urman RD (2014) Liposomal bupivacaine and clinical outcomes. Best Pract Res Clin Anaesthesiol 28(1):15–27
121. Tumber PS (2014) Optimizing perioperative analgesia for the complex pain patient: medical and interventional strategies. Can J Anaesth 61(2):131–140
122. Ujiki MB, Gitelis ME, Carbray J, Lapin B, Linn J, Haggerty S, Wang C, Tanaka R, Barrera E, Butt Z, Denham W (2015) Patient-centered outcomes following laparoscopic inguinal hernia repair. Surg Endosc 29(9):2512–2519
123. VanDenKerkhof EG, Peters ML, Bruce J (2013) Chronic pain after surgery: time for standardization? A framework to establish core risk factor and outcome domains for epidemiological studies. Clin J Pain 29:2–8
124. van Rijckevorsel DC, de Vries M, Schreuder LT, Wilder-Smith OH, van Goor H (2015) Risk factors for chronic postsurgical abdominal and pelvic pain. Pain Manag 5(2):107–116
125. Walsh M, Woodhouse LJ, Thomas SG, Finch E (2011) Strategies aimed at preventing chronic post-surgical pain: comprehensive perioperative pain management after total joint replacement surgery. Physiother Can 63:289–304
126. Wang C, Cai XZ, Yan SG (2015) Comparison of periarticular multimodal drug injection and femoral nerve block for postoperative pain management in total knee arthroplasty: a systematic review and meta-analysis. J Arthroplasty 30(7):1281–1286
127. Werner MU, Mjöbo HN, Nielsen PR, Rudin A (2010) Prediction of postoperative pain: a systematic review of predictive experimental pain studies. Anesthesiology 112:1494–1502

128. Wildgaard K, Ravn J, Nikolajsen L, Jakobsen E, Jensen TS, Kehlet H (2011) Consequences of persistent pain after lung cancer surgery: a nationwide questionnaire study. Acta Anaesthesiol Scand 55:60–68
129. Yin JB, Cui GB, Mi MS, Du YX, Wu SX, Li YQ, Wang W (2014) Local infiltration analgesia for postoperative pain after hip arthroplasty: a systematic review and meta-analysis. J Pain 15(8):781–799
130. Yu N, Long X, Lujan-Hernandez JR, Succar J, Xin X, Wang X (2014) Transversus abdominis-plane block versus local anesthetic wound infiltration in lower abdominal surgery: a systematic review and meta-analysis of randomized controlled trials. BMC Anesthesiol 14:121
131. Yu L, Ran B, Li M, Shi Z (2013) Gabapentin and pregabalin in the management of postoperative pain after lumbar spinal surgery: a systematic review and meta-analysis. Spine 38(22): 1947–1952
132. Zhao X, Tong Y, Ren H, Ding XB, Wang X, Zong JY, Jin SQ, Li Q (2014) Transversus abdominis plane block for postoperative analgesia after laparoscopic surgery: a systematic review and meta-analysis. Int J Clin Exp Med 7(9):2966–2975

MIX
Papier aus verantwortungsvollen Quellen
Paper from responsible sources
FSC® C105338

If you have any concerns about our products,
you can contact us on
ProductSafety@springernature.com

In case Publisher is established outside the EU,
the EU authorized representative is:
Springer Nature Customer Service Center GmbH
Europaplatz 3, 69115 Heidelberg, Germany

Printed by Libri Plureos GmbH
in Hamburg, Germany